T0350908

EMERGING TECHNOLOGIES FOR 3D VIDEO

EMERGING TECHNOLOGIES FOR 3D VIDEO

CREATION, CODING, TRANSMISSION AND RENDERING

Edited by

Frédéric Dufaux
Télécom Paris Tech, CNRS, France

Béatrice Pesquet-Popescu
Télécom Paris Tech, France

Marco Cagnazzo
Télécom Paris Tech, France

A John Wiley & Sons, Ltd., Publication

Library of Congress Cataloging-in-Publication Data

Emerging technologies for 3D video : creation, coding, transmission, and
rendering / Frederic Dufaux, Beatrice Pesquet-Popescu, Marco Cagnazzo.
      pages cm
  Includes bibliographical references and index.
   ISBN 978-1-118-35511-4 (cloth)
  1.   3-D video–Standards. 2.   Digital video–Standards.   I. Dufaux, Frederic,
1967- editor of compilation. II. Pesquet-Popescu, Beatrice, editor of
compilation. III. Cagnazzo, Marco, editor of compilation. IV. Title: Emerging
technologies for three dimensional video.
   TK6680.8.A15E44 2013
   006.6′96–dc23

                                                          2012047740

A catalogue record for this book is available from the British Library.

ISBN: 9781118355114

Set in 10/12pt, Times by Thomson Digital, Noida, India.

Printed and bound in Singapore by Markono Print Media Pte Ltd

Contents

PART VI APPLICATIONS AND IMPLEMENTATION

20 Interactive Omnidirectional Indoor Tour

Jean-Charles Bazin, Olivier Saurer, Friedrich Fraundorfer, and Marc Pollefeys

Preface

The underlying principles of stereopsis have been known for a long time. Stereoscopes to see photographs in 3D appeared and became popular in the nineteenth century. The first demonstrations of 3D movies took place in the first half of the twentieth century, initially using anaglyph glasses, and then with polarization-based projection. Hollywood experienced a first short-lived golden era of 3D movies in the 1950s. In the last 10 years, 3D has regained significant interests and 3D movies are becoming ubiquitous. Numerous major productions are now released in 3D, culminating with *Avatar*, the highest grossing film of all time.

In parallel with the recent growth of 3D movies, 3DTV is attracting significant interest from manufacturers and service providers. This is obvious by the multiplication of new 3D product announcements and services. Beyond entertainment, 3D imaging technology is also seen as instrumental in other application areas such as video games, immersive video conferences, medicine, video surveillance, and engineering.

With this growing interest, 3D video is often considered as one of the major upcoming innovations in video technology, with the expectation of greatly enhanced user experience.

This book intends to provide an overview of key technologies for 3D video applications. More specifically, it covers the state of the art and explores new research directions, with the objective to tackle all aspects involved in 3D video systems and services. Topics addressed include content acquisition and creation, data representation and coding, transmission, view synthesis, rendering, display technologies, human perception of depth, and quality assessment. Relevant standardization efforts are reviewed. Finally, applications and implementation issues are also described.

More specifically, the book is composed of six parts. Part One addresses different aspects of *3D content acquisition and creation*. In Chapter 1, Lee presents depth cameras and related applications. The principle of active depth sensing is reviewed, along with depth image processing methods such as noise modelling, upsampling, and removing motion blur. In Chapter 2, Kirmani, Colaço, and Goyal introduce the space-from-time imaging framework, which achieves spatial resolution, in two and three dimensions, by measuring temporal variations of light intensity in response to temporally or spatiotemporally varying illumination. Chapter 3, by Vazquez, Zhang, Speranza, Plath, and Knorr, provides an overview of the process generating a stereoscopic video (S3D) from a monoscopic video source (2D), generally known as 2D-to-3D video conversion, with a focus on selected recent techniques. Finally, in Chapter 4, Zone* provides an overview of numerous contemporary strategies for shooting narrow and variable interaxial baseline for stereoscopic cinematography. Artistic implications are also discussed.

A key issue in 3D video, Part Two addresses *data representation, compression, and transmission*. In Chapter 5, Kaaniche, Gaetano, Cagnazzo, and Pesquet-Popescu address the

---

* It is with great sadness that we learned that Ray Zone passed away on November 13, 2012.

problem of disparity estimation. The geometrical relationship between the 3D scene and the generated stereo images is analyzed and the most important techniques for disparity estimation are reviewed. Cagnazzo, Pesquet-Popescu, and Dufaux give an overview of existing data representation and coding formats for 3D video content in Chapter 6. In turn, in Chapter 7, Mora, Valenzise, Jung, Pesquet-Popescu, Cagnazzo, and Dufaux consider the problem of depth map coding and present an overview of different coding tools. In Chapter 8, Vetro and Müller provide an overview of the current status of research and standardization activity towards defining a new set of depth-based formats that facilitate the generation of intermediate views with a compact binary representation. In Chapter 9, Cheung and Cheung consider interactive media streaming, where the server continuously and reactively sends appropriate subsets of media data in response to a client's periodic requests. Different associated coding strategies and solutions are reviewed. Finally, Gürler and Tekalp propose an adaptive P2P video streaming solution for streaming multiview video over P2P overlays in Chapter 10.

Next, Part Three of the book discusses *view synthesis and rendering*. In Chapter 11, Wang, Lang, Stefanoski, Sorkine-Hornung, Sorkine-Hornung, Smolic, and Gross present image-domain warping as an alternative to depth-image-based rendering techniques. This technique utilizes simpler, image-based deformations as a means for realizing various stereoscopic post-processing operators. Gilliam, Brookes, and Dragotti, in Chapter 12, examine the state of the art in plenoptic sampling theory. In particular, the chapter presents theoretical results for uniform sampling based on spectral analysis of the plenoptic function and algorithms for adaptive plenoptic sampling. Finally, in Chapter 13, Klose, Lipski, and Magnor present a complete end-to-end framework for stereoscopic free viewpoint video creation, allowing one to viewpoint-navigate through space and time of complex real-world, dynamic scenes.

As a very important component of a 3D video system, Part Four focuses on *3D display technologies*. In Chapter 14, Konrad addresses digital signal processing methods for 3D data generation, both stereoscopic and multiview, and for compensation of the deficiencies of today's 3D displays. Numerous experimental results are presented to demonstrate the usefulness of such methods. Borel and Doyen, in Chapter 15, present in detail the main 3D display technologies available for cinemas, for large-display TV sets, and for mobile terminals. A perspective of evolution for the near and long term is also proposed. In Chapter 16, Arai focuses on integral imaging, a 3D photography technique that is based on integral photography, in which information on 3D space is acquired and represented. This chapter describes the technology for displaying 3D space as a spatial image by integral imaging. Finally, in Chapter 17, Kovács and Balogh present light-field displays, an advanced technique for implementing glasses-free 3D displays.

In most targeted applications, humans are the end-users of 3D video systems. Part Five considers *human perception of depth* and *perceptual quality assessment*. More specifically, in Chapter 18, Watt and MacKenzie focus on how the human visual system interacts with stereoscopic 3D media, in view of optimizing effectiveness and viewing comfort. Three main issues are addressed: incorrect spatiotemporal stimuli introduced by field-sequential stereo presentation, inappropriate binocular viewing geometry, and the unnatural relationship between where the eyes fixate and focus in stereoscopic 3D viewing. In turn, in Chapter 19, Hanhart, De Simone, Rerabek, and Ebrahimi consider mechanisms of 3D vision in humans, and their underlying perceptual models, in conjunction with the types of distortions that today's and tomorrow's 3D video processing systems produce. This complex puzzle is examined with a focus on how to measure 3D visual quality, as an essential factor in the success of 3D technologies, products, and services.

In order to complete the book, Part Six describes *target applications for 3D video*, as well as *implementation issues*. In Chapter 20, Bazin, Saurer, Fraundorfer, and Pollefeys present a semi-automatic method to generate interactive virtual tours from omnidirectional video. It allows a user to virtually navigate through buildings and indoor scenes. Such a system can be applied in various contexts, such as virtual tourism, tele-immersion, tele-presence, and e-heritage. Daniyal and Cavallaro address the question of how to automatically identify which view is more useful when observing a dynamic scene with multiple cameras in Chapter 21. This problem concerns several applications ranging from video production to video surveillance. In particular, an overview of existing approaches for view selection and automated video production is presented. In Chapter 22, Bourge and Bellon present the hardware architecture of a typical mobile platform, and describe major stereoscopic 3D applications. Indeed, smartphones bring new opportunities to stereoscopic 3D, but also specific constraints. Chapter 23, by Le Feuvre and Mathieu, presents an integrated system for displaying interactive applications on multiview screens. Both a simple GPU-based prototype and a low-cost hardware design implemented on a field-programmable gate array are presented. Finally, in Chapter 24, Tseng and Chang propose an optimized disparity estimation algorithm for high-definition 3DTV applications with reduced computational and memory requirements.

By covering general and advanced topics, providing at the same time a broad and deep analysis, the book has the ambition to become a reference for those involved or interested in 3D video systems and services. Assuming fundamental knowledge in image/video processing, as well as a basic understanding in mathematics, this book should be of interest to a broad readership with different backgrounds and expectations, including professors, graduate and undergraduate students, researchers, engineers, practitioners, and managers making technological decisions about 3D video.

Frédéric Dufaux
Béatrice Pesquet-Popescu
Marco Cagnazzo

List of Contributors

Jun Arai, NHK (Japan Broadcasting Corporation), Japan

Tibor Balogh, Holografika, Hungary

Jean-Charles Bazin, Computer Vision and Geometry Group, ETH Zürich, Switzerland

Alain Bellon, STMicroelectronics, France

Thierry Borel, Technicolor, France

Arnaud Bourge, STMicroelectronics, France

Mike Brookes, Department of Electrical and Electronic Engineering, Imperial College London, UK

Marco Cagnazzo, Département Traitement du Signal et des Images, Télécom ParisTech, France

Andrea Cavallaro, Queen Mary University of London, UK

Tian-Sheuan Chang, Department of Electronics Engineering, National Chiao Tung University, Taiwan

Gene Cheung, Digital Content and Media Sciences Research Division, National Institute of Informatics, Japan

Ngai-Man Cheung, Information Systems Technology and Design Pillar, Singapore University of Technology and Design, Singapore

Andrea Colaço, Media Lab, Massachusetts Institute of Technology, USA

Fahad Daniyal, Queen Mary University of London, UK

Francesca De Simone, Multimedia Signal Processing Group (MMSPG), Ecole Polytechnique Fédérale de Lausanne (EPFL), Switzerland

Didier Doyen, Technicolor, France

Pier Luigi Dragotti, Department of Electrical and Electronic Engineering, Imperial College London, UK

Frédéric Dufaux, Département Traitement du Signal et des Images, Télécom ParisTech, France

Touradj Ebrahimi, Multimedia Signal Processing Group (MMSPG), Ecole Polytechnique Fédérale de Lausanne (EPFL), Switzerland

Friedrich Fraundorfer, Computer Vision and Geometry Group, ETH Zürich, Switzerland

Raffaele Gaetano, Département Traitement du Signal et des Images, Télécom ParisTech, France

Christopher Gilliam, Department of Electrical and Electronic Engineering, Imperial College London, UK

Vivek K. Goyal, Research Laboratory of Electronics, Massachusetts Institute of Technology, USA

Markus Gross, Disney Research Zurich, Switzerland

C. Göktuğ Gürler, College of Engineering, Koç University, Turkey

Philippe Hanhart, Multimedia Signal Processing Group (MMSPG), Ecole Polytechnique Fédérale de Lausanne (EPFL), Switzerland

Alexander Sorkine-Hornung, Disney Research Zurich, Switzerland

Joël Jung, Orange Labs, France

Mounir Kaaniche, Département Traitement du Signal et des Images, Télécom ParisTech, France

Ahmed Kirmani, Research Laboratory of Electronics, Massachusetts Institute of Technology, USA

Felix Klose, Institut für Computergraphik, TU Braunschweig, Germany

Sebastian Knorr, imcube labs GmbH, Technische Universität Berlin, Germany

Janusz Konrad, Department of Electrical and Computer Engineering, Boston University, USA

Péter Tamás Kovács, Holografika, Hungary

Manuel Lang, Disney Research Zurich, Switzerland

Seungkyu Lee, Samsung Advanced Institute of Technology, South Korea

Jean Le Feuvre, Département Traitement du Signal et des Images, Telecom ParisTech, France

Christian Lipski, Institut für Computergraphik, TU Braunschweig, Germany

Kevin J. MacKenzie, Wolfson Centre for Cognitive Neuroscience, School of Psychology, Bangor University, UK

Marcus Magnor, Institut für Computergraphik, TU Braunschweig, Germany

Yves Mathieu, Telecom ParisTech, France

Elie Gabriel Mora, Orange Labs, France; Département Traitement du Signal et des Images, Télécom ParisTech, France

Karsten Müller, Fraunhofer Institute for Telecommunications, Heinrich-Hertz-Institut, Germany

Béatrice Pesquet-Popescu, Département Traitement du Signal et des Images, Télécom ParisTech, France

Nils Plath, imcube labs GmbH, Technische Universität Berlin, Germany

Marc Pollefeys, Computer Vision and Geometry Group, ETH Zürich, Switzerland

Martin Rerabek, Multimedia Signal Processing Group (MMSPG), Ecole Polytechnique Fédérale de Lausanne (EPFL), Switzerland

Olivier Saurer, Computer Vision and Geometry Group, ETH Zürich, Switzerland

Aljoscha Smolic, Disney Research Zurich, Switzerland

Olga Sorkine-Hornung, ETH Zurich, Switzerland

Filippo Speranza, Communications Research Centre Canada (CRC), Canada

Nikolce Stefanoski, Disney Research Zurich, Switzerland

A. Murat Tekalp, College of Engineering, Koç University, Turkey

Yu-Cheng Tseng, Department of Electronics Engineering, National Chiao Tung University, Taiwan

Giuseppe Valenzise, Département Traitement du Signal et des Images, Télécom ParisTech, France

Carlos Vazquez, Communications Research Centre Canada (CRC), Canada

Anthony Vetro, Mitsubishi Electric Research Labs (MERL), USA

Simon J. Watt, Wolfson Centre for Cognitive Neuroscience, School of Psychology, Bangor University, UK

Oliver Wang, Disney Research Zurich, Switzerland

Liang Zhang, Communications Research Centre Canada (CRC), Canada

Ray Zone, The 3-D Zone, USA

Acknowledgements

We would like to express our deepest appreciation to all the authors for their invaluable contributions. Without their commitment and efforts, this book would not have been possible.

Moreover, we would like to gratefully acknowledge the John Wiley & Sons Ltd. staff, Alex King, Liz Wingett, Richard Davies, and Genna Manaog, for their relentless support throughout this endeavour.

Frédéric Dufaux
Béatrice Pesquet-Popescu
Marco Cagnazzo

Part One

Content Creation

1

Consumer Depth Cameras and Applications

Seungkyu Lee

Samsung Advanced Institute of Technology, South Korea

1.1 Introduction

Color imaging technology has been advanced to increase spatial resolution and color quality. However, its sensing principle limits three-dimensional (3D) geometry and photometry information acquisition. Many computer vision and robotics researchers have tried to reconstruct 3D scenes from a set of two-dimensional (2D) color images. They have built a calibrated color camera network and employed visual hull detection from silhouettes or key point matching to figure out the 3D relation between the set of 2D color images. These techniques, however, assume that objects seen from different views have identical color and intensity. This photometry consistency assumption is valid only for Lambertian surfaces that are not common in the real world. As a result, 3D geometry data capture using color cameras shows limited performance under limited lighting and object environments.

3D sensing technologies such as digital holograph, interferometry, and integral photography have been studied. However, they show limited performance in 3D geometry and photometry acquisition. Recently, several consumer depth-sensing cameras using near-infrared light have been introduced in the market. They have relatively low spatial resolution compared with color sensors and show limited sensing range and accuracy. Thanks to their affordable prices and the advantage of direct 3D geometry acquisition, many researchers from graphics, computer vision, image processing, and robotics have employed this new modality of data for many applications. In this chapter, we introduce two major depth-sensing principles using active IR signals and state of the art applications.

1.2 Time-of-Flight Depth Camera

In active light sensing technology, if we can measure the flight time of a fixed-wavelength signal emitted from a sensor and reflected from an object surface, we can calculate the

Emerging Technologies for 3D Video: Creation, Coding, Transmission and Rendering, First Edition.
Frédéric Dufaux, Béatrice Pesquet-Popescu, and Marco Cagnazzo.
© 2013 John Wiley & Sons, Ltd. Published 2013 by John Wiley & Sons, Ltd.

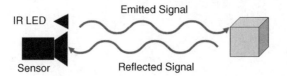

Figure 1.1 Time-of-flight depth sensor

distance of the object from the sensor based on the speed of light. This is the principle of a time-of-flight (ToF) sensor (Figure 1.1). However, it is not simple to measure the flight time directly at each pixel of any existing image sensor. Instead, if we can measure the phase delay of the reflected signal compared with the original emitted signal, we can calculate the distance indirectly. Recent ToF depth cameras in the market measure the phase delay of the emitted infrared (IR) signal at each pixel and calculate the distance from the camera.

1.2.1 Principle

In this section, the principle of ToF depth sensing is explained in more detail with simplified examples. Let us assume that we use a sinusoidal IR wave as an active light source. In general, consumer depth cameras use multiple light-emitting diodes (LEDs) to generate a fixed-wavelength IR signal. What we can observe using an existing image sensor is the amount of electrons induced by collected photons during a certain time duration. For color sensors, it is enough to count the amount of induced electrons to capture the luminance or chrominance of the expected bandwidth. However, a single shot of photon collection is not enough for phase delay measurement. Instead, we collect photons multiple times at different time locations, as illustrated in Figure 1.2.

Q_1 through Q_4 in Figure 1.2 are the amounts of electrons measured at each corresponding time. Reflected IR shows a phase delay proportional to the distance from the camera. Since we have reference emitted IR and its phase information, electron amounts at multiple time locations (Q_1 through Q_4 have a 90° phase difference to each other) can tell us the amount of delay as follows:

$$t(d) = \arctan\left(\frac{Q_3 - Q_4}{Q_1 - Q_2}\right) = \arctan\left(\frac{\alpha \sin(\phi_3) - \alpha \sin(\phi_4)}{\alpha \sin(\phi_1) - \alpha \sin(\phi_2)}\right) \tag{1.1}$$

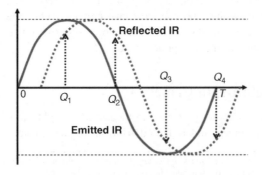

Figure 1.2 Phase delay measurement

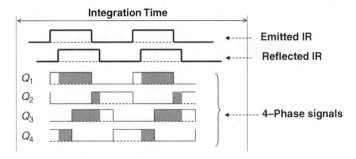

Figure 1.3   Four-phase depth sensing

where α is the amplitude of the IR signal and ϕ_1 through ϕ_4 are the normalized amounts of electrons.

In real sensing situations, a perfect sine wave is not possible to produce using cheap LEDs of consumer depth cameras. Any distortion on the sine wave causes miscalculation of the phase delay. Furthermore, the amount of electrons induced by the reflected IR signal at a certain moment is very noisy due to the limited LED power. In order to increase the signal-to-noise ratio, sensors collect electrons from multiple cycles of reflected IR signal, thus allowing some dedicated integration time.

For a better understanding of the principle, let us assume that the emitted IR is a square wave instead of sinusoidal and we have four switches at each sensor pixel to collect Q_1 through Q_4. Each pixel of the depth sensor consists of several transistors and capacitors to collect the electrons generated. Four switches alter the on and off states with 90° phase differences based on the emitted reference IR signal as illustrated in Figure 1.3. When a switch is turned on and reflected IR goes high, electrons are charged as indicated by shaded regions.

In order to increase the signal-to-noise ratio, we repeatedly charge electrons through multiple cycles of the IR signal to measure Q_1 through Q_4 during a fixed integration time for a single frame of depth image acquisition. Once Q_1 through Q_4 are measured, the distance can be calculated as follows:

$$
\text{Distance} = \frac{c}{2} t(d) = \frac{c}{2} \arctan\left(\frac{Q_3 - Q_4}{Q_1 - Q_2}\right)
$$
$$
= \frac{c}{2} \arctan\left(\frac{\alpha q_3 - \alpha q_4}{\alpha q_1 - \alpha q_2}\right) = \frac{c}{2} \arctan\left(\frac{q_3 - q_4}{q_1 - q_2}\right)
$$
(1.2)

where c is the speed of light (3×10^8 m/s) and $t(d)$ is the flight time. Note that q_1 through q_4 are normalized electric charges and α is the amplitude of reflected IR that does not affect the distance calculation. In other words, depth can be calculated correctly regardless of IR amplitude. The emitted IR should be modulated with a high enough frequency to estimate the flight time.

As indicated in Figure 1.4, what we have calculated is the distance R from the camera to an object surface along the reflected IR signal. This is not necessarily the distance along the z-direction of the 3D sensor coordinate. Based on the location of each pixel and field of view information, Z in Figure 1.4 can be calculated from R to obtain an undistorted 3D geometry. Most consumer depth cameras give calculated Z distance instead of R for user convenience.

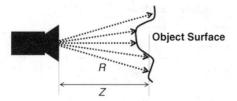

Figure 1.4 Relation between R and Z

1.2.2   Quality of the Measured Distance

Even though the ToF principle allows the distance imaging within the sensing range decided by the IR modulation frequency, the quality of measured depth suffers from various sensor systematic or nonsystematic noises (Huhle *et al.*, 2008; Edeler *et al.*, 2010; Foix *et al.*, 2011; Matyunin *et al.*, 2011). Owing to the limited power of IR light, incoming reflected IR into each image sensor pixel induces a limited number of electrons for depth calculation. The principle of ToF depth sensing can calculate correct depth regardless of the power of IR light and the amplitude of reflected IR. However, a lower absolute amount of electrons suffers from electronic noises such as shot noise. To resolve this problem, we increase the integration time to collect a sufficient number of electrons for higher accuracy of depth calculation. However, this limits the frame rate of sensors. Increase of modulation frequency also increases sensor accuracy under identical integration time, because it allows more cycles of modulated IR waves for a single depth frame production. However, this also limits the maximum sensing range of depth sensors. The left image in Figure 1.5 shows a 3D point cloud collected by a ToF depth sensor. The viewpoint is shifted to the right of the camera showing occluded regions by foreground chairs. Note that the 3D data obtained from a depth sensor are not the

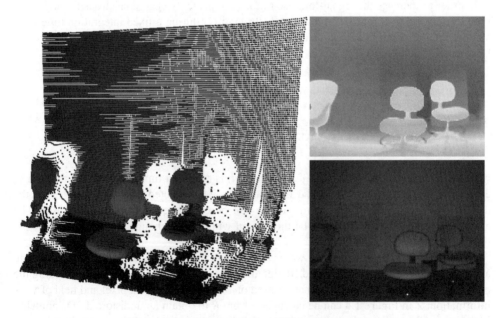

Figure 1.5 Measured depth and IR intensity images

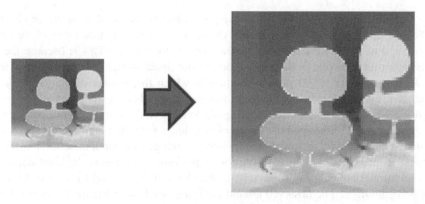

Figure 1.6 Depth image 2D super-resolution

complete volumetric data. Only the 3D locations of the 2D surface seen from the camera's viewpoint are given. The right images in Figure 1.5 are depth and IR intensity images.

In active light sensors, the incoming light signal-to-noise ratio is still relatively low compared with the passive light sensors like a color camera due to the limited IR signal emitting power. In order to increase the signal-to-noise ratio further, depth sensors merge multiple neighbor sensor pixels to measure a single depth value, decreasing depth image resolution. This is called pixel binning. Most consumer depth cameras perform the pixel binning and sacrifice image resolution to guarantee a certain depth accuracy. Therefore, many researchers according to their applications perform depth image super-resolution (Schuon *et al.*, 2008; Park *et al.*, 2011; Yeo *et al.*, 2011) before the use of raw depth images, as illustrated in Figure 1.6.

The left image in Figure 1.6 is a raw depth image that is upsampled on the right. Simple bilinear interpolation is used in this example. Most interpolation methods, however, consider the depth image as a 2D image and increase depth image resolution in the 2D domain. On the other hand, if depth is going to be used for 3D reconstruction, upsampling only in two axes is not enough. The left image in Figure 1.7 shows an example of 2D depth image

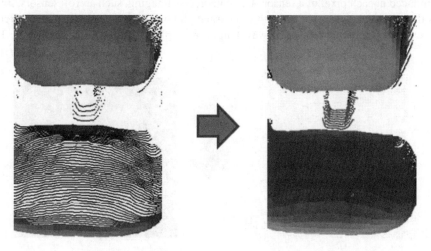

Figure 1.7 Depth image 2D versus 3D super-resolution

super-resolution. The upper part of the chair is almost fronto-parallel to the depth camera and shows a dense super-resolution result. On the other hand, the lower part of the chair is perpendicular to the depth camera and shows lots of local holes. This is because the point gaps between depth pixels in the z-direction are not filled enough. The right image in Figure 1.7 is an example of 3D super-resolution in which the number of points going to be inserted is adaptively decided by the maximum distance in between depth pixels.

Figure 1.7 is an example of 3D super-resolution where the original depth point cloud is upsampled around 25 times and the IR intensity value is projected onto each point. A sensor pixel taking the boundary region of the foreground chair gives a depth in between foreground and background. Once the super-resolution is performed, this miscalculated depth point makes additional boundary noise pixels, as can be seen in the right image of Figure 1.7 where odd patterns can be observed around the foreground chair. Figure 1.8 shows this artifact more clearly.

Figure 1.8 shows upsampled depth point cloud where the aligned color value is projected onto each pixel. The left image in Figure 1.8 shows lots of depth points in between the foreground chair and background. The colors projected onto these points are from either the foreground or background of the aligned 2D color image. This is a huge artifact, especially for the 3D reconstruction application, where random view navigation will see this noise more seriously, as shown in Figure 1.8.

The right image in Figure 1.9 is an example of boundary noise point elimination. Depth points away from both the foreground and background point cloud can be eliminated by outlier elimination methods.

When we take a moving object in real time, another artifact like motion blur comes out in depth images (Lindner and Kolb, 2009; Hussmann *et al.*, 2011; Lee *et al.*, 2012). Motion blur is a long-standing issue of imaging devices because it leads to a wrong understanding and information of real-world objects. For distance sensors in particular, motion blur causes distortions in the reconstructed 3D geometry or totally different distance information. Various techniques have been developed to mitigate the motion blur in conventional sensors. Current technology, however, either requires a very short integration time to avoid motion blur or adopts computationally expensive post-processing methods to improve blurred images.

Different from passive photometry imaging devices collectively using the amount of photons induced in each pixel of a sensor, active geometry imaging, such as ToF sensors, investigates the relation between the amount of charged photons to figure out the phase difference of an emitted light source of fixed wavelength, as explained earlier in this chapter. These

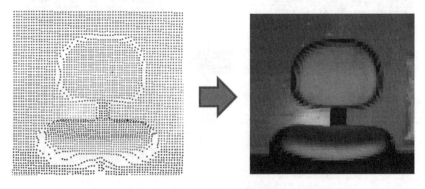

Figure 1.8 Depth point cloud 3D super-resolution

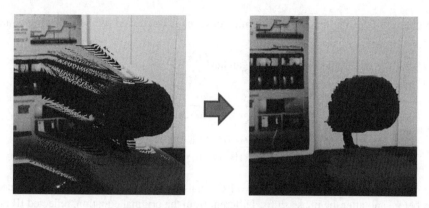

Figure 1.9 Boundary noise elimination

sensors investigate the flight time of an emitted and reflected light source to calculate the distance. The phase difference of the reflected light in these principles represents the difference in distance from the camera. The distance image, as a result, is an integration of phase variation over a pixel grid. When there is any movement of an object or the camera itself, a phase shift will be observed at the corresponding location, shaping the infrared wavefront. The phase shift observed by a sensor pixel causes multiple reflected IR waves to capture different phases within the integration time. This gives the wrong distance value calculation. As a result, the phase shift within a photon integration time produces motion blur that is not preferable for robust distance sensing. The motion blur region is the result of a series of phase shifts within the photon integration time. Reducing the integration time is not always a preferable solution because it reduces the amount of photons collected to calculate a distance-decreasing signal-to-noise ratio. On the other hand, post-processing after distance calculation shows limited performance and is a time-consuming job.

When we produce a distance image that includes motion blur, we can detect the phase shift within an integration time by investigating the relation between the separately collected amounts of photons by multiple control signals. Figure 1.10 shows what happens if there is any phase shift within an integration time in a four-phase ToF sensing mechanism. From their definitions, the four electric charges Q_1–Q_4 are averages of total cumulated electric charges over multiple "on" phases. Without any phase shift, distance is calculated by equation (1.2) by obtaining the phase difference between emitted and reflected IR waves. When there is

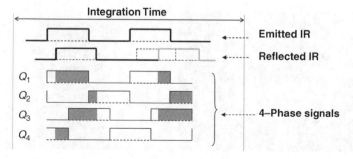

Figure 1.10 Depth motion blur

single phase shift within the integration time, distance is calculated by the following equation:

$$
\begin{aligned}
\text{Distance} &= \frac{c}{2}t(d) = \frac{c}{2}\arctan\left(\frac{Q_3 - Q_4}{Q_1 - Q_2}\right) \\
&= \frac{c}{2}\arctan\left[\frac{(\alpha_1 q_{13} + \alpha_2 q_{23}) - (\alpha_1 q_{14} + \alpha_2 q_{24})}{(\alpha_1 q_{11} + \alpha_2 q_{21}) - (\alpha_1 q_{12} + \alpha_2 q_{22})}\right] \\
&= \frac{c}{2}\arctan\left[\frac{\alpha_1(q_{13} - q_{14}) + \alpha_2(q_{23} - q_{24})}{\alpha_1(q_{11} - q_{12}) + \alpha_2(q_{21} - q_{22})}\right]
\end{aligned}
\tag{1.3}
$$

where α_1 and α_2 are the amplitudes of two IR signals and q_{1x} and q_{2x} are normalized electron values before and after the phase shifts. Different from the original equation, reflected IR amplitudes α_1 and α_2 cannot be eliminated from the equation and affect the distance calculation.

Figure 1.10 shows what happens in a depth image with a single phase shift. During the integration time, the original (indicated in black) and phase-shifted (indicated in grey) reflected IR come in sequentially and will be averaged to calculate a single depth value. Motion blurs around moving objects will be observed, showing quite different characteristics from those of conventional color images, as shown in Figure 1.11. Note that the motion blur regions (indicated by dashed ellipses) have nearer or farther depth values than both foreground and background neighbor depth values. In general with multiple or continuous phase shifts, miscalculated distance is as follows:

$$
\text{Distance} = \frac{c}{2}\arctan\left(\frac{Q_3 - Q_4}{Q_1 - Q_2}\right) = \frac{c}{2}\arctan\left(\frac{\sum_i \alpha_i q_{i3} - \sum_i \alpha_i q_{i4}}{\sum_i \alpha_i q_{i1} - \sum_i \alpha_i q_{i2}}\right)
\tag{1.4}
$$

Each control signal has a fixed phase delay from the others that gives a dedicated relation of collected electric charges. A four-phase ToF sensor makes a 90° phase delay between control signals, giving the following relations: $Q_1 + Q_2 = Q_3 + Q_4 = Q^{sum}$ and $|Q_1 - Q_2| + |Q_3 - Q_4| = Q^{sum}$. Q^{sum} is the total amount of electric charge delivered by the reflected IR. In principle, every regular pixel has to meet with these conditions if there is no significant noise (Figure 1.12). With an appropriate sensor noise model and thresholds, these relations can be used to see if the pixel has regular status. A phase shift causes a very significant distance error exceeding the common sensor noise level and is effectively detected by testing whether either of the relations is violated.

Figure 1.11 Depth motion blur examples

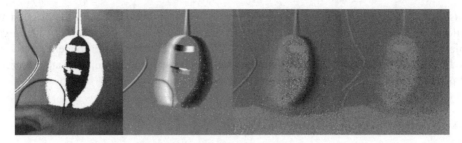

Figure 1.12 Boundary noise elimination

There are several other noise sources (Foix *et al.*, 2011). The emitted IR signal amplitude attenuates while traveling in proportion to the reciprocal of the square of the distance. Even though this attenuation should not affect the depth calculation of the ToF sensor in principle, the decrease of signal-to-noise ratio will degenerate the repeatability of the depth calculation. The uniformity assumption of the emitted IR onto the target object also causes spatial distortion of the calculated depth. In other words, each sensor pixel will collect reflected IR of different amplitudes even though reflected from surfaces of identical distance. Daylight interference is another very critical issue of the practicality of the sensor. Any frequency of IR signal emitted from the sensor will exist in daylight, which works as a noise with regard to correct depth calculation. Scattering problems (Mure-Dubois and Hugli, 2007) within the lens and sensor architecture are a major problem with depth sensors owing to their low signal-to-noise ratio.

1.3   Structured Light Depth Camera

Kinect, a famous consumer depth camera in the market, is a structured IR light-type depth sensor which is well-known 3D geometry acquisition technology. It is composed of an IR emitter, IR sensor, and color sensor, providing an IR amplitude image, depth map, and color image. Basically, this technology utilizes conventional color sensor technology with relatively higher resolution. Owing to the limit of the sensing range of the structured light principle, the operating range of this depth sensor is around 1–4 m.

1.3.1   Principle

In this type of sensor, a predetermined IR pattern is emitted onto the target objects (Figure 1.13). The pattern can be a rectangular grid or a set of random dots. A calibrated IR

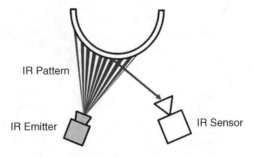

Figure 1.13   Structured IR light depth camera

sensor reads the pattern reflected from the object surfaces. Different from the original IR pattern, the IR sensor reads a distorted pattern due to the geometric variation of the target objects. This allows information to be obtained on the pixel-wise correspondence between the IR emitter and IR sensor. Triangulation of each point between the projected and observed IR patterns enables calculation of the distance from the sensor.

1.4 Specular and Transparent Depth

Figure 1.14 illustrates how the light ray arrives at a Lambertian surface and travels back to the sensor. Lambertian objects distribute incoming light equally in all directions. Hence, sensors can receive reflected light regardless of their position and orientation. Most existing 3D sensing technologies assume Lambertian objects as their targets. Both the ToF sensor in the left in Figure 1.14 and the structured light sensor on the right in Figure 1.14 can see emitted light sources.

However, most real-world objects have non-Lambertian surfaces, including transparency and specularity. Figure 1.15 shows what happens if active light depth sensors see a specular surface and why the specular object is challenging to handle. Unlike the Lambertian material, specular objects reflect the incoming light into a limited range of directions. Let us assume that the convex surfaces in Figure 1.15 are mirrors, which reflect incoming light rays toward a very narrow direction. If the sensor is not located on the exact direction of the

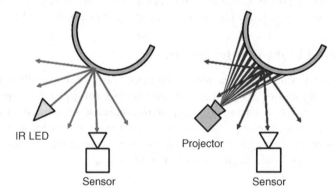

Figure 1.14 Lambertian surface

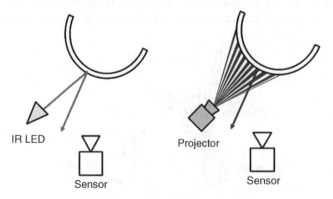

Figure 1.15 Depth of a specular surface

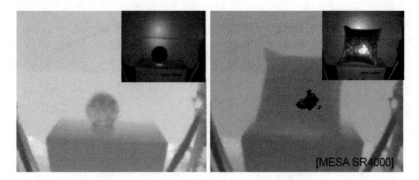

Figure 1.16 Specular object examples of ToF sensor

reflected ray from the mirror surface, the sensor does not receive any reflected IR and it is impossible to calculate any depth information. On the other hand, if the sensor is located exactly on the mirror reflection direction, the sensor will receive an excessive amount of concentrated IR light, causing saturation in measurement. Consequently, the sensors fail to receive the reflected light in a sensible range. Such a phenomenon results in missing measurements for both types of sensors.

Figure 1.16 shows samples of specular objects taken by a ToF depth sensor. The first object is a mirror where the flat area is all specular surfaces. The second object shows specularity in a small region. The sensor in the first case is not on the mirror reflection direction and no saturation is observed. However, depth within the mirror region is not correct. The sensor in the second case is on the mirror reflection direction and saturation is observed in the intensity image. This leads to wrong depth calculation.

Figure 1.17 shows samples of specular objects taken by a structured light depth sensor. The mirror region of the first case shows the depth of the reflected surface. The second case also shows a specular region and leads to miscalculation of depth.

In Figure 1.18 we demonstrate how a transparent object affects the sensor measurement. Considering transparent objects with background, depth sensors receive reflected light from both the foreground and background (Figure 1.19). The mixture of reflected light from foreground and background misleads the depth measurement. Depending on the sensor types, however, the characteristics of the errors vary. Since a ToF sensor computes the depth of a transparent object based on the mixture of reflected IR from foreground and background, the depth measurement includes some bias toward the background. For the structured light

Figure 1.17 Specular object examples of structured IR sensor

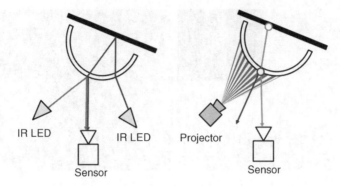

Figure 1.18 Depth of transparent object

Figure 1.19 Transparent object examples

sensor, the active light patterns are used to provide the correspondences between the sensor and projector. With transparent objects, the measurement errors cause the mismatch on correspondences and yield data loss.

In general, a multipath problem (Fuchs, 2010) similar to transparent depth occurs in concave objects, as illustrated in Figure 1.20. In the left image in Figure 1.20, two different IR paths having different flight times from the IR LEDs to the sensor can arrive at the same sensor pixel. The path of the ray reflected twice on the concave surface is a spurious IR signal and distracts from the correct depth calculation of the point using the ray whose path is reflected only once. In principle, a structured light sensor suffers from a similar problem.

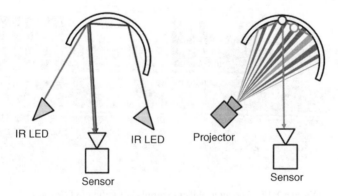

Figure 1.20 Multipath problem

The sensor can observe an unexpected overlapped pattern from multiple light paths if the original pattern is emitted onto a concave surface, as shown on right in Figure 1.20.

1.5 Depth Camera Applications

The recent introduction of cheap depth cameras affects many computer vision and graphics applications. Direct 3D geometry acquisition makes 3D reconstruction easier with a denser 3D point cloud compared with using multiple color cameras and a stereo algorithm. Thus, we do not have to employ the sparse feature matching method with photometry consistency of the Lambertian surface assumption that is fragile in reflective or transparent materials of real-world objects.

1.5.1 Interaction

One of the most attractive applications of early stage of depth cameras is body motion tracking, gesture recognition, and interaction (Shotton *et al.*, 2011). Interactive games or human–machine interaction can employ a depth camera supported by computer vision algorithms as a motion sensor. In this case, higher pixel-wise accuracy and depth precision is not expected. Instead, a faster frame rate and higher spatial resolution are more important for interactive applications such as games with body motion controlling. The system can be trained based on the given input depth data with noise. Furthermore, they are mostly indoor applications that are free from the range ambiguity (Choi *et al.*, 2010; Droeschel *et al.*, 2010) and daylight interference problems of depth cameras.

1.5.2 Three-Dimensional Reconstruction

Another important application of depth cameras is 3D imaging (Henry *et al.*, 2010). This application puts more strict conditions on the depth image. Not just higher frame rate and spatial resolution, but every pixel also has to capture highly accurate depths for precise 3D scene acquisition (Figure 1.21). Furthermore, we have to consider the case of outdoor usage, where a much longer sensing range has to be covered under daylight conditions. Therefore, the range ambiguity now matters in this application. Depth motion blur also has to be removed before the 3D scene reconstruction in order to avoid 3D distortion around edges. A

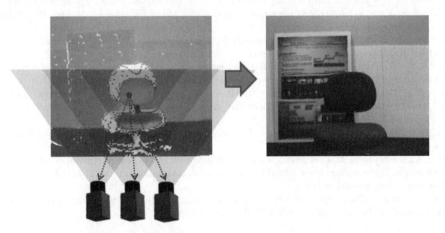

Figure 1.21 A 3D reconstruction example from multiple cameras

reconstructed 3D scene can generate as many view images as required for future 3D TV, such as glassless multiview displays or integral imaging displays (Dolson *et al.*, 2008; Shim *et al.*, 2012). Digital holography displays accept the 3D model and corresponding texture. A mixed reality of the virtual and real worlds is another example of 3D reconstruction application (Ryden *et al.*, 2010; Newcombe *et al.*, 2011).

References

Choi, O., H. Lim, B. Kang, *et al.* (2010) Range unfolding for time-of-flight depth cameras, in *2010 17th IEEE International Conference on Image Processing (ICIP 2010)*, IEEE, pp. 4189–4192.

Dolson, J., J. Baek, C. Plagemann, and S. F Thrun (2008) Fusion of time-of-flight depth and stereo for high accuracy depth maps, in *IEEE Conference on Computer Vision and Pattern Recognition, 2008. CVPR 2008*, IEEE.

Droeschel, D., D. Holz, and S. Behnke (2010) Multi-frequency phase unwrapping for time-of-flight cameras, in *2010 IEEE/RSJ International Conference on Intelligent Robots and Systems (IROS)*, IEEE, pp. 1463–1469.

Edeler, T., K. Ohliger, S. Hussmann, and A. Mertins (2010) Time-of-flight depth image denoising using prior noise information, in *2010 IEEE 10th International Conference on Signal Processing (ICSP)*, IEEE, pp. 119–122.

Foix, S., G. Alenya, and C. Torras (2011) Lock-in time-of-flight (ToF) cameras: a survey. *IEEE Sens. J.*, **11** (9): 1917–1926.

Fuchs, S. (2010) Multipath interference compensation in time-of-flight camera images, in *2010 20th International Conference on Pattern Recognition (ICPR)*, IEEE, pp. 3583–3586.

Henry, P., M. Krainin, E. Herbst *et al.* (2010) RGB-D mapping: using depth cameras for dense 3D modeling of indoor environments. RGB-D: Advanced Reasoning with Depth Cameras Workshop in conjunction with RSS.

Huhle, B., T. Schairer, P. Jenke, and W. Strasser (2008) Robust non-local denoising of colored depth data, in *IEEE Computer Society Conference on Computer Vision and Pattern Recognition Workshops, 2008. CVPR Workshops 2008*, IEEE, pp. 1–7.

Hussmann, S., A. Hermanski, and T. Edeler (2011) Real-time motion artifact suppression in ToF camera systems. *IEEE Trans. Instrum. Meas.*, **60** (5): 1682–1690.

Lee, S., B. Kang, J. Kim, and C. Kim (2012) Motion blur-free time-of-flight range sensor, in *Sensors, Cameras, and Systems for Industrial and Scientific Applications XIII*, (eds R. Widenhorn, V. Nguyen, and A. Dupret), Proceedings of the SPIE, Vol. **8298**, SPIE, Bellingham, WA.

Lindner, M. and A. Kolb (2009) Compensation of motion artifacts for time-of-flight cameras, in *Dynamic 3D Imaging* (eds A. Kolb and R. Koch), Lecture Notes in Computer Science, Vol. **5742**, Springer, pp. 16–27.

Matyunin, S., D. Vatolin, Y. Berdnikov, and M. Smirnov (2011) Temporal filtering for depth maps generated by Kinect depth camera, in *3DTV Conference: The True Vision – Capture, Transmission and Display of 3D Video (3DTV-CON), 2011*, IEEE.

Mure-Dubois, J. and H. Hugli (2007) Real-time scattering compensation for time-of-flight camera, in *Proceedings of the ICVS Workshop on Camera Calibration Methods for Computer Vision Systems – CCMVS2007*, Applied Computer Science Group, Bielefeld University, Germany.

Newcombe, R. A., S. Izadi, O. Hilliges *et al.* (2011) Kinectfusion: real-time dense surface mapping and tracking, in *Proceedings of the 2011 10th IEEE International Symposium on Mixed and Augmented Reality*, IEEE Computer Society, Washington, DC.

Park, J., H. Kim, Y.-W. Tai *et al.* (2011) High quality depth map upsampling for 3D-ToF cameras, in *2011 IEEE International Conference on Computer Vision (ICCV)*, IEEE.

Ryden, F., H. Chizeck, S. N. Kosari *et al.* (2010) Using Kinect and a haptic interface for implementation of real-time virtual mixtures. RGB-D: Advanced Reasoning with Depth Cameras Workshop in conjunction with RSS.

Schuon, S., C. Theobalt, J. Davis, and S. Thrun (2008) High-quality scanning using time-of-flight depth superresolution, in *IEEE Computer Society Conference on Computer Vision and Pattern Recognition Workshops, 2008. CVPR Workshops 2008*, IEEE.

Shim, H., R. Adels, J. Kim *et al.* (2012) Time-of-flight sensor and color camera calibration for multi-view acquisition. *Vis. Comput.*, **28** (12), 1139–1151.

Shotton, J., A. Fitzgibbon, M. Cook, and A. Blake (2011) Real-time human pose recognition in parts from single depth images, in *Proceedings of the 2011 IEEE Conference on Computer Vision and Pattern Recognition*, IEEE Computer Society, Washington, DC.

Yeo, D., E. ul Haq, J. Kim *et al.* (2011) Adaptive bilateral filtering for noise removal in depth upsampling, in *2010 International SoC Design Conference (ISOCC)*, IEEE, pp. 36–39.

2

SFTI: Space-from-Time Imaging

Ahmed Kirmani, Andrea Colaço, and Vivek K. Goyal
Massachusetts Institute of Technology, USA

2.1 Introduction

For centuries, the primary technical meaning of image has been a visual representation or counterpart, formed through the interaction of light with mirrors and lenses, and recorded through a photochemical process. In digital photography, the photochemical process has been replaced by a sensor array, but the use of optical elements is unchanged. Thus, the spatial resolution in this traditional imaging is limited by the quality of the optics and the number of sensors in the array; any finer resolution comes from multiple frames or modeling that will not apply to a generic scene (Milanfar, 2011). A dual configuration is also possible in which the sensing is omnidirectional and the light source is directed, with optical focusing (Sen *et al.*, 2005). The spatial resolution is then limited by the illumination optics, specifically the spot size of the illumination.

Our *space-from-time imaging* (SFTI) framework provides methods to achieve spatial resolution, in two and three dimensions, rooted in the temporal variations of light intensity in response to temporally- or spatiotemporally-varying illumination. SFTI is based on the recognition that parameters of interest in a scene, such as bidirectional reflectance distribution functions at various wavelengths and distances from the imaging device, are embedded in the impulse response (or transfer function) from a light source to a light sensor. Thus, depending on the temporal resolution, any plurality of source–sensor pairs can be used to generate an image; the spatial resolution can be finer than both the spot size of the illumination and the number of sensors.

The use of temporal resolution in SFTI is a radical departure from traditional imaging, in which time is associated only with fixing the period over which light must be collected to achieve the desired contrast. Very short exposures or flash illuminations are used to effectively "stop time" (Edgerton and Killian, 1939), but these methods could still be called *atemporal* because even a microsecond integration time is enough to combine light from a large range of transport paths involving various reflections. No temporal variations are

Emerging Technologies for 3D Video: Creation, Coding, Transmission and Rendering, First Edition.
Frédéric Dufaux, Béatrice Pesquet-Popescu, and Marco Cagnazzo.
© 2013 John Wiley & Sons, Ltd. Published 2013 by John Wiley & Sons, Ltd.

present at this time scale (nor does one attempt to capture them), so no interesting inferences can be drawn.

SFTI introduces inverse problems and is thus a collection of computational imaging problems and methods. Since light transfer is linear, the inverse problems are linear in the intensity or reflectance parameters; however, propagation delays and radial falloff of light intensity cause the inverse problems to be nonlinear in geometric parameters. Formulating specific inverse problem parameterizations that have numerically stable solutions is at the heart of SFTI. This chapter summarizes two representative examples: one for reflectance estimation first reported by Kirmani *et al.* (2011b, 2012) and the other for depth acquisition first reported by Kirmani *et al.* (2011a). We first review basic terminology and related work and then demonstrate how parameters of a static scene are embedded into the impulse response of the scene.

2.2 Background and Related Work

2.2.1 *Light Fields, Reflectance Distribution Functions, and Optical Image Formation*

Michael Faraday was the first to propose that light should be interpreted as a field, much like a magnetic field. The phrase *light field* was coined by Alexander Gershun (Gershun, 1939) to describe the amount of light traveling in every direction, through every point in space, at any wavelength and any time. It is now synonymous with the *plenoptic function* $L(x, y, z, \theta, \lambda, t)$, which is a function of three spatial dimensions, two angular dimensions, wavelength, and time. The time parameter is usually ignored since all measurements are made in steady state. Moreover, light travels in straight lines in any constant-index medium. The macroscopic behavior at a surface is described by the *bidirectional reflectance distribution function* (BRDF) (Nicodemus, 1965). This function of wavelength, time, and four geometric dimensions takes an incoming light direction parameterized by (θ_{in}, ϕ_{in}) and outgoing (viewing) direction $(\theta_{out}, \phi_{out})$, each defined with respect to the surface normal, and returns the scalar ratio of reflected radiance exiting along the viewing angle to the irradiance along the incoming ray's direction. The BRDF determines how a scene will be perceived from a particular viewpoint under a particular illumination. Thus, the scene geometry, BRDF, and scene illumination determine the plenoptic function.

To explain the light field and BRDF concepts, we consider image formation in flatland as shown in Figure 2.1a. A broad wavelength light source (shown in light gray) is radially transmitting pulses of light uniformly in all directions using a pulse intensity modulation function $s(t)$. The scene plane is completely characterized by the locations of its end points and the BRDF of its surface. Assuming that our scene plane is opaque, the BRDF is a scalar function of geometric and material parameters, $b(\alpha, x, y, \theta, \lambda)$. To explain the BRDF, consider a unit-intensity light ray incident on the scene point (x, y) at an incidence angle α. The scene point scatters the different wavelengths composing the light ray differently in all directions θ. The intensity of attenuated light observed at a particular wavelength and at a particular observation angle is equal to the BRDF function value $b(\alpha, x, y, \theta, \lambda)$. The light from the illumination source reaches different points on the scene plane at different incidence angles as well as at different time instances due to the time delay in light propagation. The light field $L(x, y, z, \theta, \lambda, t)$ is then defined as a function describing the intensity of light rays originating from the scene points under a given illumination setting.

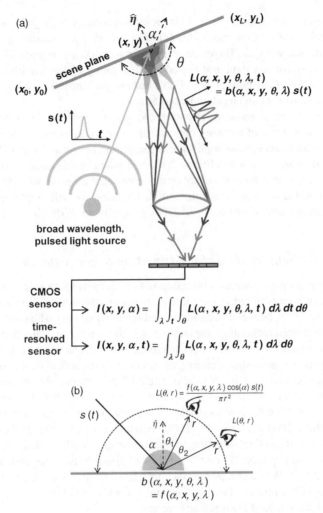

Figure 2.1 Light fields and reflectance distribution functions. (a) The image formation process in flatland. (b) The Lambertian assumption for surface reflectance distribution

In traditional optical imaging, light reflected from the scene plane is focused on a sensor pixel using focusing optics such as lenses. This focusing is equivalent to integration of the light field along θ. Furthermore, a gray-scale CMOS sensor forms an image of the scene $I(x, y, \alpha)$ by integrating along wavelength and time. Thus, traditional image formation loses a lot of valuable scene information contained in the light field. For example, it is impossible to recover scene geometry from a single image $I(x, y, \alpha)$.

Generalizing to a three-dimensional scene, an ordinary digital camera captures a two-dimensional (2D) color image $I(x, y, \alpha, \lambda)$ by marginalizing $L(\alpha, x, y, z, \theta, \lambda, t)$ and sampling the wavelength as

$$I(x, y, \alpha, \lambda) = \int_0^{T_{\exp}} \iiint L(\alpha, x, y, z, \theta, \phi, \lambda, t) \, d\phi \, d\theta \, dz \, dt$$

where T_{exp} represents an exposure time. This projection ignores the angular and time dimensions of the plenoptic function and trades spatial resolution for sampling multiple wavelengths; for example, with the Bayer pattern. Recent work in the area of computational photography has extensively explored the angular sampling of light fields and its applications (Ng *et al.*, 2005; Georgeiv *et al.*, 2006), but the possibilities arising from temporal sampling of light fields remain largely unexplored.

In SFTI, our goals are to reconstruct both scene geometry and reflectance from a single viewpoint, with limited use of focusing optics, from time samples of the incident light field. A practical and widely accepted simplifying assumption about the scene BRDF, called the Lambertian assumption, is shown in Figure 2.1b. According to this assumption, the BRDF is independent of the viewing angle θ and we observe the same intensity of light from all directions at the same radial distance. For simplicity and clarity, we will employ the Lambertian model; incorporating some other known BRDF is generally not difficult.

2.2.2   Time-of-Flight Methods for Estimating Scene Structure

The speed of light enables distances (longitudinal resolution) to be inferred from the time difference between a transmitted pulse and the arrival of a reflection from a scene or, similarly, by the phase offset between transmitted and received signals when these are periodic. This is a well-established technology for ranging, which we refer to as depth map acquisition when spatial (transverse) resolution is also acquired. A light detection and ranging (LIDAR) or laser radar system uses raster scanning of directed illumination of the scene to obtain spatial resolution (Schwarz, 2010). A time-of-flight (TOF) camera obtains spatial resolution with an array of sensors (Gokturk *et al.*, 2004; Foix *et al.*, 2011). These TOF-based techniques have better range resolution and robustness to noise than using stereo disparity (Forsyth and Ponce, 2002; Seitz *et al.*, 2006; Hussmann *et al.*, 2008) or other computer vision techniques – including structured-light scanning, depth-from-focus, depth-from-shape, and depth-from-motion (Forsyth and Ponce, 2002; Scharstein and Szeliski, 2002; Stoykova *et al.*, 2007). While companies such as Canesta, MESA Imaging, 3DV, and PMD offer commercial TOF cameras, these systems are expensive and have low spatial resolution when compared with standard 2D imaging cameras.

All previous techniques using TOF have addressed only the estimation of structure (e.g. three-dimensional (3D) geometry of scenes and shapes of biological samples). SFTI provides methods for estimation of both reflectance and structure. Furthermore, unlike previous methods – including notably the method of Kirmani *et al.* (2009) using indirect illumination and sensing – SFTI is based on intensity variation as a function of time rather than collections of distance measurements.

2.2.3   Synthetic Aperture Radar for Estimating Scene Reflectance

The central contribution of SFTI is to use *temporal* information, in conjunction with appropriate post-measurement signal processing, to form images whose *spatial* resolution exceeds the spatial resolution of the illumination and light collection optics. In this connection it is germane to compare SFTI with synthetic aperture radar (SAR), which is a well-known microwave approach for using time-domain information plus post-measurement signal processing to form images with high spatial resolution (Kovaly, 1976; Munson *et al.*, 1985; Cutrona, 1990). In stripmap mode, an airborne radar transmits a sequence of high-bandwidth pulses on a fixed slant angle toward the ground. Pulse-compression reception of individual pulses provides across-track spatial resolution superior to that of the radar's antenna pattern as the range response of the compressed pulse sweeps across the ground plane. Coherent integration over many pulses provides along-track spatial resolution by forming a synthetic aperture whose diffraction limit is much smaller than that of the radar's antenna pattern.

SAR differs from SFTI in two general ways. First, SAR requires the radar to be in motion, whereas SFTI does not require sensor motion. Second, SAR is primarily a microwave technique, and most real-world objects have specular BRDFs at microwave wavelengths. With specular reflections, an object is directly visible only when the angle of illumination and angle of observation satisfy the law of reflection, and multiple reflections – which are not accounted for in first-order SAR models – are strong. On the other hand, most objects are Lambertian at optical wavelengths, so optical SFTI avoids these sources of difficulty.

2.3 Sampled Response of One Source–Sensor Pair

Broadly stated, SFTI is any computational imaging in which spatial resolution is derived from time-resolved sensing of the response to time-varying illumination. The key observation that enables SFTI is that scene information of interest can be embedded in the impulse response (or transfer function) from a light source to a light sensor. Here, we develop an example of an impulse response model and sampling of this response to demonstrate the embedding that enables imaging. The following sections address the specific inverse problems of inferring scene reflectance and structure.

2.3.1 Scene, Illumination, and Sensor Abstractions

Consider a static 3D scene to be imaged in the scenario depicted in Figure 2.2. We assume that the scene is contained in a cube of dimensions $L \times L \times L$ for some finite L. Further assume that the scene surfaces are all Lambertian so that their perceived brightnesses are invariant to the angle of observation (Oren and Nayar, 1995); incorporation of any known BRDF would not add insight to our model. Under these assumptions, the scene at one wavelength or collection of wavelengths can be completely represented as a 3D function $f : [0, L]^3 \to [0, 1]$. For any $\mathbf{x} = (x_1, x_2, x_3) \in [0, L]^3$, the value of the function $f(\mathbf{x})$ represents the radiometric response; $f(\mathbf{x}) = 0$ implies that there is no scene point present at position $\mathbf{x}$, and $f(\mathbf{x}) > 0$ implies that there is a scene point present with a reflectance value given by $f(\mathbf{x})$. We assume that all surface points have nonzero reflectance to avoid ambiguity. To incorporate dependence on wavelength, the codomain of f could be made multidimensional.

The single illumination source is monochromatic and omnidirectional with time-varying intensity denoted by $s(t)$. The single time-resolved sensor collects light from all directions uniformly; that is, it lacks any spatial resolution. We model the sensor as a linear, time-invariant (LTI) filter that has a known impulse response $h(t)$. The source and the sensor are assumed to be synchronized in time and their positions are assumed to be known with respect to a global coordinate system.

Every ray of light originating at the source either does not interact with the scene at all or is reflected off a scene point, resulting in its attenuation and diffusion. This diffused light is collected by the sensor. We assume no occlusions in the light path starting from the source to any scene point and ending at the sensor.

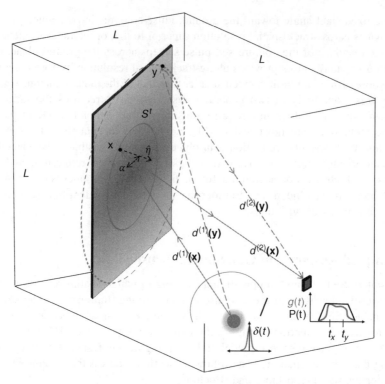

Figure 2.2 Space-from-time imaging setup. An impulsive illumination source transmits a spherical wavefront towards a scene plane with nonconstant reflectance. The light reflected from the plane is time sampled at a sensor, which has no focusing optics and, therefore, receives contributions from all the scene points. The scene impulse response at any time, shown as the gray waveform, is determined by the elliptical radon transform (ERT) defined in (2.3). For a scene plane with constant reflectance, the scene impulse response (shown in black) is a parametric signal; it is a piecewise linear function as described in (2.14)

2.3.2   Scene Response Derivation

At any given time, the light incident at the sensor is a linear combination of the time-delayed reflections from all points on the scene surface. For any point $\mathbf{x}$, let $d^{(1)}(\mathbf{x})$ denote the distance from illumination source to $\mathbf{x}$ and let $d^{(2)}(\mathbf{x})$ denote the distance from $\mathbf{x}$ to the sensor (see Figure 2.2). Then, $d(\mathbf{x}) = d^{(1)}(\mathbf{x}) + d^{(2)}(\mathbf{x})$ is the total distance traveled by the contribution from $\mathbf{x}$. This contribution is attenuated by the reflectance $f(\mathbf{x})$, square-law radial falloff, and $\cos(\alpha(\mathbf{x}))$ to account for foreshortening of the surface with respect to the illumination as per the Lambertian reflectance model (see Figure 2.1b), where $\alpha(\mathbf{x})$ is the angle between the surface normal at $\mathbf{x}$ and a vector from $\mathbf{x}$ to the illumination source. Thus, when the intensity of the omnidirectional illumination is a unit impulse at time 0, denoted $s(t) = \delta(t)$, the contribution from point $\mathbf{x}$ is the light intensity signal $a(\mathbf{x})f(\mathbf{x})\delta(t - d(\mathbf{x}))$, where we have normalized to unit speed of light and

$$a(\mathbf{x}) = \frac{\cos(\alpha(\mathbf{x}))}{\left(d^{(1)}(\mathbf{x})d^{(2)}(\mathbf{x})\right)^2}. \tag{2.1}$$

Combining contributions over the plane, the total light incident at the sensor is

$$g(t) = \int_0^L \int_0^L \int_0^L a(\mathbf{x}) f(\mathbf{x}) d(t - d(\mathbf{x})) \, dx_1 \, dx_2 \, dx_3. \tag{2.2}$$

Thus, evaluating $g(t)$ at a fixed t amounts to integrating over $\mathbf{x} \in [0, L]^3$ with $d(\mathbf{x}) = t$. Define the *isochronal surface* $S^t = \{x \in [0, L]^3 : d(\mathbf{x}) = t\}$. Then

$$g(t) = \iint a(\mathbf{x}(u_1, u_2)) f(\mathbf{x}(u)) \, du_1 \, du_2 \tag{2.3}$$

where $\mathbf{x}(u_1, u_2)$ is a parameterization of $S^t \cap [0, L]^3$ with unit Jacobian. The intensity of light incident at the sensor $g(t)$ thus contains the surface integrals over S^t of the desired function f. Each S^t is a level surface of $d(\mathbf{x})$ for a fixed t; as illustrated in Figure 2.2, these are concentric, 3D ellipsoids with varying eccentricities that are governed by t and foci fixed at the source and sensor locations. Note that, under our aforementioned assumptions, $g(t)$ is the impulse response of the scene because light transport is an LTI phenomenon.

A digital computational recovery can use only samples rather than a continuous-time function directly. We measure samples of the scene response $r(t)$, which is the convolution of source signal $s(t)$, scene impulse response $g(t)$, and sensor impulse response $h(t)$ (see Figure 2.3). The sensor noise is represented by $\eta(t)$. Except at very low photon flux, $\eta(t)$ is modeled well as signal independent, zero mean, white and Gaussian.

We now see how uniform sampling of $r(t)$ relates to linear functional measurements of f. This also establishes the foundations of a Hilbert space view of SFTI. Suppose discrete samples are obtained at the sensor using a sampling interval of T:

$$r[n] = (\{s(t) * h(t)\} * g(t))|_{t=nT}, \qquad n = 1, 2, \ldots, N. \tag{2.4}$$

To simplify notation, we combine the effect of the source and sensor impulse responses into an LTI prefilter denoted by $\tilde{h}$. Then, a sample $r[n]$ can be seen as a standard $L^2(\mathbb{R})$ inner product between g and a time-reversed and shifted $\tilde{h}$ (Unser, 2000; Vetterli *et al.*, 2013):

$$r[n] = \langle g(t), \tilde{h}(nT - t) \rangle. \tag{2.5}$$

Using (2.2), we can express (2.5) in terms of f using the standard $L^2([0, L]^2)$ inner product:

$$r[n] = \langle f, \varphi_n \rangle$$

where

$$\varphi_n(\mathbf{x}) = a(\mathbf{x}) \tilde{h}(nT - d(\mathbf{x})).$$

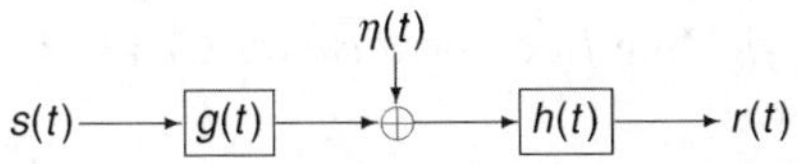

Figure 2.3   Block diagram abstraction for signal sampled at sensor

2.3.3 Inversion

Our general inverse problem from this formulation is to estimate the scene structure and reflectance $f(\mathbf{x})$ from measured samples. It is clearly not possible to do this when the samples are obtained from a single sensor and the illumination has a single source. Even if we had $g(t)$ rather than samples of a lowpass filtered version of $g(t)$, the collapse of all information from isochronal surface S^t to a single time instant is irreversible. Thus, we must have a plurality of source–sensor pairs.

It is worthwhile to note that (2.2) and (2.3) are a one-dimensional projection of an *elliptical Radon transform* (ERT) of the spatial function $f(\mathbf{x})$, as often defined in integral geometry (Gouia, 2011; Krishnan *et al.*, 2012). A full ERT would require one or both ellipsoid foci to be varied; this corresponds to a moving source or sensor. ERTs are generally not invertible and are notoriously ill-conditioned (Coker and Tewfik, 2007; Gouia-Zarrad and Ambartsoumian, 2012).

In the next two sections, we discuss accurate and robust imaging despite noisy observations and limited diversity in source and sensor locations. In the first case we assume that the scene geometry (i.e., support of $f(\mathbf{x})$) is known, and we use SFTI principles to recover scene reflectance using only diffuse light; we call this *diffuse imaging*. In the second case, the scene reflectance is assumed to be constant (e.g., $f(\mathbf{x}) = 1$ on its support, but the support is unknown), and we demonstrate how to use SFTI for estimating scene structure of piecewise planar scenes; we call this *compressive depth acquisition* (CoDAC).

2.4 Diffuse Imaging: SFTI for Estimating Scene Reflectance

Consider the imaging of a plane with known position, orientation, and dimensions ($L \times L$). Since the plane's geometry is known, the SFTI problem reduces to the recovery of the reflectance pattern on the plane's surface, which is a gray-scale image. The 3D function $f(\mathbf{x})$ can now be reduced to a 2D function $f : [0, L]^2 \rightarrow [0, 1]$, with coordinate system now defined on the plane.

As discussed in Section 2.3.2, time-resolved measurements of the scene impulse response $g(t)$ using a single source–sensor pair is not enough to uniquely recover the scene reflectance function $f(\mathbf{x})$. Therefore, we add a small number of identical, omnidirectional, time-resolved sensors indexed by $k \in \{1, 2, \ldots, K\}$ at a regular grid spacing as depicted in Figure 2.4. As before, we have a single, monochromatic, omnidirectional illumination source, with time-varying intensity $s(t)$.

2.4.1 Response Modeling

Let $\mathbf{x} = (x_1, x_2) \in [0, L]^2$ be a point on the plane to be imaged. Generalizing the distances defined in Section 2.3.2, let $d_k^{(2)}(\mathbf{x})$ denote the distance from $\mathbf{x}$ to sensor k and let $d_k(\mathbf{x}) = d^{(1)}(\mathbf{x}) + d_k^{(2)}(\mathbf{x})$ denote the total distance traveled by the contribution from $\mathbf{x}$. Since the scene is restricted to a plane, instead of entire isochronal surfaces, only the intersections with the plane are relevant, yielding *isochronal curves* $C_k^t = \{x \in [0, L]^2 : d_k(\mathbf{x}) = t\}$. Each isochronal curve is an ellipse.

Assume for the moment $s(t) = \delta(t)$; then, following our derivation of scene impulse response from Section 2.3.2, we arrive at the following formula for the light incident at

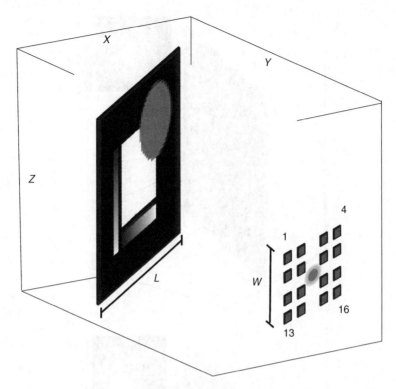

Figure 2.4 An omnidirectional source illuminates a planar scene, and the diffused reflected light is measured at a 4 × 4 array of unfocused omnidirectional sensors

sensor k at time t:

$$
\begin{aligned}
g_k(t) &= \int_0^L \int_0^L a_k(\mathbf{x})f(\mathbf{x})\delta(t - d_k(\mathbf{x}))\, dx_1\, dx_2 \\
&= \int_{C_k^t} a_k(\mathbf{x})f(\mathbf{x})\, ds = \int a_k(\mathbf{x}(u))f(\mathbf{x}(u))\, du
\end{aligned}
\tag{2.6}
$$

where

$$
a_k(\mathbf{x}) = \frac{\cos(\alpha(\mathbf{x}))}{(d^{(1)}(\mathbf{x})d_k^{(2)}(\mathbf{x}))^2}.
\tag{2.7}
$$

Examples of distance functions and geometric attenuation factors for different sensors are shown in Figure 2.5.

Now suppose that discrete samples are obtained at sensor k with sampling prefilter $h_k(t)$ and sampling interval T. Following the derivation in Section 2.3.2 and defining $\tilde{h}_k = s * h_k$, we obtain the following expression for discrete-time samples of the scene response at sensor k:

$$
r_k[n] = (g_k(t) * \tilde{h}_k(t))\big|_{t=nT_k} = \langle g_k(t), \tilde{h}_k(nT_k - t) \rangle
\tag{2.8}
$$

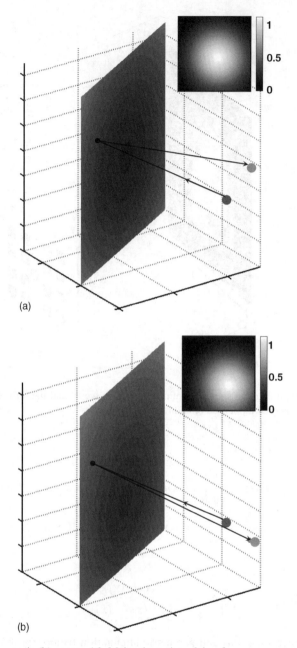

(a)

(b)

Figure 2.5 (a) Sensor 4; (b) sensor 16. Main plots: time delay from source to scene to sensor is a continuous function of the position in the scene. Insets: the geometric attenuation of the light is also a continuous function of the position in the scene in (2.7)

for $n = 1, 2, \ldots, N$. Using (2.6), we can express (2.8) in terms of f using the standard $L^2([0, L]^2)$ inner product:

$$r_k[n] = \langle f, \varphi_{k,n} \rangle \tag{2.9a}$$

where

$$\varphi_{k,n}(\mathbf{x}) = a_k(\mathbf{x})\tilde{h}_k(nT_k - d_k(\mathbf{x})). \tag{2.9b}$$

Over a set of sensors and sample times, $\{\varphi_{k,n}\}$ will span a subspace of $L^2([0, L]^2)$, and a sensible goal is to form a good approximation of f in that subspace.

A common practical model for the prefilter impulse response is the "integrate and dump" or time-gated sampling:

$$h(t) = \begin{cases} 1, & \text{for } 0 \le t \le T \\ 0, & \text{otherwise} \end{cases} . \tag{2.10}$$

Now, since $h(t)$ is nonzero only for $t \in [0, T]$, by (2.8), the sample $r_k[n]$ is the integral of $g_k(t)$ over $t \in [(n-1)T, nT]$. Thus, under the idealization $s(t) = \delta(t)$, by (2.6), $r_k[n]$ is an α-weighted integral of f between the contours $C_k^{(n-1)T}$ and C_k^{nT}. To interpret this as an inner product with f as in (2.9), we see that $\varphi_{k,n}(\mathbf{x})$ is $a_k(\mathbf{x})$ between $C_k^{(n-1)T}$ and C_k^{nT} and zero otherwise. Figure 2.6a shows a single representative $\varphi_{k,n}$. The functions $\{\varphi_{k,n}\}_{n \in \mathbb{Z}}$

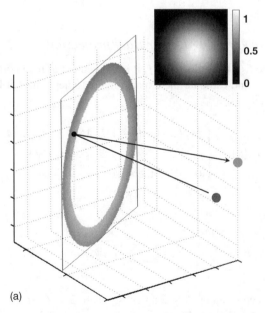

(a)

Figure 2.6 (a) Sensor 4. A single measurement function $\varphi_{k,n}(\mathbf{x})$ when $h(t)$ is the box function given in (2.10). Inset is unchanged from Figure 2.5a. (b) Sensor 16. When $h(t)$ is the box function given in (2.10), the measurement functions for a single sensor $\{\varphi_{k,n}(\mathbf{x})\}_{n=1}^{N}$ partition the plane into elliptical annuli. Shading is by discretized delay using the scale of Figure 2.5. Also overlaid is the discretization of the plane of interest into an $M \times M$ pixel array

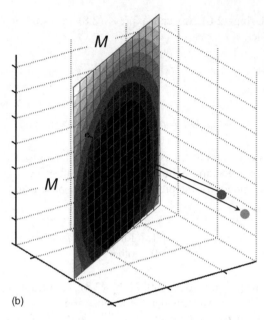

(b)

Figure 2.6 (*Continued*)

for a single sensor have disjoint supports; their partitioning of the domain $[0, L]^2$ is illustrated in Figure 2.6b.

2.4.2 Image Recovery using Linear Backprojection

To express an estimate $\hat{f}$ of the reflectance f, it is convenient to fix an orthonormal basis for a subspace of $L^2([0, L]^2)$ and estimate the expansion coefficients in that basis. For an $M \times M$ pixel representation, let

$$\psi_{i,j}(\mathbf{x}) = \begin{cases} M/L, & \text{for } (i-1)L/M \leq x_1 \leq iL/M, \\ & (j-1)L/M \leq x_2 \leq jL/M \\ 0, & \text{otherwise} \end{cases} \tag{2.11}$$

so that

$$\hat{f} = \sum_{i=1}^{M} \sum_{j=1}^{M} c_{i,j} \psi_{i,j}$$

in the span of $\{\psi_{i,j}\}$ is constant on $\Delta \times \Delta$ patches, where $\Delta = L/M$.

For $\hat{f}$ to be consistent with the value measured by sensor k at discrete time n, we must have

$$r_k[n] = \langle \hat{f}, \varphi_{k,n} \rangle = \sum_{i=1}^{M} \sum_{j=1}^{M} c_{i,j} \langle \psi_{i,j}, \varphi_{k,n} \rangle. \tag{2.12}$$

Note that the inner products $\{\langle \psi_{i,j}, \varphi_{k,n} \rangle\}$ exclusively depend on Δ, the positions of illumination and sensors, the plane geometry, the effective sampling prefilter $\tilde{h}(t)$, and the sampling interval T – not on the unknown reflectance of interest f. Hence, we have a system of linear equations to solve for the coefficients $\{c_{i,j}\}$. (In the case of the basis given in (2.11), the coefficients are the pixel values multiplied by Δ.)

When we specialize to the box sensor impulse response in (2.10) and the basis in (2.11), many inner products $\langle \psi_{i,j}, \varphi_{k,n} \rangle$ are zero and so the linear system is sparse. The inner product $\langle \psi_{i,j}, \varphi_{k,n} \rangle$ is nonzero when reflection from the (i,j) pixel affects the light intensity at sensor k within time interval $[(n-1)T, nT]$. Thus, for a nonzero inner product the (i,j) pixel must intersect the elliptical annulus between $C_k^{(n-1)T}$ and C_k^{nT}. With reference to Figure 2.6a, this occurs for a small fraction of (i,j) pairs unless M is small or T is large. The value of a nonzero inner product depends on the fraction of the square pixel that overlaps with the elliptical annulus and the geometric attenuation factor $a_k(\mathbf{x})$.

To express the linear system given in (2.12) with a matrix multiplication, replace double indexes with single indexes (i.e., vectorize, or reshape) as

$$\mathbf{y} = \mathbf{A}\,\mathbf{c} \tag{2.13}$$

where $\mathbf{y} \in \mathbb{R}^{KN}$ contains the data samples $\{r_k[n]\}$, the first N from sensor 1, the next N from sensor 2, and so on; and $\mathbf{c} \in \mathbb{R}^{M^2}$ contains the coefficients $\{c_{i,j}\}$, varying i first and then j. Then $\langle \psi_{i,j}, \varphi_{k,n} \rangle$ appears in row $(k-1)N + n$, column $(j-1)M + i$ of $\mathbf{A} \in \mathbb{R}^{KN \times M^2}$. Figure 2.7 illustrates an example of the portion of $\mathbf{A}$ corresponding to sensor 1 for the scene in Figure 2.4.

Assuming that $\mathbf{A}$ has a left inverse (i.e., rank $(\mathbf{A}) = M^2$), one can form an image by solving (2.13). The portion of $\mathbf{A}$ from one sensor cannot have full column rank because of the collapse of information along elliptical annuli depicted in Figure 2.6a. Full rank can be achieved with an adequate number of sensors, noting that sensor positions must differ to increase rank, and greater distance between sensor positions improves conditioning, as shown in our simulations in Kirmani *et al.* (2011b, 2012).

Impulsive illumination severely limits the illumination energy, leading to poor SNR, especially due to the two radial falloff attenuations in (2.7). In our recent work (Kirmani *et al.*, 2011b, 2012), we have generalized the above example of imaging planar scene reflectance using impulsive illumination to bandlimited sources. Our simulations show that diffuse

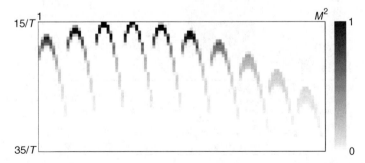

Figure 2.7 Visualizing linear system representation $\mathbf{A}$ in (2.13). Contribution from sensor 1 is shown. The matrix has 100 columns (since $M = 10$) and 40 rows (through choice of T). All-zero rows arising from times prior to first arrival of light and after the last arrival of light are omitted

imaging can be implemented with practical opto-electronic hardware used in optical communications. Our initial work on diffuse imaging also inspires detailed study of how pixel size, sensor locations, sampling prefilters, and sampling rates affect the conditioning of the inverse problem.

2.5 Compressive Depth Acquisition: SFTI for Estimating Scene Structure

We now turn our attention to the problem of scene structure recovery. For simplicity in explaining the basic principles, we initially consider a scene containing a single planar surface, and we assume the reflectance on that surface is constant; in the abstraction developed in Section 2.3, $f(\mathbf{x})$ is nonzero only on one planar facet in $[0, L]^3$, and the intensity of response from that facet does not vary with $\mathbf{x}$ within the facet. Our goal is to estimate the plane's position, orientation, and dimensions.

The response modeling in Section 2.3.2 applies, but as was made more explicit in Section 2.4.2, the difficulty is in the form of the inverse problem. The forward operator $\mathbf{A}$ in (2.13) is a nonlinear function of the plane's geometric parameters that we are trying to estimate. Thus, estimating the geometric parameters – even if the reflectances in $\mathbf{c}$ are known – is generally a difficult nonlinear inverse problem.

In order to make this problem tractable, we need to modify our response modeling. We will also develop a way to acquire spatial information distinct from the use of a small array of sensors as in Section 2.4.

2.5.1 Single-Plane Response to Omnidirectional Illumination

Consider a scene containing a single planar facet occupying solid angle $[\theta_1, \theta_2] \times [\phi_1, \phi_2]$ as shown in Figure 2.8a. Assume a unit-intensity illumination pulse $s(t) = \delta(t)$ with the source at the origin, and assume a single photodetector is also at the origin.

For any point $\mathbf{x} \in \mathbb{R}^3$ on the facet, light originating at the source will be reflected from $\mathbf{x}$, attenuated due to scattering, and arrive back at the detector delayed in time by an amount

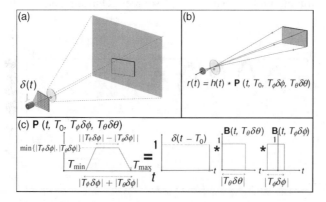

Figure 2.8 (a) A scene containing a single planar surface. (b) The photodetector output (prior to sampling) is the detector response $h(t)$ convolved with the scene impulse response $P(t, T_0, T_\phi \delta\phi, T_\theta \delta\theta)$. (c) Diagrammatic explanation of the parametric signal $P(t, T_0, T_\phi \delta\phi, T_\theta \delta\theta)$

proportional to the distance $2\|\mathbf{x}\|$. Thus, the signal incident on the photodetector in response to impulse illumination of a patch of differential area $d\mathbf{x}$ centered at $\mathbf{x}$ is

$$g_{\mathbf{x}}(t) = a\delta(t - 2\|\mathbf{x}\|)\,d\mathbf{x}$$

where a is the total attenuation (transmissivity), which is assumed to be constant across the facet. (Ignoring variations in radial falloff is a far-field assumption.) Denoting the photodetector impulse response by $h(t)$, the electrical output of the photodetector due only to this patch is

$$r_{\mathbf{x}}(t) = h(t) * g_{\mathbf{x}}(t) = ah(t - 2\|\mathbf{x}\|)\,d\mathbf{x}.$$

The response from the whole facet is obtained by integrating $r_{\mathbf{x}}(t)$ over $\theta \in [\theta_1, \theta_2]$ and $\phi \in [\phi_1, \phi_2]$. A calculation detailed in Kirmani $et\ al.$ (2011a) shows that, under an appropriate small-angle approximation, the full response is

$$r(t) = \frac{a}{T_\phi T_\theta} h(t) * P(t, T_0, T_\phi\delta\phi, T_\theta\delta\theta) \tag{2.14}$$

where $P(t, T_0, T_\phi\delta\phi, T_\theta\delta\theta)$ is defined graphically in Figure 2.8c with

$$T_\phi = \frac{2d_\perp}{\gamma(\phi_1, \theta_1)}\tan\phi_1\sec^2\phi_1, \qquad T_\theta = \frac{2d_\perp}{\gamma(\phi_1, \theta_1)}\tan\theta_1\sec^2\theta_1,$$

$T_0 = 2d_\perp\gamma(\phi_1, \theta_1)$, $\gamma(\phi_1, \theta_1) = \sqrt{\sec^2\phi_1 + \tan^2\theta_1}$, $\delta\theta = \theta_2 - \theta_1$, $\delta\phi = \phi_2 - \phi_1$, and $d_\perp$ is the distance between the illumination source and the nearest point on the plane containing the facet.

Recall now the general principle of SFTI that scene information of interest can be embedded in the scene impulse response. Here, that impulse response is specifically a scalar multiple of the parametric function $P(t, T_0, T_\phi\delta\phi, T_\theta\delta\theta)$ – a trapezoid with corners specified by the dimensions of the facet and its distance from the imaging device.

Estimating the function $P(t, T_0, T_\phi\delta\phi, T_\theta\delta\theta)$ by processing the digital samples $r[k]$ of function $r(t)$ enables estimation of various parameters including $T_{\min}$ and $T_{\max}$, which correspond to the nearest and farthest scene points. The detector impulse response $h(t)$ is generally modeled as a lowpass filter. Thus, the general deconvolution problem of obtaining $P(t, T_0, T_\phi\delta\phi, T_\theta\delta\theta)$ from samples $r[k]$ is ill-posed and highly sensitive to noise. However, our modeling shows that the light transport function $P(t, T_0, T_\phi\delta\phi, T_\theta\delta\theta)$ is piecewise linear. This knowledge makes the recovery of $P(t, T_0, T_\phi\delta\phi, T_\theta\delta\theta)$ a $para$-$metric\ deconvolution$ problem that we solve using the parametric signal processing framework described in Blu $et\ al.$ (2008). Because of the use of parametric processing, the estimation of distance parameters $T_{\min}$ and $T_{\max}$ can be very accurate relative to the bandwidth of $h(t)$.

Note now what is not contained in $P(t, T_0, T_\phi\delta\phi, T_\theta\delta\theta)$: the position of the planar facet within the field of view; that is, any spatial resolution. This is intuitive because the illumination and sensing are omnidirectional. To obtain spatial resolution we introduce spatially patterned illumination or sensing.

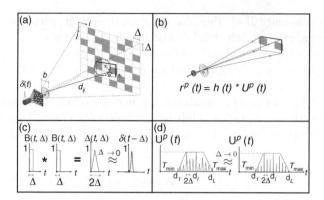

Figure 2.9 (a) Binary spatial patterning. (b) Scene response with spatial patterning. (c) Diagrammatic explanation of the response from one small pixel (small Δ). (d) Modeling of the parametric signal $U^p(t)$ as a weighted sum of equally spaced Dirac deltas. Note that $U^p(t)$ has the same time envelope as the signal $P(t, T_0, T_\phi \delta \phi, T_\theta \delta \theta)$

2.5.2 Spatially-Patterned Measurement

We have thus far discussed inferences from samples of the filtered full scene response $r(t)$. It lacks spatial information because $r(t)$ contains the aggregate contribution of $r_x(t)$ for all points x in the scene. If we were to integrate only over a subset of points in the scene and had such measurements for several different subsets, it would be possible to make some spatial inferences. We have experimentally achieved such partial spatial integration in two complementary ways: in Kirmani *et al.* (2011a) we illuminated subsets of the scene by using a spatial light modulator (SLM), still collecting light at a single omnidirectional sensor; in Colaço *et al.* (2012) we collected light from subsets of the scene by focusing onto a digital micromirror device (DMD), still using omnidirectional illumination. Similar mathematical modeling applies in both cases.

Figure 2.9a illustrates spatially patterned measurement by highlighting a subset of a pixelated version of the field of view (FOV). The pixels could be generated by either patterned illumination or patterned sensing; we will henceforth make no distinction between these. The pixels discretize the FOV into an $N \times N$ array of small squares of size $\Delta \times \Delta$. We index both the pixels by (i,j). The binary spatial pattern is indexed by p, giving $\{c_{i,j}^p\}_{i,j=1}^N$, where each $c_{i,j}^p$ is either zero or one. Let $d_{i,j}$ denote the depth in the direction of pixel (i,j). The depths for all (i,j) associated with the scene form the *depth map* $\mathbf{D} = \{\mathbf{D}_{i,j}\}_{i,j=1}^N$, where $\mathbf{D}_{i,j} = d_{i,j}$ if rays along pixel (i,j) intersect the rectangular facet and $\mathbf{D}_{i,j} = 0$ otherwise.

Analysis of the scene response is detailed in Kirmani *et al.* (2011a). To summarize, since the response of a subset of the scene is measured and the light intensity signals are nonnegative, the response to pattern p is a function $U^p(t)$ that lies under $P(t, T_0, T_\phi \delta \phi, T_\theta \delta \theta)$. Ideally, the fraction $U^p(t)/P(t, T_0, T_\phi \delta \phi, T_\theta \delta \theta)$ equals the fraction of the scene at distance $t/2$ that is included in pattern p. With a discretization of the depth (known to be limited to $T_{\min}/2$ to $T_{\max}/2$), we can introduce auxiliary variables that ultimately yield a linear forward model for acquisition. Fix L distinct depth values $\{d_1, d_2, \ldots, d_L\}$. Then define the ℓth *depth mask* through $I_{i,j}^\ell = 1$ if $d_{i,j} = d_\ell$ and $I_{i,j}^\ell = 0$ otherwise. Clearly, $D_{i,j} = \sum_{\ell=1}^L d_\ell I_{i,j}^\ell$.

As detailed in Kirmani *et al.* (2011a), samples of $r^p(t) = h(t) * U^p(t)$ can be used to compute the relative intensity $U^p(t)/P(t, T_0, T_\phi\delta\phi, T_\theta\delta\theta)$ at $t/2 \in \{d_1, d_2, \ldots, d_L\}$. By combining the data from M spatial patterns, after normalization, we obtain the linear system of equations

$$\underbrace{\mathbf{y}}_{M\times L} = \underbrace{\mathbf{C}}_{M\times N^2} \underbrace{[I^1 \quad I^2 \quad \cdots \quad I^L]}_{N^2\times L} \tag{2.15}$$

where $\mathbf{y}$ holds the processed, acquired data; $\mathbf{C}$ the binary spatial patterns, reshaped into $1 \times N^2$ row vectors; and each I^ℓ the binary auxiliary variables $\{I^\ell_{i,j}\}^N_{i,j=1}$, reshaped into an $N^2 \times 1$ column vector. This system of equations has N^2L unknowns and ML constraints.

2.5.3 Algorithms for Depth Map Reconstruction

The acquisition methodology with spatial patterning is attractive when the number of patterns M is smaller than the number of pixels in the depth map N^2. However, this causes (2.15) to be underdetermined. To be able to find a solution, we exploit the structure of scene depth.

When a scene contains only planar facets, its depth map $\mathbf{D}$ has a sparse second-order finite difference. Thus, it is possible to recover $\mathbf{D}$ from $\mathbf{y}$ by solving the following constrained ℓ_1-regularized optimization problem:

$$\arg\min_{\mathbf{D}}\|\mathbf{y} - \mathbf{C}[I^1 \quad I^2 \quad \cdots \quad I^L]\|^2_{\mathrm{F}} + \|\Phi\mathbf{D}\|_1 \tag{2.16a}$$

subject to

$$\sum_{\ell=1}^{L} d_\ell I^\ell = \mathbf{D} \tag{2.16b}$$

$$\sum_{\ell=1}^{L} I^\ell_{i,j} = 1 \quad \text{for} \quad (i,j) \in \{1, 2, \ldots, N\}^2 \tag{2.16c}$$

and

$$I^\ell_{i,j} \in \{0,1\} \text{ for } \ell \in \{1, 2, \ldots, L\} \quad \text{and} \quad (i,j) \in \{1, 2, \ldots, N\}^2 \tag{2.16d}$$

where $\|\cdot\|_{\mathrm{F}}$ is the Frobenius norm and Φ is the second-order finite-difference operator matrix.

While this optimization problem already contains a convex relaxation in its use of $\|\Phi\mathbf{D}\|_1$, it is nevertheless computationally intractable because of the integrality constraints in (2.16d). Using a further relaxation of replacing (2.16d) with

$$I^\ell_{i,j} \in [0,1] \text{ for } \ell \in \{1, 2, \ldots, L\} \quad \text{and} \quad (i,j) \in \{1, 2, \ldots, N\}^2 \tag{2.16e}$$

yields a tractable formulation. We solved this convex optimization problem using CVX, a package for specifying and solving convex programs (Grant and Boyd, 2011).

To summarize this section on estimating planar scene structure using SFTI, we have presented a method for acquiring 2D depth maps of piecewise-planar scenes using samples measured by a single photodetector with spatially-patterned integration. In contrast to the 2D laser scanning in LIDAR systems and the focused 2D sensor array in TOF cameras, our acquisition architecture consists of a single photodetector and a nonscanning 2D SLM or DMD. In Kirmani *et al.* (2011a) we experimentally demonstrated the acquisition of scene depth at both high range resolution and high spatial resolution with significantly reduced device complexity and hardware cost as compared with state-of-the-art LIDAR systems and TOF cameras. In Colaço *et al.* (2012) we experimentally demonstrated depth acquisition with a highly sensitive Geiger-mode avalanche photodiode detector, including in the presence of an unknown partially-transmissive occluder. These novel capabilities are achieved by developing an SFTI depth acquisition framework based on parametric signal processing that exploits the sparsity of natural scene structures. We refer the reader to Kirmani *et al.*, (2011a) and Colaço *et al.* (2012) for additional details on theory, extensions, and proof-of-concept experiments.

2.6   Discussion and Future Work

SFTI is a novel framework for optical imaging where spatial resolution is obtained through computational processing of time-varying light intensities. While TOF has commonly been used to measure *distances*, the key innovation is to use *intensity variation with time* to create new computational optical imaging modalities. The linearity of a forward model and the plausibility of inversion depend on what is unknown and how it is parameterized. The most straightforward – but still novel – setting has a planar scene in a known position with unknown reflectance. The reflectance can be computed from measurements by a few omnidirectional sensors in response to a single time-varying omnidirectional illumination (Kirmani *et al.*, 2011b, 2012).

Our first experimental verification of this form of SFTI was in a setting where the omnidirectional illumination came from shining a focused, pulsed light source (laser) on a diffuse surface, and the omnidirectional sensing came from focusing a streak camera on the diffuse surface; this produces the seemingly magical effect of using a diffuse surface as if it were a mirror. A more challenging problem is to form a depth map of a piecewise-planar scene in an unknown configuration. With spatially-patterned integration and more sophisticated signal processing, we have achieved this as well in Kirmani *et al.* (2011a) and Colaço *et al.* (2012).

A traditional camera uses lenses to form an optical image of the scene and thus obtain spatial correspondences between the scene and film or a sensor array. The film or sensor array has an integration time, but only one scalar value per spatial location – no time resolution. SFTI can produce high spatial resolution with a handful of source–sensor pairs, where every source and sensor is omnidirectional (i.e., completely lacks spatial resolution). Between these opposite extremes lie myriad possible trade-offs between time resolution, number and spacing of sensors, and spatial resolution. This is a rich area for both exploration of theoretical foundations and development of prototype systems. Both theory and practice should incorporate optophysical details (such as diffraction) and device realities (such as source power and pulse width constraints, sensor speed and jitter, thermal noise, and shot noise).

Our proof-of-concept experiments show radically new capabilities, but they are far from the end of the story. Rather, they motivate us to work toward a complete theory while they also demonstrate that our theory can have technological impact. Of course, we are not suggesting that standard, economical cameras will be displaced, but we believe that SFTI may become part of competitive solutions, especially in 3D acquisition.

Acknowledgments

This material is based on work supported in part by the National Science Foundation under grants 0643836 and 1161413, the DARPA InPho program through the US Army Research Office Award W911-NF-10-1-0404, a Qualcomm Innovation Fellowship, and an Esther and Harold E. Edgerton Career Development Chair.

References

Blu, T., Dragotti, P.-L., Vetterli, M. *et al.* (2008) Sparse sampling of signal innovations. *IEEE Signal Process. Mag.*, **25** (2), 31–40.

Coker, J. D. and Tewfik, A. H. (2007) Multistatic SAR image reconstruction based on an elliptical-geometry Radon transform, in *International Waveform Diversity and Design Conference*, IEEE, pp. 204–208.

Colaço, A., Kirmani, A., Howland, G. A. *et al.* (2012) Compressive depth map acquisition using a single photon-counting detector: parametric signal processing meets sparsity, in *Proceedings of IEEE International Conference on Computer Vision and Pattern Recognition, Providence, RI*, IEEE Computer Society, Washington, DC, pp. 96–102.

Cutrona, L. J. (1990) Synthetic aperture radar, in *Radar Handbook*, (ed. M. I. Skolnik), McGraw-Hill, New York, NY, ch 21.

Edgerton, H. E. and Killian, Jr., J. R. (1939) *Flash! Seeing the Unseen by Ultra High-Speed Photography*, Hale, Cushman and Flint, Boston, MA.

Foix, S., Alenyà, G., and Torras, C. (2011) Lock-in time-of-flight (ToF) cameras: A survey. *IEEE Sensors J.*, **11** (9), 1917–1926.

Forsyth, D. A. and Ponce, J. (2002) *Computer Vision: A Modern Approach*. Prentice Hall.

Georgeiv, T., Zheng, K. C., Curless, B. *et al.* (2006) Spatio-angular resolution tradeoff in integral photography, in *Proceedings of the 17th Eurographics Symposium on Rendering. Vienna, Austria* (eds T. Akenine-Moller and W. Heidrich), Eurographics Association, Aire-la-Ville, Switzerland, pp. 263–272.

Gershun, A. (1939) The light field. *J. Math. Phys.*, **18**, 51–151. (Translation from 1936 Russian paper Svetovoe Pole to English by P. Moon and G. Timoshenko.)

Gokturk, S. B., Yalcin, H., and Bamji, C. (2004) A time-of-flight depth sensor – system description, issues and solutions, in *Computer Vision and Pattern Recognition Workshop*, IEEE, p. 35.

Gouia, R. (2011) Some problems of integral geometry in advanced imaging. Doctoral dissertation, University of Texas–Arlington.

Gouia-Zarrad, R. and Ambartsoumian, G. (2012) Approximate inversion algorithm of the elliptical Radon transform, in *2012 8th International Symposium on Mechatronics and its Applications (ISMA)*, IEEE, pp. 1–4.

Grant, M. and Boyd, S. (2011) CVX: Matlab software for disciplined convex programming, version 1.21. http://cvxr.com/cvx.

Hussmann, S., Ringbeck, R., and Hagebeuker, B. (2008) A performance review of 3D TOF vision systems in comparison to stereo vision systems, in *Stereo Vision* (ed. A. Bhatti), InTech, pp. 103–120.

Kirmani, A., Hutchison, T., Davis, J., and Raskar, R. (2009) Looking around the corner using transient imaging, in *2009 IEEE 12th International Conference on Computer Vision*, IEEE, pp. 159–166.

Kirmani, A., Colaço, A., Wong, F. N. C., and Goyal, V. K. (2011a) Exploiting sparsity in time-of-flight range acquisition using a single time-resolved sensor. *Opt. Express*, **19** (22), 21485–21507.

Kirmani, A., Jeelani, H., Montazerhodjat, V., and Goyal, V. K. (2011b) Diffuse imaging: replacing lenses and mirrors with omnitemporal cameras, in *Wavelets and Sparsity XIV* (eds M. Papadakis, D. Van De Ville, and V. K. Goyal), Proceedings of the SPIE, Vol. **8138**, SPIE,. Bellingham, WA.

Kirmani, A., Jeelani, H., Montazerhodjat, V., and Goyal, V. K. (2012) Diffuse imaging: creating optical images with unfocused time-resolved illumination and sensing. *IEEE Signal Process. Lett.*, **19** (1), 31–34.

Kovaly, J. J. (1976) *Synthetic Aperture Radar*, Artech House, Dedham, MA.

Krishnan, V. P., Levinson, H., and Quinto, E. T. (2012) Microlocal analysis of elliptical Radon transforms with foci on a line, in *The Mathematical Legacy of Leon Ehrenpreis* (eds I. Sabadini and D. C. Struppa), Springer Proceedings in Mathematics, Book **16**, Springer, Milan, pp. 163–182.

Milanfar, P. (ed.) (2011) *Super-Resolution Imaging*, CRC Press, Boca Raton, FL.

Munson, Jr., D. C., O'Brien, J. D., and Jenkins, W. K. (1985) A topographic formulation of spotlight-mode synthetic aperture radar. *Proc. IEEE*, **71** (8), 917–925.

Ng, R., Levoy, M., Brédif, M. *et al.* (2005) Light field photography with a hand-held plenoptic camera, Stanford University Computer Science Technical Report CSTR 2005-02, http://graphics.stanford.edu/papers/lfcamera/lfcamera-150dpi.pdf (accessed December 2012).

Nicodemus, F. E. (1965) Directional reflectance and emissivity of an opaque surface. *Appl. Opt.*, **4** (7), 767–773.

Oren, M. and Nayar, S. K. (1995) Generalization of the Lambertian model and implications for machine vision. *Int. J. Comput. Vision*, **14** (3), 227–251.

Scharstein, D. and Szeliski, R. (2002) A taxonomy and evaluation of dense two-frame stereo correspondence algorithms. *Int. J. Comput. Vision*, **47** (1–3), 7–42.

Schwarz, B. (2010) LIDAR: mapping the world in 3D. *Nat. Photonics*, **4** (7), 429–430.

Seitz, S. M., Curless, B., Diebel, J. *et al.* (2006) A comparison and evaluation of multi-view stereo reconstruction algorithms, in *2006 IEEE Computer Society Conference on Computer Vision and Pattern Recognition*, IEEE, pp. 519–528.

Sen, P., Chen, B., Garg, G. *et al.* (2005) Dual photography. *ACM Trans. Graph. – Proc. ACM SIGGRAPH 2005*, **24** (3), 745–755).

Stoykova, E., Alatan, A. A., Benzie, P. *et al.* (2007) 3-D time-varying scene capture technologies – a survey. *IEEE Trans. Circ. Syst. Vid. Technol.*, **17** (11), 1568–1586.

Unser, M. (2000) Sampling – 50 years after Shannon. *Proc. IEEE*, **88** (4), 569–587.

Vetterli, M., Kovačević, J., and Goyal, V. K. (2013) *Foundations of Signal Processing*, Cambridge University Press.

3

2D-to-3D Video Conversion: Overview and Perspectives

Carlos Vazquez[1], Liang Zhang[1], Filippo Speranza[1], Nils Plath[2], and Sebastian Knorr[2]

[1]*Communications Research Centre Canada (CRC), Canada*
[2]*imcube labs GmbH, Technische Universität Berlin, Germany*

3.1 Introduction

Stereoscopic three-dimensional (S3D) imaging technologies provide an enhanced sensation of depth by presenting the viewer with two slightly different perspectives of the same scene as captured, for example, by two horizontally aligned cameras; the left view is presented to the left eye only and the right view is presented to the right eye only. The various S3D technologies differ mainly on how such fundamental eye–view separation is achieved. The benefits of S3D imaging are not limited to an enhanced sensation of depth, but also include an increased sense of immersion, presence, and realism (Yano and Yuyama, 1991; IJsselsteijn *et al.*, 1998; Freeman and Avons, 2000; Seuntiëns *et al.*, 2005; Häkkinen *et al.*, 2008; Kaptein *et al.*, 2008).

There has been a significant effort by the video and entertainment industries to foster adoption of S3D, particularly for the development of three-dimensional television (3DTV). As a result, in the last few years there has been progress in several key technology areas, including display, encoding and transmission, and content generation technologies. Technological advances are very important to build the eco-system required for 3D services. However, such advances would be pointless without the availability of a wide variety of program content. Currently, the availability of 3D content is still limited, but there are signs of a steady increase in production which could be attributed to several factors. First, a growing number of equipment manufacturers are including in their line-up 3D production and post-production tools. This should decrease production costs and thus generate more content production. The second factor is an increase in education about 3D technologies. One of the

Emerging Technologies for 3D Video: Creation, Coding, Transmission and Rendering, First Edition.
Frédéric Dufaux, Béatrice Pesquet-Popescu, and Marco Cagnazzo.
© 2013 John Wiley & Sons, Ltd. Published 2013 by John Wiley & Sons, Ltd.

positive results of such effort is that more content creators are embracing 3D attracted by its intrinsic artistic and entertainment value. The third factor, and the focus of this chapter, is the rapid development of 2D-to-3D video conversion techniques that allow the generation of stereoscopic video from a standard monoscopic video source.

In general, the 2D-to-3D conversion process consists of two interrelated steps. In the first step, the depth structure of the scene is defined. This step usually implies the generation of a depth map or a 3D model for each frame of a scene; a depth map is a matrix containing the depth value of each point of light, or pixel, in the image, while a 3D model will approximate the spatial position in the 3D space of each feature in the scene. In the second step, the generated depth information and the original 2D texture content are used in conjunction with rendering techniques to generate a new perspective image to form a stereoscopic pair. The new virtual perspectives do not contain information for objects that were occluded in the original scene; therefore, a potentially critical process of this second step is hole-filling; that is, the reconstruction and painting of those empty areas.

The techniques to define the depth of the scene and to render the new views can be approximately classified as semi-automatic or fully automatic. The semi-automatic approach usually involves a great deal of manual intervention, often object by object. This approach has been used extensively by the 3D cinema industry because it provides the highest quality conversions; however, it is very time consuming, very expensive, and not applicable to live content. For these reasons, there is a growing interest in developing depth generation and rendering techniques for automatic conversion. The depth information can be generated in different ways; the most common approach is to extract the depth information contained in the monocular depth cues, such as motion parallax, image blur, linear perspective, and so on.

Automatic rendering techniques are generally based on well-known image processing techniques, such as depth image-based rendering (DIBR) and warping, whose efficiency and accuracy are constantly improving. Thus, even though the automatic approach does not currently produce 2D-to-3D conversions as good as those obtained by the semi-automatic approach, the gap is closing rapidly.

In the next sections we will provide an overview of depth generation and rendering processes with a focus on recent techniques. In Section 3.2 we discuss the 2D-to-3D conversion problem; in Section 3.3 we present the most recent techniques for the definition of the depth structure of the scene, and in Section 3.4 we outline the main approaches used to render novel views; in Section 3.5 we discuss quality of experience issues; finally, in Section 3.6 we discuss future perspectives and research challenges.

3.2 The 2D-to-3D Conversion Problem

3.2.1 General Conversion Approach

The conversion of video material from 2D format to S3D consists of a set of operations that result in the generation of two or more video streams for stereoscopic viewing (Zhang *et al.*, 2011). Figure 3.1 shows a general block diagram of the process in which three main operations are highlighted: (a) the generation of the scene's depth information by analyzing the 2D image; (b) the mapping of depth information into parallax between the original 2D image and the new image to be generated, which depends on the viewing conditions; (c) the synthesis of the new views for stereoscopic or auto-stereoscopic viewing. The definition of a

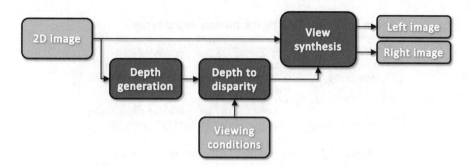

Figure 3.1   General scheme of 2D-to-3D conversion process

2D-to-3D conversion system consists in selecting the set of techniques used to perform these three main operations.

It is worth noting that the output of a 2D-to-3D system might consist of an arbitrary number of novel views. For stereoscopic viewing, which requires only two views, only one novel view is actually needed since the other image of the stereoscopic pair is usually the original itself. For auto-stereoscopic viewing, the number of required views depends on the display employed. Of course, it is worth noting that the underlying principle in stereoscopic and auto-stereoscopic systems, at least for those currently in use (e.g., parallax barrier and lenticular systems), is the same: at any given moment and position only two views are necessary and sufficient for the 3D effect.

For clarity, in the next sections we will examine algorithms and processes assuming a stereoscopic viewing situation, with the understanding that the same algorithms and processes apply to auto-stereoscopic viewing as well.

3.2.2   Depth Cues in Monoscopic Video

The first question that arises when one talks about 2D-to-3D conversion is: How is it possible to generate a stereoscopic representation from a monoscopic image? The answer to this question lies in the way the human visual system (HVS) perceives depth. The perception of depth is an active process that involves analyzing the scene and extracting every bit of information that helps the brain identify the depth position of objects. There are several depth cues that the HVS uses in this process, and all of them contribute to the perception of depth in the scene (Ostnes *et al.*, 2004).

Figure 3.2 shows a summary of the depth cues that are used by the HVS; for a more detailed description, please see Chapter 18. Some of those cues are binocular in nature, such as stereopsis and vergence. They use information from both eyes to generate depth information. Stereopsis, for example, is a very strong depth cue exploited by 3D viewing systems to provide a strong sense of depth. The HVS extracts the parallax information by analyzing the differences between the retinal projections of the scene in both eyes and translating those differences into depth information.

Other depth cues, such as linear perspective, relative size of objects, relative height in the image, accommodation, atmospheric scattering, motion parallax, and interposition, are monocular in nature. In other words, these depth cues are contained in the information provided by a single eye.

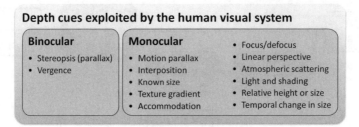

Figure 3.2 Depth cues used by the human visual system

Most of these monocular cues are also contained in images or sequences. Accordingly, they can be used to extract the depth information of the scene and thus provide the missing parallax information for the 2D-to-3D conversion process. This is indeed the rationale behind 2D-to-3D conversion: to use the monocular depth cues in the image to infer the corresponding parallax and create a second image using that information. An example of a 2D image in which many monocular depth cues are visible is shown in Figure 3.3a. As can be seen, the relative size of the poles, the linear perspective provided by the walkway, the gradient of the bolts' texture on the structure, and the partial occlusion of the bike and car on the right are all powerful indicators of the depth in the scene. In contrast, in Figure 3.3b only the shadings and the blur provide some depth information, making it more difficult to interpret the depth in the scene.

3.2.2.1 Description of Main Problems: Depth Structure Definition and Image Rendering

The two main operations of the 2D-to-3D conversion process are (a) the definition of the depth structure of the scene and (b) the generation of the second view needed for stereoscopic viewing.

The definition of the depth structure is based on the analysis of the scene and the extraction of as many monocular depth cues as possible to place objects in depth. The process used by the HVS to extract the depth information from monocular depth cues is complex and often based on learned features, like normal sizes of objects and relative location in the image. Reproducing this process in a computer system for automation is difficult and prone to errors. As a consequence, most of the 2D-to-3D conversions done today involve a human analysis of the scene and require a manual definition of the depth structure of the scene.

The rendering of the second image, on the other hand, can involve more automatic operations. The process involves displacing objects from the position in the original image to

(a)   (b)

Figure 3.3 Example of images with (a) several and (b) few monocular depth cues

generate parallax between corresponding objects in the two images forming the stereoscopic pair. The displacement is related to the depth of objects in the scene, which, as noted, is computed in the first operation. This displacement can be carried out by reprojection of the scene into a new virtual camera or by direct shift of pixels to new positions. There is, however, one major obstacle: the new point of view reveals details of objects and the background that are not visible in the original image and which need to be inferred from neighboring regions in the same image or neighboring images in the same sequence. This process is difficult to automate. In fact, high-quality conversions are done mostly manually by artists.

The inferential nature of the 2D-to-3D video conversion process – in which the depth cannot be directly measured and is generated either manually or by analyzing the 2D image, as well as the need to incorporate information that is not available in the original 2D image – leads to some trade-offs driving the techniques used for the conversion.

3.2.2.2 Quality versus Cost Issue

There are two main approaches to the conversion process: fully automatic and semi-automatic. There is a clear trade-off between the level of automation and the quality of the resulting conversion. The more human labor used, the more control is possible in the quality of the final product and the more flexibility is available for artistic creativity in the conversion. For feature film conversions a semi-automatic approach with a large amount of human intervention is applied (Seymour, 2012), leading to high-quality results and full control of the artistic dimension, but with high associated costs. Fully automatic approaches, on the other hand, could allow for reduced costs, but at the expense of less control on the quality of the final product and no space for artistic creativity. Automatic algorithms for the extraction of depth and for synthesizing new views are still prone to errors and introduce artifacts that are hard to remove.

3.2.2.3 On-Line versus Off-Line Issue

The second important trade-off often discussed in 2D-to-3D conversion is the on-line versus off-line conversion. The real-time on-line constraint introduces limitations in terms of processing time that makes it difficult to apply complex algorithms and methods to the generation of the depth and the view synthesis processes. Simpler techniques, which rely more on the perception of depth by the HVS and are tailored to specific types of content, are applied in this case. The quality obtained from these types of systems usually suffers from the restrictions imposed.

3.3 Definition of Depth Structure of the Scene

In this section we will explore the different methods currently used to define the depth structure of the scene for 2D-to-3D conversion of video content. We have classified the methods into two main categories:

- **Depth creation methods:** These methods attempt to generate a consistent but not necessarily truthful representation of the depth in the original image (Section 3.3.1).
- **Depth recovery methods:** These methods attempt to generate a consistent and truthful representation of the depth in the original image by analyzing and measuring parameters of particular depth cues (Section 3.3.2).

3.3.1   Depth Creation Methods

Methods in this category aim at defining a depth structure for the scene that provides a good viewing experience without attempting to reflect exactly the real underlying depth. These kinds of methods are usually employed in automatic real-time systems that are limited in resources and, thus, cannot use complex algorithms for extracting the depth.

3.3.1.1   Depth Models

One of the simplest ways of generating the depth information is to assume that most natural and artificial environments have a general depth structure that can be approximated with a predefined depth model. By classifying the scene into several classes, one can pick the depth model that best describes the scene and use it as the underlying depth structure in the conversion process. Several conversion methods are either based on this approach or use this approach as one of the strategies in the definition of the depth.

A commonly used depth model assumes that objects in the vicinity of the bottom of the image are closer to the camera than objects in the vicinity of the top (Yamada *et al.*, 2005). This broad assumption surprisingly holds true for many natural scenes in which the close ground is captured in the lower part of the image and the far sky in the upper part. The depth is then gradually increased from bottom to top to account for the change in depth across the vertical dimension of the image. Another popular depth model assumes that the scene depicts a room in which the floor, the ceiling, and the lateral walls are all close to the viewer while the back wall is farther away, leading to a box or spherical model for the depth. Examples of these two models are shown in Figure 3.4a and b.

Depth models are efficient and effective to some extent. They are easy to implement and can be effective for many natural scenes if the depth effect is not excessively stressed. These models are, however, very limited with respect to the complexity of the scene and the amount of depth that can be obtained from their use. A proper classification of the scene in order to select the appropriate model is needed as a first step (Battiato *et al.*, 2004a). It is worth noting that depth models are primarily used as an initial and rough approximation to the depth representation, which is then refined using other depth cues (Huang *et al.*, 2009; Yamada and Suzuki, 2009).

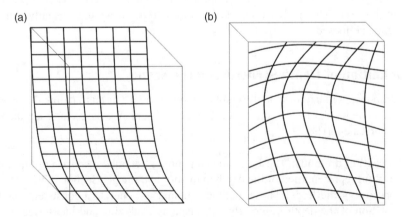

(a)   (b)

Figure 3.4   Examples of depth models used in 2D-to-3D conversion

3.3.1.2   Surrogate Depth Maps

Surrogate depth-map solutions are based on general assumptions about how the HVS perceives depth. The general idea is to use some feature of the 2D image as if it were depth; that is, the feature becomes a "surrogate" for depth. As a consequence, the resulting depth information might not be directly related to the actual depth structure of the scene. Nonetheless, this depth information has to provide enough depth elements in the form of parallax to ensure that the scene is perceived as having some stereoscopic depth. The robustness of the depth perception process of the HVS is then responsible for building a consistent and logical depth structure that is compatible with previous knowledge and normal assumptions in image analysis. In particular, the ability of the HVS to support sparse depth information and fill whatever could be missing or even tolerate a number of discrepancies between visual depth cues is essential for this kind of approach (Tam and Zhang, 2006; Tam *et al.*, 2009a). Methods in this category normally rely on luminance intensity, color distribution, edges, or saliency features that help provide some amount of depth distribution in the scene (Nothdurft, 2000; Tam *et al.*, 2005, 2009b; Kim *et al.*, 2010).

The term "surrogate" highlights the fact that these methods are not pretending to extract the depth information from the scene but are merely trying to provide a depth perception that is consistent with the scene and comfortable for viewers. The efficacy of these methods might be explained by the fact that depth perception is an active process that depends not only on binocular depth cues, but also on monocular depth cues and accumulated knowledge from past experiences. What is provided when a stereoscopic pair of images is presented to a viewer in addition to what is available from a single 2D image is basically parallax information. If this information is inconsistent with information extracted from other depth cues or with what is expected from previous experiences, then the HVS will, under certain circumstances, override the (inconsistent) parallax information to preserve the naturalness of the scene in the depth dimension. If, on the other hand, the parallax information is consistent with all other sources of depth information, then the depth perception is reinforced and improved. The main condition for these methods to work is that the parallax information provided by the "surrogate depth" is not strong enough to dominate other depth cues in the scene.

There are several methods reported in the literature that exploit this approach. Methods that rely on edges, for example, generate the depth information by analyzing edges in the image (Ernst, 2003; Tam *et al.*, 2005) or video (Chang *et al.*, 2007). Depth edges separate objects in the scene from the local background and are normally correlated to color or luminance edges in the 2D image. By assigning depth to image edges, objects are separated from the background and, as a consequence, the depth is improved.

Other methods rely on color information for the generation of the depth. Tam *et al.*, (2009b) used the red component (Cr) of the YCbCr color space as a surrogate for the depth in a 2D-to-3D conversion system (Vázquez and Tam, 2010). The Cr component has the characteristic that green and blue tints are represented as dark whereas red tints are represented as bright regions in the Cr image. By using such image as depth, reddish regions will appear closer to the viewer than greenish and bluish ones do. This is often in correspondence with natural images in which the red tints of skin are close to the viewer while the green and blue tints of sky and landscapes are farther from the viewer. The Cr component has some additional advantages: the shadows in the image will correspond to crevasses in the 3D objects and, thus, will correctly represent farther away features. Small features, like snowflakes or raindrops, are easily separated from the surrounding background because of the difference in color.

Saliency maps or visual attention maps (Kim *et al.*, 2010) are also used as surrogates for the depth. The rationale behind this approach is that humans tend to concentrate their attention on objects that are salient in the scene or that have salient features (Nothdurft, 2000). By moving those objects closer to the viewer and separating them from the background the scene acquires a depth volume that is, in most cases, consistent with the underlying scene structure.

Surrogate depth methods are generally used in fast and automatic conversion systems because of the simplicity in the generation of the depth information. By being based on the robustness of the HVS for the perception of depth, they provide images that look very natural and, as such, are very effective for converting 2D content containing humans and naturally occurring environments.

3.3.2 Depth Recovery Methods

The second main category of methods for the generation of depth information is based on the extraction of this information from the 2D image by analyzing the depth cues that are available and converting them into depth values. An example of an image and its corresponding depth map is shown in Figure 3.5. Recovering the depth information provided by the depth map is the main objective of algorithms covered in this section.

3.3.2.1 Artistic Definition of Depth

The first set of methods to recover the depth information from 2D depth cues is based on specialized human intervention. In other words, the methods require an expert observer (e.g., an artist) to analyze the image and determine the depth structure of the scene. This approach offers very good results since humans are very good at interpreting images and defining the relative position of objects in depth. However, on the downside, this process could be very labor intensive (Seymour, 2012).

The subjective nature of the method might lead to some differences in the techniques actually used in the field. However, such techniques all share the same underlying common approach, which consists of segmenting the image into its main components and assigning a depth to each of those components. The segmentation operation is a major issue for these methods. It normally requires many artists working in parallel on several key frames, delineating objects and separating the images in objects or depth layers. This operation is normally carried out by a rotoscoping team and could involve deconvolution at objects edges to ensure that the foreground objects are correctly separated from the background.

(a)   (b)

Figure 3.5   A 2D image and corresponding depth map

There are two main approaches for the depth assignment operation:

- Use a 3D computer graphics (CG) model to represent the object and "project" the texture over this model. This approach is also called rotomation (a combination of rotoscoping and CG animation). Rotomation is extremely complex because it requires the generation and animation of a 3D model; the latter is rather labor intensive and, thus, it might not be a suitable solution for short and complex action shots; for example, animation of hair or clothes. However, for long shots with still or tracking cameras capturing simple geometric objects (like houses, cars, etc.), this technique might be adequate since it offers very accurate and realistic results.
- Assign a depth to every pixel in the object in the form of a depth map and then adjust the depth so that the real depth of the object is captured by the pixel-based depth information (Vázquez and Tam, 2011). This approach is normally preferred, but it requires the artists to do a detailed work of depth verification and adjustment for every frame in the video.

The output of this process is a very well defined depth for every pixel in the scene and sometimes even more than just a depth value. For example, in the case of transparencies and reflections, more than one depth value can be associated with a given pixel that belongs to more than one object.

One additional issue to address in this depth assignment operation is the amount of depth that is assigned to each object and the distribution of the depth in the scene. Since the assignment is done by an artist with little to no absolute reference from real objects' sizes and depth positioning, it could be difficult to ensure the consistency of the depth distribution between objects and the varying depth within complex objects; for example, a face. Cardboard, miniaturization, and excessive depth are some of the effects that need to be avoided by proper assignment of depth to objects in the scene.

3.3.2.2 Structure from Motion

A 3D scene structure is a scene representation that consists of a number of 3D real world coordinates. As stated in Section 3.2.3, a complete scene structure reconstruction is normally impracticable or even impossible for an automatic 2D-to-3D video conversion. Fortunately, a sparse 3D scene structure is generally sufficient to achieve satisfactory results for 2D-to-3D conversion. A primary technique for the reconstruction of a sparse 3D scene structure is structure-from-motion (SfM) (Jebara *et al.*, 1999; Hartley and Zisserman, 2004; Pollefeys *et al.*, 2004; Imre *et al.*, 2006).

A sparse 3D scene structure is represented by a set of 3D feature points in a reconstructed 3D world. The first step of the process is the initial 3D scene structure reconstruction and camera track estimation from the first pair of images, which uses the tracked feature points to estimate the fundamental matrix $\mathbf{F}$. Then, the projection matrices $\mathbf{P}$ for these two video frames are determined with singular value decomposition (SVD). If the internal calibration parameters are unknown, which in general is the case for TV broadcast, home videos, or cinema movies, a self-calibration procedure has to be carried out (Luong and Faugeras, 1997; Mendonca and Cipolla, 1999; Pollefeys, 1999). Once the projection matrices $\mathbf{P}$ are known, the 3D points of the tracked feature points can be found via optimal triangulation as described in Hartley and Sturm (1997).

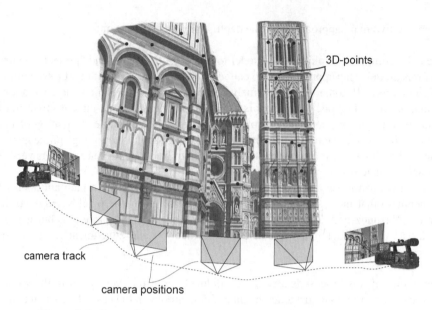

Figure 3.6   Sparse 3D scene structure and camera track determined by SfM

The next step is to update the initial 3D scene structure and camera tracks using consecutive video frames. First, the camera projection matrix for the next selected video frame is determined using already existing correspondences between the previously reconstructed 3D points and the tracked 2D points in the current frame. Then, the 3D scene structure and the camera tracks are refined with all tracked feature points between the current frame and the previous ones. This refinement is done via global nonlinear minimization techniques, also known as bundle adjustment (Triggs *et al.*, 2000; Lourakis and Argyros, 2004). This procedure is repeated for all selected frames until the refined estimates of the sparse 3D scene structure and camera tracks reach the desired accuracy.

Figure 3.6 shows the resulting sparse 3D point cloud and the computed camera track using SfM. The generation of the stereoscopic view for each frame of the camera track will be described in Section 3.4.2.4.

The sparse 3D scene structure reconstruction is a very computationally intensive process if performed with all the frames in a video sequence. The reconstruction is also heavily dependent on the initial structure computation. Camera motion degeneracy could cause the estimation of the fundamental matrix to fail (Torr *et al.*, 1999), and consecutive frame pairs may have baselines that are too short for accurate triangulation. To address these problems, the reconstruction of the sparse 3D scene structure and camera tracks is usually performed only on selected video frames, called key frames, chosen in order to reduce computational complexity, to improve triangulation accuracy, and to avoid camera motion degeneracy. An alternative approach to deal with these problems is to use a prioritized sequential 3D reconstruction approach, as proposed in Imre *et al.* (2006). The main element of such an approach is that key frames are selected according to a priority metric, and the frame pair with the highest priority metric is then used for the initial reconstruction.

3.3.2.3 Structure from Pictorial Cues

In addition to motion, there are several pictorial depth cues that allow us to perceive depth in a 2D representation of the scene, as noted in Section 3.2.2. These pictorial depth cues have been extensively applied in visual arts to enhance the perception of depth. In the following subsections, we will discuss two categories of pictorial cues commonly used to extract depth information from 2D images.

Depth from Focus/Defocus

Focus/defocus is related to the depth of field of real cameras and is reflected in 2D images in the form of objects being focused or defocused. In the real world we use the accommodation mechanism of the eye to keep objects in focus. We tend to accommodate to the distance of an object of interest, leaving all other objects at other depths out of focus. This mechanism is reproduced in video sequences because of the limited depth of field of the acquisition system's optics, such that objects of interest appear in focus in the 2D image while all other objects appear out of focus. This difference in focus information between objects in the image can be used to determine the depth relationship between objects.

Pentland (1987) studied how to extract depth maps from the degree of defocus in a thin lens system and developed a relationship between the distance D to an imaged point and the amount of defocus by the following equation:

$$D = \frac{F v_0}{v_0 - F - \sigma k f} \tag{3.1}$$

where v_0 is the distance between the lens and the image plane, f is the f-number of the lens system, F is the focal length of the lens system, k is a constant, and σ is the radius of the imaged point's "blur circle." From Equation 3.1, the problem of depth extraction is converted into a task of estimating camera parameters and the blur parameter σ. Given camera parameters which can be obtained from camera calibration, the depth distance D is determined by the blur parameter σ.

Several approaches have been developed to extract the blur parameter σ from a single image. Inverse filtering in the frequency domain (Pentland, 1987) was one of the first methods proposed to recover the blur parameter. It was later improved (Ens and Lawrence, 1993) by the introduction of regularization to solve instabilities related to inverse filtering in the frequency domain. Elder and Zucker (1998) developed a local-scale control method to detect edges at different levels of blur and to compute the blur associated with those edges. Wavelet decomposition (Valencia and Rodríguez-Dagnino, 2003; Guo *et al.*, 2008) and higher order statistics maps (Ko *et al.*, 2007) are two other techniques used in the extraction of blur information from 2D image content.

The approach of recovering depth from focus/defocus is relatively simple, but it can produce inaccurate results. Specifically, if the focus point is located in the center of the scene, objects in front or behind that point might have the same amount of blur; accordingly, it becomes impossible to distinguish an out-of-focus region in the foreground from an out-of-focus region in the background.

Depth from Geometry-Related Pictorial Cues

Geometry-related pictorial depth cues include linear perspective, known size, relative size, height in picture, and interposition. Some of these cues are stronger than others. The interposition, for example, can teach us the order of objects in depth, but not the distance in depth between them. Some cues might be hard to use in an application for 2D-to-3D conversion. The information related to the size of objects, such as known size and relative size, is difficult to use since it requires the previous knowledge of normal sizes for those objects and the identification of objects. The most commonly used geometric cues are the linear perspective and the height in picture.

Linear perspective refers to the property of parallel lines converging at infinite distance or, equivalently, a fixed-size object producing a smaller visual angle when more distant from the eye. Human-constructed scene structures frequently have linear features that run from near to far. This characteristic is used for depth estimation by detecting parallel lines in the images and identifying the point where these lines converge (vanishing point). Then a suitable assignment of depth can be derived based on the position of the lines and the vanishing point (Battiato et al., 2004b; Huang et al., 2009).

The height in picture denotes that objects that are closer to the bottom of the images are generally closer than objects at the top of the picture. Outdoor and landscape scenes mainly contain this pictorial depth cue. To extract this depth cue, horizontal lines usually have to be identified so that the image can be divided into stripes that go from the left border to the right border. For this purpose, a line-tracing algorithm is applied to recover the optimal dividing lines subject to some geometric constraints (Jung et al., 2009). A depth-refining step is further applied to improve the quality of the final depth map.

3.4   Generation of the Second Video Stream

The next step in the 2D-to-3D video conversion is related to the generation of the second video stream for stereoscopic visualization. The view synthesis process takes as input the original 2D image and the depth information generated in the previous step; the output consists of one or more new images corresponding to new virtual views needed for stereoscopic viewing. The main steps in the view generation process are converting the depth information into disparity values in order to displace pixels from the original image to their positions in the new image, the actual displacements of pixels, and the correction of artifacts that could have been introduced in the previous process. The latter typically requires the filling of newly exposed regions that were not visible (i.e., available) in the original 2D image.

3.4.1   Depth to Disparity Mapping

Depth information gathered from the original image can be represented in the form of a depth map, a 3D point cloud, or as mesh surfaces in a 3D space. This information, in conjunction with the position of pixels in the original 2D image, is used to obtain the position of corresponding pixels in the new image to be generated. One suggested solution is to back-project the pixels from the original image to their positions in the 3D space and then reproject them into the new virtual image plane (Hartley and Zisserman, 2004). Another approach is to convert the depth of pixels into disparities in the 2D plane and apply a 2D transformation to the original image by displacing each pixel to a new position in a 2D plane (Konrad, 1999). In most cases a parallel camera configuration is assumed, implying that the disparity has only horizontal components, greatly simplifying the rendering process.

The mapping from the desired depth to disparity between pixels in the left and right images of a stereoscopic image pair allows achieving the desired depth effect. The perceived

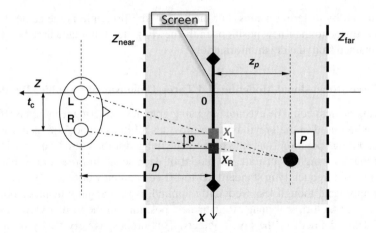

Figure 3.7 Model of a stereoscopic viewing system

volume of the scene can be controlled directly by adjusting the farthest and closest distances Z_{far} and Z_{near}, which define the depth range of the scene. These distances are better defined relative to the width of the screen by the two parameters k_{near} and k_{far}, which represent the distance (as a percentage of the display width W) that is in front of and at the back of the screen, respectively (ISO/IEC, 2007):

$$R = Z_{near} + Z_{far} = W(k_{near} + k_{far}) \qquad (3.2)$$

The depth of each pixel in the image should be contained in this range in order to respect the desired volume. Disparity values for each pixel are derived from the depth value and used to render the new view. Figure 3.7 is a schematic of a model of a stereoscopic viewing system shown with a display screen and the light rays from a depicted object being projected to the left eye and the right eye.

In this figure, the following expression provides the horizontal disparity p presented by the display according to the perceived depth of a pixel at z_p:

$$p = x_R - x_L = t_c \left(1 - \frac{D}{D - z_p} \right) \qquad (3.3)$$

where t_c corresponds to the inter-pupillary distance (i.e., distance between the human eyes), which is usually assumed to be about 65 mm. D represents the viewing distance from the display. Hence, the pixel disparity is expressed as

$$p_{pix} = \frac{pN_{pix}}{W} \qquad (3.4)$$

where N_{pix} is the horizontal pixel resolution of the display and W its width in the same units as p.

3.4.2 View Synthesis and Rendering Techniques

Once the desired range of disparities is known, the next step in the view synthesis process is the generation of the new image as it would be seen by a different perspective. The process can be achieved by different means, but the underlying principle is the same: to shift the pixels in one image to their new positions in the newly generated image based on the disparity value obtained from the depth information.

3.4.2.1   Three-Dimensional Modeling and Three-Dimensional Graphics Rendering

The rendering process could be executed by using graphics rendering techniques. If the depth recovery method used in the previous stage provides a full 3D model of each scene, then a classical rendering operation is used to recover the image for the virtual camera. This type of approach requires a highly accurate 3D model of each scene in order to be successful; as such, it is not used frequently in standard 2D-to-3D conversion work.

Another approach, though less frequently applied, is to manually displace objects from their positions in the original image to their new positions in the virtual view based on the amount of depth defined for the given objects. This process is very costly because of the involvement of human operators. Automatic operations are preferred whenever possible, and most post-production houses providing 2D-to-3D conversion services are equipped with tools to do this view synthesis process in a semi-automatic way (Seymour, 2012).

3.4.2.2   Depth-Image-Based Rendering (DIBR)

DIBR permits the creation of novel images, using information from depth maps, as if they were captured with a camera from different viewpoints (Fehn, 2003). The virtual camera is usually positioned to form a parallel stereoscopic setup with the original one and has the same intrinsic parameters and camera optical direction.

The depth information contained in the depth map is converted into disparity – that is, difference in horizontal position of pixels between the original and virtual cameras – by using the camera configuration. This disparity is then used in the DIBR-based view synthesis process to shifts pixels to their new positions in the virtual image. The process of shifting can be interpreted theoretically as a reprojection of 3D points onto the image plane of the new virtual camera. The points from the original camera, for which the depth is available, are positioned in the 3D space and then reprojected onto the image plane of the new virtual camera. In practice, however, pixels are shifted horizontally to their new positions without having to find their real positions in the 3D space (Mancini, 1998). The shift is only horizontal because of the parallel configuration of the camera pair. There are two schemes for the pixel projection shifting: backward and forward. Examples of backward and forward mappings shifts are shown in Figure 3.8. The backward pixel mapping generates the new image I_v from the color image I_o and the disparity map d_v which corresponds to the image to be rendered (Figure 3.8a):

$$I_v(x) = I_o(x - d_v) \tag{3.5}$$

where x is the pixel position in the rendered color image. In this case, since the depth is normally defined in the original image sampling grid, a process of interpolation is required to obtain the disparity in the grid of the rendered image.

The forward pixel mapping generates the new image I_v from the color image I_o and its associated disparity map d_o (Figure 3.8b):

$$I_v(x + d_o) = I_o(x) \tag{3.6}$$

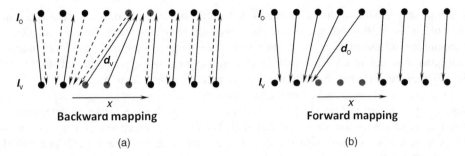

Figure 3.8 Pixel projection shifting based on 2D disparity map: (a) backward; (b) forward. I_o original image, I_v rendered image, d_v disparity of I_v, and d_o disparity of I_o

where x designates the pixel position in the original color image. In this case, generated image samples are not necessarily located on points of the regular sampling grid; and a resampling process is required for visualization and processing purposes.

Backward mapping is generally faster than forward mapping. However, the interpolation step needed to transform the disparity map could result in additional errors, especially at object borders and large depth discontinuities. This problem is illustrated in Figure 3.8a, where the disparity map, defined by vectors going from points in the original image sampling grid to arbitrary points in the rendered image plane (dashed lines in Figure 3.8a), is to be transformed into vectors going from points in the sampling grid of the rendered image to arbitrary points in the original image plane (solid lines in Figure 3.8a).

3.4.2.3 Image Warping

Another approach to synthesize views can be realized by nonlinear warping. Recently, warping has received great attention in applications such as image retargeting (Wang *et al.*, 2008; Krähenbühl *et al.*, 2009). These transformations are called nonhomogeneous to highlight the fact that not every image point undergoes an identical distortion (Shamir and Sorkine, 2009).

The ability to hide distortions in less salient image regions while preserving the aspect ratio of important parts of an image, such as persons or structure of buildings, is a major advantage. In practice, this is achieved by applying insights from research on human visual attention (Goferman *et al.*, 2010; Cheng *et al.*, 2011b).

The transformation itself is usually determined by use of meshes which are superimposed over an image. The spatial configuration of pixels already suggests that regular mesh grids are a suitable representation, but, of course, other mesh structures are also possible (Guo *et al.*, 2009; Didyk *et al.*, 2010). The grid size may be chosen depending on computational availability and time affordability.

Following the notation from Shamir and Sorkine (2009) and Wang *et al.* (2011), let $\Omega : \mathbb{R}^2 \to \mathbb{R}^2$ denote a mapping that shifts a node $v_i \in \mathbb{R}^2$ of an input mesh grid to a new position v'_i. Since these mappings are not necessarily bijective, it is imperative to be able to handle possible overlaps or holes which may occur during rendering; that is, multiple input nodes occupy the same pixel position after warping or pixel positions in the output that are not associated with any input node.

Didyk *et al.* (2010) initially divided the input image into large sections by superimposing a regular mesh grid (quads). Then, each section was divided into four subsections, if the minimum and maximum disparity values that occur inside a section exceeded a specified threshold. This division process was iterated until disparity extrema no longer exceeded the threshold or until the quads reached a minimum area of one pixel. Finally, the four corners of each quad were deformed using $\Omega = D$; that is, they are mapped from the left to the right view. Here, D refers to the disparity map that is deduced from a depth map.

Wang et al. (2011) followed a different approach. While images are again represented as mesh grids, warping is formulated as an optimization problem in order to determine the parameters that govern the warping function Ω. The objective function that is minimized has two purposes: (1) it ensures that the disparity at each pixel is correct; (2) it prevents the texture from tearing.

For pixel pairs at depth discontinuities that form an overlap, the energy allows the folding and ensures that the foreground is not folded behind the background. However, in the case of a hole the position of the pixel with less depth is fixed, so that the objective function pulls the background to the fixed foreground, effectively closing the hole while still preserving the specified disparity between the left and right frames.

Errors introduced by this approach are directed into less dominant regions of the image using a saliency measure during the process. In order to avoid a negative viewer experience, the warping transformation introduces only horizontal disparities.

For a more detailed description of this approach, please consult Chapter 11 (Image Domain Warping for Stereoscopic 3D Applications).

3.4.2.4 Approaches Based on Sparse 3D Scene Structure

Knorr et al. (2008) presented a system for generation of super-resolution stereo and multi-view video from monocular video based on *realistic stereo view synthesis* (RSVS) (Knorr and Sikora, 2007). It combines both the powerful algorithms of SfM described in Section 3.3.2.2 and the idea of image-based rendering to achieve photo-consistency without relying on dense depth estimation. Figure 3.9 shows the image-based rendering approach of RSVS. First, a virtual stereo camera with a user-defined stereo baseline is placed into the calculated camera track. The 3D scene points computed with SfM are reprojected into a virtual stereo frame and into surrounding views of the original monocular video sequence. Then, 2D point correspondences are utilized to estimate perspective homographies between the virtual and neighboring original views of the camera track. Finally, each pixel of a neighboring original view is warped into the stereo frame according to the perspective transformation equation

$$m_i = \mathbf{H}_i m_{\text{stereo}} \tag{3.7}$$

where $\mathbf{H}_i$ is the homography between the stereo frame and the right original view in Figure 3.9.

However, the approach is limited to rigid scene structures and slight horizontal camera motion. Kunter et al. (2009) extended this system to deal with independently moving objects within a rigid scene.

Owing to the homography warping involved during the view generation, the view rendering approach based on a sparse 3D scene structure does not have, contrary to the classic DIBR approach, "holes" within the synthesized images (Knorr and Sikora, 2007). In addition, super-resolution techniques can be exploited to enhance the image resolution of the reconstructed stereoscopic views since additional information from consecutive frames of the monocular video sequence is available (Knorr et al., 2008). However, such an approach could become error prone due to the assumption of a global planar homography, which is

Figure 3.9 Virtual view reconstruction using SfM and perspective transformations

basically valid only for rotating cameras (virtual and original views share the same camera center) or for a scene of one plane. One possible solution to this restriction is to apply local homographies on segmented image regions.

Finally, Rotem *et al.* (2005) applied only planar transformations on temporal neighboring views to virtually imitate a parallel stereo camera rig. This approach does not utilize 3D information from SfM, but time consistency along the sequence is heavily dependent on the 3D scene and camera motion. Thus, a stereo rig has to be modeled correctly to ensure a realistic 3D performance.

3.4.3 Post-Processing for Hole-Filling

The most significant problem in DIBR is how to deal with newly exposed areas appearing in the new images. This situation arises when the closest object is shifted, uncovering a previously occluded region. These unfilled pixels, commonly called "holes," can be classified into two categories: (i) cracks and (ii) disocclusions.

Cracks in the image are created by an object surface deformation since the object projected in the virtual view may not have the same area as in the source image. Normally, cracks are one or two pixels wide and can be easily filled by linear interpolation, within the virtual view, along the horizontal dimension.

Disocclusions are created by severe depth discontinuities in the disparity map. Whenever a pixel with a large disparity is followed (in the horizontal direction) by a pixel with a smaller disparity, the difference in disparity will create either an occlusion (superposition of objects) or a disocclusion (lack of visual information). Disocclusions are located along the boundaries of objects as well as at the edges of the images, and they correspond to areas behind objects that cannot be seen from the viewpoint of the source image (Fehn, 2003). Disocclusions are more concerning since they are more difficult to address and can have a significant negative effect on the picture quality of the virtual view. In the next sections, we review some of the solutions proposed to minimize the negative effects of disocclusions.

3.4.3.1 Depth Map Smoothing

One solution to disocclusion is to preprocess depth maps (Fehn, 2004; Tam *et al.*, 2004; Zhang *et al.*, 2004). Since the disocclusion problem arises where there are severe depth discontinuities (sharp edges) in the depth map, smoothing edge transitions in the depth map will help avoid the appearance of disocclusions. Consequently, it avoids artifacts related to filling disocclusions, producing images with high pictorial quality and free of annoying inpainting artifacts. This quality is obtained, however, at the expense of a loss in depth contrast and the introduction of object geometrical distortions.

One undesirable effect of smoothing edges is the geometrical distortion; for example, vertical straight lines in the left image could be rendered as curves in the right image if they are close to a sharp depth edge. To reduce this geometry distortion, asymmetric smoothing of depth map with more strength in the vertical direction than in the horizontal direction was proposed by Zhang and Tam (2005), which is consistent with known characteristics of the binocular system of the human eyes. Different from the constant smoothing strength for the whole depth map, some more sophisticated methods for adaptive smoothing of depth maps are further proposed in the literature; for example, edge-dependent depth filtering (Chen *et al.*, 2005), distance-dependent depth filtering (Daribo T*et al.*, 2007), discontinuity-adaptive smoothing filtering (Lee and Ho, 2009), and layered depth filtering (Wang *et al.*, 2007). Recently, an adaptive edge-oriented smoothing process was proposed (Lee and Effendi, 2011) in which the asymmetric smoothing filter is performed in the hole regions with vertical lines that belong to the background and the horizontal smoothing filter is done in the hole regions without vertical lines. Nevertheless, smoothing of depth maps, while attenuating some artifacts, will lead to a reduction of the depth contrast contained in the rendered stereoscopic views. Future studies will be required to examine this trade-off more closely, although based on the studies conducted so far the benefits appear to outweigh this disadvantage (Tam and Zhang, 2004; Zhang and Tam, 2005).

3.4.3.2 Spatial and Temporal Inpainting

Originally used for restoration of old paintings, inpainting has been introduced to the domain of digital image processing by Bertalmío *et al.* (2000) and Cheng *et al.* (2011) in order to repair damaged or missing parts of images by generating new and plausible content from available data in the image. In the case of stereo view generation, inpainting algorithms are concerned with approximating texture data for disoccluded areas by using image content available in the original images.

In order to avoid fatigue and nausea arising from vertical disparities, pixels are usually only shifted in the horizontal direction (Allison, 2007; Choi *et al.*, 2012). As a consequence, the shapes of holes that occur mainly extend in the vertical direction. Current literature typically mentions inpainting approaches that are based on (1) structure continuation, (2) texture replication, or (3) combinations of both (Smolic *et al.*, 2011).

Using boundary continuation, the image is modeled as a function in terms of curvature or gradients. Forcing this model to span over corrupted or missing data can be used to restore the image. This can be achieved by solving partial differential equations (Bertalmío *et al.*, 2000, 2001) or variational problems (Chan and Shen, 2005) where "healthy" pixels along the boundary of corrupted regions are set as boundary constraints. This has the effect that pixel information is propagated into the holes. Since each pixel has only a direct influence on its direct neighbors, this method can be regarded as local (Caselles, 2011).

Texture-based methods represent images as sets of overlapping patches, separated into patches containing holes and those that are free of holes. Partially disoccluded patches are completed with the best matching hole-free patch according to a prespecified matching criterion. The final results depend greatly on the order in which patches are filled. So, in order to control this effect, prioritization schemes have been devised (Criminisi *et al.*, 2004). Commonly, patches along the boundary of the holes, which have an overlap with both known and hole pixels, have a higher fill-in priority and are filled in first. Methods in this category can be regarded as global, since the entire image or image cube can be searched to find the best matching patch (Caselles, 2011).

One of the first works on inpainting for videos is described by Bertalmío *et al.* (2001). They used a system of partial differential equations to close holes frame by frame. However, temporal coherence is not accounted for, which results in artifacts and flickering in the filled-in output. In Patwardhan, *et al.* (2007), each frame is roughly segmented in fore- and background, and this is used to produce time-consistent image mosaics. These mosaics are then used to fill as many holes as possible. All remaining holes are filled with an exemplar-based method. Cheung *et al.* (2008) expressed inpainting as a probabilistic problem. For a set of patches generated from input images a set of probabilistic models is trained. These models are then exploited to synthesize data in order to fill holes. An extension of Efros and Leung (1999) to video can be found in Wexler *et al.* (2004), where hole filling is posed as an optimization problem in which holes are represented as patch cubes. These so-called space–time cubes extend spatially in the current frame as well as temporally into the neighboring frames. An objective function ensures that for each space–time cube a similarly textured hole-free cube with a similar trajectory is found in the image sequence in order to fill holes.

Another approach (Cheng *et al.*, 2011a) based on the seminal work of Criminisi *et al.* (2004) regards a video as an image stack consisting of N frames. A frame at time step t has a number of reference frames in the forward and backward directions assigned to it, which will serve as reservoir of image content to fill holes in the current frame. Pixels at time step t, which are identified as holes, are stored in a priority queue. The priority of a hole pixel x is determined by the ratio of the sum of image gradient magnitudes in a 2D patch centered around the pixel and its corresponding depth value. As long as the priority queue is not empty, the pixel with the highest priority value is picked and compared with all source pixels in the specified range of reference frames. At each reference frame the best matching source pixel according to a similarity measure is chosen. And finally, all matches are combined as a weighted sum to yield the new fill value. This process is repeated for all frames in the image stack until all holes have been filled.

With the algorithms described so far, it is possible to repair image content in still images as well as in image sequences. However, especially for stereo view synthesis, not only is it important to be spatially and temporally coherent but also to be view consistent between the left and right video frames. Raimbault and Kokaram (2012) describe a procedure related to the method of Komodakis (2006) that addresses this shortcoming. Their probabilistic framework employs exemplar-based inpainting, motion vectors, and disparity values of the left and right views to fill missing image content. However, for several methods, limitations exist due to motion constraints or assumption (Jia, Y. *et al.*, 2005; Jia, J. *et al.*, 2006; Patwardhan *et al.*, 2007) or restrictions in the size of input data that can be processed due to computational complexities (Wexler *et al.*, 2004; Cheung *et al.*, 2008). More detailed overviews can be found in Shih and Chang (2005), Tauber *et al.* (2007), and Fidaner (2008).

3.5 Quality of Experience of 2D-to-3D Conversion

Ideally, 2D-to-3D conversion should produce content that has no picture artifacts, that has good depth, and that respects the artistic composition of the original video. In other words, 2D-to-3D conversion should produce stereoscopic video comparable, in terms of quality of experience, to any other original stereoscopic video. There are three basic perceptual dimensions which could affect the quality of experience provided by stereoscopic imaging: picture quality, depth quality, and visual comfort. In addition, some researchers have also considered more general dimensions such as naturalness and sense of presence (Seuntiëns *et al.*, 2005). Incorrect 2D-to-3D conversion might introduce many types of errors that could affect any of these dimensions.

Picture quality refers to the technical quality of the picture provided by the system. Generally, picture quality is mainly affected by errors introduced by encoding and/or transmission processes. However, the process of 2D-to-3D conversion could introduce additional artifacts, either through inaccurate depth maps or incorrect rendering, which could further reduce the picture quality of the converted video.

Depth quality refers to the degree to which the image conveys a sensation of depth. The presence of monocular cues, such as linear perspective, blur, gradients, and so on, permits one to convey some sensation of depth even in standard monoscopic images. By adding disparity information, the 2D-to-3D conversion process provides an enhanced sense of depth. However, inaccurate depth information might introduce errors, either locally or globally, which could also negatively affect the perceived sensation of depth. Furthermore, incorrect depth values might produce visible depth distortions, such as the so-called "cardboard effect." In this effect, an individual object appears unnaturally thin, as if it has only one depth layer; that is, it looks like a cardboard model that has been cut and pasted into the image.

Visual discomfort refers to the subjective sensation of discomfort often associated with the viewing of stereoscopic images. The issue of comfort is relevant since some researchers have expressed concerns related to the safety and health of viewing stereoscopic images. It is widely believed that the main determinant of such discomfort is the presence of excessive disparity/parallax, possibly because it worsens the conflict between accommodation and vergence. Therefore, it has been suggested that the disparities in the stereoscopic image should be small enough so that the perceived depths of objects fall within a "comfort zone." Several approaches have been proposed to define the boundaries of the comfort zone (Yano *et al.*,

2002; Nojiri *et al.*, 2004; Speranza *et al.*, 2006; Lambooij *et al.*, 2009; Mendiburu, 2009). In general, these approaches converge to similar boundaries, which when expressed in terms of retinal disparities correspond to about $\pm 1°$ of visual angle for both positive and negative disparities. Clearly, it is very important to ensure that the 2D-to-3D conversion does not result in disparities outside the comfort zone.

Naturalness refers to the perception of the stereoscopic image as being a truthful representation of reality (i.e. perceptual realism). The stereoscopic image may exhibit different types of distortion which could make it to appear less natural. We have already considered one example of such distortion with the cardboard effect. Another example, which could also originate from an incorrect selection of conversion parameters, is the puppet theater effect. In this effect, the stereoscopic objects are sometimes perceived as unnaturally large or small, compared with the rest of the image. Finally, any of the distortions or artifacts mentioned above could decrease the sense of presence, which refers to the subjective experience of being in one place or environment even when one is situated in another.

3.6 Conclusions

This chapter has provided an overview of depth generation and rendering processes with a focus on the most recent techniques for the definition of the depth structure of the scene and the main approaches used to render new virtual views. It is clear that the research on 2D-to-3D video conversion is progressing at a very rapid pace and with very encouraging results. Such progress is simply the consequence of the increasing attention that this research has been receiving from both industry and academia.

Despite some initial skepticism, 2D-to-3D video conversion is quickly becoming one of the most important means of generating stereoscopic content for the nascent 3DS cinema and 3DTV industries. Semi-automatic 2D-to-3D video conversion techniques are already being used, quite successfully, to convert both old and new video material originally captured in 2D. The same techniques are also used to fix problems in content natively captured in 3D; for example, when 3D capture was not possible or when a captured scene was incorrect and in need of modifications or adjustments. The success of these semi-automatic techniques has demonstrated that, when done properly, 2D-to-3D video conversion is undistinguishable from original S3D content.

The true potential of 2D-to-3D conversion will be reached only with fully automatic (real-time) conversion techniques, however. Such techniques could make 3D content truly ubiquitous and leave the choice of 2D or 3D content selection to the viewers. This vision is already becoming a reality. Some television manufacturers and online video providers have begun incorporating real-time 2D-to-3D conversion into the player system in order to provide the viewer with the capability to switch at will between 2D and 3D content.

However, to be successful these fully automatic techniques must eventually provide a quality of 3D experience that parallels that provided by actual stereoscopic content. This will require higher accuracy both in the definition of a scene's depth and in the rendering. For example, we noted that current depth generation techniques are mostly based on single monocular cues, which generally provide ambiguous depth information and, thus, result in less accurate depth maps. To achieve better accuracy, it might be necessary to consider multiple monocular cues, albeit not all with the same relevance. Better depth information might also help addressing those situations in which depth estimation has thus far proven particularly challenging; for example, reflections, transparencies, and so on. In addition to improving depth estimation, it will be necessary to improve the rendering processes as well. We have seen that, despite significant progress, much work still needs to be done in this area of research.

Considering its great potential, it is reasonable to predict that the development of fast, reliable, and computationally efficient 2D-to-3D video conversion techniques will significantly alter the S3D video industry.

References

Allison, R. (2007) Analysis of the influence of vertical disparities arising in toed-in stereoscopic cameras. *J. Imaging Sci. Technol.*, **51**, 317–327.

Battiato, S., Capra, A., Curti, S., and La Cascia, M. (2004a) 3D Stereoscopic image pairs by depth-map generation, in *2nd International Symposium on 3D Data Processing, Visualization and Transmission, Thessaloniki, Greece*, IEEE Computer Society, pp. 124–131.

Battiato, S., Curti, S., La Cascia, M. *et al.* (2004b) Depth map generation by image classification, in *Three-Dimensional Image Capture and Applications VI* (eds B. D. Corner, P. Li, and R. P. Pargas), Proceedings of the SPIE, Vol. 5302, SPIE, Bellingham, WA, pp. 95–104.

Bertalmío, M., Sapiro, G., Caselles, V., and Ballester, C. (2000) Image inpainting, in *SIGGRAPH, 2000, Proceedings of the 27th Annual Conference on Computer Graphics and Interactive Techniques* (eds J. S. Brown and K. Akeley), ACM Press/Addison-Wesley, New York, NY, pp. 417–424.

Bertalmío, M., Bertozzi, A. L., and Sapiro, G. (2001) Navier–Stokes, fluid dynamics, and image and video inpainting, in *IEEE Computer Society Conference on Computer Vision and Pattern Recognition (CVPR)*, Vol. 1, IEEE Computer Society, Los Alamitos, CA, p. 355.

Caselles, V. (2011) Exemplar-based image inpainting and applications. International Congress on Industrial and Applied Mathematics – ICIAM, Vancouver, BC, Canada.

Chan, T. and Shen, J. (2005) Variational image inpainting. *Commun. Pure Appl. Math.*, **58** (5), 579–619.

Chang, Y.-L., Fang, C.-Y., Ding, L.-F. *et al.* (2007) Depth map generation for 2D-to-3D conversion by short-term motion assisted color segmentation, in *2007 IEEE International Conference on Multimedia and Expo, ICME 2007 Proceedings*, IEEE, pp. 1958–1961.

Chen, W., Chang, Y., Lin, S. *et al.* (2005) Efficient depth image based rendering with edge dependent depth filter and interpolation, in *IEEE International Conference on Multimedia and Expo, 2005. ICME 2005*, IEEE, pp. 1314–1317.

Cheng, C., Lin, S., and Lai, S. (2011a) Spatio-temporally consistent novel view synthesis algorithm from video-plus-depth sequences for autostereoscopic displays. *IEEE Trans. Broadcast.*, **57** (2), 523–532.

Cheng, M., Zhang, G., Mitra, N. *et al.* (2011b) Global contrast based salient region detection, in *CVPR '11 Proceedings of the 2011 IEEE Conference on Computer Vision and Pattern Recognition*, IEEE Computer Society, Washington, DC, pp. 409–416.

Cheung, V., Frey, B., and Jojic, N. (2008) Video epitomes. *Int. J. Comput. Vision*, **76** (2), 141–152.

Choi, J., Kim, D., Choi, S., and Sohn, K. (2012) Visual fatigue modeling and analysis for stereoscopic video. *Opt. Eng.*, **51**, 017206.

Criminisi, A., Pérez, P., and Toyama, K. (2004) Region filling and object removal by exemplar-based image inpainting. *IEEE Trans. Image. Process.*, **13** (9), 1200–1212.

Daribo, I., Tillier, C., and Pesquet-Popescu, B. (2007) Distance dependent depth filtering in 3D warping for 3DTV, in *IEEE 9th Workshop on Multimedia Signal Processing*, IEEE, pp. 312–315.

Didyk, P., Ritschel, T., Eisemann, E. *et al.* (2010) Adaptive image-space stereo view synthesis, in Proceedings of the Vision, Modeling, and Visualization Workshop 2010, Siegen, Germany, Eurographics Association, pp. 299–306.

Efros, A., and Leung, T. (1999) Texture synthesis by non-parametric sampling, in *The Proceedings of the Seventh IEEE International Conference on Computer Vision, 1999*, Vol. 2, IEEE Computer Society, Washington, DC, pp. 1033–1038.

Elder, J. H., and Zucker, S. W. (1998) Local scale control for edge detection and blur estimation. *IEEE Trans. Pattern Anal.*, **20** (7), 699–716.

Ens, J., and Lawrence, P. (1993) An investigation of methods for determining depth from focus. *IEEE Trans. Pattern Anal.*, **15** (2), 97–107.

Ernst, F. E. (2003) 2D-to-3D video conversion based on time-consistent segmentation. Proceedings of the Immersive Communication and Broadcast Systems Workshop, Berlin, Germany.

Fehn, C. (2003) A 3D-TV approach using depth-image-based rendering (DIBR), in *Visualization, Imaging, and Image Processing*, Vol. I (ed. M. H. Hamza), ACTA Press, I-396-084.

Fehn, C. (2004) Depth-image-based rendering (DIBR), compression, and transmission for a new approach on 3D-TV, in *SPIE Stereoscopic Displays and Virtual Reality Systems XI* (eds A. J. Woods, J. O. Merritt, S. A. Benton, and M. T. Bolas), Proceedings of the SPIE, Vol. 5291, SPIE, Bellingham, WA, pp. 93–104.

Fidaner, I. (2008) A survey on variational image inpainting, texture synthesis and image completion. Technical Report, Boğaziçi University, Istanbul, Turkey.

Freeman, J., and Avons, S. (2000) Focus group exploration of presence through advanced broadcast services, in *Human Vision and Electronic Imaging V* (eds B. E. Rogowitz and T. N. Pappas), Proceedings of the SPIE, Vol. **3959**, SPIE, Bellingham, WA, pp. 530–539.

Goferman, S., Zelnik-Manor, L., and Tal, A. (2010) Context-aware saliency detection, in *2010 IEEE Conference on Computer Vision and Pattern Recognition (CVPR)*, IEEE, pp. 2376–2383.

Guo, G., Zhang, N., Huo, L., and Gao, W. (2008) 2D to 3D conversion based on edge defocus and segmentation, in *IEEE International Conference on Acoustics, Speech and Signal Processing, 2008. ICASSP 2008*, IEEE, pp. 2181–2184.

Guo, Y., Liu, F., Shi, J. *et al.* (2009) Image retargeting using mesh parametrization. *IEEE Trans. Multimedia*, **11** (5), 856–867.

Häkkinen, J., Kawai, T., Takatalo, J. *et al.* (2008) Measuring stereoscopic image quality experience with interpretation based quality methodology, in *Image Quality and System Performance V* (eds S. P. Farnand and F. Gaykema), Proceedings of the SPIE, Vol. 6808, SPIE, Bellingham, WA, p. 68081B.

Hartley, R., and Sturm, P. (1997) Triangulation. *Comput. Vision Image Und.*, **68** (2), 146–157.

Hartley, R. and Zisserman, A. (2004) *Multiple View Geometry in Computer Vision*, Cambridge University Press.

Huang, X., Wang, L., Huang, J. *et al.* (2009) A depth extraction method based on motion and geometry for 2D to 3D conversion. Third International Symposium on Intelligent Information Technology Application.

IJsselsteijn, W., de Ridder, H., Hamberg, R. *et al.* (1998) Perceived depth and the feeling of presence in 3DTV. *Displays*, **18** (4), 207–214.

Imre, E., Knorr, S., Alatan, A. A., and Sikora, T. (2006) Prioritized sequential 3D reconstruction in video sequences of dynamic scenes, in *2006 IEEE International Conference on Image Processing (ICIP)*, IEEE, pp. 2969–2972.

ISO/IEC (2007) 23002-3:2007 *Information Technology – MPEG video Technologies – Part 3: Representation of Auxiliary Video and Supplemental Information*, International Organization for Standardization/International Electrotechnical Commission, Lausanne.

Jebara, T., Azarbayejani, A., and Pentland, A. (1999) 3D structure from 2D motion. *IEEE Signal Proc. Mag.*, **16** (3), 66–84.

Jia, J., Tai, Y., Wu, T., and Tang, C. (2006) Video repairing under variable illumination using cyclic motions. *IEEE Trans. Pattern Anal. (PAMI)*, **28** (5), 832–839.

Jia, Y., Hu, S., and Martin, R. (2005) Video completion using tracking and fragment merging. *Vis. Comput.*, **21** (8), 601–610.

Jung, Y. J., Baik, A., Kim, J., and Park, D. (2009) A novel 2D-to-3D conversion technique based on relative height-depth cue, in *Stereoscopic Displays and Applications XX* (eds A. J. Woods, N. S. Holliman, and J. O. Merritt), Proceedings of the SPIE, Vol. 7237, SPIE, Bellingham, WA, p. 72371U.

Kaptein, R., Kuijsters, A., Lambooij, M. *et al.* (2008) Performance evaluation of 3D-TV systems, in *Image Quality and System Performance V* (eds S. P. Farnand and F. Gaykema), Proceedings of the SPIE, Vol. 6808, SPIE, Bellingham, WA, pp. 1–11.

Kim, J., Baik, A., Jung, Y. J., and Park, D. (2010) 2D-to-3D conversion by using visual attention analysis, in *Stereoscopic Displays and Applications XXI* (eds A. J. Woods, N. S. Holliman, and N. A. Dodgson), Proceedings of the SPIE, Vol. 7524, SPIE, Bellingham, WA, p. 752412.

Knorr, S., and Sikora, T. (2007) An image-based rendering (IBR) approach for realistic stereo view synthesis of TV broadcast based on structure from motion, in *IEEE International Conference on Image Processing, 2007. ICIP 2007*, IEEE, pp. VI-572–VI-575.

Knorr, S., Kunter, M., and Sikora, T. (2008) Stereoscopic 3D from 2D video with super-resolution capability. *Signal Process. Image Commun.*, **23** (9), 665–676.

Ko, J., Kim, M., and Kim, C. (2007) 2D-to-3D stereoscopic conversion: depth-map estimation in a 2D single-view image, in *Applications of Digital Image Processing XXX* (ed. A. G. Tescher), Proceedings of the SPIE, Vol. 6696, SPIE, Bellingham, WA, p. 66962A.

Komodakis, N. (2006) Image completion using global optimization, in *2006 IEEE Conference on Computer Vision and Pattern Recognition*, Vol. 1, IEEE, pp. 442–452.

Konrad, J. (1999) View reconstruction for 3-D video entertaiment: Issues, algorithms and applications, in *Seventh International Conference on Image Processing and Its Applications, 1999*, Vol. 1, IEEE, pp. 8–12.

Krähenbühl, P., Lang, M., Hornung, A., and Gross, M. (2009) A system for retargeting of streaming video. *ACM Trans. Graph. (TOG)*, **28** (5), 126:1–126:10.

Kunter, M., Knorr, S., Krutz, A., and Sikora, T. (2009) Unsupervised object segmentation for 2D to 3D conversion, in *Stereoscopic Displays and Applications XX* (eds A. J. Woods, N. S. Holliman, and J. O. Merritt), Proceedings of the SPIE, Vol. 7237, SPIE, Bellingham, WA, p. 72371B.

Lambooij, M., IJsselsteijn, W., Fortuin, M., and Heynderickx, I. (2009) Visual discomfort and visual fatigue of stereoscopic displays: a review. *J. Imaging Sci. Technol.*, **53** (3), 030201.

Lee, P.-J. and Effendi. (2011) Nongeometric distortion smoothing approach for depth map preprocessing. *IEEE Trans. Multimedia*, **13** (2), 246–254.

Lee, S. B. and Ho, Y. S. (2009) Discontinuity-adaptive depth map filtering for 3D view generation. The 2nd International Conference on Immersive Telecommunications, Berkeley, CA, USA.

Lourakis, M. and Argyros, A. (2004) The design and implementation of a generic sparse bundle adjustment software package based on the Levenberg–Marquardt algorithm. Tech. Rep., Institute of Computer Science – FORTH, Heraklion, Crete, Greece.

Luong, Q.-T. and Faugeras, O. D. (1997) Self-calibration of a moving camera from point correspondences and fundamental matrices. *Int. J. Comput. Vision*, **22** (3), 261–289.

Mancini, A. (1998) Disparity estimation and intermediate view reconstruction for novel applications in stereoscopic video. Master's thesis, McGill University, Department Electrical Engineering, Montréal, QC, Canada.

Mendiburu, B. (2009) *3D Movie Making: Stereoscopic Digital Cinema from Script to Screen*, Focal Press/Elsevier.

Mendonça, P., and Cipolla, R. (1999) A simple technique for self-calibration. IEEE Computer Society Conf. on Computer Vision and Pattern Recognition (CVPR).

Nojiri, Y., Yamanoue, H., Hanazato, A. *et al.* (2004) Visual comfort/discomfort and visual fatigue caused by stereoscopic HDTV viewing, in *Stereoscopic Displays and Virtual Reality Systems XI* (eds A. J. Woods, J. O. Merritt, S. A. Benton, and M. T. Bolas), Proceedings of the SPIE, Vol. 5291, SPIE, Bellingham, WA, pp. 303–313.

Nothdurft, H. (2000) Salience from feature contrast: additivity across dimensions. *Vision Res.*, **40**, 1183–1201.

Ostnes, R., Abbott, V., and Lavender, S. (2004) Visualisation techniques: an overview – Part 1. *Hydrogr. J.* (113) 4–7.

Patwardhan, K., Sapiro, G., and Bertalmío, M. (2007) Video inpainting under constrained camera motion. *IEEE Trans. Image Process.*, **16** (2), 545–553.

Pentland, A. P. (1987) A new sense for depth of field. *IEEE Trans. Pattern Anal.*, **9**, 523–531.

Pollefeys, M. (1999) Self-calibration and metric 3D reconstruction from uncalibrated image sequences. PhD thesis, Katholieke Universiteit Leuven.

Pollefeys, M., Gool, L. V., Vergauwen, M., Verbiest, F., Cornelis, K., Tops, J. *et al.* (2004) Visual modeling with a hand-held camera. *Int. J. Comput. Vision*, **59** (3), 207–232.

Raimbault, F., and Kokaram, A. (2012) Stereo-video inpainting. *J. Electron. Imaging*, **21**, 011005.

Rotem, E., Wolowelsky, K., and Pelz, D. (2005) Automatic video to stereoscopic video conversion, in *Stereoscopic Displays and Virtual Reality Systems XII* (eds A. J. Woods, M. T. Bolas, J. O. Merritt, and I. E. McDowall), Proceedings of the SPIE, Vol. 5664, SPIE, Bellingham, WA, pp. 198–206.

Seuntiëns, P., Heynderickx, I., IJsselsteijn, W. *et al.* (2005) Viewing experience and naturalness of 3D images, in *Three-Dimensional TV, Video, and Display IV* (eds B. Javidi, F. Okano, and J.-Y. Son), Proceedings of the SPIE, Vol. 6016, SPIE, Bellingham, WA, pp. 43–49.

Seymour, M. (2012, May 08) Art of Stereo Conversion: 2D to 3D. Retrieved May 30, 2012, from http://www.fxguide.com/featured/art-of-stereo-conversion-2d-to-3d-2012/.

Shamir, A. and Sorkine, O. (2009) Visual media retargeting, in *SIGGRAPH ASIA '09*, ACM, New York, NY, pp. 11:1–11:13.

Shih, T. and Chang, R. (2005) Digital inpainting – survey and multilayer image inpainting algorithms, in *Third International Conference on Information Technology and Applications, 2005. ICITA 2005*, Vol. 1, IEEE, pp. 15–24.

Smolic, A., Kauff, P., Knorr, S. *et al.* (2011) Three-dimensional video postproduction and processing. *Proc. IEEE*, **99** (4), 607–625.

Speranza, F., Tam, W. J., Renaud, R., and Hur, N. (2006) Effect of disparity and motion on visual comfort of stereoscopic images, in *Stereoscopic Displays and Virtual Reality Systems XIII* (eds A. J. Woods, N. A. Dodgson, J. O. Merritt *et al.*), Proceedings of the SPIE, Vol. 6055, SPIE, Bellingham, WA, pp. 94–103.

Tam, W. J. and Zhang, L. (2004) Nonuniform smoothing of depth maps before image-based rendering, in *Three-Dimensional TV, Video and Display III* (eds B. Javidi and F. Okano), Proceedings of the SPIE, Vol. 5599, SPIE, Bellingham, WA, pp. 173–183.

Tam, W. J. and Zhang, L. (2006) 3D-TV content generation: 2D-to-3D conversion. IEEE International Conference on Multimedia and Expo, Toronto, Canada.

Tam, W. J., Alain, G., Zhang, L. *et al.* (2004) Smoothing depth maps for improved stereoscopic image quality, in *Three-Dimensional TV, Video and Display III* (eds B. Javidi and F. Okano), Proceedings of the SPIE, Vol. 5599, SPIE, Bellingham, WA, pp. 162–172.

Tam, W. J., Yee, A. S., Ferreira, J. *et al.* (2005) Stereoscopic image rendering based on depth maps created from blur and edge information, in *Stereoscopic Displays and Virtual Reality Systems XII* (eds A. J. Woods, M. T. Bolas, J. O. Merritt, and I. E. McDowall), Proceedings of the SPIE, Vol. 5664, SPIE, Bellingham, WA, pp. 104–115.

Tam, W. J., Speranza, F., and Zhang, L. (2009a) Depth map generation for 3-D TV: importance of edge and boundary information, in *Three-Dimensional Imaging, Visualization and Display* (eds B. Javidi, F. Okano, and J.-Y. Son), Springer, New York, NY.

Tam, W. J., Vázquez, C., and Speranza, F. (2009b) 3D-TV: a novel method for generating surrogate depth maps using colour information, in *Stereoscopic Displays and Applications XX* (eds A. J. Woods, N. S. Holliman, and J. O. Merritt), Proceedings of the SPIE, Vol. 7237, SPIE, Bellingham, WA, p. 72371A.

Tauber, Z., Li, Z., and Drew, M. (2007) Review and preview: disocclusion by inpainting for image-based rendering. *IEEE Trans. Syst. Man Cyb. C*, **37** (4), 527–540.

Torr, P., Fitzgibbon, A., and Zisserman, A. (1999) The problem of degeneracy in structure and motion recovery from uncalibrated images. *Int. J. Comput. Vision*, **32** (1), 47–53.

Triggs, B., McLauchlan, P., Hartley, R., and Fitzgibbon, A. (2000) Bundle adjustment – a modern synthesis, in *Vision Algorithms: Theory and Practice* (eds W. Triggs, A. Zisserman, and R. Szeliski), Springer Verlag, pp. 298–375.

Valencia, S. A., and Rodríguez-Dagnino, R. M. (2003) Synthesizing stereo 3D views from focus cues in monoscopic 2D images, in *Stereoscopic Displays and Virtual Reality Systems X* (eds A. J. Woods, M. T. Bolas, J. O. Merritt, and S. A. Benton), Proceedings of the SPIE, Vol. 5006, SPIE, Bellingham, WA, pp. 377–388.

Vázquez, C., and Tam, W. J. (2010) CRC-CSDM: 2D to 3D conversion using colour-based surrogate depth maps. International Conference on 3D Systems and Applications (3DSA 2010), Tokyo, Japan.

Vázquez, C. and Tam, W. J. (2011) A nonconventional approach to the conversion of 2D video and film content to stereoscopic 3D. *SMPTE Mot. Imag. J.*, **120** (4), 41–48.

Wang, O., Lang, M., Frei, M. *et al.* (2011) StereoBrush: interactive 2D to 3D conversion using discontinuous warps, in *Proceedings of the Eighth Eurographics Symposium on Sketch-Based Interfaces and Modeling*, ACM, New York, NY, pp. 47–54.

Wang, W., Huo, L., Zeng, W. *et al.* (2007) Depth image segmentation for improved virtual view image quality in 3-DTV, in *2007 International Symposium on Intelligent Signal Processing and Communication Systems, 2007. ISPACS 2007*, IEEE, pp. 300–303.

Wang, Y., Tai, C., Sorkine, O., and Lee, T. (2008) Optimized scale-and-stretch for image resizing. *ACM Trans. Graph. (TOG)*, **27** (5), 118:1–118:8.

Wexler, Y., Shechtman, E., and Irani, M. (2004) Space–time video completion, in *IEEE Conference on Computer Vision and Pattern Recognition*, Vol. 1, IEEE Computer Society, Los Alamitos, CA, pp. 120–127.

Yamada, K. and Suzuki, Y. (2009) Real-time 2D-to-3D conversion at full HD 1080P resolution, in *13th IEEE International Symposium on Consumer Electronics*, IEEE, pp. 103–106.

Yamada, K., Suehiro, K., and Nakamura, H. (2005) Pseudo 3D image generation with simple depth models, in *International Conference in Consumer Electronics, 2005. ICCE. 2005 Digest of Technical Papers*, IEEE, pp. 277–278.

Yano, S. and Yuyama, I. (1991) Stereoscopic HDTV: experimental system and psychological effects. *SMPTE J.*, **100** (1), 14–18.

Yano, S., Ide, S., Mitsuhashi, T., and Thwaites, H. (2002) A study of visual fatigue and visual comfort for 3D HDTV/HDTV images. *Displays*, **23** (4), 191–201.

Zhang, L. and Tam, W. J. (2005) Stereoscopic image generation based on depth images for 3D TV. *IEEE Trans. Broadcast.*, **51**, 191–199.

Zhang, L., Tam, W. J., and Wang, D. (2004) Stereoscopic image generation based on depth images, in *International Conference on Image Processing, 2004. ICIP '04*, Vol. 5, IEEE, pp. 2993–2996.

Zhang, L., Vázquez, C., and Knorr, S. (2011) 3D-TV content creation: automatic 2D-to-3D video conversion. *IEEE Trans. Broadcast.*, **57** (2), 372–383.

4

Spatial Plasticity: Dual-Camera Configurations and Variable Interaxial

Ray Zone[†]
The 3-D Zone, USA

4.1 Stereoscopic Capture

For most of its 150-year history, the capture of stereoscopic images has necessitated the use of two optical systems each with a separate lens and double recording media, whether film, video, or digital data storage. This binary dataset is necessary for the separate left- and right-eye points of view that are used to display the stereoscopic images, whether moving or still.

A fundamental problem in this binocular capture is the physical size or form factor of both the optical lenses used for photography and the camera bodies themselves. To create stereography that is comfortable in three-dimensional (3D) viewing, especially with projection and magnification on the motion picture screen, it is necessary to reduce the distance between the optical centers of the two lenses, the interaxial (IA), to values slightly smaller than the average 65 mm separating the two human eyes, or interpupillary distance. As a result, much of the invention for stereoscopic cinematography has reflected technological strategies and attempts to create smaller IA values with the horizontal parallax during image capture. Throughout the history of stereoscopic cinema, various optical devices have been employed, from arrangements with mirrors to the use of front-surface 50/50 half-silvered (transmission/reflectance) mirrors more commonly known as "beam-splitters."

Reduction and control over precise IA values has been the first goal of 3D "rig" creation. With this primary goal achieved, other elaborations of and techniques for the stereoscopic moving image have also been devised, such as variable IA and convergence, or "toe-in," of

---

[†] Ray Zone sadly died during the production of this book. He passed away on November 13, 2012, at his home in Los Angeles. He was 65 years old.

Emerging Technologies for 3D Video: Creation, Coding, Transmission and Rendering, First Edition.
Frédéric Dufaux, Béatrice Pesquet-Popescu, and Marco Cagnazzo.
© 2013 John Wiley & Sons, Ltd. Published 2013 by John Wiley & Sons, Ltd.

the optical axes of the two lenses/cameras. These kinds of control over the stereoscopic image can render its spatial character with greater plasticity as an artistic variable.

Vision scientists for centuries have speculated about the nature of spatial perception and the fundamental fact that it may be predominantly psychological in nature. George Berkeley, for example, in 1709 in his *Essay on a New Theory of Vision* attempted "to show the manner wherein we perceive by sight, the distance, magnitude, and situation of *objects*." He acknowledged "that the estimate we make of the distance of *objects* considerably remote, is rather an act of judgment grounded on experience than of *sense*" (Berkeley, 1910). Similarly, Julian Hochberg, discussing the classic structuralist theory of vision, noted that it "considered all depth cues to be symbols: the results of learned associations that have been formed between particular patterns of visual sensations and particular tactual-kinesthetic memories. What makes each of the depth cues effective," he emphasized, "is merely the fact that it has been associated with the other depth cues, and with movement, touching, etc., in the prior history of the individual" (Gombrich *et al.*, 1972).

There remains considerable controversy over this issue among vision scientists and stereoscopic motion picture professionals. Daniele Siragusano, a stereoscopic postproduction professional, for example, in his technical paper "Stereoscopic volume perception" writes quite simply, "By increasing the interaxial distance, the depth-to-width ratio gets bigger, and subsequently the perceived volume." He adds that "The depth volume is directly proportional to the interaxial distance" (Siragusano, 2012, p. 47).

This chapter provides an overview of 3D camera arrangements and "rigs" available to the stereoscopic filmmaker as of 2012. These 3D toolsets reflect various strategies for dynamic control over stereoscopic variables such as IA and convergence. As stereographic toolsets evolve, the creators of stereoscopic motion pictures can increasingly use the z-axis rendering of physical space as a fundamental parameter of cinematic storytelling, much in the same way that artistic control is similarly used with sound and music as well as color and composition.

4.2   Dual-Camera Arrangements in the 1950s

When the 3D movie boom of the 1950s was launched in Hollywood during 1952 and 1953, studio cameramen had to hurriedly devise two-camera systems for stereoscopic filming. With a 1953 article, veteran stereoscopist John Norling surveyed the various dual-camera arrangements for stereoscopic cinematography that had been in use up to that time. "The major effort in 3-D camera design," wrote Norling (1953), "is expended in providing the camera with facilities for making pictures having acceptable interaxial spacing and for convergence of the lens axes so that stereoscopic windows can be formed at desired planes." He added that "Most of today's stereoscopic motion pictures have been made with cameras not specifically designed for the purpose." That is a condition that, to a great extent, remains true in the twenty-first century.

The most obvious configuration was camera arrangement number one with two forward-facing and parallel cameras side by side. This was not a commonly used arrangement because of the bulkiness of the camera bodies and it would produce IA values more excessive than were desired for natural-looking stereography.

Camera arrangement two came into most common use in Hollywood in the 1950s (Figure 4.1). Sometimes one of the cameras was inverted or the film had to be printed "flopped" or reversed, as with all of the 3D movies shot with the "Natural Vision" rig (Figure 4.2), films such as *Bwana Devil*, which launched the 3D boom, and others such as Warner Bros.'

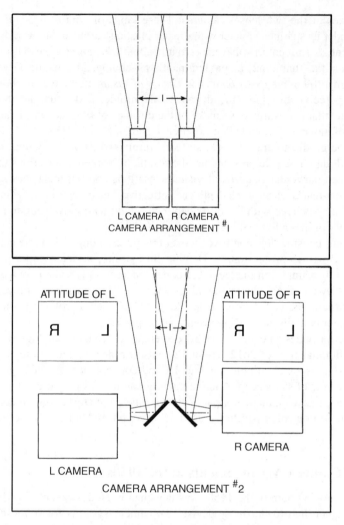

Figure 4.1 Norling's camera arrangement number two was the most commonly used in the 1950s during the 3D movie boom in Hollywood (Collection of Ray Zone)

House of Wax and *Charge at Feather River*. With camera arrangement number two, the IA was fixed and could not be varied.

One of the characteristic results of camera arrangement two was production of geometric errors, or "keystoning" differences between the left and right eye images. This convergence control was inherent in camera arrangement two but did have the advantage of lifting the stereoscopic image with emergence and negative parallax "off the screen" into the audience space, one of the visual leitmotifs of 3D films of the era.

If lenses of 50 mm focal lengths were used, Norling stated that the IA distance could be as small as 1.5 inches with camera arrangement number three (Figure 4.3). He estimated that 35 mm focal-length lenses would necessitate an IA spacing of 2.75 inches and a 25 mm focal-length lens a distance of 4 inches.

Figure 4.2 The Natural Vision system used two Mitchell cameras lens-to-lens with mirrors at 45° angles to the lens. Precise micrometer adjustments could be made to change the angle of the mirrors in relation to the lens (Collection of Ray Zone)

Camera arrangement number four shows the use of a beam-splitting transmission–reflection mirror. This arrangement permits the use of IA values going down to zero, where the two lenses overlap each other nearly identically. Less commonly used in the early 1950s in Hollywood 3D movies, camera arrangement number four, with the use of a half-silvered mirror, heralded the use of technology that would become commonplace by 2012.

4.3 Classic "Beam-Splitter" Technology

With their US Patent (No. 2,916,962) for "Optical Systems for Stereoscopic Cameras," Nigel and Raymond Spottiswoode presciently foretold the primary strategy to be used in the twenty-first century for stereoscopic cinematography (Figure 4.4). Two different camera configurations are shown in the patent drawing. The first is the so-called "over–through" configuration with a downward facing camera shooting the front-surface reflection off the half-silvered mirror. The second camera, on a corkscrew rail for adjustable IA, shoots through the mirror in a position 90° from the first camera.

A second drawing in the patent shows a variation of Norling's camera arrangement number four with a slightly more complex optical path in the camera configuration. Both configurations allow for adjustable IA that can go down to zero. "Whether extremely small interaxial spacings are necessary," observed Norling, "depends largely on the point of view of individuals proposing them, and their preferences are not necessarily based upon practical conditions."

Norling's comment reveals the highly subjective nature of stereographic decision making in the creation of spatial plasticity. Though he made commentary on the psychological effects of variable IA, Norling's observation is somewhat surprising. Nevertheless, the impact of beam-splitter technology on twenty-first century stereoscopic cinematography is widespread. And even though new 3D camera systems are built in a variety of means for the use of narrower and variable IA and convergence, or "toe-in" of optical axes, Spottiswoode's beam-splitter strategy was very much in use in 2012. A survey of a number of stereoscopic

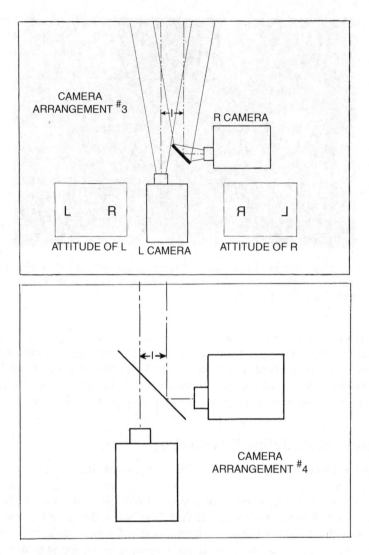

Figure 4.3 Camera arrangement number three permitted IA distance as small as 1.5 inches. Camera arrangement number four shows the use of a transmission–reflection mirror (Collection of Ray Zone)

camera systems developed for use in 2012 revealed IA reduction and control strategies that are not without precedent in stereographic history.

4.4 The Dual-Camera Form Factor and Camera Mobility

Even though digital 3D rigs could be built quite small by 2012, a rig fully loaded with onboard monitors and software tools for metadata and the "stereo image processing" could become quite bulky and heavy. An example of this is the 3ality Technica rig with all of the support technology attached, from stereoscopic metadata capture tools, to monitors for previewing of the stereographic image (Figure 4.5). A portion of the bulk with this large toolset is dedicated to correct monitoring and capture of the stereoscopic motion images

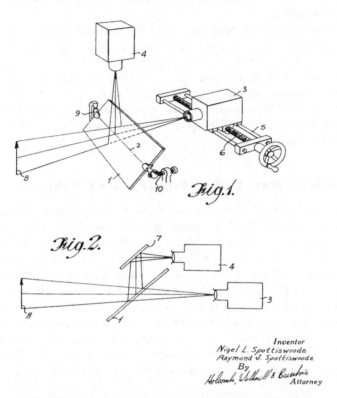

Dec. 15, 1959   N. L. SPOTTISWOODE ET AL   2,916,962
OPTICAL SYSTEMS FOR STEREOSCOPIC CAMERAS
Filed May 24, 1954

Inventor
Nigel L. Spottiswoode
Raymond J. Spottiswoode
By
Holcombe, Wellnill & Brisebois
Attorney

Figure 4.4   Nigel Spottiswoode's 1959 US Patent was foundational in clarifying the use of beam-splitter technology for stereoscopic cinematography

Figure 4.5   This fully loaded 3ality Technica dual-camera rig with two Red cameras shows how big a contemporary 3D rig can get (Photo by Ray Zone)

so that broadcast in real time is a possibility without the inducement of eyestrain or headaches in the audience. And, as the stereoscopic capture goes into the postproduction workflow, minimal corrections, ideally, will be necessary to adjust the imagery for comfortable 3D viewing and finishing.

An additional advantage of this fully loaded 3D rig is that of dynamic variable IA. This capability is especially useful for real-time 3D sportscasting where control over the IA values will render the stereographic images more pleasing and comfortable to view over a period of several hours. As of 2012, the computer-generated (CG) animated 3D movies that have been released to cinemas have had a distinct advantage over live-action 3D capture with the use of dynamic variable IA. With CG imagery, dynamic variable IA is an inherent part of the animation toolset. With live-action 3D capture, it is a factor which can add considerable size and weight to the 3D rig.

4.5 Reduced 3D Form Factor of the Digital CCD Sensor

The smaller form factor of the digital sensor and the inauguration of digital capture with "raw" recording of digital files have permitted viable use of two sensors side by side in a parallel configuration for stereoscopic cinematography after Norling's camera arrangement number one (Figure 4.6 and 4.7). It is the photographic optics and the lenses that then will dictate how small the IA can be. With smaller diameter lenses and compact housings for the CCD sensors, an acceptably narrow IA can be used for parallel shooting (Figure 4.8).

Though it is possible to shoot side by side in parallel position with two CCD digital sensors, it is nevertheless still advantageous to create stereoscopic cinematography with IA values that are extremely small. For that reason, several companies have developed very small beam-splitters that can be used with digital sensors (Figure 4.9). A few of them have even developed systems so that dynamic variable IA can be used while the cameras are rolling.

Figure 4.6 A front view of a device to join camera optics to digital sensor backs. Lenses use a fixed IA with no convergence or camera "toe-in" capability (Photo by Ray Zone)

Figure 4.7   A back view of the same device (Photo by Ray Zone)

Another strategy for variable IA has been used by cinematographer Sean Fairburn in creating the MIO (multiple interocular) 3D system using three Drift Innovation HD170 "Stealth" cameras on his "Multiplate" mount (Figure 4.10). With this system a range of different IA values is possible and greater versatility is given to the stereoscopic cinematographer in capturing subjects at varying distances from the cameras.

Figure 4.8   A reduced form factor for the beam-splitter itself is possible with the use of CCD digital sensors and raw capture of video (Photo by Ray Zone)

Figure 4.9 The back of a very compact beam-splitter device shows the two small digital sensors in place on the unit. This particular beam-splitter can also utilize dynamic variable IA (Photo by Ray Zone)

Figure 4.10 Sean Fairburn demonstrates his MIO 3D system with three cameras on his "Multiplate" mount (Photo by Ray Zone)

4.6 Handheld Shooting with Variable Interaxial

Smaller digital cameras and lighter beam-splitters have made it possible to do handheld and mobile stereoscopic cinematography and the use of variable IA at the same time. Using two Canon XF105 digital cameras on a "Hurricane" rig, director of photography Andrew Parke was able to do handheld shooting for the feature film *Night of the Living Dead 3-D 2: Reanimated* in 2011 (Figure 4.11). With rigs of this kind, of course, the weight must be sufficiently light so that a single operator can carry the entire unit.

While different IA spacings were possible with the Hurricane rig, dynamically changing the IA settings during photography was not possible. The IA range on the Hurricane rig varied from zero to slightly more than 2 inches. For handheld 3D shooting the two cameras must be in the "under–through" configuration on the beam-splitter. See Figure 4.12 and 4.13.

While the long focal-length lenses add the greatest part of the weight to this P+S Technik "Freestyle" 3D rig, dynamic variable IA is possible with handheld photography when two digital sensors, like the Silicon Imaging heads, are used. The IA will vary from zero to just over 2 inches, but the fluidity of z-space rendering, done "on the fly" during photography, can create a greater sense of immersion in the stereoscopic image. It is also a powerful yet subtle form of "damage control" to make the 3D easy to view as distances from camera to subject change over the course of the shot or scene.

Two Canon HD cameras are used with this very light 3D Film Factory "Bullet" rig in an "under–through" configuration for handheld shooting. Ease of access to clean dust off the lenses and front-silvered glass is also provided with this compact 3D rig. Variable IA will go from 0 to 3.5 inches and convergence adjustment is capable of 2.5°. The unit itself weighs 14 lbs and can be broken down in under 10 min.

Figure 4.11 Director of photography Andrew Parke does handheld 3D shooting during the filming of *Night of the Living Dead 3D 2: Reanimated* (Photo by Ray Zone)

Figure 4.12   Handheld 3D shooting is possible with dynamic variable IA using two CCD sensors with a beam-splitter rig (Photo by Ray Zone)

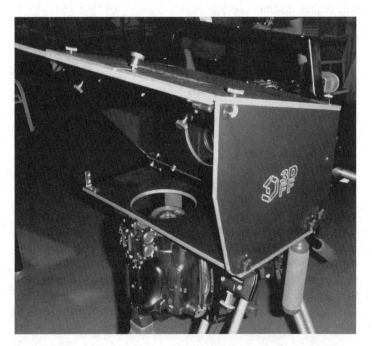

Figure 4.13   A variety of cameras can be used with many of the lighter beam-splitter rigs (Photo by Ray Zone)

4.7 Single-Body Camera Solutions for Stereoscopic Cinematography

Numerous single-body 3D camera solutions have been devised for stereoscopic cinematography over the years. John Norling himself built a dual-band 35 mm single-body 3D film camera that used variable IA and convergence "toe-in" in the 1930s. Contemporary manufacturers have also attempted to create a unified solution for 3D photography.

Panasonic has manufactured two single-body 3D cameras, the AG3D A1 in 2011 and, more recently, the HDC Z10000 for close-up stereo photography (Figure 4.14). The A1 has a fixed IA of 60 mm and the Z10000 has a fixed IA of 32 mm. The smaller IA means that the Z10000 is more fitting for getting the camera close to the subject. Weighing only 5 lbs, the Z10000 also offers convergence capabilities so that the camera can get as close as a foot away from the subject and still produce stereo that is easy to view.

Weighing 11 lbs and only 12 inches in length, the Titan 3D camera manufactured by Meduza Systems features fully motorized dynamic IA that will travel from 38 mm to 110 mm in 4 s (Figure 4.15 and Figure 4.16). These motor controls are integral to the camera, but additional wireless control can be used with the 3ality Technica THC which will fully motorize iris, focus, zoom, IA, and convergence. Various focal lengths are offered with the prime lenses and the zoom ranges from 7.1 to 28.4 mm. Two 2/3-inch CMOS sensors capture image resolution at 2048 × 1080 for 1080p. Standard monitors like the Transvideo and Marshall displays are supported.

Using the IA base within a single large lens, the single-mode 3D system built by K2E Inc. of Korea is capable of delivering a 2-inch IA with convergence capability for macro shooting down to 45 cm to subject (Figure 4.17). A zoom magnifying 36 times power maintains

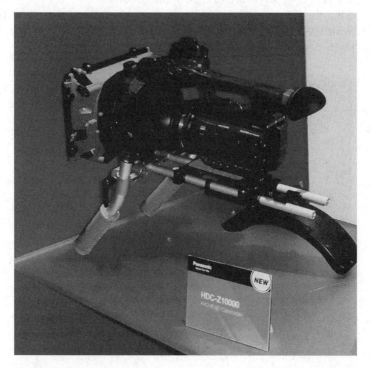

Figure 4.14 The Panasonic HDC Z10000 3D camera is built for close-up stereo photography (Photo by Ray Zone)

Figure 4.15 The Titan single-body 3D camera, seen here from the front, provides dynamic variable IA and 5° of convergence travel (Photo by Ray Zone)

binocular symmetries between left and right eye views (Figure 4.18). An additional advantage of a single lens system is synchronization of image pairs.

Weighing 35 lbs, the 3D can be shot with frame-by-frame sequential mode at 1080 60p or in dual mode with 1080 60i plus 1080 60i. With error-free capture of stereoscopic motion pictures this single-mode system is useful for live 3D broadcasting.

Figure 4.16 Seen from the back, the Titan 3D camera utilizes HDMI out, USB ports, Dual Link HD-SDI and advanced camera control ports (Photo by Ray Zone)

Figure 4.17   A "single mode" 3D system uses only one large lens to pull stereo pairs of images (Photo by Ray Zone)

With a 65 mm digital sensor, the Phantom 65 camera (v642) is a digital equivalent of large format cinematography (Figure 4.19). With fixed IA at 1.5 inches, two separate left and right eye apertures are captured on the single large sensor. Running at 140 frames per second, the Phantom will record full 4k resolution. There is a drop in resolution but the Phantom will

Figure 4.18   A side view of the single mode 3-D system shows the long lens used to produce left and right eye images (Photo by Ray Zone)

Figure 4.19   The Phantom 65 camera can record 4k 3D at 140 fps (Photo by Ray Zone)

record at higher frame rates. Recording at 1920×1080, for example, the Phantom will record at 320 fps. Weighing 12 lbs, the Phantom 65 camera is the world's only high-speed digital 3D camera.

4.8   A Modular 3D Rig

Designed specifically for the RED Epic and Scarlet cameras, the Arri Alexa M and Canon C500 and C300 cameras, the twenty-first century 3D BX4 beam-splitter weighs 11 lbs stripped down and can be fitted with additional motorized IA and convergence controls for remote operation (Figure 4.20). With variable IA up to 2 inches, the BX4 uses a proprietary mirror formulation to eliminate the need for quarter-wave retarder filtration. Two optional filter holders can be attached or removed so that directors of photography can now use linear polarizing filters, or other kinds of filters, for the first time with 3D shooting.

Twenty-first century 3D CEO Jason Goodman has designed a highly modular rig that can be adapted or expanded for additional capabilities for stereoscopic cinematography. It is a rig that will also work for steadicam, handheld, and remote head applications.

4.9   Human Factors of Variable Interaxial

John Norling commented on the perceptual differences visible in the stereoscopic image with varying degrees of IA separation. "Misinterpretation of a 3-D picture with regard to depth has been all too common among observers," wrote Norling. "Some scenes have appeared deeper than normal and well-known objects have taken on strange forms. These conflict with the individual's experience and misinterpretation is the result" (Norling, 1953).

Figure 4.20 The twenty-first century 3D BX4 stereoscopic beam-splitter with two RED Epic cameras (Photo by Ray Zone)

Norling comments about wide-base IA, so-called "hyperstereo," that "if it is too large the 3-D picture will suffer from 'miniaturization'" and that "objects and people will look smaller than they should and the depth of the scene will be abnormally increased in relation to the real spacing of objects." This "puppet theater" or "lilliputian" effect is a by-product of hyper-stereoscopic capture.

When Raymond and Nigel Spottiswoode photographed their dual-band 35 mm 3D film *The Distant Thames* in 1951, they used two 3-strip Technicolor cameras, side by side shooting parallel. Because of the wide camera bodies, this necessitated an IA distance of about 9 inches, far in excess of the normal 65 mm human interocular. With an exit poll at the theater, the Spottiswoode brothers queried the audience. "The unprompted answers to a questionnaire revealed that 75 per cent of the audience was unaware of this shrinking of the scene to model size" (Spottiswoode and Spottiswoode, 1953).

Conversely, Norling also commented about "hypostereo" whereby the IA distance is reduced. "If the interaxial spacing is too small the 3-D picture will suffer from 'giantism,'" he observed, and that "objects and people will look larger than they should, and the depth of the scene in relation to the spacing of objects becomes drastically reduced." Norling added that "the phenomenon of giantism may be interpreted differently by different people. With some, the giantism may pass unnoticed if the true sizes of the objects photographed are known. With others, it may be immediately apparent and sometimes annoying."

These perceptual variables in the stereoscopic image indicate the need for precision in the use of IA values for 3D cinematography. The plasticity of images rendered in Z-space through stereoscopic cinematography represents an artistic opportunity for filmmakers. Dynamic variable IA is a means for the cinematic storyteller to shape the depth of the world pictured on the motion picture screen, in real time, as the eyes of the audience bear witness to the transformation of visual space.

References

Berkeley, G. (1910) *A New Theory of Vision and Other Select Philosophical Writings*, E.P. Dutton & Co., New York, NY, p. 13.

Gombrich, E.H., Hochberg, J., and Black, M. (1972) *Art, Perception, and Reality*, Johns Hopkins University Press, Baltimore, MD, p. 51.

Norling, J. (1953) Basic principles of 3-D photography and projection, in *New Screen Techniques* (ed. M. Quiqley Jr.), Quigley Publishing Company, New York, NY, pp. 34–53.

Siragusano, D. (2012) Stereoscopic volume perception. *SMPTE Mot. Imag. J.*, **121** (4), 44–53.

Spottiswoode, R. and Spottiswoode, N. (1953) *The Theory of Stereoscopic Transmission and Its Application to the Motion Picture*, University of California Press, Berkeley, CA, p. 153.

Part Two

Representation, Coding and Transmission

5

Disparity Estimation Techniques

Mounir Kaaniche, Raffaele Gaetano, Marco Cagnazzo,
and Béatrice Pesquet-Popescu
Département Traitement du Signal et des Images, Télécom ParisTech, France

5.1 Introduction

The main idea behind the extraction of three-dimensional (3D) information consists of estimating the spatial displacement between two images generated by recording two slightly different view angles of the same scene. The computation of such displacement across all the pixels leads to the so-called *disparity map*. The disparity estimation problem, also known as the stereo matching problem, has been extensively studied in computer vision, and surveys of the different techniques proposed in the literature can be found in Scharstein and Szeliski (2002) and Brown *et al.* (2003).

Traditionally, two approaches, block-based or pixel-based, can be used to estimate the disparity field. In the first class of approaches, it is assumed that the disparity is block-wise constant. However, this assumption does not always hold, especially around object discontinuities. Furthermore, using the block-matching technique yields inaccurate results in texture-less regions. Pixel-based approaches attempt to overcome this drawback by assigning a disparity value to each pixel, thus producing a dense disparity map. The richness of such a result makes pixel-based approaches preferred to block-based ones in many applications, such as view synthesis and 3D reconstruction.

In turn, pixel-based algorithms are classified into two categories: local and global methods. Local methods, in which the disparity at each pixel only depends on the intensity values within a local window, perform well in highly textured regions. However, they often produce noisy disparities in untextured regions and fail in occluded areas, unless additional regularization constraints are introduced. To overcome this problem, global methods have been developed. These aim to find the disparity map by minimizing a global energy function over the whole image.

The remainder of this chapter is organized as follows. In Section 5.2 we present the geometrical models for stereoscopic imaging. We first introduce the basic concepts of camera

Emerging Technologies for 3D Video: Creation, Coding, Transmission and Rendering, First Edition.
Frédéric Dufaux, Béatrice Pesquet-Popescu, and Marco Cagnazzo.
© 2013 John Wiley & Sons, Ltd. Published 2013 by John Wiley & Sons, Ltd.

models, and then we establish the geometrical relation between the two views, laying the base for setting up the stereo matching problem, presented in Section 5.3. In the latter section, we also recall the most common hypotheses for simplifying the problem. After that, we summarize the fundamental steps involved in stereo matching algorithms. Next, a survey of the state-of-the-art disparity estimation methods is provided in Section 5.4. Finally, in Section 5.5, we draw conclusions and we end the chapter.

5.2 Geometrical Models for Stereoscopic Imaging

Before discussing the stereo imaging systems, we first introduce the pinhole camera model, which is considered as the simplest camera model. However, experiences have shown that such a simple model can accurately approximate the geometry and optics of most modern cameras (Faugeras, 1994). Thus, it defines the geometric relationship between a 3D object point and its two-dimensional (2D) corresponding projection onto the image plane. This geometric mapping from 3D to 2D is called a perspective projection.

5.2.1 The Pinhole Camera Model

Acquisition of an image using the pinhole camera model is illustrated in Figure 5.1.

This geometric representation consists of a retinal plane, a focal plane, and an image plane. The point C is called the *optical center* or *center of projection*. The distance between the focal plane and the retinal one is called the *focal length* of the optical system and is denoted by f. The line going through the optical center and perpendicular to the retinal plane is called the *optical axis*, and it intersects the image plane at a point c called the *principal point*.

We should note that, for each object in the 3D world, the camera model considered forms an inverted image of that object on the retinal plane located behind the focal plane. However, it appears preferable to place the image plane at a distance f in front of the focal plane in order to obtain a noninverted image of the object.

Now, let us explore the equations for the perspective projection describing the relation between a point M in the 3D world and its projection m onto the image plane (Figure 5.2).

This relationship is composed of the three following transformations:

- A perspective projection that defines the relation between the 3D-point M expressed in the camera coordinate system and its projection m expressed in the same coordinate system.

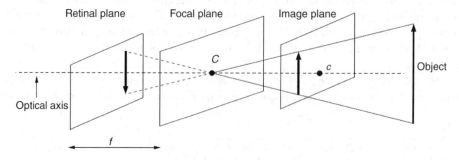

Figure 5.1 Pinhole camera model

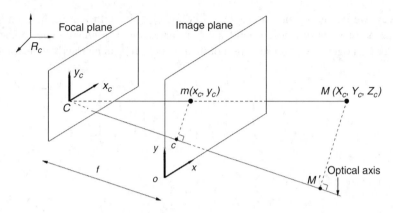

Figure 5.2 Perspective projection

- A transformation from camera to image that consists in converting the coordinates of m from the camera coordinate system to a new coordinate system attached to the image.
- A transformation from world to camera that expresses the relation between the 3D world coordinate system and the 3D camera coordinate system.

5.2.1.1 Perspective Projection Using Homogeneous Coordinates

Let $\mathcal{R}_c$ be the camera coordinate system originated at the optical center C and whose Z-axis coincides with the optical axis of the camera. For the sake of simplicity, we impose that the horizontal and vertical axes are parallel to those of the image plane.

Consider the point M in the 3D world with coordinates (X_c, Y_c, Z_c) in $\mathcal{R}_c$ and its projection m with coordinates (x_c, y_c) in $\mathcal{R}_c$. Let m' and M' be the projections of m and M, respectively, onto the optical axis. Using similarities of triangles (C, m, c) and (C, M, M'), we can deduce that

$$x_c = f \frac{X_c}{Z_c}$$
$$y_c = f \frac{Y_c}{Z_c}$$

(5.1)

Since the above equations are not linear with respect to the Euclidean coordinates X_c, Y_c, and Z_c, homogeneous coordinates will be introduced in order to obtain linear equations. These coordinates are defined as follows: if (X, Y, Z) are the *Euclidean coordinates* of a point in $\mathbb{R}^3$, its *homogeneous coordinates* are defined in the projective space $\mathbb{P}^3 \subset \mathbb{R}^4 \backslash \{0\}$ as $(\lambda X, \lambda Y, \lambda Z, \lambda)$ for any nonzero real value of λ.

More precisely, the projective space $\mathbb{P}^3$ is a subset of $\mathbb{R}^4 \backslash \{0\}$ defined by the relation of equivalence:

$$(X, Y, Z, W) \simeq (X', Y', Z', W') \Leftrightarrow$$
$$\exists \lambda \neq 0 | (X, Y, Z, W) = \lambda \cdot (X', Y', Z', W')$$

(5.2)

Let us now denote by $\tilde{\mathbf{M}} = (X, Y, Z, 1) \simeq (\lambda X, \lambda Y, \lambda Z, \lambda) \in \mathbb{P}^3$, with $\lambda \neq 0$, the homogeneous coordinates of the point M whose Euclidean coordinates are (X, Y, Z) in $\mathbb{R}^3$. In a similar way, we denote by $\tilde{m} = (x, y, 1) \simeq (\lambda x, \lambda y, \lambda) \in \mathbb{P}^2$, with $\lambda \neq 0$, the homogeneous coordinates of the point m whose Euclidean coordinates are (x, y) in the image plane.

Using the homogeneous representation of the points, (5.1) can be rewritten as follows:

$$\begin{pmatrix} s \cdot x_c \\ s \cdot y_c \\ s \end{pmatrix} = \begin{pmatrix} f & 0 & 0 & 0 \\ 0 & f & 0 & 0 \\ 0 & 0 & 1 & 0 \end{pmatrix} \begin{pmatrix} X_c \\ Y_c \\ Z_c \\ 1 \end{pmatrix} \tag{5.3}$$

where s is a nonzero factor.

5.2.1.2   Transformation from Camera to Image

Most of the current imaging systems define the origin O of the image coordinate system at the top left pixel of the image. This coordinate system will be denoted by $\mathcal{R}_O$. However, the previous image coordinates (x_c, y_c) are expressed in $\mathcal{R}_c$ with the origin in c. Thus, a transformation between these two coordinate systems is required. If we denote by (x, y) the corresponding pixel in the digitized image (i.e., expressed in $\mathcal{R}_O$), this position is related to the camera coordinate system by the following transformation:

$$\begin{pmatrix} x \\ y \\ 1 \end{pmatrix} = \begin{pmatrix} k_x & 0 & x_0 \\ 0 & k_y & y_0 \\ 0 & 0 & 1 \end{pmatrix} \begin{pmatrix} x_c \\ y_c \\ 1 \end{pmatrix} \tag{5.4}$$

where:

- k_x and k_y are the horizontal and vertical scale factors (pixels/mm);
- x_0 and y_0 are the pixel coordinates of the principal point c in $\mathcal{R}_O$.

These parameters are called the camera *intrinsic parameters* since they do not depend on the position and orientation of the camera (Faugeras, 1994; Pedersini *et al.*, 1999).

Therefore, the perspective projection equation (5.3) becomes

$$\begin{pmatrix} s \cdot x \\ s \cdot y \\ s \end{pmatrix} = \begin{pmatrix} k_x & 0 & x_0 \\ 0 & k_y & y_0 \\ 0 & 0 & 1 \end{pmatrix} \begin{pmatrix} f & 0 & 0 & 0 \\ 0 & f & 0 & 0 \\ 0 & 0 & 1 & 0 \end{pmatrix} \begin{pmatrix} X_c \\ Y_c \\ Z_c \\ 1 \end{pmatrix} \tag{5.5}$$

By defining $\alpha_x = k_x f$ and $\alpha_y = k_y f$, this equation is often rewritten as

$$\begin{pmatrix} s \cdot x \\ s \cdot y \\ s \end{pmatrix} = \begin{pmatrix} \alpha_x & 0 & x_0 \\ 0 & \alpha_y & y_0 \\ 0 & 0 & 1 \end{pmatrix} \begin{pmatrix} 1 & 0 & 0 & 0 \\ 0 & 1 & 0 & 0 \\ 0 & 0 & 1 & 0 \end{pmatrix} \begin{pmatrix} X_c \\ Y_c \\ Z_c \\ 1 \end{pmatrix} \tag{5.6}$$

5.2.1.3   Transformation from World to Camera

Let $\mathcal{R}$ be the 3D world coordinate system and (X, Y, Z) the coordinates of the point M in $\mathcal{R}$. These coordinates are related to those expressed in $\mathcal{R}_c$ by a rotation $\mathbf{R}$ followed by a translation $\mathbf{t}$. Thus, we obtain

$$\begin{pmatrix} X_c \\ Y_c \\ Z_c \end{pmatrix} = \mathbf{R} \begin{pmatrix} X \\ Y \\ Z \end{pmatrix} + \mathbf{t} \tag{5.7}$$

Using the homogeneous coordinates, (5.7) can also be expressed as

$$\begin{pmatrix} X_c \\ Y_c \\ Z_c \\ 1 \end{pmatrix} = \begin{pmatrix} \mathbf{R} & \mathbf{t} \\ 0 & 1 \end{pmatrix} \begin{pmatrix} X \\ Y \\ Z \\ 1 \end{pmatrix} \tag{5.8}$$

The matrix $\mathbf{R}$ of size 3×3 and the vector $\mathbf{t}$ of size 3×1 describe the orientation and position of the camera with respect to the world coordinate system. They are called the *extrinsic parameters* of the camera (Faugeras, 1994; Pedersini *et al.*, 1999).

Finally, combining the three transformations described above yields the general form of the perspective projection matrix $\mathbf{P}$ of the camera:

$$\mathbf{P} = \mathbf{A}\mathbf{P}_0\mathbf{K} \tag{5.9}$$

where

$$\mathbf{A} = \begin{pmatrix} \alpha_x & 0 & x_0 \\ 0 & \alpha_y & y_0 \\ 0 & 0 & 1 \end{pmatrix}, \quad \mathbf{P}_0 = \begin{pmatrix} 1 & 0 & 0 & 0 \\ 0 & 1 & 0 & 0 \\ 0 & 0 & 1 & 0 \end{pmatrix} \quad \text{and} \quad \mathbf{K} = \begin{pmatrix} \mathbf{R} & \mathbf{t} \\ 0 & 1 \end{pmatrix} \tag{5.10}$$

This expression can also be rewritten as

$$\mathbf{P} = \mathbf{A}[\mathbf{R}\,\mathbf{t}] \tag{5.11}$$

Consequently, the relationship between the homogeneous coordinates of a world point $\tilde{\mathbf{M}}$ and its projection $\tilde{\mathbf{m}}$ onto the image plane is described by

$$\tilde{\mathbf{m}} = \mathbf{P}\,\tilde{\mathbf{M}} \tag{5.12}$$

5.2.2   Stereoscopic Imaging Systems

The principle of a binocular stereoscopic imaging system consists of generating two views, called the left image and right image, by recording two slightly different view angles of the same scene (see Figure 5.3). The left and right cameras are respectively represented by their optical centers $C^{(l)}$ and $C^{(r)}$, and their perspective projection matrix $\mathbf{P}^{(l)}$ and $\mathbf{P}^{(r)}$.

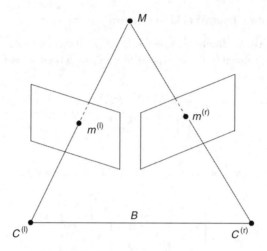

Figure 5.3 Stereoscopic imaging system

Assume that the object point M is projected onto two points $m^{(l)}$ and $m^{(r)}$ in the left and right images to create the so-called *homologous points*. Using (5.12), this projection can be expressed as

$$\tilde{\mathbf{m}}^{(l)} = \mathbf{P}^{(l)}\,\tilde{\mathbf{M}}$$
$$\tilde{\mathbf{m}}^{(r)} = \mathbf{P}^{(r)}\tilde{\mathbf{M}}$$

(5.13)

5.2.2.1 Epipolar Geometry

The epipolar geometry describes the relation between the two generated images. The different entities involved in this geometry are illustrated in Figure 5.4.

The plane passing through the object point M and the centers of projection $C^{(l)}$ and $C^{(r)}$ is called the *epipolar plane* Π. The intersection of the line joining the optical centers (called *baseline*) with the two images plane results in two points $e^{(l)}$ and $e^{(r)}$ called the *epipoles*.

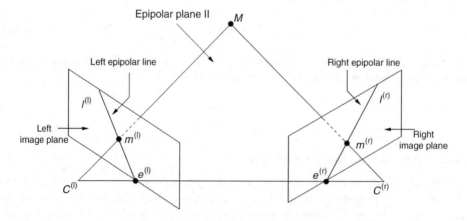

Figure 5.4 Epipolar geometry between a pair of images

The intersection of the left (right) image plane with the line joining the point M and the left (right) optical center $C^{(l)}$ ($C^{(r)}$) corresponds to $m^{(l)}$ ($m^{(r)}$).

Suppose now that we only know $m^{(l)}$ and we have to search for its corresponding point $m^{(r)}$. The plane Π is determined by the baseline and the ray defined by $m^{(l)}$. From the above, we know that the ray passing through $m^{(r)}$ lies in Π. Hence, the point $m^{(r)}$ lies on the line of intersection $l^{(r)}$ of Π with the right image plane. This line is called the *epipolar line $l^{(l)}$* associated with $m^{(l)}$.

The knowledge of the epipolar geometry leads to an important observation. Given an image point $m^{(l)}$ lying on the left epipolar line $l^{(l)}$, its corresponding point $m^{(r)}$ must lie on the conjugate epipolar line $l^{(r)}$. This corresponds to the epipolar constraint, which can be expressed mathematically as

$$(\tilde{\mathbf{m}}^{(r)})^{\mathrm{T}}\mathbf{F}\tilde{\mathbf{m}}^{(l)} = 0 \tag{5.14}$$

where $\mathbf{F}$ is the *fundamental matrix*. This matrix, of size 3×3, depends on the intrinsic and extrinsic parameters of the cameras (Zhang and Xu, 1997). An overview of techniques to find $\mathbf{F}$ can be found in Zhang (1998).

Thus, such a constraint allows reducing the search of homologous points to a one-dimensional (1D) search problem rather than a 2D one. In practice, this problem can be further simplified by using an appropriate processing known as *epipolar rectification*.

5.2.2.2 Epipolar Rectification

There are two primary camera configurations used in stereoscopic vision. The first configuration, which is considered at the beginning of Section 5.2.2, is called the *converging camera configuration*. In this configuration, cameras are rotated towards each other by a small angle and, thus, the epipolar lines appear inclined. The second configuration is called the *parallel camera configuration* (Figure 5.5), and it is composed of two cameras with parallel optical axes.

It is important to note that the parallel camera configuration has several advantages. Indeed, it simplifies the search for homologous points, which will be situated on parallel horizontal lines. Furthermore, in this particular case, the epipolar lines are horizontal and coincide with the image scan lines. Hence, two corresponding pixels are located at the same line in the two images as shown in Figure 5.5. In the other case (i.e., converging camera configuration), the epipolar lines can be made horizontal by using an appropriate processing.

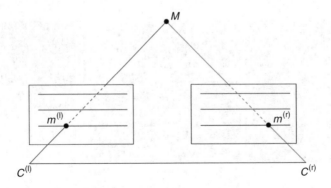

Figure 5.5   Epipolar rectification

This procedure is called rectification. It consists of determining a transformation that projects each image onto a common image plane. A survey of different rectification methods proposed in the literature can be found in Hartley and Gupta (1993), Fusiello *et al.* (2000b), and Papadimitriou and Dennis (1996).

5.3 Stereo Matching Process

5.3.1 Disparity Information

Assume that a point **M** in the 3D scene is projected onto two points $m^{(l)}(x^{(l)}, y^{(l)})$ and $m^{(r)}(x^{(r)}, y^{(r)})$ in the images $I^{(l)}$ and $I^{(r)}$. One image (generally the left one) is considered as a reference image. In order to estimate the disparity values for all the pixels in the reference image, a stereo matching process is performed by searching for each point $m^{(l)}$ in the reference image its homologous point $m^{(r)}$ in the other image. Thus, the stereo matching problem can be mathematically formulated as the search for a function u that associates a disparity vector to each pixel $m^{(l)}(x^{(l)}, y^{(l)})$ in the reference image:

$$u : \mathbb{R}^2 \rightarrow \mathbb{R}^2$$
$$(x^{(l)}, y^{(l)}) \mapsto u(x^{(l)}, y^{(l)}) = (x^{(l)} - x^{(r)}, y^{(l)} - y^{(r)}) \tag{5.15}$$

When the two views $I^{(l)}$ and $I^{(r)}$ are rectified, the displacement between the homologous points is purely horizontal. In this case, (5.15) becomes

$$u : \mathbb{R}^2 \rightarrow \mathbb{R}$$
$$(x^{(l)}, y^{(l)}) \mapsto u(x^{(l)}, y^{(l)}) = x^{(l)} - x^{(r)} \tag{5.16}$$

Figure 5.6 shows as an example of the "cones" stereo image pair and the corresponding disparity map when the left image is selected as the reference one (Scharstein and Szeliski, 2003).

In the following, the stereo images are assumed to be geometrically rectified.

5.3.2 Difficulties in the Stereo Matching Process

Independently of the scene geometry, the stereo matching process is based on comparing the intensity values of the images. For this reason, it is considered as a difficult problem in stereo

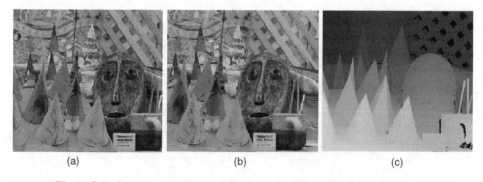

| (a) | (b) | (c) |

Figure 5.6 Cones stereo pairs: (a) left image; (b) right image; (c) disparity map

vision due to the presence of occlusions, illumination variations, and lack of texture (Dhond and Aggarwal, 1989).

- **Occlusion problem (Dhond and Aggarwal, 1989):** In general, stereo or multiview images contain nearly similar contents since they correspond to the same scene. However, there are some areas in one image that are absent in the other image, and they are referred to as *occluded areas*. This occlusion effect is due to the different viewpoints of the cameras and the presence of discontinuities in the scene. Therefore, the disparity is undefined in occlusion areas because such areas cannot be found in the other image.
- **Illumination variations:** In real stereoscopic imaging systems, the characteristics of the cameras may be slightly different. Consequently, some illumination changes may appear between the captured images. This illumination variation will cause a serious problem in the correspondence process. In order to overcome this problem, a preprocessing step, like the histogram equalization, is often applied to the original stereo image pair.
- **Untextured regions:** Obviously, a pixel is less discriminating in areas with repetitive structure or texture. Indeed, it will be difficult to distinguish pixels in the same area having similar intensity. Thus, such homogeneously textured regions result in ambiguities in the stereo matching process, simply because of the presence of multiple possible matches.

In summary, stereo matching is an ill-posed problem with inherent ambiguities.

5.3.3 Stereo Matching Constraints

In order to overcome the ambiguities mentioned above and make the problem more tractable, a variety of constraints and assumptions are typically made. The most commonly used constraints are the following ones (Boufama and Jin, 2002; Miled *et al.*, 2006):

- **Epipolar constraint:** Given an image point in one image, the corresponding point must lie on an epipolar line in the other image. The main advantage of this constraint is that it reduces the matching problem from a 2D search problem to a 1D one. Furthermore, when the stereo images are rectified, the 1D search problem is further simplified since the epipolar line will coincide with the image scan line.
- **Uniqueness constraint:** This imposes that a given pixel in one image can match to no more than one pixel in the other image. It is often used to identify occlusions by enforcing one-to-one correspondences for visible pixels across images. However, this constraint fails in the presence of transparent objects.
- **Ordering constraint:** This ensures that the order of two pixels in a line of one image is preserved in its homologous line of the other image. The advantage of using this assumption is that it allows the detection of occlusions.
- **Smoothing constraint:** This imposes that disparity varies smoothly across homogeneous areas (inside objects). It is generally satisfied everywhere except at depth boundaries.

5.3.4 Fundamental Steps Involved in Stereo Matching Algorithms

Before presenting the different disparity estimation techniques reported in the literature, we should note that stereo matching algorithms are generally based on the following steps (Scharstein and Szeliski, 1998):

- matching cost computation and support aggregation;
- disparity computation and optimization;
- post-processing of disparity.

5.3.4.1 Matching Cost Computation and Support Aggregation

To solve the stereo correspondence problem, great attention is paid to the matching cost that can be defined locally (at the pixel level), or over a certain area of support (window). In this respect, many cost functions have been used in the literature to measure the similarity between a given pixel and each possible candidate matching sample. The most common pixel-based matching costs are absolute intensity differences AD (Kanade, 1994) and squared intensity differences SD (Matthies *et al.*, 1989), given by

$$
\begin{aligned}
\mathrm{AD}(x, y, u) &= \left| I^{(\mathrm{l})}(x, y) - I^{(\mathrm{r})}(x - u, y) \right| \\
\mathrm{SD}(x, y, u) &= \left[I^{(\mathrm{l})}(x, y) - I^{(\mathrm{r})}(x - u, y) \right]^2
\end{aligned}
\tag{5.17}
$$

However, AD and SD measures are known to be very sensitive to outliers.

Furthermore, other matching costs such as normalized cross-correlation (Ryan *et al.*, 1980), gradient-based, and nonparametric measures have also been proposed because they are more robust to changes in cameras (Scharstein, 1994; Zabih and Woodfill, 1994). Measures based on information theory, like entropy and mutual information, have also been used (Fookes *et al.*, 2002). Finally, Birchfield and Tomasi (1998, 1999) have defined a similarity measure that is insensitive to image sampling. More precisely, they proposed using the linearly interpolated intensity functions surrounding the pixel in the reference image and the possible candidate pixel in the other image. While these latter matching costs have the advantage of alleviating some problems (like noise, changes in cameras, and image sampling), they entail a higher computational cost and, therefore, are not appropriate for real-time applications. To overcome this drawback, some workers have proposed applying a down-sampling step to the stereo images before performing the matching cost computation. For instance, in Chapter 24, Y.-C. Tseng and T.-S. Chang review the main issues on computational complexity and hardware implementation of disparity estimation algorithms.

Consequently, the choice of a matching cost depends on both the complexity of the algorithm that can be tolerated and on the properties of the images.

Once the similarity measure is defined, the next step is to sum or to average the matching cost over a support region. For this reason, this step is known as *support aggregation*. A typical aggregation can be implemented by using 2D convolution with a support window, square windows, and windows with adaptive size (Kanade and Okutomi, 1994).

5.3.4.2 Disparity Computation and Optimization

Once the cost aggregation step has been applied to the possible candidate matching pixels, the candidate leading to the minimum cost will be considered as the best match and, therefore, the disparity value can be computed as defined in (5.16). For this purpose, and based on the optimization strategy, we can mainly classify the disparity estimation techniques into *local* and *global* optimization methods. Describing these methods will be the objective of Section 5.4.

5.3.4.3   Post-Processing of Disparity

In order to improve the accuracy of the disparity map, the following steps can be applied to the estimated field:

- **Left/Right Consistency (LRC) Check:** This step consists of comparing the computed disparity map when the left image has been selected as a reference image with that obtained by using the right image as a reference image. Since the left and right images correspond to the same scene, the relation between the two resulting disparity maps should only differ by sign for pixels visible in the stereo pairs (Fua, 1993). In practical terms, for each pixel p of the left-referenced disparity map, the matching point p' of the other map is computed, and the two corresponding disparity values (one for each view) are compared. If the absolute difference between these values is over a certain threshold (typically fixed to one) the LRC check fails and the corresponding disparity value is invalidated. This consistency check fails for all mismatched pixels, including the occluded ones, which makes this post-processing step particularly advantageous.
- **Interpolation (Cochran and Medioni, 1992):** Since the disparity is not defined for the occluded pixels, the main question that arises is how the positions of these pixels, referred also to as *holes*, can be filled. A simple and frequently used solution to fill in these holes is to spread the neighboring disparity values: for each invalidated pixel p, the closest valid pixels to the left and to the right are searched, and the lowest of the disparity values corresponding to the two locations is chosen for p. Selecting the lower disparity is motivated by the fact that occlusions occur at the background. Another efficient filtering technique consists of applying a mirror effect to the disparity values adjacent to the holes.
- **Filtering (Birchfield and Tomasi, 1998):** Generally, a median filter can be applied to the estimated map to smooth it and to further remove the false matches. This filtering may become necessary if some hole-filling procedure has been performed, since the latter typically generates horizontal streaks in the disparity map. In the general case, this step is especially interesting in the presence of weakly textured regions, where some pixels may have been mismatched although the disparity can be correctly estimated in the neighborhood.

5.4   Overview of Disparity Estimation Methods

5.4.1   Local Methods

Local methods are based on comparing the similarity between two sets of pixels. According to the type of the matching sets, these methods can be mainly classified into two categories: feature- and area-based approaches.

5.4.1.1   Feature-Based Methods

First, this class of methods aims to convert both images' intensities into a set of features like edges (Otha and Kanade, 1985), segments (Medioni and Nevatia, 1985), or curves (Schmid and Zisserman, 1998). Then, a correspondence step between the extracted features is established. Recently, a scale-invariant feature transform (SIFT)-based method has been proposed to estimate the disparity field and tested in the context of multiview video coding (Liu *et al.*, 2010). These methods have the advantage that features are less sensitive to noise and photometric distortions and, therefore, are fewer and more easily distinguishable than a pixel intensities-based description. While feature-based methods yield generally accurate results, their main drawback is the sparsity of the resulting disparity map. In this case, an interpolation step is required if a dense disparity field is desired (Konrad and Lan, 2000). Furthermore, they result in an additional complexity since they require a prior feature extraction step by using some conventional techniques such as the Harris-point and the Canny edge detectors.

5.4.1.2   Area-Based Methods

In contrast with feature-based methods, area-based ones have the advantage of directly generating a dense disparity map, but they are more sensitive to locally ambiguous regions like the occlusion and the untextured ones. This class of methods, where pixels or regions are matched to find the corresponding pixels, is also known as *correlation-based methods*. The basic idea of this method is illustrated in Figure 5.7. It consists firstly of defining a window around the pixel $m^{(1)}$ to be matched, and then comparing it with the window surrounding each possible candidate corresponding pixel located in a search area.

A major difficulty of this method is the choice of the window size and shape. Indeed, the first works have been developed by simply using a rectangular (or a square) window with a fixed size. However, the limitation of this technique is that it fails at object edges and boundaries. Furthermore, using small windows in untextured regions do not capture enough intensity variations to make reliable the correspondence task, whereas the use of large windows tends to blur the depth boundaries. To overcome these problems, more sophisticated approaches have been proposed by varying the size and the shape of the window according to the intensity variations (Kanade and Okutomi, 1994; Woo and Ortega, 2000). Fusiello *et al.* (2000a) proposed to try a set of windows and then select the one giving the highest correlation measure.

More recently, an alternative method has been proposed (Yoon and Kweon, 2006) in which, instead of searching for the optimal window shape, the rectangular supports around the two pixels to be matched are adaptively weighted, according to color/intensity similarity and geometric distance of each pixel of the window with respect to the central one. This method *de facto* allows the use of arbitrarily shaped windows for support matching, hence proving particularly effective around depth discontinuities. However, to provide an accurate matching, the base support window needs to be bigger than in other methods, resulting in a significant increase of the computational complexity. Anyway, as for other area-based

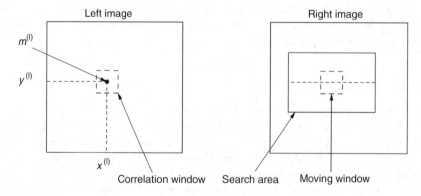

Figure 5.7   Correlation-based stereo matching method

methods, owing to the intrinsic data-parallel nature of this algorithm, a considerable speed up may be expected from implementations on parallel computing devices.

It is worth pointing out that these techniques are very popular for their simplicity, and their variants have been widely employed in many applications such as motion estimation (Alkanhal *et al.*, 1999) and image registration (Zitova and Flusser, 2003). Motivated by its simplicity, many research studies in stereovision have been devoted to the extension of the correlation-based methods to the context of color stereo images. Generally, there are two ways of incorporating color information into disparity estimation algorithms. The first one consists of computing the correlation separately with each of the color components and then of merging the results by means of a weighting operation, for example (Belli *et al.*, 2000). The second one aims at defining a new color-based correlation function that takes into account color information available in the stereo pair (Koschan, 1993; El Ansari *et al.*, 2007).

5.4.2 Global Methods

Contrary to local methods, the global ones aim at finding the disparity map that minimizes a global energy function over the entire image. Generally, such a function is composed of two terms and in its most general form can be written as

$$E(\boldsymbol{u}) = E_{\mathrm{d}}(\boldsymbol{u}) + \alpha E_{\mathrm{s}}(\boldsymbol{u}) \tag{5.18}$$

where $E_{\mathrm{d}}(\boldsymbol{u})$ is a *data term* that measures the distance between corresponding pixels, $E_{\mathrm{s}}(\boldsymbol{u})$ is a *regularization term*, also called a *smoothing prior*, that enforces smoothness of the disparity field $\boldsymbol{u} = (u(m, n))_{(m,n) \in \mathcal{D}}$, where $\mathcal{D}$ denotes the image support, and α is a positive constant called a regularization parameter that controls the amount of smoothness in the resulting solution.

In order to estimate the disparity map, the above global cost function is first reformulated in an appropriate way and then minimized by resorting to some specific algorithms. The best known ones are graph cuts (Boykov *et al.*, 2001; Kolmogorov and Zabih, 2001) and variational methods (Slesareva *et al.*, 2005; Kosov *et al.*, 2009; Miled *et al.*, 2009).

5.4.2.1 Graph Cuts

The idea of these methods is to cast the stereo matching problem as a pixel labeling one. To specify such a problem, we need to define a set of *sites* and a set of *labels*. In the following, the spatial domain (m, n) in each image and its associated disparity value will be simply denoted by s and $u(s)$ for notation concision. More precisely, sites represent all the pixels in the reference image, and labels correspond to a finite set of their possible disparity values. Note that in this case the precision of the estimated disparity is imposed by the chosen set of values. These two sets will be designated respectively by $\mathcal{P}$ and $\mathcal{L}$:

$$\mathcal{P} = \{1, 2, \ldots, N\} \tag{5.19}$$
$$\mathcal{L} = \{l_1, l_2, \ldots, l_k\}$$

where N and k respectively represent the number of all the pixels which need to be matched and the number of candidate disparity values.

Now, the goal is to find a labeling $\mathcal{U} = \{u(1), u(2), \ldots, u(N)\}$ that assigns to each site s in the set $\mathcal{P}$ a label (i.e., disparity) $u(s)$ from the set $\mathcal{L}$. Thus, the labeling problem can be seen as a mapping from $\mathcal{P}$ to $\mathcal{L}$. In this case, the data term of (5.18) is often expressed as

$$E_d(\boldsymbol{u}) = \sum_{s \in \mathcal{P}} D_s(u(s)) \qquad (5.20)$$

where D_s measures the similarity between the pixel s in the left image and the pixel s shifted by $u(s)$ in the right image by using, for example, one of the standard cost functions described in Section 5.3.4.1.

Concerning the smoothness term $E_s(\boldsymbol{u})$, it aims at modeling how each pixel s interacts with its neighboring pixels. The set of the neighbors of s will be denoted by $\mathcal{N}_s$. Let $\mathcal{N}$ be the set of all neighboring pairs $\{s, t\}$. Given this set $\mathcal{N}$, the smoothness term can be rewritten in the following form:

$$E_s(u) = \sum_{\{s,t\} \in \mathcal{N}} V_{\{s,t\}}(u(s), u(t)) \qquad (5.21)$$

where $V_{\{s,t\}}(u(s), u(t))$, called a *neighbor interaction function*, is used to enforce the neighboring pixels to have similar disparity values. It can take different types of smoothness priors like everywhere smooth prior, piecewise constant prior, and piecewise smooth prior.

Once the stereo correspondence problem is reformulated as a labeling one, the disparity can be determined by finding the minimal cut through a constructed graph.

Note that the first method using graph cuts to solve the stereo correspondence problem was introduced by Roy and Cox (1998). Later, Boykov *et al.* (2001) developed algorithms that approximate the minimization of the energy function for two general classes of interactions $V_{\{s,t\}}(u(s), u(t))$: semi-metric and metric. These metrics include the truncated ℓ_2 distance and the Potts model (Boykov *et al.*, 1998). Inspired by the developed algorithms, Kolmogorov and Zabih (2001) proposed to minimize an energy function composed of the two aforementioned terms and an *occlusion term* that imposes a penalty to take into account the occlusion problems. Moreover, such graph-cuts-based algorithms have also been employed to estimate the disparity map of color stereo images after a segmentation step that decomposes the images into homogeneous color regions (Hong and Chen, 2004; Bleyer and Gelautz, 2005).

5.4.2.2 Variational Methods

Variational approaches have also been extensively used to compute a consistent disparity map by solving the associated partial differential equation. Indeed, such techniques have already attracted much interest in the computer vision community where they were first introduced by Horn and Schunck (1981) for the purpose of estimating a dense flow from a sequence of images. In opposition to the previous approaches which are performed in a discrete manner and produce integer disparity values, variational techniques work in a continuous space and, therefore, have the advantage of producing a disparity field with ideally infinite precision.

While the data term of (5.18) is often defined by using a classical cost function like (5.17), most existing studies have mainly focused on the choice of the regularization term. In this context, the regularization step is based on a diffusion process which is isotropic in the homogeneous areas and anisotropic around the discontinuities. For this reason, the smoothness term depends generally on the norm of the gradient of u, denoted by $|\nabla u|$, and can be written as

$$E_s(u) = \Phi(\nabla u) \tag{5.22}$$

where Φ is a given regularization functional.

One of the most basic regularization techniques takes a quadratic form and is referred to as *Tikhonov* regularization (Tikhonov and Arsenin, 1977). However, the use of an ℓ_2 term favors the smoothness over the whole disparity field, which will result in the destruction of important details located around the discontinuities. In order to produce a smooth disparity map while preserving edges, another regularization type has been introduced by considering the total variation of u (Rudin *et al.*, 1992). This functional, denoted by $\mathrm{tv}(u)$, can be defined as the sum over the whole spatial domain of the norm of the spatial gradient of u. After discretization, various expressions of the total variation can be found. For example, one can use the following form of the isotropic total variation which has the advantage to be invariant by rotation:

$$\mathrm{tv}(u) = \sum_{x=1}^{M-1}\sum_{y=1}^{N-1} \left(|u(x+1,y) - u(x,y)|^2 + |u(x,y+1) - u(x,y)|^2\right)^{1/2} \tag{5.23}$$

where $M \times N$ corresponds to the size of the disparity map.

Moreover, an anisotropic variant of the total variation has also been used:

$$\mathrm{tv}(u) = \sum_{x=1}^{M-1}\sum_{y=1}^{N-1} |u(x+1,y) - u(x,y)| + \sum_{x=1}^{M-1}\sum_{y=1}^{N-1} |u(x,y+1) - u(x,y)| \tag{5.24}$$

The advantage of this form is that it leads to a reduced computational complexity of the total variation. Furthermore, it is closely related to the sparse ℓ_1 criterion of the detail coefficients of a nonredundant Haar wavelet transform.

Moreover, Alvarez *et al.* (2002) have proposed to exploit the edges of the reference image in order to preserve the discontinuities in the resulting disparity field. This can be achieved by using a regularization based on the Nagel–Enkelmann operator (Nagel and Enkelmann, 1986). This function depends in this case on the gradient of the disparity ∇u and the gradient of the reference image $\nabla I^{(1)}$, and is given by

$$\Phi(\nabla u, \nabla I^{(1)}) = (\nabla u)^T D(\nabla I^{(1)})(\nabla u) \tag{5.25}$$

where $D(\nabla I^{(1)})$ is a symmetric matrix of size 2×2 given by

$$D(\nabla I^{(1)}) = \frac{1}{|\nabla I^{(1)}|^2 + 2v^2} \left[\begin{pmatrix} \dfrac{\partial \nabla I^{(1)}}{\partial y} \\ -\dfrac{\partial \nabla I^{(1)}}{\partial x} \end{pmatrix} \begin{pmatrix} \dfrac{\partial \nabla I^{(1)}}{\partial y} \\ -\dfrac{\partial \nabla I^{(1)}}{\partial x} \end{pmatrix}^T + v^2 I_{2,2} \right] \tag{5.26}$$

where $I_{2,2}$ denotes the identity matrix of size 2×2 and v is a weighting parameter.

Once the regularization terms have been defined, the aforementioned approaches minimize the energy function by solving the associated Euler–Lagrange partial differential equation. For this purpose, several iterative methods can be used, such as Jacobi, Gauss–Seidel, and gradient descent methods (Young, 1971).

However, these techniques often are computationally demanding and they require a careful study for the discretization of the associated partial differential equations. In addition, they require the determination of the Lagrange parameter α, which may be a difficult task. The latter problem becomes even more involved when a sum of regularization terms has to be considered to address multiple constraints, which may arise in the problem.

The work by Miled *et al.* (2009) attempted to overcome these difficulties by formulating the stereo matching problem as a constrained convex optimization problem. The proposed approach incorporates different constraints corresponding to a *prior* knowledge and relying on various properties of the disparity field to be estimated. Each constraint is represented by a convex set, and the intersection of these sets constitutes the feasibility set. Then, an appropriate convex quadratic objective function is minimized on the feasibility set by using an efficient block-iterative algorithm. The optimization algorithm used, based on the computation of subgradient projections, has the ability to handle a wide range of convex constraints. This work has also been extended to the context of color stereo images where a color-based objective function, which is the sum of intensity differences over the three color channels, is minimized under specific convex constraints (Miled *et al.*, 2008).

However, in order to guarantee the convergence of the algorithm, the cost function must be quadratic and strictly convex. To overcome this problem, a more general convex optimization framework based on proximal methods has recently been developed and applied to disparity estimation (El Gheche *et al.*, 2011; Chaux *et al.*, 2012). More precisely, the proposed work enables the use of nondifferentiable cost functions, offering a great flexibility in the choice of the criterion involved, which, for example, can be based on the ℓ_1-norm, or ℓ_p-norm with $p \geq 1$ or the Kullback–Leibler divergence.

A particular advantage related to these methods is due to their block-iterative structure, which directly provides the possibility of developing parallel implementations. More precisely, at each iteration these algorithms first perform the independent activation of the criteria involved, each criterion generating a contribution for the update of the current solution, and then combine the different results. Adopting a task-parallel approach, the different contributions may be associated, for instance, with the different cores of a multi-core CPU device.

Alternatively, the reduced inter-task dependencies can be indirectly exploited by adopting a data-parallel approach, justified by the massive presence of matrix operations. In particular, most of the operations required to compute the partial solutions are point-wise (e.g., computation of the proximity operator related to the chosen cost function) or have close-range dependencies (e.g., gradients, Haar decompositions). These operations exhibit a significant performance enhancement potential when implemented on massively data-parallel architectures like graphics processing units. In some cases, the computation of fast Fourier transforms is needed, which has also been shown to achieve significant speed-ups on such devices (Volkov and Kazian, 2008). Finally, note also that a theoretical framework for computing global solutions of variational models with different convex regularizers has been addressed by Pock *et al.* (2010).

5.4.2.3　Other Methods

In addition to the above, other optimization algorithms have been developed based on dynamic programming and belief propagation.

Dynamic programming techniques use the ordering and smoothness constraints to optimize correspondences in each scan line (Otha and Kanade, 1985). In this context, the disparity estimation problem consists of the following two steps. Given two scan lines of length N, disparity values are first calculated for each possible pairing of pixels, yielding a path cost matrix of size $N \times N$. Then, the matching pixel with the minimum path cost is selected through the scan lines. While this method has the advantage of enforcing the smoothness constraint of the disparity along the epipolar line, its main limitation is its inability to enforce smoothness in both horizontal and vertical directions. To solve this problem, a two-pass dynamic programming technique has been developed by Kim *et al.* (2005). Moreover, Veksler (2005) proposed to perform this algorithm on a 2D tree structure instead of individual 1D scan lines.

Concerning the belief propagation (BP) algorithm, it has mainly been developed for performing inference on graphical models such as Bayesian networks. In particular, this algorithm was also found to be efficient for handling Markov random fields (MRFs) which have been used to formulate the disparity estimation problem (Sun *et al.*, 2003; Tappen and Freeman, 2003). Indeed, an MRF can be seen as a graph model in which nodes represent random variables. In this context, the joint probability $p(\mathbf{u}) = p(u_1, \ldots, u_N)$ of the MRF model takes the following general form:

$$p(\mathbf{u}) = \prod_{s \in \mathcal{P}} \Psi(u(s)) \prod_{\{s,t\} \in \mathcal{N}} \Phi(u(s), u(t)) \tag{5.27}$$

where Ψ is a function that measures the similarity between the pixel s in the left image and the pixel s shifted by $u(s)$ in the right image, $\mathcal{P}$ is the set of all the pixels, Φ is a neighbor interaction function, and $\mathcal{N}$ corresponds to the set of all neighboring pairs $\{s, t\}$.

The disparity optimization step aims at finding an estimator for $\mathbf{u} = (u_1, \ldots, u_N)$. The most common estimators are maximum a posteriori and minimum mean squared error (MMSE). The first one finds the best estimate of $(u_1, \ldots, u_N)$ that maximizes (5.27). According to the notation defined in Section 5.4.2.1, by setting $\Psi(u(s)) = e^{-D_s(u(s))}$, $\Phi(u(s), u(t)) = e^{-V_{\{s,t\}}(u(s), u(t))}$, and taking the log of (5.26), it has been shown by Tappen and Freeman (2003) that maximizing the probability in (5.27) is equivalent to minimizing a global energy function similar to that defined in the context of graph-cut algorithms. However, it has been noticed that this estimation technique may yield to disparity maps with stair-stepping effects (i.e., effects of having large flat regions with sudden jumps). To effectively deal with this problem, the MMSE has been used. This estimate weights the discrete disparity levels according to their marginal probabilities, resulting in sub-pixel disparities being assigned and a smooth disparity map.

Note that many researchers have formulated the stereo matching problem as a Markov field and solved it using BP (Sun *et al.*, 2003). However, applying BP on large images leads to a heavy computational cost. To overcome this problem, other techniques have been proposed, such as hierarchical BP (Felzenszwalb and Huttenlocher, 2006) and block-based BP (Tseng *et al.*, 2007).

5.5   Conclusion

Before concluding this chapter, it is worth pointing out that an exhaustive quantitative comparison between many disparity estimation algorithms is available on the Middlebury website at http://vision.middlebury.edu/stereo/ (Scharstein and Szeliski, 2002). This website also includes a large database of stereo images with their associated ground truth disparities. The database has been more recently extended with additional datasets containing images with illumination variations, for testing disparity estimation algorithms in the presence of structured lights. These datasets are publicly available at http://vision.middlebury.edu/stereo/data/scenes2005 and http://vision.middlebury.edu/stereo/data/scenes2006.

The Middlebury website also contains benchmarks for other imaging techniques (http://vision.middlebury.edu/flow, http://vision.middlebury.edu/mview, http://vision.middlebury.edu/MRF, http://vision.middlebury.edu/color), including the optical flow estimation benchmark (/flow) that, owing to its closeness to the disparity estimation problem, provides further instruments and figures useful for the evaluation of disparity estimation techniques. On the original stereo matching benchmark, intensive experiments have shown that graph cuts and variational methods are among the most powerful techniques that produce consistent and smooth disparity maps while preserving the depth discontinuities.

In summary, we have reviewed the main issues on stereovision and disparity estimation algorithms. More precisely, we first studied the relationship between an object point in the 3D space and its 2D corresponding projections onto the image planes. After that, we introduced the disparity estimation problem and addressed the most important algorithms to solve it.

The common idea behind all the existing disparity estimation methods consists of first defining a matching cost function to measure the similarity between a given pixel (in the reference image) and each possible candidate matching sample (in the other image). Then, the disparity map is estimated by minimizing the retained cost function. Generally, methods that produce dense disparity maps are preferred in computer vision to those which yield sparse disparity results based on the matching of some features. Indeed, producing dense disparity maps has attracted a great deal of attention in many application fields, such as view-synthesis and 3D-reconstruction.

To this end, two main classes of methods have been developed: local and global. The first class performs well in highly textured regions, but it fails in occluded areas and often produces a noisy disparity map in textureless regions. To overcome these limitations, many research studies have been developed by resorting to global methods, involving regularization aspects. In this context, some approaches perform the optimization in a discrete manner, while others work in a continuous space, allowing the generation of a dense disparity field with ideally infinite precision.

References

Alkanhal, M., Turaga, D., and Chen, T. (1999) Correlation based search algorithms for motion estimation. Proceedings of Picture Coding Symposium, PCS'99, Portland, OR, pp. 99–102.

Alvarez, L., Deriche, R., Sanchez, J., and Weickert, J. (2002) Dense disparity map estimation respecting image discontinuities: a PDE and scale-space based approach. *J. Vis. Commun. Image R.*, **13** (1–2), 3–21.

Belli, T., Cord, M., and Philipp-Foliguet, S. (2000) Color contribution for stereo image matching. Proceedings of 2000 International Conference on Color in Graphics and Image Processing, Saint-Etienne, France, pp. 317–322.

Birchfield, S. and Tomasi, C. (1998) A pixel dissimilarity measure that is insensitive to image sampling. *IEEE Trans. Pattern Anal.*, **20** (4), 401–406.

Birchfield, S. and Tomasi, C. (1999) Depth discontinuities by pixel to pixel stereo. *Int. J. Comput. Vision*, **35** (3), 269–293.

Bleyer, M. and Gelautz, M. (2005) Graph-based surface reconstruction from stereo pairs using image segmentation, in *Videometrics VIII* (eds J.-A. Beraldin, S. F. El-Hakim, A. Gruen, and J. S. Walton), Proceedings of the SPIE, Vol. **5665**, SPIE, Bellingham, WA, pp. 288–299.

Boufama, B. and Jin, K. (2002) Towards a fast and reliable dense matching algorithm. Proceedings of 2002 International Conference on Vision Interface, Calgary, Canada, pp. 178–185.

Boykov, Y. Veksler, O. and Zabih, R. (1998) Markov random fields with efficient approximations, in *Proceedings. 1998 IEEE Computer Society Conference on Computer Vision and Pattern Recognition*, IEEE Computer Society, Washington, DC, pp. 648–655.

Boykov, Y., Veksler, O., and Zabih, R. (2001) Fast approximate energy minimization via graph cuts. *IEEE Trans. Pattern Anal.*, **23** (11), 1222–1239.

Brown, M. Z., Burschka, D., and Hager, G. D. (2003) Advances in computational stereo. *IEEE Trans. Pattern Anal.*, **25** (8), 993–1008.

Chaux, C., El-Gheche, M., Farah, J. *et al.* (2012) A parallel proximal splitting method for disparity estimation under illumination variation. *J. Math. Imaging Vision*, doi: 10.1007/s10851-012-0361-z.

Cochran, S. D. and Medioni, G. (1992) 3-D surface description from binocular stereo. *IEEE Trans. Pattern Anal.*, **14** (10), 981–994.

Dhond, V. R. and Aggarwal, J. K. (1989) Structure from stereo – a review. *IEEE Trans. Syst. Man Cyb.*, **19** (6), 1489–1510.

El Ansari, M., Masmoudi, L., and Bensrhair, A. (2007) A new regions matching for color stereo images. *Pattern Recogn. Lett.*, **28** (13), 1679–1687.

El Gheche, M., Pesquet, J.-C., Farah, J. *et al.* (2011) Proximal splitting methods for depth estimation. Proceedings of 2011 International Conference on Speech and Signal Processing, Prague, Czech Republic, pp. 853–856.

Faugeras, O. (1994) *Three-Dimensional Computer Vision: A Geometric Viewpoint*, MIT Press, Cambridge, MA.

Felzenszwalb, P. F. and Huttenlocher, D. P. (2006) Efficient belief propagation for early vision. *Int. J. Comput. Vision*, **70** (1), 41–54.

Fookes, C., Bennamoun, M. and LaManna, A. (2002) Improved stereo image matching using mutual information and hierarchical prior probabilities, in *Proceedings. 16th International Conference on Pattern Recognition, 2002*, Vol. **2**, IEEE, pp. 937–940.

Fua, P. (1993) A parallel stereo algorithm that produces dense depth maps and preserves image features. *Mach. Vision Appl.*, **6** (1), 35–49.

Fusiello, A., Roberto, V., and Trucco, E. (2000a) Symmetric stereo with multiple windowing. *Int. J. Pattern Recogn.*, **14** (8), 1053–1066.

Fusiello, A., Trucco, E., and Verri, A (2000b) A compact algorithm for rectification of stereo pairs. *Int. J. Mach. Vision Appl.*, **12** (1), 16–22.

Hartley, R. and Gupta, R. (1993) Computing matched-epipolar projections. Proceedings of 1993 International Conference on Computer Vision and Pattern Recognition, New York, United States, pp. 549–555.

Hong, L. and Chen, G. (2004) Segment-based stereo matching using graph cuts, in *CVPR2004. Proceedings of the 2004 IEEE Computer Society Conference on Computer Vision and Pattern Recognition, 2004*, Vol. **1**, IEEE Computer Society, Washington, DC, pp. I-74–I-81.

Horn, K. P. and Schunck, G. (1981) Determining optical flow. *Artif. Intell.*, **17** (1–3), 185–203.

Kanade, T. (1994) Development of a video-rate stereo machine. In Proceedings of Image Understanding Workshop, Monterey, CA, pp. 549–557.

Kanade, T. and Okutomi, M. (1994) A stereo matching algorithm with an adaptive window: theory and experiment. *IEEE Trans. Pattern Anal.*, **16** (9), 920–932.

Kim, C., Lee, K., Choi, B., and Lee, S. (2005) A dense stereo matching using two-pass dynamic programming with generalized ground control points, in *Proceedings of 2005 IEEE Computer Society Conference on Computer Vision and Pattern Recognition*, Vol. **2**, IEEE Computer Society, Washington, DC, pp. 1075–1082.

Kolmogorov, V. and Zabih, R. (2001) Computing visual correspondence with occlusions using graph cuts, in *Proceedings. Eighth IEEE International Conference on Computer Vision, 2001. ICCV 2001*, IEEE, pp. 508–515.

Konrad, J. and Lan, Z. D. (2000) Dense disparity estimation from feature correspondences, in *Stereoscopic Displays and Virtual Reality Systems VII* (eds J. O. Merritt, S. A. Benton, A. J. Woods, and M. T. Bolas), Proceedings of the SPIE, Vol. **3957**, SPIE, Bellingham, WA, pp. 90–101.

Koschan, A. (1993) Dense stereo correspondence using polychromatic block matching, in *Computer Analysis of Images and Patterns* (eds D. Chetverikov and W. G. Kropatsch), Lecture Notes in Computer Science, Vol. **719**, Springer, Berlin, pp. 538–542.

Kosov, S., Thormahlen, T., and Seidel, H.-P. (2009) Accurate real-time disparity estimation with variational methods, in *Advances in Visual Computing* (eds G. Bebis, R. Boyle, B. Parvin*et al.*), Lecture Notes in Computer Science, Vol. **5875**, Springer, Berlin, pp. 796–807.

Liu, K. H., Liu, T. J., and Liu, H. H. (2010) A SIFT descriptor based method for global disparity vector estimation in multiview video coding, *2010 IEEE International Conference on Multimedia and Expo (ICME)*, IEEE, pp. 1214–1218.

Matthies, L., Szeliski, R., and Kanade, T. (1989) Kalman filter-based algorithms for estimating depth from image sequences. *Int. J. Comput. Vision*, **3** (3), 209–236.

Medioni, G. and Nevatia, R. (1985) Segment-based stereo matching. *Comput. Vis. Graph. Image Process.*, **31** (1), 2–18.

Miled, W., Pesquet, J.-C., and Parent, M. (2006) Dense disparity estimation from stereo images. Proceedings of 2006 International Symposium on Image/Video Communications, Hammamet, Tunisia.

Miled, W., Pesquet-Popescu, B., and Pesquet, J.-C. (2008) A convex programming approach for color stereo matching, in *2008 IEEE 10th Workshop on Multimedia Signal Processing*, IEEE, pp. 326–331.

Miled, W., Pesquet, J.-C., and Parent, M. (2009) A convex optimization approach for depth estimation under illumination variation. *IEEE Trans. Image Process.*, **18** (4), 813–830.

Nagel, H. and Enkelmann, W. (1986) An investigation of smoothness constraints for the estimation of displacement vector fields from image sequences. *IEEE Trans. Pattern Anal.*, **8** (5), 565–593.

Otha, Y. and Kanade, T. (1985) Stereo by intra and inter-scanline search using dynamic programming. *IEEE Trans. Pattern Anal.*, **7** (2), 139–154.

Papadimitriou, D. V. and Dennis, T. J. (1996) Epipolar line estimation and rectification for stereo image pairs. *IEEE Trans. Image Process.*, **5** (4), 672–677.

Pedersini, F., Sarti, A., and Tubaro, S. (1999) Multicamera systems. *IEEE Signal Process. Mag.*, **16** (3), 55–65.

Pock, T., Cremers, D., Bischof, H., and Chambolle, A. (2010) Global solutions of variational models with convex regularization. *SIAM J. Imaging Sci.*, **3** (4), 1122–1145.

Roy, S. and Cox, I. (1998) A maximum-flow formulation of the N-camera stereo correspondence problem, in *6th International Conference on Computer Vision, 1998*, IEEE Computer Society, Washington, DC, pp. 492–499.

Rudin, L. I., Osher, S., and Fatemi, E. (1992) Nonlinear total variation based noise removal algorithms. *Phys. D*, **60** (1–4), 259–268.

Ryan, T. W., Gray, R. T., and Hunt, B. R. (1980) Prediction of correlation errors in stereo pair images. *Opt. Eng.*, **19** (3), 312–322.

Scharstein, D. (1994) Matching images by comparing their gradient fields, in *Proceedings of the 12th IAPR International Conference on Pattern Recognition, 1994*, Vol. **1**, IEEE, pp. 572–575.

Scharstein, D. and Szeliski, R. (1998) Stereo matching with nonlinear diffusion. *Int. J. Comput. Vision*, **28** (2), 155–174.

Scharstein, D. and Szeliski, R. (2002) A taxonomy and evaluation of dense two-frames stereo correspondence algorithms. *Int. J. Comput. Vision*, **47** (1), 7–42.

Scharstein, D. and Szeliski, R. (2003) High-accuracy stereo depth maps using structured light, in *Proceedings of 2003 IEEE Computer Society Conference on Computer Vision and Pattern Recognition*, Vol. **1**, IEEE Computer Society, Washington, DC, pp. 195–202.

Schmid, C. and Zisserman, A. (1998) The geometry and matching of curves in multiple views, in *ECCV '98 Proceedings of the 5th European Conference on Computer Vision*, Vol. **I**, Springer-Verlag, London, pp. 394–409.

Slesareva, N., Bruhn, A., and Weickert, J. (2005) Optic flow goes stereo: a variational method for estimating discontinuity-preserving dense disparity maps, in *PR'05 Proceedings of the 27th DAGM conference on Pattern Recognition*, Springer-Verlag, Berlin, pp. 33–40.

Sun, J., Zheng, N. N., and Shum, H. Y. (2003) Stereo matching using belief propagation. *IEEE Trans. Pattern Anal.*, **25** (7), 787–800.

Tappen, M. F. and Freeman, W. T. (2003) Comparison of graph cuts algorithms with belief propagation for stereo using identical MRF parameters, in *Proceedings of 2003 IEEE International Conference on Computer Vision*, Vol. **2**, IEEE, pp. 900–906.

Tikhonov, A. and Arsenin, A. (1977) *Solution of Ill-Posed Problems*, John Wiley & Sons, Inc., New York, NY.

Tseng, Y.-C., Chang, N., and Chang, T.-S. (2007) Low memory cost block-based belief propagation for stereo correspondence, in *2007 IEEE International Conference on Multimedia and Expo*, IEEE, pp. 1415–1418.

Veksler, O. (2005) Stereo correspondence by dynamic programming on a tree, in *Proceedings of 2005 IEEE Computer Science Conference on Computer Vision and Pattern Recognition*, Vol. **2**, IEEE, pp. 384–390.

Volkov, V. and Kazian, B. (2008) Fitting FFT onto the G80 architecture. *Methodology*, 1–6.

Woo, O. and Ortega, A. (2000) Overlapped block disparity compensation with adaptive windows for stereo image coding. *IEEE Trans. Circ. Syst. Vid.*, **10** (2), 194–200.

Yoon, K. and Kweon, I. (2006) Adaptive support-weight approach for correspondence search. *IEEE Trans. Pattern Anal.*, **28** (4), 650–656.

Young, D. (1971) *Iterative Solution of Large Linear Systems*, Academic Press, New York, NY.

Zabih, R. and Woodfill, J. (1994) Non-parametric local transforms for computing visual correspondence, in *ECCV'94 Proceedings of the Third European Conference on Computer Vision*, Vol. **II**, Springer-Verlag, New York, NY, pp. 151–158.

Zhang, Z. (1998) Determining the epipolar geometry and its uncertainty: a review. *Int. J. Comput. Vision*, **27** (2), 161–195.

Zhang, Z. and Xu, G. (1997) A general expression of the fundamental matrix for both perspective and affine cameras, in *IJCAI'97 Proceedings of the Fifteenth International Joint Conference on Artificial Intelligence*, Vol. **2**, Morgan Kaufmann, San Francisco, CA, pp. 1502–1507.

Zitova, B. and Flusser, J. (2003) Image registration methods: a survey. *Image Vis. Comput.*, **21** (11), 977–1000.

6

3D Video Representation and Formats

Marco Cagnazzo, Béatrice Pesquet-Popescu, and Frédéric Dufaux

Département Traitement du Signal et des Images, Télécom ParisTech, France

6.1 Introduction

Three-dimensional (3D) video is increasingly considered as the next major innovation in video technology, with the goal to provide a leap forward in quality of experience (QoE). As a consequence, research and development in 3D video is attracting sustained interest. In parallel, announcements and deployments of new 3D video products and services are gaining momentum.

It is obvious that efficient representation is needed for successful 3D video applications. This element is closely intertwined with the other components of a 3D video system: content acquisition, transmission, rendering, and display. It also has a significant impact on the overall performance of the system, including bandwidth requirement and end-user visual quality, as well as constraints such as backward compatibility with existing equipment and infrastructure.

Several attributes are highly desirable. First, it is important to decouple data representation from content acquisition and display. More specifically, 3D video formats should allow stereo devices to cope with varying display characteristics and viewing conditions by means of advanced stereoscopic display processing. For instance, controlling the baseline stereo distance to adjust depth perception is helpful to improve QoE and visual comfort. With the emergence of high-quality auto-stereoscopic multiview displays, 3D video formats should enable the synthesis of many high-quality views. Moreover, the required bit rate should remain decoupled from the number of views, calling for efficient compression. Free viewpoint video, which allows for interactively varying the viewpoint, also demands many views.

In the current eco-system, a variety of 3D video formats, coding schemes, and display technologies coexist. In this context, standardization is key to guarantee interoperability and support mass adoption.

Emerging Technologies for 3D Video: Creation, Coding, Transmission and Rendering, First Edition.
Frédéric Dufaux, Béatrice Pesquet-Popescu, and Marco Cagnazzo.
© 2013 John Wiley & Sons, Ltd. Published 2013 by John Wiley & Sons, Ltd.

Given the above considerations, several 3D video representation and formats have been proposed, as summarized in Vetro *et al.* (2008), Mueller *et al.* (2010), and Vetro *et al.* (2011a). Products and services based on standardized formats, frame-compatible stereoscopic formats (Vetro, 2010) and multiview video coding (MVC) (ITU-T and ISO/IEC JTC 1, 2010; Vetro *et al.*, 2011b), have already been brought to market.

We review a variety of 3D video representations in Section 6.2 and existing standardized formats in Section 6.3.

6.2 Three-Dimensional Video Representation

In this section, we describe 3D video representations, including stereoscopic 3D (S3D) video, multiview video (MVV), video-plus-depth, multiview video-plus-depth (MVD), and layered depth video (LDV). We also discuss and analyze their respective strengths and shortcomings.

6.2.1 Stereoscopic 3D (S3D) Video

S3D video is the oldest, simplest, and most widely used representation for 3D video. For instance, 3D films and 3D TV services to date rely on this representation.

S3D is based on the principle of stereopsis, in which two views with a disparity, referred to as left and right views, are respectively received by the left and right eyes of an observer. The resulting binocular disparity is then exploited by the brain to create a perception of depth. An example of a left and right view pair is illustrated in Figure 6.1 for the sequence Breakdancers (Zitnick *et al.*, 2004).

One of the advantages of S3D video representation is that the data acquisition is relatively easy. More precisely, two side-by-side cameras fixed on a rig can be used, as illustrated in Figure 6.2a, similar to the arrangement of human eyes. In such a setting, the distance between the two cameras, also referred to as the baseline, determines the disparity between the left and right views and accordingly the amount of perceived depth. An alternative setting, which allows for smaller baseline than the side-by-side setting, is to use a beam splitter and two perpendicular cameras as shown in Figure 6.2b. An extensive discussion on S3D video acquisition is presented in Chapter 4.

Another advantage of S3D video representation is that it is effective in terms of required bandwidth and storage. Straightforwardly, the raw data rate for an S3D video is doubled

Figure 6.1 Stereoscopic 3D video representation: example of left and right views

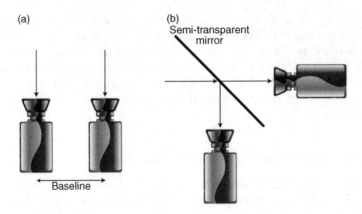

Figure 6.2   Stereoscopic 3D video acquisition

when compared with a regular two-dimensional (2D) video. When bandwidth is not critical, simulcast transmission of both views can be used. In this case, both views are independently encoded and the redundancies between the left and right views are not exploited. This solution is currently used in 3D Cinema, which is based on JPEG 2000 (Rabbani and Joshi, 2002). In 3D TV services, where backward compatibility is paramount, frame-compatible interleaving formats are preferred (Vetro, 2010). Finally, in order to improve compression, MVC (ITU-T and ISO/IEC JTC 2010; Vetro *et al.*, 2011b) takes advantage of the redundancies between the left and right views. These different formats will be further discussed in Section 6.3.

In addition, S3D video representation is well suited for current stereoscopic display technologies, including passive systems based on polarized glasses or active systems based on liquid-crystal shutter glasses.

However, the S3D representation also suffers from a number of severe drawbacks. First, it only adds one depth cue: binocular disparity. Therefore, it still falls short of providing a full 3D experience. Second, it is obviously desirable to decouple the data representation from the display characteristics. However, with S3D the video content is tuned for one specific display dimension and viewing condition, as determined by the fixed baseline. This lack of flexibility is an important shortcoming. In particular, it is desirable to be able to adjust the depth perception according to the screening environment and user preference.

6.2.2   Multiview Video (MVV)

As a straightforward extension of the S3D representation, several texture videos are acquired in a synchronous manner by a system of cameras in MVV representation. An illustration of the resulting representation is given in Figure 6.3.

Figure 6.3   Multiview video representation: example of N views

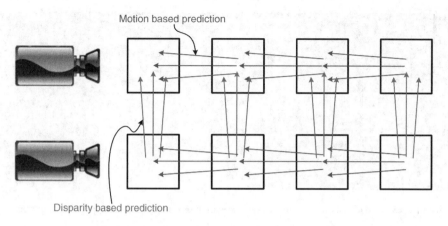

Figure 6.4 Disparity-based prediction in a stereo pair

Clearly, in MVV with N views, the raw data rate is multiplied by N when compared with a monoview 2D video. This typically generates a tremendous amount of data and, therefore, powerful compression is required. Fortunately, as all the cameras are capturing the same scene from varying viewpoints, MVV is characterized by significant inter-view statistical redundancies.

Similar to S3D video, views in MVV can be independently encoded (i.e., *simulcast*), or jointly encoded by taking advantage of the correlations existing amongst the views.

Typically, hybrid video coding schemes perform temporal motion-compensated prediction. Namely, already encoded frames are used in order to predict the current frame, only encoding the prediction residual. This mechanism can be straightforwardly extended to inter-view prediction from neighboring views (Merkle *et al.*, 2007). More specifically, as illustrated in Figure 6.4 for two views, temporal predictions inside each view (motion-compensated predictions) are combined with predictions between frames at the same temporal instant, but in different views (disparity-compensated predictions).

MVC (ITU-T and ISO/IEC JTC 1, 2010; Vetro *et al.*, 2011b), an extension of H.264/ advanced video coding (AVC), addresses the encoding of MVV content, combining conventional intra-view motion-based prediction and inter-view disparity-based prediction. It provides an efficient compression at the expense of high encoding complexity. It will be discussed further in Section 6.3.6.

MVV is especially suited for emerging auto-stereoscopic displays, which require a large number of views. Furthermore, it allows one to preserve the full resolution of the video sequence. In addition, difficulties associated with view synthesis can be avoided. Finally, the representation can easily be made backward compatible with legacy 2D displays by simply extracting one of the views.

The major shortcoming of the representation is that, even when using an efficient coding scheme such as MVC, the bit rate essentially grows linearly with the number of encoded views.

6.2.3 Video-Plus-Depth

Video-plus-depth is another useful representation. More specifically, a 2D video is augmented with an extra channel conveying depth information. Depth information results in a display-independent representation that enables synthesis of a number of views.

Figure 6.5   Video-plus-depth representation: example of video and corresponding depth map

The depth map is a grayscale image where each pixel is assigned a depth value representing the distance between the corresponding 3D point in the scene and the camera. The distance is restricted to a range $[Z_{near}, Z_{far}]$, corresponding to the two extreme depths which can be represented. Typically, this interval is quantized to 8 bits, and thus the closest point (at distance Z_{near}) is associated with the value 255 and the farthest point (at distance Z_{far}) with the value 0.

The video-plus-depth representation is illustrated in Figure 6.5, with an example of a 2D video frame and its associated depth map.

One of the difficulties of this representation is to obtain an accurate depth map. Depth data can be derived from stereo video content by means of disparity estimation (Scharstein and Szeliski, 2002; Kauff *et al.*, 2007; see also Chapter 5).

Another approach is to compute depth information from monoview video sequences by 2D-to-3D video conversion (Zhang *et al.*, 2011; see also Chapter 3). However, automatic 2D-to-3D conversion remains a challenging problem.

Finally, range cameras can be used to directly acquire depth information (see also Chapter 1). For instance, devices based on the time-of-flight principle can be used for this purpose (Foix *et al.*, 2011), with a light pulse and a dedicated sensor to capture a whole scene in three dimensions. While developed as a gesture-based user interface for gaming, the Kinect (Microsoft, 2010) features a depth sensor combining an infrared laser with a monochrome CMOS sensor. It operates within a range of a few meters and produces a 640×480 VGA resolution depth map with 11-bit depth.

Two straightforward advantages of the video-plus-depth representation are that the 2D video stream provides backward compatibility with legacy devices and it is independent of underlying coding formats for both the 2D video and depth streams.

Additionally, this representation supports advanced stereoscopic display processing technologies. More specifically, new views can be synthesized using depth-image-based-rendering (Zhang and Chen, 2004; Kauff *et al.*, 2007). In this way, depth perception can be adapted at the receiver end, depending on the display characteristics and viewing conditions. This is an important feature to improve QoE and avoid fatigue. However, the shortcoming of this approach is the quality of the synthesized views. In particular, dealing efficiently with occlusions and disocclusions remains a challenge.

Finally, the transmission of the extra depth map channel only incurs a limited bandwidth increase. Indeed, depth data have different characteristics when compared with natural video

Figure 6.6 Multiview video-plus-depth representation: example of N views and corresponding depth maps

data. It usually results in increased compression performance; nevertheless, conventional coding methods may not be optimal. MPEG-C Part 3, more formally referred to as ISO/IEC 23002-3 (ISO/IEC JTC 1, 2007), is a standardized format for auxiliary video data including depth map. It will be discussed in more detail in Section 6.3. Other more advanced coding techniques better suited for depth map compression have also been proposed (Maitre and Do, 2009; Merkle *et al.*, 2009; Oh *et al.*, 2009; see also Chapter 7). When compressing depth information, an important aspect to consider is that coding artifacts impact the quality of synthesized views. More specifically, coding distortions in the depth map lead to erroneous pixel shifts in synthesized views. In particular, it is critical to precisely represent edges in the depth map.

6.2.4 Multiview Video-Plus-Depth (MVD)

In MVD, each of the N views is acquired with its associated depth, as illustrated in Figure 6.6. This can be further exploited to reconstruct the 3D geometry of the scene with much better accuracy than the one obtained from MVV or video-plus-depth only.

The two sequences, video texture and depth, can then be encoded and transmitted independently. Alternatively, texture and depth can be jointly encoded, to exploit the redundancies between them, resulting in better coding performance. Given the great flexibility of this representation, a new, dedicated compression standard is under development. Chapter 8 covers this subject thoroughly.

6.2.5 Layered Depth Video (LDV)

A 3D scene can be described using the layered depth image (LDI), first introduced by Shade *et al.* (1998). An LDI is a representation of the 3D scene from a given viewpoint. However, for each pixel, besides the color and the depth that can be measured from the viewpoint, information about the occluded pixels is given in subsequent layers, allowing a fast image synthesis at viewpoints close to the starting one. LDV is straightforwardly defined as a sequence of LDIs.

More formally, an LDI is an $N \times M$ bi-dimensional array of so-called *layered depth pixels*, plus the camera parameters. In turn, each layered depth pixel is composed of a certain

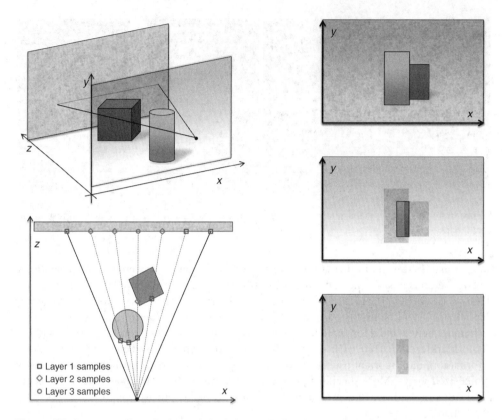

Figure 6.7 Representation of a layered depth image. Left column, top: 3D scene; bottom: scheme of lines of sight. Right column, top: front layer; middle: second layer; bottom: last layer. Depth layers are not shown

number of layers, and for each layer the color and the depth information are provided. As a consequence, the color and the depth data belonging to the first layer are identical to those of the image plus depth representation shown in Figure 6.5. In other words, the first layer (or front layer) is obtained by sampling the first surface seen along the line of sight. Subsequent layers (or back layers) are related to the next surfaces encountered on the same line. This is depicted in Figure 6.7.

An LDI thus contains information about occluded pixels in the current viewpoint. Therefore, it can be seen as an extension of the image-plus-depth representation (Section 6.2.3). It allows a more accurate synthesis of viewpoints near to the original one, since the previously occluded regions may now be rendered from layers other than the front one. Moreover, the LDI representation is usually more compact than MVD, since back layers are often quite sparse. On the other hand, LDIs usually have tighter limitations than MVD on the viewpoints that can be reliably reconstructed.

There are several ways to generate LDV. For synthetic scenes, where geometric information is perfectly known, the layers can easily be obtained using standard or ad-hoc modified ray-tracing algorithms. A similar approach is possible if MVD data are available. Finally, the case of MVV can be brought back to the previous one by computing depth maps by means of disparity estimation, as explained in Section 6.2.3 and in Chapter 5.

LDV can be compressed by adapting multiview video coding algorithms. In particular, the front layer is identical to a video-plus-depth sequence, while the compression of back layers requires some modifications to manage the holes. For example, Yoon and Ho (2007) proposed to fill the holes with the pixels from the front layer, or, as an alternative, to aggregate the pixel horizontally. In both cases, one obtains a regularly shaped image, which can be compressed by an ordinary video encoder such as H.264/AVC.

6.3 Three-Dimensional Video Formats

Given the variety of 3D video representations presented in the Section 6.2, the next step is to consider compression issues and associated formats. The three most widely used standardized formats to date are simulcast, frame-compatible stereo interleaving, and MVC, as illustrated in Figure 6.8a–c respectively. Hereafter, we also consider three other formats for the sake of completeness: MPEG-4 multiple auxiliary components (MAC), MPEG-C Part 3, and MPEG-2 multiview profile (MVP). All these formats are discussed in more detail in this section. We also provide an analysis of their strengths and weaknesses.

6.3.1 Simulcast

Even though the views of an MVV sequence are correlated, for the simplicity of coding and transmission they may be encoded independently of each other (Figure 6.8a and Figure 6.9).

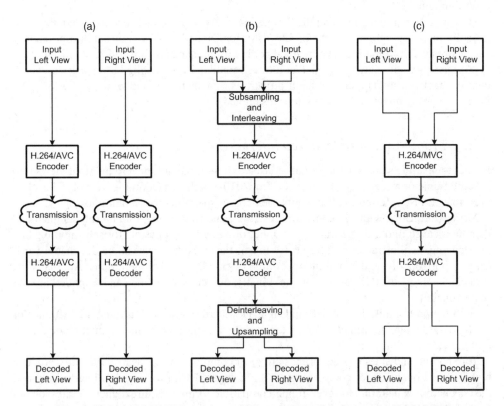

Figure 6.8 (a) H.264/AVC simulcast; (b) frame-compatible format; (c) H.264/MVC

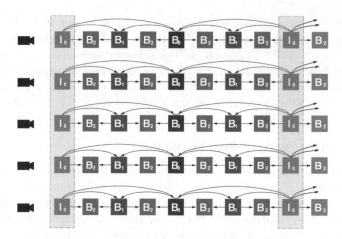

Figure 6.9 Simulcast encoding of a multiview video sequence

The independent transmission of views is called simulcast. This solution has the obvious advantage of simplicity. In addition, no synchronization between views is required.

Simulcast offers a number of advantages. First, state-of-the-art standard video coders can be used. Moreover, it is straightforward to decode one view for legacy 2D displays, hence guaranteeing backward compatibility. Finally, computational complexity and delay are kept to a minimum.

However, owing to the fact that the correlation between cameras is not exploited, this coding solution is not optimal in terms of rate distortion. Nonetheless, significant bit-rate savings can be obtained by applying asymmetrical coding (Stelmach and Tam, 1998; Stelmach *et al.*, 2000). Namely, one view is encoded at a lower quality, with a negligible impact on overall quality thanks to human stereo vision properties. For instance, one view can be low-pass filtered, coarsely quantized, or spatially subsampled.

6.3.2 Frame-Compatible Stereo Interleaving

Frame-compatible stereoscopic formats consist of a multiplex of the left and right views into a single frame or a sequence of frames (Vetro, 2010), as illustrated in Figure 6.8b. This class of formats is also referred to as stereo interleaving or spatial/temporal multiplexing.

More specifically, with spatial multiplexing, left and right views are first sub-sampled and then interleaved into a single frame. A number of patterns are possible for sub-sampling and interleaving, as illustrated in Figure 6.10. For instance, both views can be decimated vertically or horizontally, and stored in top–bottom or side-by-side configurations. Alternatively, data can be interleaved in row or column. Finally, quincunx sampling can be applied instead of decimation.

With temporal multiplexing, left and right views are interleaved as alternating frames or fields. By reducing the frame/field rate, the sample data rate can be made equal to the rate of a single view.

Top–bottom and side-by-side formats are most commonly used, as they typically produce higher visual quality after compression, when compared with row/column interleaving and the checkerboard pattern. However, frame-compatible formats require higher bit rates, owing to the spurious high frequencies introduced. Moreover, they may result in cross-talk and

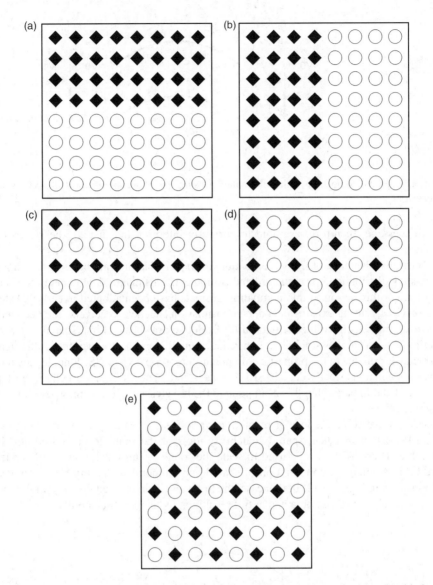

Figure 6.10 Frame-compatible formats with spatial multiplexing: (a) top–bottom, (b) side-by-side, (c) row interleaved, (d) column interleaved, (e) checkerboard (◆ denotes samples from the first view, ○ denotes samples from the second view)

color bleeding artifacts. In the case of interlaced content, the top–bottom format should be avoided as it further reduces the vertical resolution.

In order to correctly interpret frame-compatible formats and to de-interleave the data, auxiliary information is needed. For this purpose, supplementary enhancement information (SEI) has been standardized in the framework of H.264/MPEG-4 AVC (ITU-T and ISO/IEC JTC 1, 2010). Note, however, that SEI messages are not a normative part of the decoding process.

In particular, the frame packing arrangement SEI message is capable of signaling the various interleaving schemes described in Figure 6.10. Decoders understanding

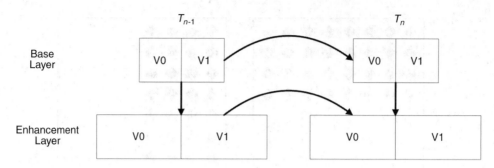

Figure 6.11 Frame-compatible with full-resolution using spatial scalability tools of SVC (side-by-side format)

this SEI message, therefore, are able to correctly interpret and display the stereo video content.

The obvious advantage of frame-compatible formats is that they are backward compatible with existing distribution infrastructure and equipment. This makes deployment much easier and faster. For this reason, frame-compatible formats have been adopted in early 3D stereoscopic services, reusing existing encoders, transmission channels, receivers, and decoders.

However, frame-compatible formats suffer from two drawbacks. First, spatial or temporal resolution is lost. This possibly results in reduced quality and user experience, although the impact may lessen due to the properties of human stereoscopic vision. Second, legacy receivers may not able to correctly decode SEI messages and, hence, fail to correctly interpret the interleaved data. In particular, this is an issue in the broadcast environment, where it is costly to upgrade devices.

In order to overcome the loss of resolution, various schemes have been proposed (Vetro *et al.*, 2011a). For instance, spatial scalability tools of the scalable video coding (SVC) extension of H.264/AVC can be used to scale the frame to full resolution, as illustrated in Figure 6.11. Similarly, a combination of spatial and temporal scalability tools can also be used, as shown in Figure 6.12. In this case, the first view is first scaled up by spatial scalability. The second view is then obtained at full resolution by temporal scalability.

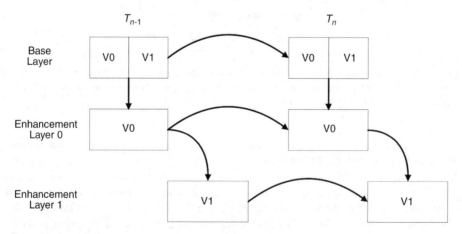

Figure 6.12 Frame-compatible with full-resolution using spatial and temporal scalability tools of SVC (side-by-side format)

6.3.3 MPEG-4 Multiple Auxiliary Components (MAC)

MPEG-4 Part 2 (ISO/IEC JTC 1, 2004) defines a specific tool which can be used for encoding the video-plus-depth data, namely MAC. Initially, MAC was designed to represent the grayscale shape describing the transparency of a video object, known as the *alpha channel*. However, it can also be defined in more general ways to describe shape, depth shape, or other secondary texture components. Therefore, the depth video can be encoded as one of the auxiliary components (Karim *et al.*, 2005). In MAC, the encoding of the auxiliary components follows a similar scheme as the one for the texture encoding, which employs motion compensation and block-based discrete cosine transform. It uses the same motion vectors as the 2D video for motion compensation of the auxiliary components.

Furthermore, this coding scheme can be also be used for stereo video coding as proposed in Cho *et al.* (2003). In this case, the disparity vector field, the luminance, and the chrominance data of the residual texture are assigned as three components of MAC.

6.3.4 MPEG-C Part 3

MPEG standardized another format, called MPEG-C Part 3 or ISO/IEC 23002-3 (ISO/IEC JTC 1, 2007), which specifies a video-plus-depth coding format. This standardized solution responds to the broadcast infrastructure needs. The specification is thus based on the encoding of a 3D content inside a conventional MPEG-2 transport stream, which includes the texture video, the depth video, and some auxiliary data. It provides interoperability of the content, display technology independence, capture technology independence, backward compatibility, compression efficiency, and the ability of the user to control the global depth range. MPEG-C Part 3 does not introduce any specific coding algorithms for depth or texture. It supports already existing coding formats, like MPEG-2 and H.264/AVC, and is used only to specify high-level syntax that allows a decoder to interpret two incoming video streams correctly as texture and depth data.

A key advantage of this solution is that it provides backward compatibility with legacy 2D devices and allows the evolution of 2D video bitstream normalization inside the high-level syntax.

Owing to the nature of the depth data, its smooth gray-level representation leads to a much higher compression efficiency than the texture video. Thus, only a small extra bandwidth is needed for transmitting the depth map. The total bandwidth for video-plus-depth data transmission is then reduced compared with the stereo video data. Some depth coding experiments (Fehn, 2003) have shown that the depth bitstream bit rate required in this setting for good depth representation is approximately 10–20% of the texture bitstream bit rate.

6.3.5 MPEG-2 Multiview Profile (MVP)

Already in the MPEG-2 standard (ITU-T and ISO/IEC JTC 1, 2000), the MVP was defined in order to transmit two video signals for stereoscopic TV applications. One of the main features of the MVP is the use of scalable coding tools to guarantee backward compatibility with the MPEG-2 main profile. The MVP relies on a multilayer representation such that one view is designed as the base layer and the other view is assigned as the enhancement layer. The base layer is encoded in conformance with the main profile, while the enhancement layer is encoded with the scalable coding tools. Also, the MVP conveys the camera parameters (i.e., geometry information, focal length, etc.) in the bitstream. An example

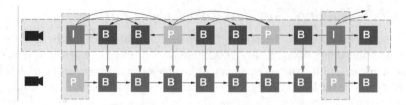

Figure 6.13   Prediction structure in MPEG-2 multiview profile

of the prediction structure for the MVP is shown in Figure 6.13, using a group of pictures structure IBBP.

As can observed, temporal prediction only is used in the base layer. As a consequence, backward compatibility with legacy 2D decoders is achieved, since the base layer represents a conventional 2D video sequence. In the enhancement layer, temporal prediction and interview prediction are simultaneously performed to achieve higher efficiency.

6.3.6   Multiview Video Coding (MVC)

The MVC extension (ITU-T and ISO/IEC JTC 1, 2010; Vetro *et al.*, 2011b) of the H.264/AVC standard was first introduced in 2008 to support a standard representation of S3D and MVV representations, described respectively in Sections 6.2.1 and 6.2.2. It is illustrated in Figure 6.8c.

In a nutshell, the MVC extension allows one to encode multiple view videos while keeping as much as possible the same structure of H.264/AVC. In particular, efficient compression is achieved by simply allowing pictures from different views to be used as prediction references for the current block of pixels. As a consequence, the bitstream of MVC is very similar to that of the original standard, and it benefits as well from the flexible prediction structure of H.264/AVC. Finally, the profiles and levels introduced for the extension are tightly related to those of H.264/AVC. All these issues are addressed in this section, which is concluded by a few experimental results about compression performance.

6.3.6.1   Requirements and Bitstream Structure

Just like other video standards, MVC was designed with the primary target of achieving a high *compression efficiency*, with low delay, and small consumption of memory and computational power. However, more specific requirements were considered for MVC. In addition to temporal random access, *view-switching random access* should be provided. Similarly, the concept of scalability should be extended in order to include *view scalability* – that is, the ability to decode only some of the encoded views without accessing the whole encoded bitstream. Another requirement was *backward compatibility*: it should be possible decode a single view (called *base view*) by just feeding an ordinary H.264/AVC decoder with a subset of the MVC bitstream.

Therefore, the part of the MVC bitstream carrying the base view is necessarily identical to an H.264/AVC stream. The other views' stream has a similar structure, but a flag (more precisely the *network abstraction layer (NAL) unit type* defined in the H.264/AVC syntax) signals that it is not a part of the base view. This allows a legacy decoder to properly ignore the corresponding data units, while an MVC decoder will recognize the NAL unit type and will

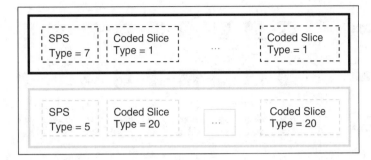

Figure 6.14   MVC bitstream structure, made up of the SPS and the encoded slices. First row: base view. Second row: non-base view. A legacy decoder will recognize and decode the NAL units of the base view thanks to the type, and discard those of the other views

be able to decode both types of encoded data. The resulting MVC bitstream structure is shown in Figure 6.14.

As far as the MVC stream syntax is concerned, it mainly consists of some extension of the H.264/AVC sequence parameter set (SPS). The extended SPS is signaled by a different NAL unit type (as shown in Figure 6.14) and conveys a view identification index, the information about the view dependency (see the *inter-view prediction* issue in Section 6.3.6.2), and some further information related to view scalability.

6.3.6.2   Inter-view Prediction

The main tool introduced in MVC in order to achieve efficient compression is the *inter-view prediction*. As shown in Sections 6.2.1 and 6.2.2, stereoscopic videos or, more generally, MVVs are captured by nearby cameras; therefore, a large amount of redundancy between views exists. Just like in MPEG-2 MVP, a picture can be predicted not only from pictures captured by the same camera, but also from pictures belonging to other views. For each block to be encoded, a different prediction can be performed; for example, a first block in a picture can be encoded by temporal prediction while a second one from the same picture can be encoded using inter-view prediction. The choice of the actual predictor is up to the encoder (it is not dictated by the norm); for example, it could be performed by minimizing a rate-distortion cost function.

In order to enable the inter-view prediction, the reference picture lists of H.264/AVC are used. For the base view, they only contain temporal references, since this view must be compatible with a single view decoder. For the other views, reference pictures from other views are added to the reference list, where they can be inserted at any position. The standard provides the tools needed to signal this information to the decoder. In MVC, it should be noted that only pictures corresponding to the same capture (or display) time can be added to the reference lists. Using the terminology of the standard, inter-view reference pictures must belong to the same *access unit* as the current picture (an access unit is the set of all the NAL units related to the same temporal instant).

Different prediction orders can be envisaged, depending on the content characteristics and the application at hand. For example, in Figure 6.15, a view-progressive encoding is presented, in which the view dependencies are exploited only for the first frame of each group of pictures. More specifically, the first frame of the central view (denoted by I_0 in the figure)

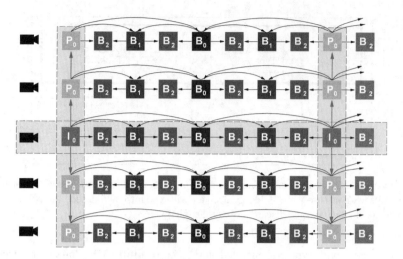

Figure 6.15 View-progressive encoding

is encoded as Intra. Then, the first frame of neighboring views is predicted from I_0, while for further views the first frame is predicted from the nearest encoded view, as shown in the figure. Finally, for other frames within each view, only temporal dependencies are exploited.

Another prediction configuration is shown in Figure 6.16, and called "fully hierarchical encoding," since bidirectional predictions are allowed along both the time and the view dimensions. This configuration aims at better exploiting all the existing dependencies in time and view. It also provides a scalable representation in time and in the number of cameras.

The inter-view prediction should not prevent random access, in time and view. This functionality is achieved in MVC by means of *instantaneous refresh decoding* (IDR) pictures. As in 2D standards, an IDR picture is encoded without temporal references. However, inter-view prediction (within the same access unit) can be used, thus reducing the coding rate without affecting the temporal random access functionality. The MVC standard also provides the *anchor picture* type, which is very similar to IDR pictures. The availability of IDR and anchor pictures allows an efficient random access in both time and view axes.

6.3.6.3 Profiles and Levels

The MVC extension defines two profiles, called *stereo high* and *multiview high*, and based on the *high* profile of H.264/AVC. The stereo high profile only supports two different views, while the multiview profile supports multiple views but does not support interlaced video. The MVC profiles and their relationship with H.264/AVC profiles are depicted in Figure 6.17.

For stereoscopic video, we remark that a bitstream can be compatible with both profiles. To this end, it suffices to avoid using the interlaced coding tools of the stereo high profile. Moreover, a flag can be set in the bitstream in order to signal this compatibility. The stereo high profile is currently used in stereoscopic Blu-ray 3D discs.

As far as *levels* are concerned, many of them impose just the same limits as the corresponding levels in the single-view H.264/AVC decoder (e.g., the overall bit rate). Other limits are increased with a fixed factor of two (e.g., the decoder buffer capacity and throughput), allowing then to decode a stereoscopic video having a given resolution with

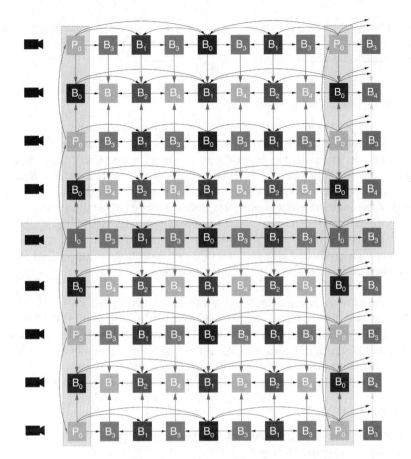

Figure 6.16   Fully hierarchical encoding

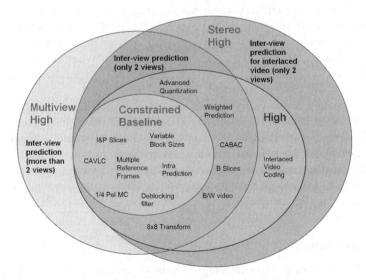

Figure 6.17   MVC profiles

the same level used for the corresponding single-view video. On the other hand, if more than two views are to be decoded, a higher level than the one that would be used for the single-view case is needed.

6.3.6.4   Compression Performance

Several studies have been carried out in order to compute the compression performance improvement of MVC with respect to independent coding of each view (simulcast).

A synthetic and increasingly common way to report compression performance is the Bjøntegaard metric (Bjøntegaard, 2001), which estimates the rate reduction of a given codec with respect to the reference one for the same quality over a given rate interval. This rate reduction is usually referred to as Bjontegaard delta rate (BDR). Likewise, it is possible to define a Bjontegaard delta peak signal-to-noise ratio (PSNR) as the average PSNR increase for a same rate in a given PSNR interval.

Using inter-view prediction allows an average rate reduction of 20% (BDR) for a multi-view video with eight views (Tian *et al.*, 2007). In some specific cases, gains as large as 50% (BDR) have been reported (Merkle *et al.*, 2007).

Subjective quality assessment results are presented in Vetro *et al.* (2011a) for HD stereo content. It is shown that, on the whole, the subjective quality obtained by MVC (stereo high profile) is equally good to that obtained using H.264/AVC simulcast, as long as the dependent-view bit rate is larger or equal to 25% of the base-view bit rate. In other words, MVC results in a 37.5% bit-rate savings compared with H.264/AVC simulcast.

6.4   Perspectives

In this chapter we have reviewed a number of representations for 3D video, as well as existing standardized compression formats.

With the goal to go beyond existing standards, MPEG has been undertaking a new phase of standardization for 3D video coding (3DVC). Two major objectives are targeted: to support advanced stereoscopic display processing and to improve support for high-quality auto-stereoscopic multiview displays. It is expected that 3DVC enhances the 3D rendering capabilities, when compared with the limited depth range supported with frame-compatible formats. Moreover, 3DVC does not trade off the resolution. When compared with MVC, 3DVC is expected to significantly reduce the bit rate needed to generate the views required for display. 3DVC will be more thoroughly discussed in Chapter 8.

In the development of 3D video formats, quality assessment is a significant challenge (Huynh-Thu *et al.*, 2010). While subjective evaluation of 2D video quality has reached some maturity, with several methodologies recommended by ITU, the subjective assessment of 3D video raises new issues. In particular, the viewing experience becomes multidimensional and involves not only visual quality, but also depth perception and viewing comfort. Finally, the specific 3D display technology also has a significant impact. 3D video quality assessment will be considered in Chapter 19.

Acknowledgments

We would like to acknowledge the Interactive Visual Media Group of Microsoft Research for providing the Breakdancers data set (Zitnick *et al.*, 2004).

References

Bjøntegaard, G. (2001) Calculation of average PSNR differences between RD curves, ITU-T SG16/Q.6, Doc. VCEG-M033, Austin, TX.

Cho, S., Yun, K., Bae, B., and Hahm, Y. (2003) Disparity compensated coding using MAC for stereoscopic video, in *2003 IEEE International Conference on Consumer Electronics (ICCE)*, IEEE, pp. 170–171.

Fehn, C. (2003) A 3D-TV approach using depth-image-based rendering (DIBR), in *Visualization, Imaging, and Image Processing (VIIP 2003)* (ed. M. H. Hamza), ACTA Press, paper 396-084.

Foix, S., Alenyà, G., and Torras, C. (2011) Lock-in time-of-flight (ToF) cameras: a survey. *IEEE Sensors J.*, **11** (9), 1917–1926.

Huynh-Thu, Q., Le Callet, P. and Barkowsky, M. (2010) Video quality assessment: from 2D to 3D – challenges and future trends, in *2010 17th IEEE International Conference in Image Processing (ICIP)*, IEEE, pp. 4025–4028.

ISO/IEC JTC 1 (2004) ISO/IEC 14496-2, *Information Technology – Coding of Audio-Visual Objects – Part 2: Visual*, ISO, Geneva.

ISO/IEC JTC 1 (2007) ISO/IEC 23002-3, *Information Technology – MPEG Video Technologies – Part 3: Representation of Auxiliary Video Streams and Supplemental Information*, ISO, Geneva.

ITU-T and ISO/IEC JTC 1 (2000) ITU-T Recommendation H.262 and ISO/IEC 13818-2 (MPEG-2 Video), Generic coding of moving pictures and associated audio information: video.

ITU-T and ISO/IEC JTC 1 (2010) ITU-T Recommendation H.264 and ISO/IEC 14496-10 (MPEG-4 AVC), Advanced video coding for generic audiovisual services.

Karim, H.A., Worrall, S., Sadka, A.H., and Kondoz, A.M. (2005) 3-D video compression using MPEG4-multiple auxiliary component (MPEG4-MAC). Proceedings of the Visual Information Engineering (VIE).

Kauff, P., Atzpadin, N., Fehn, C. *et al.* (2007) Depth map creation and image-based rendering for advanced 3DTV services providing interoperability and scalability. *Signal Process. Image Commun.*, **22** (2), 217–234.

Maitre, M. and Do, M. N. (2009) Shape-adaptive wavelet encoding of depth maps. Proceedings of Picture Coding Symposium (PCS'09), Chicago, USA.

Merkle, P., Smolic, A., Mueller, K., and Wiegand, T. (2007) Efficient prediction structures for multiview video coding. *IEEE Trans. Circ. Syst. Vid. Technol.*, **17** (11), 1461–1473.

Merkle, P., Morvan, Y., Smolic, A. *et al.* (2009) The effects of multiview depth video compression on multiview rendering. *Signal Process. Image Commun.*, **24** (1–2), 73–88.

Microsoft (2010) Kinect. http://www.xbox.com/kinect/ (accessed March 30, 2012).

Mueller, K., Merkle, P., Tech, G., and Wiegand, T. (2010) 3D video formats and coding methods, in *2010 17th IEEE International Conference on Image Processing (ICIP)*, IEEE, pp. 2389–2392.

Oh, K.-J., Yea, S., Vetro, A., and Ho, Y.-S. (2009) Depth reconstruction filter and down/up sampling for depth coding in 3-D video. *IEEE Signal Proc. Lett.*, **16** (9), 747–750.

Rabbani, M. and Joshi, R. (2002) An overview of the JPEG2000 still image compression standard. *Signal Process. Image Commun.*, **17** (1), 3–48.

Scharstein, D. and Szeliski, R. (2002) A taxonomy and evaluation of dense two-frame stereo correspondence algorithms. *Int. J. Comput. Vision*, **47** (1/2/3), 7–42.

Shade, J., Gortler, S., He, L., and Szeliski, R. (1998) Layered depth images, in *SIGGRAPH98, The 25th Annual Conference on Computer Graphics and Interactive Techniques*, ACM, New York, NY, pp. 231–242.

Stelmach, L. and Tam, W. J. (1998) Stereoscopic image coding: effect of disparate image-quality in left- and right-eye views. *Signal Process. Image Commun.*, **14**, 111–117.

Stelmach, L., Tam, W.J., Meegan, D., and Vincent, A. (2000) Stereo image quality: effects of mixed spatio-temporal resolution. *IEEE Trans. Circ. Syst. Vid. Technol.*, **10** (2), 188–193.

Tian, D., Pandit, P., Yin, P., and Gomila, C. (2007) Study of MVC coding tools, Joint Video Team (JVT) Doc, JVT-Y044, Shenzhen, China, October.

Vetro, A. (2010) Frame compatible formats for 3D video distribution, in *2010 17th IEEE International Conference on Image Processing (ICIP)*, IEEE, pp. 2405–2408.

Vetro, A., Yea, S., and Smolic, A. (2008) Towards a 3D video format for auto-stereoscopic displays, in *Applications of Digital Image Processing XXXI* (ed. A. G. Tescher), Proceedings of the SPIE, Vol. 7073, SPIE, Bellingham, WA, p. 70730F.

Vetro, A., Tourapis, A.M., Müller, K., and Chen, T., (2011a) 3D-TV content storage and transmission. *IEEE Trans. Broadcast.*, **57** (2), 384–394.

Vetro, A., Wiegand, T., and Sullivan, G. J. (2011b) Overview of the stereo and multiview video coding extensions of the H.264.MPEG-4 AVC standard. *Proc. IEEE*, **99** (4), 626–642.

Yoon, S. U. and Ho, Y. S. (2007) Multiple color and depth video coding using a hierarchical representation. *IEEE Trans. Circ. Syst. Vid. Technol.*, **17** (11), 1450–1460.

Zhang, C. and Chen, T. (2004) A survey on image-based rendering - representation, sampling and compression. *Signal Process. Image Commun.*, **19** (1), 1–28.

Zhang, L., Vazquez, C., and Knorr, S. (2011) 3D-TV content creation: automatic 2D-to-3D video conversion. *IEEE Trans. Broadcast.*, **57** (2), 372–383.

Zitnick, C.L., Kang, S.B., Uyttendaele, M. *et al.* (2004) High-quality video view interpolation using a layered representation. *ACM Trans. Graph. (TOG) – Proc. ACM SIGGRAPH 2004*, **23** (3), 600–608.

7

Depth Video Coding Technologies

Elie Gabriel Mora,[1,2] Giuseppe Valenzise,[2] Joël Jung,[1]
Béatrice Pesquet-Popescu,[2] Marco Cagnazzo,[2] and Frédéric Dufaux[2]

[1]*Orange Labs, France*
[2]*Département Traitement du Signal et des Images, Télécom ParisTech, France*

7.1 Introduction

With the recent breakthroughs in three-dimensional (3D) cinema and the emergence of new multimedia services such as 3D television or free viewpoint television (FTV), more cost-effective and lightweight 3D video formats are being explored, both in the industry and in academic research fields.

The stereoscopic 3D format was initially proposed, which consisted of two views of the same scene captured by two cameras separated by a specific baseline. It allowed stereoscopic viewing of the video by matching each view to each eye of the viewer. The mixed-resolution stereo format followed. It exploited the binocular suppression theory, which states that if the two views are of different quality, then the perceived binocular quality for a viewer would be closer to the higher quality (Brust *et al.*, 2009). This means that one of the two views can be coded in a lower resolution, hence reducing the total bit rate needed to transmit the stereo-scopic sequence.

Two views, however, are not enough to deliver a wide-ranged 3D content. Fluidity in navigation is also not achievable with two views, although it is a must in FTV applications for instance. Hence, the multiview video format was considered, consisting of N views captured by N cameras arranged and spaced in a specific way (see Chapter 6). However, this format is not cost efficient as the total bit rate needed to encode and transmit the N views is high.

Depth-based formats are the latest proposed 3D video formats, and they are currently being thoroughly explored both in standardization and in academic research fields. First we have the video+depth format, which consists of a color view and its associated depth. The color and depth videos are used to extrapolate another color view at a different position in space, through the depth image-based rendering (DIBR) technique, hence allowing stereo-scopic viewing. If multiple views are needed, then the multiview video+depth format can be

Emerging Technologies for 3D Video: Creation, Coding, Transmission and Rendering, First Edition.
Frédéric Dufaux, Béatrice Pesquet-Popescu, and Marco Cagnazzo.
© 2013 John Wiley & Sons, Ltd. Published 2013 by John Wiley & Sons, Ltd.

used, which consists of N color views with their associated depths. Multiple intermediate views can thus be synthesized using DIBR.

The advantage of depth-based formats is that they are cost effective, as the depth information requires less bit rate to encode than color data does. However, acquiring a depth map (a depth video frame) to begin with is a challenge, since current depth estimation techniques such as stereo matching algorithms are far from perfect, presenting many artifacts in the estimated depth map. Depth cameras (see Chapter 1) directly producing a depth video stream are an interesting alternative, but they are not mature at this stage. An expensive approach would be to acquire a raw depth signal and clean up each frame in a post-production process, but this is only an available option for resourceful cinema studios. Nevertheless, depth-based formats present great potential and they are currently being explored in standardization to progressively make their way in the industry.

As delivering 3D content is costly, high compression rates must be achieved in order not to exceed the bandwidth capacity of service providers. If depth-based formats are used, depth videos should be compressed as well. Depth videos have different characteristics than color videos; thus, specific coding tools are designed and used for depth contents.

The evaluation of a depth coding tool, however, should not be carried out on the depth video itself, since the depth video is not displayed on screen. Indeed, the depth data are just used to synthesize (existing or virtual) views. Hence, evaluation of depth coding tools is done on the synthesized views themselves. Subjective tests performed on synthesized views remain the most effective method to evaluate different depth coding technologies, but subjective viewing sessions are logistically difficult to organize. Objective results are faster and easier to get, although they are not always accurate and, hence, not always reliable. The best known objective metrics are the Bjontegaard delta rate (BD-Rate) and Bjontegaard delta peak signal-to-noise ratio (BD-PSNR) metrics (Bjontegaard, 2001). When reporting BD-Rate and BD-PSNR data it is common practice to use as rate the sum of bit rates of the color and depth videos used for synthesis, and as PSNR it is the PSNR of the synthesized views measured with respect to synthesized views with uncompressed original color and depth videos.

The rest of this chapter is organized as follows: Section 7.2 presents an analysis of depth maps and their special characteristics. It also presents the different redundancies that can be exploited to maximize the compression of the depth maps, hence defining three main categories of depth map coding tools. These tools will be detailed in Section 7.3. Section 7.4 gives a specific example of such a tool with its associated results. Section 7.5 concludes this chapter.

7.2   Depth Map Analysis and Characteristics

A depth map matches each pixel in a color video to its distance from the camera. It is a single component (grayscale) image, composed essentially of smooth regions separated by sharp borders, as shown in Figure 7.1. It is not affected by illumination, so it contains no texture or shadows.

Depth compression can be achieved in three ways. A first class of methods exploits the inherent characteristics of depth maps, either on a block-by-block basis or considering higher level data structures, such as a slice or a frame; the latter methods are referred to as "above block-level coding tools." Second, correlations with the color video can be exploited. Indeed, motion vector fields and intra modes between the color and depth videos are highly

Figure 7.1 A color frame and its associated depth map

correlated, especially around object borders. Special depth transforms or depth mode decisions can depend on color information as well. Finally, since depth maps are not displayed on the screen but, rather, are used to synthesize views, depth coding can be optimized for the quality of these synthesized views, meaning new distortion models only considering depth coding effects on rendered views can be designed.

Hence, we have three main categories of depth coding tools: tools that exploit the inherent characteristics of depth maps, tools that exploit the correlations with the associated color video, and tools that optimize depth map coding for the quality of the synthesis. The following section details each category by giving examples of such tools used in the literature or proposed as a response to MPEG's call for proposals (CfP) on 3D video (3DV), in November 2011 (MPEG, 2011).

7.3 Depth Map Coding Tools

7.3.1 Tools that Exploit the Inherent Characteristics of Depth Maps

Tools in this category can be further subdivided into two subcategories: above block-level coding tools and block-level coding tools. The first subcategory of tools deals with depth maps as a whole (without using blocks), while the second uses the typical block-based coding structure of video encoders and changes it slightly by introducing new coding modes more adapted to depth contents.

7.3.1.1 Above Block-Level Coding Tools

One example of a tool in this category is reduced resolution coding of depth maps, which is considered as a form of data compression (Rusanovsky and Hannuksela, 2011). Indeed, upsampling artifacts in depth maps are minimal. Hence, significant gains can be obtained if depth maps are coded in a reduced resolution. In the test model for Advanced Video Coding (AVC)-based 3DV solutions (Hannuksela, 2012) released in March 2012, depth maps are downsampled to quarter resolution (half resolution in each direction) before encoding.

Another example is reduced motion vector resolution. The eight-tap filters used for motion-compensated interpolation, as defined in the sixth working draft of high-efficiency video coding (HEVC; Bross *et al.*, 2012), produce ringing artifacts at sharp edges in the depth map. Hence, motion-compensated prediction and disparity-compensated prediction have been modified in depth map coding in such a way that no interpolation is used, meaning only full sample accuracy motion or disparity vectors are used. The removed unnecessary accuracy also reduces the bit rate needed to signal motion vector differences (Schwarz, *et al.*, 2011).

View Synthesis Prediction is another powerful above block-level coding tool (Rusanovsky and Hannuksela, 2011) used for both texture and depth. Using an already coded and reconstructed depth map $D0^*$ at view 0, a depth map $D1'$ can be synthesized at view 1, which is then used as predictor for coding the depth map D1 at the same view. Synthesis is just a matter of mapping the pixels from a source image $s(x, y)$ to a destination target image $t(x, y)$ at a desired position (view):

$$t(\lfloor x + D(x, y) \rfloor, y) = s(x, y)$$

$$D(x, y) = \frac{fB}{z(x, y)}$$

$$z(x, y) = \left[\frac{d(x, y)}{255} \left(\frac{1}{Z_{near}} - \frac{1}{Z_{far}} \right) + \frac{1}{Z_{far}} \right]^{-1}$$

(7.1)

where f is the focal length, B is the camera baseline, $D(x, y)$ is the disparity at position (x, y), and $d(x, y)$ is the depth value at position (x, y). Since we are synthesizing a depth map here, we have $s(x, y) = d(x, y)$.

Another tool also useful for prediction is the Z_{near} Z_{far} Compensation tool for weighted prediction (Domanski *et al.*, 2011). Different frames or different views in one depth sequence may have different extremal values for the depth (referred to as the Z_{near} and Z_{far} parameters). Since the depth map is usually rescaled in the $[0, 255]$ interval, different depth images may have different rescaling. In this case, using one depth map as reference for the other one will thus result in a poor prediction. Consequently, a coherent scaling should be applied among all depth maps. This can be easily achieved using the depth extremal values. For example, if a depth map with parameters Z_{near_s} and Z_{far_s} is stored in the reference picture list for another depth map with parameters Z_{near_t} and Z_{far_t}, the scaling for the current depth map is defined as follows:

$$L_T = L_S \frac{Z_{far\_s} - Z_{near\_s}}{Z_{far\_t} - Z_{near\_t}} + 255 \times \frac{Z_{near\_s} - Z_{near\_t}}{Z_{far\_t} - Z_{near\_t}}$$

(7.2)

where L_S is the original depth value and L_T is the rescaled one.

7.3.1.2 Block-Level Coding Tools

The first tool to be presented in this category is the approximation of a depth block using modelling functions. This was presented in Merkle *et al.* (2008). Basically, a depth map is composed of smooth regions separated by borders. Hence, a depth block can be approximated with either a constant function, a linear function, a piece-wise constant function (*wedgelet* function), or a piece-wise linear function (*platelet* function). However, if none of these functions provides a good approximation of the original signal, the block can be divided into four blocks in a quad-tree manner. The process is then repeated until an approximation function is found for each leaf of the quad-tree.

The same logic was refined and enhanced in a proposal to the MPEG CfP for 3DV (Schwarz *et al.*, 2011). Four new intra modes were presented in the proposal, two of which

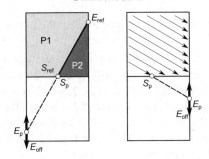

Figure 7.2 Depth modelling modes 1 and 2

(modes 3 and 4) use texture information and will be discussed in Section 7.3.2.2. In mode 1, the depth block is approximated into two constant regions, R_1 and R_2, separated by a straight line (wedgelet). For both regions, a parameter called region constant P_i is defined as the average of the values of the pixels covered by R_i and is sent in the bitstream to the decoder. Partition information is also sent, consisting in the start and end points of the straight line separating the two regions. In mode 2, P_1 and P_2 are sent but the partition information is not. Instead, it is inferred from neighboring blocks. However, the resulting partitioning may not be adequate for the block. Hence, an offset E_{off} is introduced that corrects the end point of the straight line, as shown in Figure 7.2. E_{off} is sent in the bitstream as well.

While the two previous tools involved intra predictions, the adaptive 2D block matching (2D-BM)/3D block matching (3D-BM) selection tool deals with inter predictions (Kamolrat *et al.*, 2010). Basically, 2D-BM is how a traditional encoder finds the best temporal match of a currently coded block in a temporal reference frame. The best match is the block in the reference frame that minimizes the sum of squared errors (SSE), defined as follows:

$$\text{SSE}(i,j) = \sum_{m=0}^{M-1} \sum_{n=0}^{N-1} (f(m,n) - g(m+i,n+j))^2 \tag{7.3}$$

where f is the original $M \times N$ block to be coded and g is the $M \times N$ reference block. Here, the search is done in two dimensions: horizontal and vertical. However, in depth, we may have movements in the depth direction as well, so we can increase the search accuracy if we extend the search to the third dimension, depth, yielding the 3D-BM. The best match is the one that would minimize this new formulation of the SSE:

$$\text{SSE}(i,j,k) = \sum_{m=0}^{M-1} \sum_{n=0}^{N-1} (f(m,n) - g(m+i,n+j) + k)^2 \tag{7.4}$$

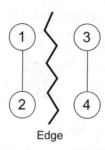

Figure 7.3 Constructing the EAT of 2 × 2 block

3D-BM is more effective than 2D-BM at high bit rates, when the savings obtained by minimizing distortion are higher than the cost of adding an additional component k to motion vectors. Inversely, 2D-BM is better than 3D-BM at low bit rates. The adaptive 2D-BM/3D-BM selection tool evaluates the rate-distortion (R-D) cost of coding each block with 2D-BM and with 3D-BM and chooses to code with the block-matching technique that yields the lower cost. The choice is signaled in the bitstream as well. This tool outperforms 2D-BM and 3D-BM at middle bit rates.

Transforms can also be adapted to depth contents. Shen *et al.* (2010) proposed edge-adaptive transforms (EATs) that take the edge structure of the depth block directly into account and, hence, produce many zero coefficients after filtering. An edge detection stage is first applied to the residual and a binary edge map is established. From the binary edge map, an adjacency matrix $A(i,j)$ is computed. $A(i,j) = A(j,i) = 1$ if pixels i and j are immediate neighbors (a pixel has four immediate neighbors: the left, right, above, and below pixels) and if they are not separated by an edge. Otherwise $A(i,j) = A(j,i) = 0$.

From the adjacency matrix, a degree matrix D can be computed, where $D(i,j) = 0$ if $i \neq j$ and $D(i,i)$ is the number of nonzero elements in the ith row of A. The Laplacian matrix L can then be computed as $L = D - A$. Then from L, the EAT matrix E^t can be computed using the cyclic Jacobi method. An example is given in Figure 7.3.

In this example, the A, D, L, and E^t matrices would equal the following:

$$A = \begin{bmatrix} 0 & 1 & 0 & 0 \\ 1 & 0 & 0 & 0 \\ 0 & 0 & 0 & 1 \\ 0 & 0 & 1 & 0 \end{bmatrix} \qquad D = \begin{bmatrix} 1 & 0 & 0 & 0 \\ 0 & 1 & 0 & 0 \\ 0 & 0 & 1 & 0 \\ 0 & 0 & 0 & 1 \end{bmatrix}$$

$$L = \begin{bmatrix} 1 & -1 & 0 & 0 \\ -1 & 1 & 0 & 0 \\ 0 & 0 & 1 & -1 \\ 0 & 0 & -1 & 1 \end{bmatrix} \qquad E^t = \begin{bmatrix} \frac{1}{\sqrt{2}} & \frac{1}{\sqrt{2}} & 0 & 0 \\ 0 & 0 & \frac{1}{\sqrt{2}} & \frac{1}{\sqrt{2}} \\ \frac{-1}{\sqrt{2}} & \frac{1}{\sqrt{2}} & 0 & 0 \\ 0 & 0 & \frac{-1}{\sqrt{2}} & \frac{1}{\sqrt{2}} \end{bmatrix}$$

It can be proven that for an $N \times N$ piece-wise constant-depth block with M constant regions, the EAT coefficients will be composed of at most M nonzero elements and $N^2 - M$

zero elements. Hence, the coding efficiency is improved compared with the discrete cosine transform (DCT). However, since the edge information should be transmitted, the use of EATs is not always the best choice. Therefore, an R-D criterion is set to choose the best transform between DCT and EAT for each block, meaning the choice for each block must be signaled as well in the bitstream.

7.3.2   Tools that Exploit the Correlations with the Associated Texture

The texture and the depth component are highly correlated, and this correlation can be exploited to achieve high depth compression gains. First, prediction modes can be selected for depth according to the information available in the texture. Moreover, the prediction information itself, such as motion vectors, reference picture indexes, or even intra modes, can be directly inherited from texture blocks. Finally, the texture information can be used to design special transforms more adapted to depth in order to increase coding efficiency.

7.3.2.1   Prediction Mode Inheritance/Selection

Depth Block Skip is a tool that was presented as part of a proposal (Lee J. *et al.*, 2011) to the MPEG CfP for 3DV. Basically, it forces to code a depth block in SKIP mode if the temporal correlation in texture is found to be high enough. Indeed, if a high temporal correlation exists in texture, it is likely that the temporal correlation will be high in depth as well. Thus, for a currently coded depth block X in frame D_i, the collocated texture block A in frame T_i, and the collocated texture block B in frame T_{i-1} are found and the sum of squared differences (SSD) between A and B is computed. If the SSD is lower than a certain threshold, it is assumed that the temporal correlation in texture – and consequently in depth – is sufficient and, in this case, the current depth block X is forced to be coded in SKIP mode, which means the collocated block C in D_{i-1} will be copied into X, as shown in Figure 7.4. Since the process can be repeated at the decoder, the decision to code the block in SKIP is not signaled. There are, however, some drawbacks to this approach. First, two texture frames are used to code a depth frame. This is a fairly large amount of side information to use, which is justified only if this method achieves significant coding gains, which is not the case. Second, the tool makes

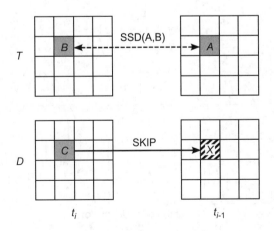

Figure 7.4   Depth block skip

depth coding prone to texture errors that occur in other access units, hence reducing coding robustness. And third, the tool violates a general design principle for the standardized bit-streams, which states that the parsing of the bitstream at the decoder should not be dependent on signal processing applied on decoded pictures.

The SKIP mode for depth can also be forced whenever the collocated texture block has also been coded in SKIP mode (Kim *et al.*, 2009). The logic behind this is that the uniformity of motion in texture is likely to also hold in depth. SKIP mode may not normally be chosen for depth because of the flickering artifacts that add "fake" motion due to bad depth map estimation. Forcing the SKIP mode allows eliminating those artifacts, hence increasing the quality of the synthesis, and achieving high compression gains as no SKIP flags are signaled. It also reduces encoder complexity as it reduces complex motion estimation/compensation processes.

7.3.2.2 Prediction Information Inheritance

Motion information (motion vectors plus reference picture indexes) is found to be highly cor-related between texture and depth, especially around borders. Seo *et al.* (2010) added a new mode, called "motion sharing," to the list of already existing modes, such as intra and inter prediction modes. In "motion sharing" mode, the motion information of the texture block corresponding to the currently coded depth block is shared with the depth block. Thus, no motion estimation is done at this stage for the depth block. The choice of "motion sharing" mode is not systematically forced as it undergoes an R-D check, just like intra or inter modes.

The Motion Parameter Inheritance tool, introduced in Schwarz *et al.* (2011), also allows sharing or inheriting the motion information of the associated texture block for the currently coded depth block, but the technique is a bit different. Here, no new modes are created. The motion parameters, as well as the entire partitioning structure of the collocated texture block, are considered for the currently coded depth block as an additional MERGE candidate in the latest HEVC standard.

The texture intra mode can also be inherited, as proposed by Bang *et al.* (2011). The collocated texture intra mode is inherited at the currently coded block in depth. As shown in Figure 7.5, if the texture block is partitioned smaller than the depth block, the intra mode of the top-left block is inherited. If the texture block is partitioned bigger than the depth block,

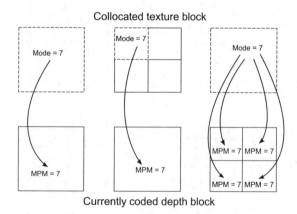

Figure 7.5 Texture intra-mode inheritance

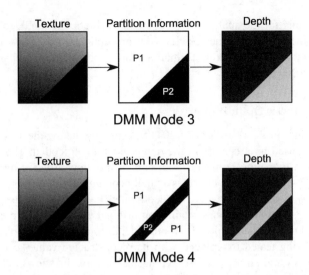

Figure 7.6 Depth modelling modes 3 and 4

the same texture intra mode is inherited for all the smaller partitions in depth. Inheriting the texture intra mode means adding it as a candidate to the most probable mode list for the currently coded depth block, where it will serve as a predictor for the depth intra mode.

Texture information can also be used to define new depth-modelling intra modes. We have already discussed in Section 7.3.1.2 of the first two depth-modelling modes proposed in Schwarz *et al.* (2011). Mode 3 still seeks to approximate the input block with two constant regions separated by a line, although here the partition information is inherited from the texture itself. Indeed, a simple thresholding of the texture Luma samples allows one to divide the texture block into two separate regions. The resulting partitioning is used as partition information for depth. No start and end points are thus computed in this mode. Since we are dealing with reconstructed and not original textures, the process is decodable and the partition information does not need to be signaled. The constants of the two regions are still sent, however. Mode 4 is exactly like mode 3 except that here the block is divided into three constant regions, as shown in Figure 7.6 (this is called contour partitioning, as opposed to wedgelet partitioning where the block is divided into two constant regions as in modes 1, 2, and 3).

7.3.2.3 Spatial Transforms

Texture information can also be used to design spatial transforms for depth. Daribo *et al.* (2008) proposed an adaptive wavelet lifting scheme wherein short filters are applied to areas in depth containing edges, and long filters applied in homogeneous depth areas. Edge detection is performed, however, on texture for decodability, and based on the observed correlation between the texture and depth edges.

7.3.3 Tools that Optimize Depth Map Coding for the Quality of the Synthesis

Since the depth map is not displayed on screen but rather used to synthesize views, the R-D model should consider distortion directly on the synthesized view in order to optimize the depth map coding for the quality of what actually matters: the rendered views. Different methods exist: some actually synthesize views during depth map coding, whereas others only estimate the distortion using texture information, and so on. We will compare these different techniques below.

7.3.3.1   View Synthesis Optimization

This method, defined in Schwarz *et al.* (2011), computes the synthesized view distortion change (SVDC) caused by a change in the depth block being coded. The SVDC metric is computed by actually synthesizing, in-loop, part of the view affected by the depth block change. Figure 7.7 shows how the metric is actually computed.

A depth frame s_D is considered where the depth block being tested has been replaced by its original values, all subsequent blocks retain their original values, and all previous blocks are already coded and reconstructed blocks. Another depth frame $\tilde{s}_D$ is formed just like s_D, only the depth block being tested is replaced with its reconstructed value according to the test being performed. s_D and $\tilde{s}_D$ are used to synthesize two texture views s'_T and $\tilde{s}'_T$. Another view, $s'_{T,Ref}$, can be synthesized using the original depth and color frames. Computing the SSE between s'_T and $s'_{T,Ref}$ gives E, and between $\tilde{s}'_T$ and $s'_{T,Ref}$ gives $\tilde{E}$. The SVDC metric is simply the difference $\tilde{E} - E$.

This method is motivated by three reasons. First, an exact distortion metric that considers occlusions and disocclusions in the synthesized views should be provided. Second, the method should be relative to a block. Third, partial distortions should be additive. This means that if the block is divided into four sub-blocks, which is a common operation in quad-tree encoders such as HEVC, the sum of the distortions caused by a change in each sub-block should equal the distortion caused by the corresponding change in the whole block. This assumption holds for SVDC. Only the change in the synthesized view distortion caused by a change in the depth block is considered, not the total synthesized view distortion itself.

7.3.3.2   Distortion Models

Other distortion models do not rely on an actual synthesis of the intermediate view to evaluate distortion. Instead, distortion on the synthesized view is estimated. Lee J.Y. *et al.* (2011)

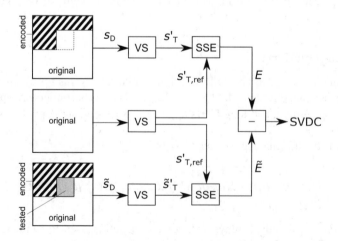

Figure 7.7   Computing the SVDC

computed the view synthesis distortion (VSD) as a depth block distortion, weighted by one-pixel texture translation differences, as defined by

$$\text{VSD} = \sum_{(x,y)} \left[\frac{\alpha}{2} |D_{x,y} - \tilde{D}_{x,y}| (|\tilde{C}_{x,y} - \tilde{C}_{x-1,y}| + |\tilde{C}_{x,y} - \tilde{C}_{x+1,y}|) \right]^2 \tag{7.5}$$

where $D_{x,y}$, $\tilde{D}_{x,y}$, and $\tilde{C}_{x,y}$ are respectively the original depth map value, the reconstructed depth map value, and the reconstructed texture value at position (x, y). The value of α is defined as

$$\alpha = \frac{fB}{255} \left(\frac{1}{Z_{\text{near}}} - \frac{1}{Z_{\text{far}}} \right) \tag{7.6}$$

where f is the focal length, B the baseline between the current view and the rendered view, and Z_{near} and Z_{far} are the values of the nearest and farthest depth values.

In Equation 7.5, it is assumed that two adjacent pixels will remain adjacent after warping, which is not always true since we may have occlusions or holes. To rectify this, $\tilde{C}_{x-1,y}$ and $\tilde{C}_{x+1,y}$ in Equation 7.5 can be replaced with $\tilde{C}_{x_L,y}$ and $\tilde{C}_{x_R,y}$, where x_L and x_R are defined as

$$\begin{aligned} x_L &= x + \arg \max_{l \geq 1} [\alpha(D_{x-l,y} - D_{x,y}) - l] \\ x_R &= x + \arg \min_{r \geq 1} [\max(\alpha(D_{x+r,y} - D_{x,y}) + r, 0)] \end{aligned} \tag{7.7}$$

Then, Equation 7.5 is modified, by considering the occlusion regions, as

$$\text{VSD} = \begin{cases} \sum_{(x,y)} \left[\frac{1}{2}\alpha |D_{x,y} - \tilde{D}_{x,y}| (|\tilde{C}_{x,y} - \tilde{C}_{x_L,y}| + |\tilde{C}_{x,y} - \tilde{C}_{x_R,y}|) \right]^2 & \text{if } x_L < x \\ 0 & \text{if } x_L \geq x \end{cases} \tag{7.8}$$

An even simpler distortion model that estimates the distortion on the synthesized views is proposed by Kim *et al.* (2009). Typically, a depth map distortion ΔD_{Depth} at position (x, y) causes a translation error ΔP in the synthesized view. There is a linear relationship between the two quantities, defined as

$$\Delta P = \alpha \Delta D_{\text{Depth}} \tag{7.9}$$

where α is defined in Equation 7.6. Moreover, it can be proven that there is a linear relationship between a global pixel displacement t_x and the distortion measured in the SSD of the original video signal translated by t_x, measured as

$$d_{\text{SSD}}(t_x) = \sum_x \sum_y \left(V_{(x,y)} - V_{(x-t_x,y)} \right)^2 \tag{7.10}$$

where $V(x, y)$ is the original video signal value at position (x, y). If multiple values of t_x are used and their corresponding distortion values $d_{\text{SSD}}(t_x)$ computed, we can derive a scale factor s as follows:

$$s = \frac{\mathbf{d}_{\text{SSD}}^{\mathbf{T}} \cdot \mathbf{t_x}}{\mathbf{t_x^T} \cdot \mathbf{t_x}} \tag{7.11}$$

where $\mathbf{d}_{\mathrm{SSD}}$ and $\mathbf{t_x}$ are the vectors formed by aggregating multiple values of $d_{\mathrm{SSD}}(t_x)$ and t_x, respectively, and T denotes vector transpose operator. For a given position error ΔP, this parameter s provides an estimation of the resulting distortion in the interpolated view.

If views are synthesized using a left and a right view, there is usually a weight for each contribution, p for the left and $1 - p$ for the right:

$$V_{\mathrm{synth}} = pV_{\mathrm{left}} + (1 - p)V_{\mathrm{right}} \tag{7.12}$$

A scale factor representing the global characteristics of V_{left} can be computed as

$$k = ps \tag{7.13}$$

Using the two parameters found above, the new distortion metric can be derived as

$$\Delta D^2_{\mathrm{synth}} = k\Delta P = k\alpha \left| \Delta D_{\mathrm{depth}} \right| \tag{7.14}$$

Finally, the Lagrangian cost used in rate distortion optimization processes can thus be written as:

$$J = \sum_x \sum_y \Delta D^2_{\mathrm{synth}}(x, y) + \lambda R_{\mathrm{depth}}$$
$$J = k\alpha \sum_x \sum_y \left| \Delta D_{\mathrm{depth}} \right| + \lambda R_{\mathrm{depth}} \tag{7.15}$$

Other depth map coding tools aim at increasing the depth map sparsity in the transform domain for increased coding efficiency. This is done by modifying the initial depth values before transform in such a way that the introduced change does not affect the synthesized views over a certain threshold. An example of such tools is given in Section 7.4.

7.4 Application Example: Depth Map Coding Using "Don't Care" Regions

As discussed previously, a key observation in coding of depth maps is that depth maps are not themselves directly viewed, but are only used to provide geometric information of the captured scene for view synthesis at the decoder. Thus, as long as the resulting geometric error does not lead to unacceptable synthesized view quality, each depth pixel only needs to be reconstructed coarsely at the decoder; for example, within a defined tolerable range. This notion of tolerable range per depth pixel can be formalized as a *"don't care region"* (DCR) using a threshold τ, by studying the synthesized view distortion sensitivity to the pixel value. Specifically, if a depth pixel's reconstructed value is inside its defined DCR, then the resulting geometric error will lead to distortion in a targeted synthesized view by no more than τ. Clearly, a sensitive depth pixel (e.g., an object boundary pixel whose geometric error will lead to confusion between background and foreground) will have a narrow DCR, and vice versa. In the rest of this section, we first provide a formal definition of DCR for disparity maps; then we discuss how the larger degree of freedom provided by DCR enables one to increase coding efficiency in transform coding of still images; finally, we describe how DCR can be embedded into motion prediction of a conventional hybrid video coder in order to increase coding efficiency.

7.4.1 Derivation of "Don't Care" Regions

"Don't care" regions were originally defined for finding the sparsest representation of transform coefficients in the spatial dimension by Cheung *et al.* (2010). An equivalent definition can be found in Valenzise *et al.* (2012). A pixel $C_n(x, y)$ in texture map n, with associated disparity value $D_n(x, y)$, can be mapped to a corresponding pixel in view $n + 1$ through a view synthesis function $s(x, y; D_n(x, y))$. In the simplest case where the views are captured by purely horizontally shifted cameras, $s(x, y; D_n(x, y))$ corresponds to a pixel in texture map C_{n+1} of view $n + 1$ displaced in the x-direction by an amount proportional to $D_n(x, y)$; that is:

$$s(x, y; D_n(x, y)) = C_{n+1}(x, y - \gamma \cdot D_n(x, y)) \tag{7.16}$$

where γ is a scaling factor depending on the camera spacing.

The view synthesis error $\varepsilon(x, y; d)$ is the absolute error between the mapped-to pixel $s(x, y; d)$ in the synthesized view $n + 1$ and the mapped-from pixel $C_n(x, y)$ in the texture C_n, given disparity value d for pixel (x, y) in C_n; that is:

$$\varepsilon(x, y; d) = |s(x, y; d) - C_n(x, y)| \tag{7.17}$$

The *don't care region* $\text{DCR}(x, y) = [\text{DCR}_{\text{low}}(x, y), \text{DCR}_{\text{up}}(x, y)]$ is defined as the *largest* contiguous interval of disparity values containing the ground-truth disparity $D_n(x, y)$ such that the view synthesis error for any point of the interval is smaller than $\varepsilon(x, y; D_n(x, y)) + \tau$, for a given threshold $\tau > 0$. The definition of DCR is illustrated in Figure 7.8. Note that, by construction, DCRs can be computed at the encoder side since both the views and the associated disparities are available. Moreover, DCR intervals are defined per pixel, thus giving precise information about how much error can be tolerated in the disparity maps. Figure 7.9 shows the width of per pixel DCR intervals for a depth frame of the *Kendo* sequence. Notice that many pixels have a very wide DCR; that is, they can be reconstructed within a broad range of depth values. This fact paves the way for the sparsification technique described next.

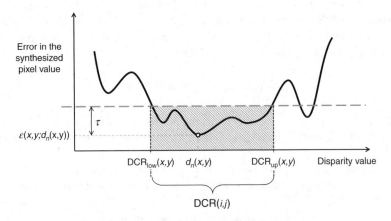

Figure 7.8 Definition of DCR for a given threshold

Figure 7.9   Width of DCR interval for frame 10 of the *Kendo* sequence ($\tau = 5$)

7.4.2   Transform Domain Sparsification Using "Don't Care" Regions

The DCR concept implies that it is not necessary to code the original depth signal with the highest possible fidelity. Instead, based on the observation that sparsity of transform coefficients is a good proxy for the bit rate, one can code the sparsest depth signal in the space defined by DCRs. By definition, the resulting synthesized view distortion will be bounded by τ (up to quantization errors).

Cheung *et al.* (2010) proposed to find the sparsest representation of a depth signal inside DCR by solving a weighted ℓ_1 minimization problem, for the case of block-based transform coding (such as in JPEG compression). Let **d** denote the pixels of a depth block and let Φ be a given transform matrix, which is used to compute transform coefficients $\mathbf{a} = \Phi^\mathrm{T}\mathbf{d}$. The sparsity optimization problem can be written as

$$\min_{\mathbf{d} \in S} \|\mathbf{a}\|_0 \quad \text{s.t.} \quad \mathbf{a} = \Phi^\mathrm{T}\mathbf{d} \tag{7.18}$$

where $\|\mathbf{a}\|_0$ is the ℓ_0-norm of **a**, which counts the nonzero elements of **a**, and S is a per block DCR; that is, a block signal **d** belongs to S if all its pixel components fall inside the per pixel DCRs of that block.

Minimizing the ℓ_0-norm in (7.18) is a hard problem. Cheung *et al.* (2010) solved the problem in (7.18) through a series of weighted ℓ_1 minimization problems such as

$$\min_{\mathbf{d} \in S} \|\mathbf{w}^\mathrm{T}\mathbf{a}\|_1 \quad \text{s.t.} \quad \mathbf{a} = \Phi^\mathrm{T}\mathbf{d} \tag{7.19}$$

where $\|\mathbf{a}\|_1 = \sum_i |a_i|$ is the ℓ_1-norm of $\mathbf{a}$ and $\mathbf{w}$ is a vector of weights which are updated at each iteration as $\mathbf{w}_i = 1/(|a_i| + \mu)$. The small constant μ is added to avoid possible division by zero. Notice that solving (7.19) corresponds to solving a linear program (LP), which can be solved in polynomial time. Since the transform coefficients have to be quantized, it is not guaranteed in general that the sparsest solution to (7.19) will be aligned to the quantization grid. This implies that, after quantization, the transform coefficient obtained may fall outside DCR. Cheung *et al.* (2010) suggested a heuristic technique to push the solution as much as possible inside the DCR before quantization, in such a way as to increase the chances that the DCR constraint is satisfied after quantizing.

Iteratively solving (7.19) can still be computationally too expensive. Therefore, Cheung *et al.* (2011) reformulated the problem by using a quadratic approximation of the view synthesis error $\varepsilon(x, y; d)$ around the ground-truth disparity value. Although the approximation of $\varepsilon(x, y; d)$ can be very coarse (see Figure 7.8), this quadratic fitting turns out to be sufficiently accurate in practice, and enables one to turn problem (7.18) into a series of iterative unconstrained quadratic programs. Finding the minimum of each of these quadratic problems corresponds to finding the solution to a set of linear equations, which is computationally simpler than solving an LP.

7.4.3 Using "Don't Care" Regions in a Hybrid Video Codec

It is possible to exploit the DCR defined in Section 7.4.1 not only in the spatial dimension, but also in the temporal dimension; that is, in the case of coding of depth video. Valenzise *et al.* (2012) have shown how the degree of freedom offered by DCR can be embedded into a state-of-the-art video encoder in order to obtain a coding gain. Specifically, they change three functional aspects of the encoder: (a) motion estimation; (b) residual coding; and (c) skip mode.

During motion estimation, the encoder searches, for a target block B, a corresponding predictor P such that it minimizes a Lagrangian cost function:

$$P^* = \arg\min_P D_{\mathrm{MV}}(B, P) + \lambda_{\mathrm{MV}} R_{\mathrm{MV}}(B, P) \tag{7.20}$$

where R_{MV} is the bit rate necessary to code the motion vector associated with P, λ_{MV} is a Lagrange multiplier, and D_{MV} is a measure of the energy of the prediction residuals $r(x, y) = P(x, y) - B(x, y)$, typically measured through the sum of absolute differences. Given per-block DCR S, it is possible to reduce the energy of the prediction residuals by computing the new residuals r' as follows:

$$r'(x, y) = \begin{cases} P(x, y) - \mathrm{DCR}_{\mathrm{up}}(x, y) & \text{if } P(x, y) > \mathrm{DCR}_{\mathrm{up}}(x, y) \\ P(x, y) - \mathrm{DCR}_{\mathrm{low}}(x, y) & \text{if } P(x, y) < \mathrm{DCR}_{\mathrm{low}}(x, y) \\ 0 & \text{otherwise} \end{cases} \tag{7.21}$$

Equation 7.21 corresponds to soft thresholding the conventional prediction residuals in the spatial domain according to defined per-pixel DCRs. The thresholded residuals r' are used to compute the discounted distortion in (7.20). An example of how prediction residuals are thresholded for a two-pixel disparity block $\mathbf{d} = [d_n(1), d_n(2)]$ is illustrated in Figure 7.10. In conventional coding, given a predictor (**pred**), one aims to reconstruct the original ground truth (**gt**). This produces the residual vectors $\mathbf{r}_1$ and $\mathbf{r}_2$. However, considering DCR, it is

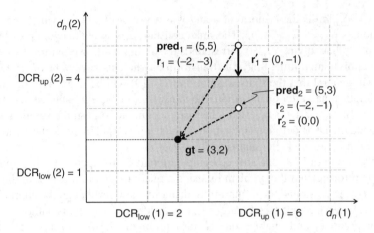

Figure 7.10 Coding the residuals using DCR with a toy example consisting of a block of two pixels ($d_n(1)$ and $d_n(2)$)

sufficient to encode a generally smaller residual ($\mathbf{r}'_1$ or $\mathbf{r}'_2$); that is, one that enables reconstruction of a value inside or on the border of the DCR (shaded area in Figure 7.10).

The soft thresholding (7.21) is used as well to modify prediction residuals in intra modes. Although the prediction technique is different from inter blocks, the larger degree of freedom entailed by DCR enables one to discount the distortion in the Lagrangian optimization in order to favor lower bit-rate solutions. Notice that the technique proposed by Valenzise *et al.* (2012) can be applied together with the transform domain sparsification approaches discussed in Section 7.4.2, where the sparsification is carried out on the prediction residuals after motion compensation. However, the joint optimization of motion vectors and of the corresponding prediction residuals is computationally not trivial; for example, it could be done by iteratively solving the LP in (7.19) for each candidate motion vector. With the current encoder architectures, this is clearly not desirable, and a more efficient solution is still under examination.

The third functional aspect to be considered in view of embedding DCR into a video encoder is how to handle the skip mode properly. The skip mode is selected during rate-distortion optimization by minimizing a Lagrangian cost function similar to (7.20). However, differently from inter and intra modes, in the skip mode no prediction residual is coded. Since the distortion evaluated during mode decision D_{MD} is computed on discounted residuals $\mathbf{r}'$, it could happen that the winner mode for a given macroblock is skip *even if the reconstructed block does not satisfy the DCR constraints*. This situation could produce arbitrarily high synthesized view distortion since the last one can increase very quickly outside DCR bounds (see Figure 7.8). A simple, though conservative, solution to this case is to prevent the skip mode whenever any pixel in the reconstructed block does not respect the DCR boundaries.

7.5 Concluding Remarks

This chapter showed different depth map coding tools used in the recent literature and in the latest proposals in response to the MPEG CfP for 3DV. Depth map coding tools are divided into three categories. Some approaches exploit the depth map inherent characteristics, such as its representation into smooth regions separated by sharp borders, its synthesis abilities, or its ability to represent motion in the depth direction. Other tools exploit the correlations between a depth map and its associated texture video. Depth mode decisions can be made, depth prediction information can be inherited, and spatial transforms can be designed specifically for depth by utilizing the associated texture information. Finally, some tools do not employ algorithms for depth map coding at all, they just propose new distortion models used in R-D depth decisions that evaluate distortion where it counts (i.e., on the synthesized views directly), since the depth map itself will not be displayed on the screen.

For instance, it is possible to fix an upper bound on the synthesized view distortion and find per-pixel intervals of depth values such that any signal reconstructed inside those intervals will produce an error within the defined bound in the synthesis. These "don't care regions" have been used to find sparse representations of depth maps in the spatial domains and to increase coding efficiency in video by enabling a larger degree of freedom in choosing motion vectors. An advantage of such approach is that DCR can be embedded into standard-compliant encoders.

More 3DV standardization meetings are likely to be held in the near future, during which new tools will be proposed, inefficient tools removed, and others further studied, until depth compression ratios reach an acceptable level. In any case, it seems that depth-based formats are starting to slowly gain the interest of the industry.

Acknowledgments

We would like to acknowledge the Interactive Visual Media Group of Microsoft Research for providing the Ballet data set (Zitnick *et al.*, 2004).

References

Bang, G., Yoo, S., and Nam, J. (2011) Description of 3D video coding technology proposal by ETRI and Kwangwoon University, ISO/IEC JTC1/SC29/WG11 MPEG2011/M22625.

Bjontegaard, G. (2001) Calculation of average PSNR differences between RD-curves. VCEG Meeting, Austin, USA.

Bross, B., Han, W., Ohm, J., Sullivan, G., and Wiegand, T. (2012) High efficiency video coding (HEVC) text specification draft 6, ITU-T SG16 WP3 and ISO/IEC JTC1/SC29/WG11 JCTVC-H1003.

Brust, H., Smolic, A., Mueller, K. *et al.* (2009) Mixed resolution coding of stereoscopic video for mobile devices, in *3DTV Conference: The True Vision – Capture, Transmission and Display of 3D Video, 2009*, IEEE, pp. 1–4.

Cheung, G., Kubota, A., and Ortega, A. (2010) Sparse representation of depth maps for efficient transform coding, in *Picture Coding Symposium (PCS), 2010*, IEEE, pp. 298–301.

Cheung, G., Ishida, J., Kubota, A., and Ortega, A. (2011) Transform domain sparsification of depth maps using iterative quadratic programming, in *2011 18th IEEE International Conference on Image Processing (ICIP)*, IEEE, pp. 129–132.

Daribo, I., Tillier, C., and Pesquet-Popescu, B. (2008) Adaptive wavelet coding of the depth map for stereoscopic view synthesis, in *2008 IEEE 10th Workshop on Multimedia Signal Processing*, IEEE, pp. 413–417.

Domanski, M., Grajek, T., Karwowski, D. *et al.* (2011) Technical description of Poznan University of Technology proposal for call on 3D video coding technology, ISO/IEC JTC1/SC29/WG11 MPEG2011/M22697.

Hannuksela, M. (2012) Test model for AVC based 3D video coding, ISO/IEC JTC1/SC29/WG11 MPEG2012/N12558.

Kamolrat, B., Fernando, W., and Mrak, M. (2010) Adaptive motion-estimation-mode selection for depth video coding, in *2010 IEEE International Conference on Acoustics Speech and Signal Processing (ICASSP)*, IEEE, pp. 702–705.

Kim, W.-S., Ortega, A., Lai, P., Tian, D., and Gomila, C. (2009) Depth map distortion analysis for view rendering and depth coding, in *2009 16th IEEE International Conference on Image Processing (ICIP)*, IEEE, pp. 721–724.

Lee, J., Oh, B.T., and Lim, I. (2011) Description of HEVC compatible 3D video coding technology by Samsung, ISO/IEC JTC1/SC29/WG11 MPEG2011/M22633.

Lee, J.Y., Wey, H.-C., and Park, D.-S. (2011) A fast and efficient multi-view depth image coding method based on temporal and inter-view correlations of texture images. *IEEE Trans. Circ. Syst. Vid. Technol.*, **21**, 1859–1868.

Merkle, P., Morvan, Y., Smolic, A. *et al.* (2008) The effect of depth compression on multiview rendering quality, in *3DTV Conference: The True Vision – Capture, Transmission and Display of 3D Video, 2008*, IEEE, pp. 245–248.

MPEG (2011) Call for proposals on 3D video coding technology, ISO/IEC JTC1/SC29/WG11 N12036.

Rusanovsky, D. and Hannuksela, M. (2011) Description of Nokia's response to MPEG 3DV call for proposals on 3DV video coding technologies', ISO/IEC JTC1/SC29/WG11MPEG2011/M22552.

Schwarz, H., Bartnik, C., and Bosse, S. (2011) Description of 3D video technology proposal by Fraunhofer HHI, ISO/IEC JTC1/SC29/WG11 MPEG2011/M22571.

Seo, J., Park, D., Wey, H.-C. *et al.* (2010) Motion information sharing mode for depth video coding, in *3DTV-Conference: The True Vision - Capture, Transmission and Display of 3D Video (3DTV-CON), 2010*, IEEE, pp. 1–4.

Shen, G., Kim, W.-S., Narang, S. *et al.* (2010) Edge-adaptive transforms for efficient depth map coding, in *Picture Coding Symposium (PCS)*, 2010, IEEE, pp. 566–569.

Valenzise, G., Cheung, G., Oliveira, R. *et al.* (2012) Motion prediction of depth video for depth-image-based rendering using don't care regions, in *Picture Coding Symposium, 2012*, IEEE, pp. 93–96.

Zitnick, C.L., Kang, S.B., Uyttendaele, M. *et al.* (2004) High-quality video view interpolation using a layered representation. *ACM Trans. Graph. (TOG) – Proc. SIGGRAPH 2004*, **23**, 600–608.

8

Depth-Based 3D Video Formats and Coding Technology

Anthony Vetro[1] and Karsten Müller[2]

[1]*Mitsubishi Electric Research Labs (MERL), USA*
[2]*Fraunhofer Institute for Telecommunications, Heinrich-Hertz-Institut, Germany*

8.1 Introduction

The primary usage scenario for three-dimensional video (3DV) formats is to support depth perception of a visual scene as provided by a 3D display system. There are many types of 3D display systems, including classic stereoscopic systems that require special-purpose glasses to more sophisticated multiview auto-stereoscopic displays that do not require glasses (Konrad and Halle, 2007). While stereoscopic systems only require two views, the multiview displays have much higher data throughput requirements since 3D is achieved by essentially emitting multiple videos in order to form view-dependent pictures. Such displays can be implemented, for example, using conventional high-resolution displays and parallax barriers; other technologies include lenticular overlay sheets and holographic screens. Each view-dependent video sample can be thought of as emitting a small number of light rays in a set of discrete viewing directions – typically between eight and a few dozen for an auto-stereoscopic display. Often, these directions are distributed in a horizontal plane, such that parallax effects are limited to the horizontal motion of the observer. A more comprehensive review of 3D display technologies is given in Chapter 15, as well as by Benzie *et al.* (2007). An overview can also be found in Ozaktas and Onural (2007).

Other goals of 3DV formats include enabling free-viewpoint video, whereby the viewpoint and view direction can be interactively changed. With such a system, viewers can freely navigate through the different viewpoints of the scene. 3DV can also be used to support immersive teleconference applications. Beyond the advantages provided by 3D displays, it has been reported that a teleconference system could enable a more realistic communication experience when motion parallax is supported.

Emerging Technologies for 3D Video: Creation, Coding, Transmission and Rendering, First Edition.
Frédéric Dufaux, Béatrice Pesquet-Popescu, and Marco Cagnazzo.
© 2013 John Wiley & Sons, Ltd. Published 2013 by John Wiley & Sons, Ltd.

Existing stereo and multiview formats are only able to support the above applications and scenarios to a limited extent. As an introduction, these formats are briefly described and the requirements and functionalities that are expected to be fully supported by depth-based formats are discussed. A more comprehensive overview can be found in Chapter 6.

8.1.1 Existing Stereo/Multiview Formats

Currently, two primary classes of multiview formats exist: frame compatible and stereo or multiview video; these are briefly reviewed in the following.

Frame-compatible formats refer to a class of stereo video formats in which the two stereo views are filtered, subsampled and arranged into a single coded frame or sequence of frames; that is, the left and right views are packed together in the samples of a single video frame (Vetro et al., 2011a). Popular arrangements include the side-by-side and top–bottom formats. Temporal multiplexing is also possible, where the left and right views would be interleaved as alternating frames or fields. The primary benefit of frame-compatible formats is that they facilitate the introduction of stereoscopic services through existing infrastructure and equipment. In this way, the stereo video can be compressed with existing encoders, transmitted through existing channels, and decoded by existing receivers. The video-level signaling for these formats is specified by the MPEG-2 Video and MPEG-4 advanced video coding (AVC) standards (Vetro et al., 2011b).

As an alternative to frame-compatible formats, direct encoding of the stereo and multiview video may also be done using multiview extensions of either MPEG-2 Video or MPEG-4 AVC standards (Vetro et al., 2011b). A key capability is the use of inter-view prediction to improve compression capability, in addition to ordinary intra- and inter-prediction modes. Another key aspect of all multiview video coding designs is the inherent support for two-dimensional (2D)/backwards compatibility with existing legacy systems. In other words, the compressed multiview stream includes a base view bitstream that is coded independently from all other views in a manner compatible with decoders for single-view profile of the standard.

8.1.2 Requirements for Depth-Based Format

Depth-based representations are another important and emerging class of 3D formats. Such formats are unique in that they enable the generation of virtual views through depth-based image rendering techniques (Kauff et al., 2007), which may be required by auto-stereoscopic or multiview displays (Müller et al., 2011). Depth-based 3D formats can also allow for advanced stereoscopic processing, such as adjusting the level of depth perception with stereo displays according to viewing characteristics such as display size, viewing distance, or user preference. The depth information itself may be extracted from a stereo pair by solving for stereo correspondences or obtained directly through special range cameras; it may also be an inherent part of the content, such as with computer-generated imagery.

ISO/IEC 23002-3 (also referred to as MPEG-C Part 3) specifies the representation of auxiliary video and supplemental information. In particular, it enables signaling for depth-map streams to support 3DV applications. The well-known 2D plus depth format (see Figure 8.1) is supported by this standard. It is noted that this standard does not specify the means by which the depth information is coded, nor does it specify the means by which the 2D video is coded. In this way, backward compatibility to legacy devices can be provided.

Figure 8.1 The 2D plus depth format. Images provided courtesy of Microsoft Research

The main drawback of the 2D plus depth format is that it is only capable of rendering a limited depth range and was not specifically designed to handle occlusions. Also, stereo signals are not easily accessible by this format; that is, receivers would be required to generate the second view to drive a stereo display, which is not the convention in existing displays.

To overcome the drawbacks of the 2D plus depth format, a multiview video plus depth (MVD) format with a limited number of original input views and associated per-pixel depth data can be considered. For instance, with two input views, high-quality stereo video is provided and the depth information would enhance 3D rendering capabilities beyond 2D plus depth. However, for high-quality auto-stereoscopic displays, wide-baseline rendering with additional views beyond the stereo range may be required. For example, formats with three or four views with associated depth-map data may be considered.

8.1.3 Chapter Organization

Depth-based 3D formats beyond 2D plus depth are a current topic of study in MPEG. This chapter provides an overview of the current status of research and standardization activity towards defining a new set of depth-based formats that facilitate the generation of intermediate views with a compact binary representation. Section 8.2 provides a brief introduction to depth-based representation and rendering techniques, which are the basis for the functionality offered by depth-based formats. Next, in Section 8.3 the different coding architectures are discussed, including those that are compatible with the existing AVC standard as well as those that are compatible to the emerging high-efficiency video coding (HEVC) standard. Hybrid coding architectures that mix coding formats are also discussed. Section 8.4 presents the various compression technologies that have been proposed and are being considered for standardization. Then, results from a large-scale experimental evaluation are presented in Section 8.5 to indicate the possible improvements over the current-state-of-the-art coding approaches. The chapter concludes with a summary of material that has been presented and discusses future research and standardization.

8.2 Depth Representation and Rendering

For the creation of a generic depth-based multiview format for various 3D displays, scene geometry needs to be provided in addition to texture information. A very general scene geometry representation is given by depth maps, which can be provided by different methods, as shown in Section 8.2.1. These depth maps can then be used to apply depth-image-based rendering (DIBR) in order to generate the required number of intermediate views for any auto-stereoscopic multiview display.

8.2.1 Depth Format and Representation

In computer graphics and computer vision, a 3D scene consists of texture and geometry information. Texture information, like color and luminance, are directly recorded by camera sensors. Geometry or 3D structure information can be obtained in different ways. For synthetic sequences, such as computer-generated scene content and animated films, scene geometry information is directly available, for example, in the form of wireframe models (ISO/IEC JTC1/SC29/WG11,1997) or 3D point coordinates (Würmlin *et al.*, 2004). For natural scenes, geometry information can be recorded as distance information by special sensors, like time-of-flight cameras (Lee *et al.*, 2010). This distance information is then recorded as a gray-value depth image. Usually, these depth images have a lower resolution than associated texture information from video cameras. Also, time-of-flight sensors currently lack accuracy for larger distances and have to be placed at slightly different positions than the video camera. Accordingly, some additional processing is required, if video and depth data are to be aligned.

For natural scenes that were only recorded by video cameras, depth data can be estimated. For this, intensive investigations of different algorithms were carried out by Scharstein and Szeliski (2002). An overview on disparity estimation techniques can also be found in Chapter 5. The basic principle of most depth estimation algorithms is to match corresponding image blocks or regions in two or more cameras with slightly different positions to obtain the displacement or disparity between corresponding texture pixels. The relation between depth information and disparity is given by projective geometry, where points of a 3D scene are projected onto camera planes with the use of projection matrices (Faugeras, 1993; Hartley and Zisserman, 2000). For parallel cameras, a simplified relation can be derived, as shown in Figure 8.2.

Here, a point $\mathbf{P}$ of a 3D scene is recorded by two cameras and projected onto their image sensors in the camera planes. The depth distance of $\mathbf{P}$ from the camera centers shall be z. Let the distance between both cameras be Δs and assume identical focal lengths f. Then, $\mathbf{P}$ is projected onto both sensors with an offset from the center of d_1 in the left and d_2 in the right

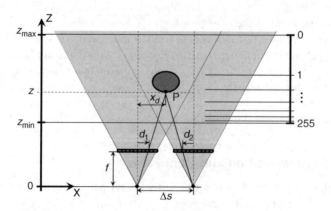

Figure 8.2 Relation between depth values z and disparity d

camera plane. This gives the following relations:

$$\frac{d_1}{x_d} = \frac{f}{z} \text{ for the left camera and } \frac{d_2}{\Delta s - x_d} = \frac{f}{z} \text{ for the right camera} \qquad (8.1)$$

Adding both equations, the following inverse relationship between depth value z and disparity value d between both projections of P can be found:

$$d = d_1 + d_2 = \frac{f x_d}{z} + \frac{f(\Delta s - x_d)}{z}$$
$$\Rightarrow d = \frac{f \Delta s}{z} \qquad (8.2)$$

The disparity value d for each pixel is obtained from matching left and right image content. Depending on the desired accuracy, disparity values are usually linearly quantized; for example, into 256 discrete values to operate with 8-bit resolution data, as done for many practical applications. According to (8.2), this refers to an inverse quantization of depth values, as indicated in the depth scale on the right side in Figure 8.2. For optimal usage of the depth range, the nearest and farthest depth values z_{min} and z_{max} are determined and inverse depth quantization is carried out in the range $[z_{min}, z_{max}]$. Therefore, the stored inverse depth values $I_d(z)$ are calculated as

$$I_d(z) = \text{round}\left[255 \times \left(\frac{1}{z} - \frac{1}{z_{max}}\right) \middle/ \left(\frac{1}{z_{min}} - \frac{1}{z_{max}}\right)\right] \qquad (8.3)$$

In practical applications, disparity estimation algorithms use a matching function (Szeliski *et al.*, 2006) with different area support and size for corresponding image blocks in left and right views (Bleyer and Gelautz, 2005). Furthermore, they apply a matching criterion; for example, the sum of absolute differences or cross-correlation. The estimation process is optimized by different means, including graph cuts (Kolmogorov and Zabih, 2002), belief propagation (Felzenszwalb and Huttenlocher, 2006), plane sweeping (Cigla *et al.*, 2007), or combined methods (Atzpadin *et al.*, 2004; Kolmogorov, 2006). Depth estimation has also been studied with special emphasis for multiview video content and temporal consistency in order to provide depth data for multiview video sequences (Tanimoto *et al.*, 2008; Lee and Ho, 2010; Min *et al.*, 2010).

Although depth estimation algorithms have been improved considerably in recent years, they can still be erroneous in some cases due to mismatches, especially for partially occluded image and video content that is only visible in one view.

8.2.2 Depth-Image-Based Rendering

If a generic format with texture and depth components of a few views is used in 3D or free viewpoint video applications, additional views have to be generated; for example, for 3D displays with a different number of views. For this view generation process, DIBR is used (Redert *et al.*, 2002; Kauff *et al.*, 2007). In this process, the texture information is projected or warped to a new viewing position, at which an intermediate view is to be synthesized. The warping distance of each texture sample is determined from the associated per-sample depth information. Current video solutions use a rectification process, in which a strictly parallel camera scenario is enforced, as shown in the stereo scheme in Figure 8.2. In such settings, the general DIBR process can be simplified to a horizontal sample shift from original to

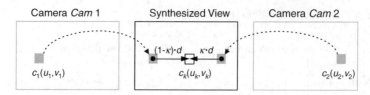

Figure 8.3 View synthesis principle with horizontal disparity-based shift from original data (*Cam* 1 and *Cam* 2) to new position in synthesized view

newly rendered views. For calculating the shift values, the disparity values between original views are obtained first by combining (8.2) and (8.3):

$$d = f\Delta s \frac{I_d(z)}{255}\left(\frac{1}{z_{\min}} - \frac{1}{z_{\max}}\right) + \frac{1}{z_{\max}} \tag{8.4}$$

In addition to the inverse depth values $I_d(z)$, focal length f, camera baseline Δs, and the nearest and farthest depth values $z_{\min}$ and $z_{\max}$ have to be known. Here, d gives the disparity or shift value between corresponding texture samples in the two original views, separated by Δs. For calculating the required horizontal shift value for a new view to be synthesized, the view position has to be known and the disparity scaled accordingly. An example is given in Figure 8.3, where two original camera positions, *Cam* 1 and *Cam* 2 with texture samples c_1 and c_2 at positions (u_1, v_1) and (u_2, v_2) are given.

Furthermore, a new view shall be synthesized between both original cameras with texture sample c_κ at position (u_κ, v_κ). Here, $\kappa \in [0 \ldots 1]$ represents the intermediate position parameter, which specifies the location between both original cameras. Accordingly, the relation between texture sample $c_1(u_1, v_1)$ in *Cam* 1 and its shifted version in the intermediate view is given by $c_{\kappa 1}(u_\kappa, v_\kappa) = c_1(u_1 + (1 - \kappa)d, v_1)$. Similarly, $c_{\kappa 2}(u_\kappa, v_\kappa) = c_2(u_2 + \kappa d, v_2)$ for *Cam* 2. In addition to shifting both original texture samples to the new position, color blending can be applied, if color variances due to different illumination in both original views occur (Müller *et al.*, 2008). This results in the final synthesized sample $c_\kappa(u_\kappa, v_\kappa)$:

$$\begin{aligned} c_\kappa(u_\kappa, v_\kappa) &= (1 - \kappa)c_{\kappa 1}(u_\kappa, v_\kappa) + \kappa c_{\kappa 2}(u_\kappa, v_\kappa) \\ &= (1 - \kappa)c_1(u_1 + (1 - \kappa)d, v_1) + \kappa c_2(u_2 - \kappa d, v_2) \end{aligned} \tag{8.5}$$

In cases, where only one original texture sample is available – for example, if scene content is occluded in one view – a synthesized texture sample $c_\kappa(u_\kappa, v_\kappa)$ is obtained by shifting the visible original sample without color blending. For an improved visual impression, additional processing steps, such as hole filling, filtering, and texture synthesis of disoccluded areas, are usually applied after sample-wise shifting and texture blending (Müller *et al.*, 2008).

Besides synthesizing new views between two original cameras, additional views can also be extrapolated (synthesized outside the viewing range of original cameras) by setting $\kappa < 0$ or $\kappa > 1$.

8.3 Coding Architectures

8.3.1 AVC-Based Architecture

Significant improvements in video compression capability have been demonstrated with the introduction of the H.264/MPEG-4 Advanced Video Coding AVC standard (Wiegand *et al.*,

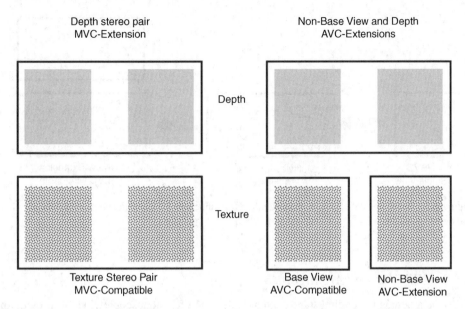

Depth stereo pair
MVC-Extension

Non-Base View and Depth
AVC-Extensions

Depth

Texture

Texture Stereo Pair
MVC-Compatible

Base View
AVC-Compatible

Non-Base View
AVC-Extension

Figure 8.4 Illustration of MVC-compatible (left) and AVC-compatible (right) architecture for depth-based 3DV coding

2003; ITU-T and ISO/IEC, 2010), which has been extensively deployed for a wide range of video products and services. An extension of this standard for multiview video coding (MVC) was first finalized in July 2009 and later amended in July 2010. The MVC format was selected by the Blu-ray Disc Association as the coding format for stereo video with high-definition resolution (BDA, 2009), and has recently been standardized for stereo broadcast as well (DVB, 2012).

Within the AVC-based framework, one target for standardization is an MVC-compatible extension including depth, where the main target is to enable 3D enhancements while maintaining MVC stereo compatibility for the texture videos. The approach would invoke an independent second stream for the representation of stereo depth maps as if they were monochrome video data, as well as high-level syntax signaling of the necessary information to express the interpretation of the depth data and its association with the video data. An illustration of the compatibility supported by this architecture is shown in Figure 8.4 (left). Macroblock-level changes to the AVC or MVC syntax, semantics, and decoding processes are not considered in this configuration in order to maintain compatibility. The standardization of this approach is currently underway and is expected to be completed by early 2013.

Considering that certain systems only maintain compatibility with monoscopic AVC, a second architecture that is being considered is an AVC-compatible extension that includes depth. In this approach, further coding efficiency gains could be obtained by improving the compression efficiency of non-base texture views and the depth data itself. However, in contrast to the MVC-compatible approach, this method requires changes to the syntax and decoding process for non-base texture views and depth information at the block level. An illustration of the compatibility supported by this architecture is shown in Figure 8.4 (right). Clearly, a notable coding efficiency benefit relative to the MVC-compatible approach would be required to justify the standardization of this approach. This is a current topic of study within the standardization committees.

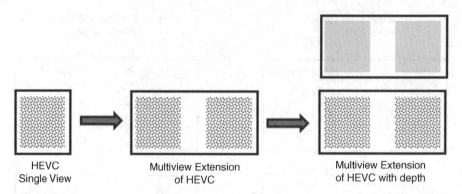

Figure 8.5 Extensions of HEVC coding standard for support of 3DV architectures with multiple views and depth

8.3.2 HEVC-Based Architecture

HEVC is a new video coding standard that is designed for monoscopic video to provide the same subjective quality at half the bit rate compared with AVC high profile. A primary usage of HEVC is to support the delivery of ultrahigh definition (UHD) video. It is believed that many UHD displays will also be capable of decoding stereo video as well. The first version of the standard was approved in January 2013.

For the highest compression efficiency, 3DV coding extensions based on the HEVC are being developed, where multiview video data with associated depth maps are coded (see Figure 8.5). In this architecture, the base view is fully compatible with HEVC in order to extract monoscopic video, while the coding of dependent views and depth maps would utilize additional tools as described in Section 8.4. It is also anticipated that there would be profiles of the standard in which stereo video can be easily extracted to support existing stereoscopic displays; in such cases, the dependency between the video data and depth data may be limited.

A subset of this 3DV coding extension would include a simple multiview extension of HEVC, utilizing the same design principles of MVC in the MPEG-4/H.264 AVC framework (i.e., providing backwards compatibility for monoscopic decoding). It is expected that this extension of HEVC will be available as a final standard by early 2014. Additionally, it is planned to develop a suite of tools for scalable coding, where both view scalability and spatial scalability would allow for backward-compatible extensions for more views. These extensions would also accommodate the depth data that would be associated with each view and/or provide for a way to enhance the resolution of views. Ideally, all of this would be achieved in such a way that decoding by legacy devices is still possible.

8.3.3 Hybrid

From a pure compression efficiency point of view, it is always best to use the most advanced codec. However, when introducing new services, providers must also consider capabilities of existing receivers and an appropriate transition plan. Considering that most terrestrial broadcast systems are based on MPEG-2 or AVC, it may not be easy to simply switch codecs in the near term.

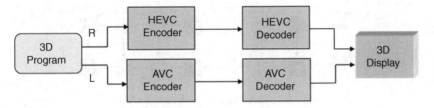

Figure 8.6 Illustration of hybrid architecture with monoscopic AVC base and HEVC enhancement program for stereo support

One solution to this problem is to transmit the 2D program in the legacy format (e.g., MPEG-2), while transmitting an additional view to support stereo services in an advanced coding format (e.g., AVC). The obvious advantage is that backward compatibility with the existing system is provided with significant bandwidth savings relative to simulcast in the legacy format. One drawback of this approach is that there is a strong dependency between the 3D program and the 2D program. Such a system would not support independent programming of 2D and 3D content programs, which may be desirable for production reasons. Also, this approach requires legacy and advanced codecs to operate synchronously, which may pose implementation challenges for certain receiver designs. Nevertheless, broadcasting trials of hybrid MPEG-2 and AVC-based systems are underway in Korea, and there are plans to standardize the transmission of such a hybrid format in ATSC.

In the context of depth-based 3D formats, there are clearly many variations that could be considered. In an AVC-compatible framework, the base view would be coded with AVC, while additional texture views and supplemental depth videos could be encoded with HEVC. A slight variation on this would be for the stereo pair of the texture to be coded with MVC and only the depth videos coded with HEVC. A simple block diagram illustrating the hybrid video coding architecture for the right and left views of a stereoscopic video program is given in Figure 8.6.

One open issue with hybrid architectures that requires further study is the degree of inter-component dependency that would be permitted across different components and different coding standards. For instance, the benefits of using decoded pictures from the AVC base view to predict pictures in the second view coded with HEVC needs to be weighed against the implementation challenges. Also, it needs to be considered whether lower level dependencies could be supported; for example, sharing of motion or mode data.

8.4 Compression Technology

For the efficient compression of 3DV data with multiple video and depth components, a number of coding tools are used to exploit the different dependencies among the components. First, one video component is independently coded by a conventional block-based 2D video coding method, such as AVC or HEVC without additional tools in order to provide compatibility with existing 2D video services. For each additional 3DV component – that is, the video component of the dependent views as well as the depth maps – additional coding tools are added on top of a 2D coding structure. Thus, a 3DV encoder can select the best coding method for each block from a set of conventional 2D coding tools and additional new coding tools, some of which are described in the following subsections.

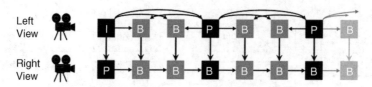

Figure 8.7 Illustration of inter-view prediction

8.4.1   Inter-View Prediction

The basic concept of inter-view prediction, which is employed in all standardized designs for efficient multiview video coding, is to exploit both inter-view and temporal redundancy for compression. Since the cameras of a multiview scenario typically capture the same scene from nearby viewpoints, substantial inter-view redundancy is present. This holds for both texture views and the corresponding depth map images associated with each view; thus, inter-prediction can be applied to both types of data independently.

A sample prediction structure is shown in Figure 8.7. In modern video coding standards such as AVC and HEVC, inter-view prediction is enabled through the flexible reference picture management capabilities of those standards. Essentially, the decoded pictures from other views are made available in the reference picture lists for use by the inter-picture prediction processing. As a result, the reference picture lists include the temporal reference pictures that may be used to predict the current picture along with the inter-view reference pictures from neighboring views. With this design, block-level decoding modules remain unchanged and only small changes to the high-level syntax are required; for example, indication of the prediction dependency across views and corresponding view identifiers. The prediction is adaptive, so the best predictor among temporal and inter-view references can be selected on a block basis in terms of rate-distortion cost.

Relative to simulcast, which does not utilize inter-view prediction, it has been shown through experiments that inter-view prediction is responsible for the majority of the coding efficiency gains. This leads to a simplified design for efficient multiview video coding (both texture and depth) with good coding efficiency capability. For additional information on the design, syntax, and coding efficiency of inter-view prediction, readers are referred to Merkle *et al.* (2007) and Vetro *et al.* (2011b).

In the following subsections, coding tools that go beyond picture-based inter-view prediction are described. Many of these require changes to lower levels of the decoding syntax and process, with the benefit of additional gains in coding efficiency.

8.4.2   View Synthesis Prediction

In addition to being used for generation of intermediate views, the DIBR techniques described in Section 8.2.2 could also be used as a unique form of inter-view prediction that is referred to as view synthesis prediction. In contrast to the inter-view prediction technique presented in Section 8.4.1, which essentially predicts a block of pixels in one view by means of a linear disparity vector, view synthesis prediction exploits the geometry of the 3D scene by warping the pixels from a reference view to the predicted view as illustrated in Figure 8.3 and Figure 8.8.

During the development of the multiview extensions of MPEG-4/H.264 AVC, depth maps were not assumed to be available or an integral part of the data format. To enable view

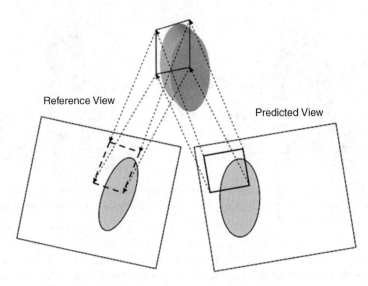

Figure 8.8   Illustration of view synthesis prediction

synthesis prediction in this framework, the depth for each block would need to be estimated and explicitly coded as side information, so that the decoder could generate the view synthesis data used for prediction. Such a scheme was first described by Martinian *et al.* (2006) and more fully elaborated on by Yea and Vetro (2009). Although coding efficiency gains were reported, the benefit of this type of prediction was limited by the overhead incurred by the additional block-based depth that was required to be sent.

In the 3DV framework, depth is available as an integral part of the data format; therefore, the generation of a synthesized view can be done without any additional side information. In this way, a synthesized view could be generated and added to the reference picture list and thus used for prediction as any other reference picture. The only requirement is to signal the appropriate reference picture index so that the decoder knows that the prediction for a particular block is done with respect to a view synthesis reference picture.

8.4.3   Depth Resampling and Filtering

Reducing the resolution of the depth maps image could provide substantial rate reductions. However, filtering and reconstruction techniques need to be carefully designed to maximize quality. Specifically, the quality of the depth map will have a direct impact on the quality of the synthesized views.

There have been several past studies on up-sampling and reconstruction techniques of reduced resolution depth. For instance, a nonlinear reconstruction filter that refines edges based on the object contours in the video data was proposed by Oh *et al.* (2009). A key advantage of this method was that the edge information was preserved. It was shown that bit-rate reductions greater than 60% relative to full-resolution coding of the depth videos could be achieved. Furthermore, improved rendering quality around the object boundaries compared with conventional filtering techniques was demonstrated.

Owing to the unique characteristics of edges within the depth image, further work in this area has verified that nonlinear approaches are generally more favorable than linear filtering

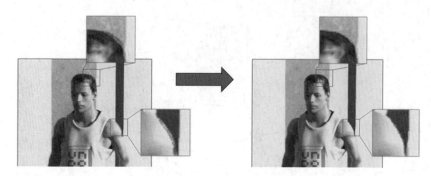

Figure 8.9 Comparison of linear interpolation for depth up-sampling (left) versus a nonlinear filtering approach (right). Original images provided courtesy of Nokia (see Plate 1 for the colored figure)

techniques; for example, as reported by Beni *et al.* (2012) and Lee *et al.* (2012). As shown in Figure 8.9 (left), linear filters will tend to blur edges and introduce artifacts in the rendered edge, while nonlinear filtering techniques (e.g., median or dilation filters) are able to better preserve the true edge characteristics and render a synthetic image with fewer artifacts, as shown in Figure 8.9 (right).

The filtering techniques can be applied either within the decoder loop or outside the decoder loop as a post-process. While there is ongoing work in this area, preliminary results suggest that most of the subjective visual benefit is achieved with post-processing techniques (Rusert, 2012). However, in-loop processing certainly has the potential to improve coding efficiency, especially for tools that rely on accurate depth values around edges to make predictions (e.g., view synthesis prediction). Additional techniques on depth video coding can also be found in Chapter 7.

8.4.4   Inter-Component Parameter Prediction

For the joint coding of video and depth data in an MVD format, dependencies between both components are identified and exploited. As each video component has an associated depth map at identical time instance and view point, a similar scene characteristic exists. This includes the collocation of scene objects with their texture and distance information in the video and depth component respectively. Furthermore, the object motion in both components is identical. Therefore, an additional coding mode can be used for depth maps, where the block partitioning into sub-blocks, as well as associated motion parameters, is inferred from the collocated block in the associated video picture (Winken *et al.*, 2012). Accordingly, it is adaptively decided for each depth block as to whether partitioning and motion information are inherited from the collocated region of the video picture or new motion data transmitted. If such information is inherited, no additional bits for partitioning and motion information are required. Note, however, that the real object motion and the motion vector, estimated by the encoder, are not necessarily identical: the estimated motion has to be coded together with other information, such as residual data. Therefore, a motion vector can be selected for a block that results in the best encoder decision but differs significantly from the true object motion. Thus, collocated blocks in the video and depth component may have different estimated motion vectors, such that the inter-component parameter prediction mode is not selected.

The common structure information in video and depth component is further used for specific depth modeling modes in a 3DV encoder, as described in Section 8.4.5.

8.4.5 Depth Modeling

For the coding of depth maps, the special characteristic and purpose of this information have to be considered. As depth maps represent the 3D geometry information of recorded or generated 3D content in the form of distance information, they are mainly characterized by unstructured constant or slowly changing areas for scene objects. In addition, abrupt value changes can occur at object boundaries between foreground and background areas. Experiments with state-of-the-art compression technology have shown that such depth maps can be compressed very efficiently. In addition, subsampling to a lower resolution prior to encoding and decoder-side up-sampling, similar to chrominance subsampling, has also been studied with good results by Oh *et al.* (2009), as discussed in Section 8.4.3. Since the purpose of depth maps is to provide scaled disparity information for texture data for view synthesis, coding methods have to be adapted accordingly. Especially sharp depth edges between foreground and background areas should be preserved during coding. A smoothing of such edges, as done by classical block-based coding methods, may lead to visible artifacts in intermediate views (Müller *et al.*, 2011). Furthermore, depth coding has to be optimized with respect to the quality of synthesized views, as the quality of the reconstructed depth data is irrelevant.

For a better preservation of edge information in depth maps, wedgelet (Merkle *et al.*, 2009) and contour-based coding modes have been introduced. During encoding, each depth block is analyzed for significant edges. If such an edge is present, a block is subdivided into two nonrectangular partitions P_1 and P_2 as shown in Figure 8.10.

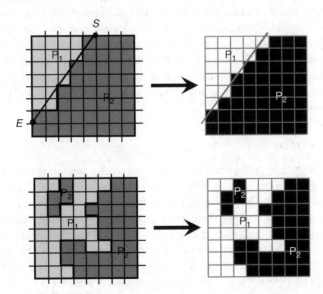

Figure 8.10 Illustration of wedgelet partition (top) and contour partition (bottom) of a depth block: original sample assignment to partitions P_1 and P_2 (left) and partition pattern (right). For the wedgelet partition, this pattern slightly differs from the original sample assignment due to the wedgelet approximation of the original contour

The partitions can be separated by a straight line as an approximation of a rather regular depth edge in this block (see Figure 8.10 top). Both partitions are then represented by a constant value. In addition to these values, the position of the separation line needs to be encoded. This is done in different ways. First, an explicit signaling is carried out, using a look-up table. This table contains all possible separation lines within a block in terms of position and orientation and provides an index for them. Second, a separation line can also be derived from neighboring blocks; for example, if an already coded neighboring block contains significant edge information which ends at the common block boundary. Then, a continuation of this edge into the current block can be assumed. Accordingly, one or both end points of the separation line (S end E in Figure 8.10, top left) can be derived from the already coded upper and left neighboring block, and thus do not need to be signaled. Third, the position of a separation line can be derived from the corresponding texture block. If a depth block contains a more complex separation between both partitions, as shown in Figure 8.10 bottom, its contour can also be derived from the corresponding texture block. In the depth encoding process, either one of the described depth modeling modes with signaling of separation information and partition values is selected or a conventional intra prediction mode is selected (Merkle *et al.*, 2012). Additional techniques on depth video coding can also be found in Chapter 7.

In addition to a good approximation of sharp depth edges by the new depth modeling modes, the rate-distortion optimization for the depth coding is adapted. As decoded depth data are used for view synthesis of the associated texture information, the distortion measure for the depth coding is obtained by comparing uncoded synthesized and reconstructed synthesized views. In this view synthesis optimization method, a block-wise processing aligned with depth data coding was introduced in order to provide a fast encoder operation (Tech *et al.*, 2012a). For an encoding optimization for one synthesized view, the method operates as follows. First, the uncoded synthesized view is rendered from uncoded texture and depth data once per frame prior to encoding. Then, the synthesized reconstructed view is rendered separately for each block from decoded texture and decoded depth data. Furthermore, only those neighboring blocks are considered in the block-wise synthesis that influence the current depth block under encoding. Examples are neighboring blocks of foreground objects in original views that occlude the current block in the synthesized view. If more synthesized views are considered, the distortion measures of each single view are averaged. More information and test results of the view synthesis optimization method with different numbers of views can be found in Tech *et al.* (2012b).

8.4.6 Bit Allocation

When coding 3DV data in the MVD format, the video component of the base view is usually coded by classical methods, such as AVC or HEVC, for providing compatibility with existing 2D video coding standards. On top of that, new coding tools are used for the video component of dependent views and for the depth data, as described in the previous subsections. These tools provide additional coding efficiency such that the dependent views and depth maps require a much smaller portion of the overall bit rate than does the base view video data. An example for the bit-rate distribution in percent for the individual components of an MVD format with two views *Cam* 1 and *Cam* 2 is shown in Figure 8.11.

In this example, a 3DV codec was used that is based on HEVC 2D video coding as developed by Schwarz *et al.* (2012). The results in Figure 8.11 show bit-rate distributions in percent at four rate points R1–R4, according to Table 8.2. For each rate point, the individual bit-rate distributions of eight different sequences from the MPEG 3DV test set were averaged.

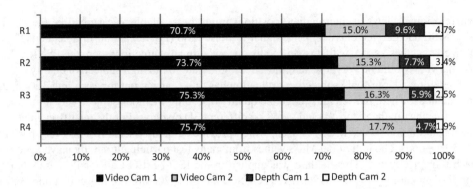

Figure 8.11 Example for average bit-rate distribution in percent of total rate over all test sets for the video and depth components in two-view 3D-HEVC-based coding with views *Cam* 1 and *Cam* 2 for four different rate points R1–R4. Here, *Cam* 1 is the independent base view and receives the largest bit-rate portion

Figure 8.11 first shows that most of the bit rate is distributed to the video component of the independent base view of *Cam* 1. In particular, the base view video component receives 71% of the bit rate at R1 and 76% at R4. Accordingly, all other components only require 29% at R4 and 24% at R1. Therefore, efficient 3DV transmission of MVD data with two views and depth maps can be achieved at approximately 1.3-times the bit rate of a 2D video transmission. Furthermore, Figure 8.11 shows that most of the bit rate is distributed to the video data. Here, the video/depth rate distribution varies from 86%/14% at the lowest rate point R1 to 93%/7% at the highest rate point R4 on average. Thus, depth data can be coded very efficiently. A comparison between MVD and pure stereo video coding showed that the perceived video quality in both cases is almost identical, even though MVD additionally provides depth data at the decoder for high-quality view synthesis at the 3D display.

8.5 Experimental Evaluation

The main objective of a 3DV coding technology is to provide high compression efficiency for a generic format that can be used to synthesize any number of views for a variety of stereoscopic and auto-stereoscopic multiview displays. For the evaluation of suitable compression methods, the quality of these synthesized views needs to be assessed. However, no original reference views exist for newly synthesized positions, such that classical objective comparison methods (e.g., mean squared error (MSE)) between decoded synthesized views and original reference cannot be applied directly. In this section, the evaluation framework that was used to assess the compression efficiency for depth-based formats is described, followed by a summary of results from a large-scale quality assessment process.

8.5.1 Evaluation Framework

As shown in Section 8.4.5 for the encoding of depth maps, a pseudo-reference can be created by synthesizing intermediate views from uncoded data and thus also applying error measures like MSE. This method, however, neglects synthesis errors that can occur due to erroneous depth maps, and thus a high-quality pseudo-reference has to be assumed. Even in cases where original views at intermediate positions might be available, measures like the pixel-wise MSE and derived classical peak signal-to-noise ratio (PSNR) are unsuitable: consider a perfectly synthesized view that is shifted by one pixel in any direction; then, the PSNR value for this view would be very low and, thus, does not relate to the high subjective quality.

Therefore, a large-scale subjective evaluation has to be carried out for assessing 3DV coding methods, where participants judge the quality of coding methods subjectively in test sessions by viewing the reconstructed views on different 3D displays. For the quality evaluation, the mean opinion score (MOS) is used, which provides a quality scale from 0 (very bad) to 10 (very good). The individual MOS values from many test participants are then averaged for each sequence tested. This method was applied in the ISO-MPEG Call for Proposals (CfP) for 3DV technology in 2011 (ISO/IEC JTC1/SC29/WG11, 2011a). For this call, a number of test parameters were considered. First, two test categories were specified: AVC compatible and HEVC compatible/unconstrained for testing the different 3DV coding proposals, based on the respective 2D video coding technology for the base view. Next, the 3DV test material was created from eight sequences with multiview video and depth components. Four of these sequences had a progressive HD resolution of 1920×1088 at 25 fps and four had a progressive resolution of 1024×768 at 30 fps, as listed in Table 8.1 and 8.2. For all test sequences, four rate points were specified, as explained in Section 8.5.2 for AVC-based technologies and in Section 8.5.3 for HEVC-based/unconstrained technologies.

All eight sequences were evaluated in two test scenarios. In the two-view scenario, video and depth components of two views {V0, V1} were coded and a stereo pair with one original and one intermediate viewing position reconstructed and synthesized as shown in Figure 8.12. This stereo pair was evaluated on a stereoscopic display. In the three-view scenario, video and depth components of three views {V0, V1, V2} were coded and different types of video data extracted. Then, a dense range of 28 views was synthesized and viewed on an auto-stereoscopic 28-view display as shown in Figure 8.13 and used for ISO/IEC JTC1/SC29/WG11 (2011a). For additional assessment, a central stereo pair in the middle of the three-view range and a random stereo pair within the viewing range were synthesized for viewing on a stereoscopic display. All results were evaluated in large-scale subjective tests with 13 international test laboratories involved (ISO/IEC JTC1/SC29/WG11, 2011b).

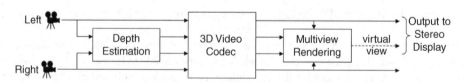

Figure 8.12   Advanced stereoscopic processing with two-view configuration.

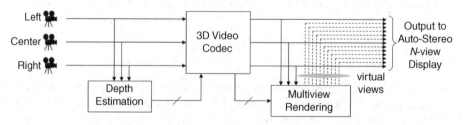

Figure 8.13   Auto-stereoscopic output with three-view configuration

8.5.2 AVC-Based 3DV Coding Results

For 3DV coding technology, that uses the AVC standard for coding the base view video, four different rate points R1–R4 (from low to high bit rate) were specified for the two-view and three-view scenarios and each test sequence in an iterative process. For finding these rate points, each sequence with video and depth component was coded at four different initial rate points. For this, the video components where jointly coded, using the MVC reference software JMVC 8.3.1. The same coding was also applied to the depth components. Then, stereo pairs and the set of 28 views for both display types were generated (ISO/IEC JTC1/SC29/WG11, 2011a). The results were subjectively assessed with the target to obtain noticeable differences between the four rate points, as well as in comparison with the uncoded synthesized data. If different rate points could not be distinguished by subjective viewing, the associated bit rates for the sequence were adapted. Finally, the bit rates for each sequence and rate point were fixed to the values shown in Table 8.1.

For the AVC-based category, 12 proposals were submitted to the CfP and subjectively tested (ISO/IEC JTC1/SC29/WG11, 2011b). Proponents had to encode the two-view and three-view MVD format with video and depth data at the four different bit rates for each sequence according to Table 8.1 and generate the required stereo pairs and the set of 28 views for both display types. Since all proposals and the coded anchors had the same bit rate, an MOS-based subjective quality comparison at each bit rate could be performed. A summary of the achieved quality improvement of the best-performing proposal in comparison with the anchor coding is shown in Figure 8.14.

Here, the viewing results of the two-view scenario on a stereoscopic display with polarized glasses (SD) as well as results of the three-view scenario on an auto-stereoscopic 28-view display (ASD) are given for each rate point. Note, that each point in Figure 8.14 represents the average MOS value over all eight sequences at that rate point and display type. The best 3DV AVC-based proposal outperforms the anchor coding results; especially at the low rate point R1. In addition, the best proposal achieves a similar or better MOS value at the next lower bit rate in comparison with the anchors; for example, the MOS for the best proposal at R1 is better than the anchor coding MOS at R2 for both display types. Comparing the associated individual bit rates in Table 8.1 of each sequence, an overall bit-rate saving of 30% could be obtained in comparison with the anchor coding.

Table 8.1 AVC-based rate points R1–R4 for two-view and three-view test scenarios for AVC-based 3DV technology

Test sequence (resolution, frame rate)	Two-view test scenario				Three-view test scenario			
	AVC bit rates (kbps)				AVC bit rates (kbps)			
	R1	R2	R3	R4	R1	R2	R3	R4
S01: Poznan_Hall2 (1920 × 1088p, 25 fps)	500	700	1000	1500	750	900	1300	2300
S02: Poznan_Street (1920 × 1088p, 25 fps)	500	700	1000	1250	750	1100	1800	4000
S03: Undo_Dancer (1920 × 1088p, 25 fps)	1000	1300	1700	2200	1380	1750	2300	2900
S04: GT_Fly (1920 × 1088p, 25 fps)	1200	1700	2100	2900	2000	2380	2900	4000
S05: Kendo (1024 × 768p, 30 fps)	400	500	800	1300	800	1000	1300	1900
S06: Balloons (1024 × 768p, 30 fps)	320	430	600	940	500	600	800	1250
S07: Lovebird1 (1024 × 768p, 30 fps)	375	500	750	1250	500	800	1250	2000
S08: Newspaper (1024 × 768p, 30 fps)	400	525	800	1300	500	700	1000	1350

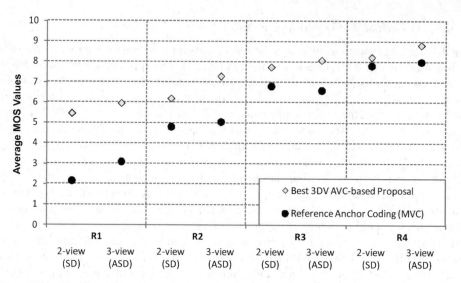

Figure 8.14 The 3D-AVC-based subjective results. Averaged MOS scores over all eight test sequences at four different bit rates R1–R4 according to Table 8.1: Evaluation of two-view scenario on stereoscopic display (SD), evaluation of three-view scenario on auto-stereoscopic 28-view display (ASD)

Comparing the two- and three-view scenarios in Figure 8.14, the subjective MOS results for each rate point are in the same quality range and thus consistent. Since two very different 3D displays were used for the viewing, the results also prove the suitability of the proposed codec for a targeted display-independent reconstruction quality based on view synthesis from a decoded generic 3DV format.

8.5.3 HEVC-Based 3DV Coding Results

For 3DV coding technology, that uses the HEVC standard for coding the base view video, four different rate points were also specified for the two-view and three-view scenarios and for each test sequence, using the same iterative procedure as described in Section 8.5.2. For the HEVC-based/unconstrained category, the anchors were produced by coding each video and depth component separately with the HEVC test Model HM2.0. Then, stereo pairs and the set of 28 views for both display types were generated similarly to the AVC-based category (ISO/IEC JTC1/SC29/WG11, 2011a). In Table 8.2, the final bit rates for the compressed MVD data at rate points R1–R4 for all sequences are shown. Note, that the HEVC-based bit rates in Table 8.2 are considerably lower than the AVC-based bit rates in Table 8.1 due to the higher compression efficiency of HEVC.

For the HEVC-based/unconstrained category, 11 proposals were submitted to the CfP and subjectively tested (ISO/IEC JTC1/SC29/WG11, 2011b). All proposals used HEVC for coding the base view. Proponents had to encode the two-view and three-view MVD format with video and depth data at the four different bit rates for each sequence according to Table 8.2 and generate the required stereo pairs and the set of 28 views for both display types respectively. Similar to the AVC-based category described in Section 8.5.2, all proposals and the coded anchors had the same bit rate, such that an MOS-based subjective quality comparison at each

Table 8.2 HEVC-based rate points R1–R4 for two-view and three-view test scenarios for HEVC-based 3DV technology

Test sequence (resolution, frame rate)	Two-view test scenario				Three-view test scenario			
	HEVC bit rates (kbps)				HEVC bit rates (kbps)			
	R1	R2	R3	R4	R1	R2	R3	R4
S01: Poznan_Hall2 (1920 × 1088p, 25 fps)	140	210	320	520	210	310	480	770
S02: Poznan_Street (1920 × 1088p@25fps)	280	480	800	1310	410	710	1180	1950
S03: Undo_Dancer (1920 × 1088p, 25 fps)	290	430	710	1000	430	780	1200	2010
S04: GT_Fly (1920 × 1088p, 25 fps)	230	400	730	1100	340	600	1080	1600
S05: Kendo (1024 × 768p, 30 fps)	230	360	480	690	280	430	670	1040
S06: Balloons (1024 × 768p, 30 fps)	250	350	520	800	300	480	770	1200
S07: Lovebird1 (1024 × 768p, 30 fps)	220	300	480	830	260	420	730	1270
S08: Newspaper (1024 × 768p, 30 fps)	230	360	480	720	340	450	680	900

bit rate could be performed as well. A summary of the quality improvement achieved of the best-performing proposal in comparison with the anchor coding is shown in Figure 8.15.

Again, the viewing results of the two-view scenario on a stereoscopic display with polarized glasses (SD) and the results of the three-view scenario on an auto-stereoscopic 28-view display (ASD) are given for each rate point as the average MOS value over all eight sequences. The best 3D-HEVC-based proposal significantly outperforms the anchor coding in both settings, such that, for example, a similar subjective quality of 5 for the anchor

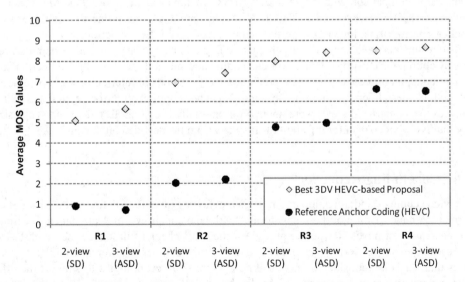

Figure 8.15 The 3D-HEVC-based subjective results. Averaged MOS scores over all eight test sequences at four different bit rates R1–R4 according to Table 8.2: evaluation of two-view scenario on stereoscopic display (SD), evaluation of three-view scenario on auto-stereoscopic 28-view display (ASD)

coding at R3 is already achieved by the best proposal at R1. Comparing the associated individual bit rates of each sequence at R3 and R1, as given in Table 8.2, average bit-rate savings of 57% and 61% for the two- and three-view scenario respectively are achieved. These are significantly higher than the bit-rate savings for the 3D-AVC-based technology for two reasons. First, the HEVC anchors were produced with single-view coding and thus no inter-view prediction was applied, in contrast to the MVC-based anchor coding. Second, a smaller random access period with a group of pictures of eight (GOP8) was used for the HEVC anchors, while the 3D-HEVC-based proposals used larger periods of GOP12 and GOP15. Accordingly, the overall bit-rate savings in the HEVC-based category also include the savings for the different random access periods as well as for the inter-view prediction. A more detailed analysis of individual results by Schwarz *et al.* (2012) showed bit-rate savings of 38% and 50% respectively for the two- and three-view scenarios in comparison with HEVC simulcast with equal random access periods. Furthermore, bit-rate savings of 20% and 23% respectively for the two- and three-view scenarios in comparison with multiview HEVC were found.

Comparing the two-view and three-view scenarios in Figure 8.15, the subjective MOS results for each rate point are in the same quality range. Thus, the 3D-HEVC-based coding technology also provides consistent results across very different 3D displays and, thus, a display-independent high-quality view synthesis from a generic 3DV format.

8.5.4   General Observations

The subjective evaluation of new 3DV coding technology showed significant improvements over the anchor coding methods in both AVC-based and HEVC-based categories. In particular, a bit-rate saving of 30% at the same subjective quality could be achieved for the 3D-AVC-based technology. For 3D-HEVC coding, bit-rate savings of more than 57% were achieved in comparison with the simulcast HEVC-anchors and still more than 20% versus a multiview HEVC-based anchor. This latter anchor has the same properties as the MVC-anchor in terms of random access period and usage of inter-view prediction.

The results obtained showed that a higher coding efficiency was achieved in both categories by optimizing existing coding tools and adding new methods, as described in Section 8.4. In particular, an improved inter-view prediction, new methods of inter-component parameter prediction, special depth coding modes, and an encoder optimization for depth data coding towards the synthesized views were applied for optimally encoding 3DV data and synthesizing multiview video data for different 3D displays from the decoded bit stream.

8.6   Concluding Remarks

The video coding standardization committees are moving quickly to define a new set of 3DV formats. A major feature of these new formats will be the inclusion of depth data to facilitate the generation of novel viewpoints of a 3D scene. This chapter introduced the depth-based representation and rendering techniques and described a number of different coding architectures that are being considered. The leading compression technologies to support the efficient representation and rendering of the depth-based 3D formats have also been described, and the results from a large-scale experimental evaluation have been summarized.

While the current 3D services are based only on stereoscopic video, the market is expected to evolve and auto-stereoscopic displays will soon mature and become cost effective. The depth-based formats and associated compression techniques described in this chapter will support such future displays and services. It is believed that a low-complexity process to generate multiple views at the receiver with high rendering quality will be an essential feature for auto-stereoscopic displays and related equipment.

Although the standardization of these formats is well underway, there are still a number of research challenges to overcome. For instance, the accurate estimation or acquisition of depth information is not considered to be fully mature at this stage, especially for real-time applications and outdoor scenes in which the depth range is large, and illumination (or other factors) could have a significant impact on the quality of the depth image. Also, while current standards will be based on the MVD format and somewhat conventional coding architectures, further research on alternative representations continues to be of interest among researchers working in this area. For instance, there have been investigations on using the transition between views as a representation of the 3D scene by Kim *et al.* (2010b), as well as work on dictionary-based representations of a 3D scene by Palaz *et al.* (2011). With these new representation formats and evaluation frameworks that consider the quality of the rendered view, there is also scope for further work on modeling and optimizing the quality of the system, as discussed, for example, by Kim *et al.* (2010a) and Tech *et al.* (2012a, 2012b).

References

Atzpadin, N., Kauff, P., and Schreer, O. (2004) Stereo analysis by hybrid recursive matching for real-time immersive video conferencing. *IEEE Trans. Circ. Syst. Vid*, **14** (3), 321–334.

Beni, P. A., Rusanovskyy, D., and Hannuksela, M. M. (2012) Non-linear depth map resampling for 3DV-ATM coding. ISO/IEC JTC1/SC29/WG11, Doc. m23721, February, San Jose, CA, USA.

Benzie, P., Watson, J., Surman, P. *et al.* (2007) A survey of 3DTV displays: techniques and technologies. *IEEE Trans. Circ. Syst. Vid.*, **17** (11), 1647–1658.

Bleyer, M. and Gelautz, M. (2005) A layered stereo matching algorithm using image segmentation and global visibility constraints. *ISPRS J. Photogramm.*, **59** (3), 128–150.

Blu-ray Disc Association (2009) Blu-ray Disc Association announces final 3D specification, http://www.blu-raydisc.com/assets/Downloadablefile/BDA-3D-Specifcation-Press-Release---Proposed-Final12-14-08version-clean-16840.pdf.

Cigla, C., Zabulis, X., and Alatan, A. A. (2007) Region-based dense depth extraction from multiview video, in *IEEE International Conference on Image Processing, 2007. ICIP 2007*, IEEE, pp. 213–216.

Digital Video Broadcast (2012) 3D moves forward: DVB Steering Board approves Phase 2a of 3DTV specification, http://www.dvb.org/news_events/press_releases/press_releases/DVB_pr227-Steering-Board-Approves-Phase-2a-3D-Specification.pdf.

Faugeras, O. (1993) *Three-Dimensional Computer Vision: A Geometric Viewpoint*, MIT Press, Cambridge, MA.

Felzenszwalb, P. F. and Huttenlocher, D. P. (2006) Efficient belief propagation for early vision. *Int. J. Comput. Vision*, **70** (1), 41–54.

Hartley, R. and Zisserman, A. (2000) *Multiple View Geometry in Computer Vision*, Cambrigde Universitity Press.

ISO/IEC JTC1/SC29/WG11 (1997) The virtual reality modeling language, DIS 14772-1, April.

ISO/IEC JTC1/SC29/WG11 (2011a) Call for proposals on 3D video coding technology, Doc. N12036, Geneva, Switzerland, March.

ISO/IEC JTC1/SC29/WG11 (2011b) Report of subjective test results from the call for proposals on 3D video coding, Doc. N12347, Geneva, Switzerland, November 2011.

ITU-T and ISO/IEC (2010) Advanced video coding for generic audiovisual services, ITU-T Rec. H.264 and ISO/IEC 14496-10 (MPEG-4 AVC), Version 10, March.

Kauff, P., Atzpadin, N., Fehn, C. *et al.* (2007) Depth map creation and image based rendering for advanced 3DTV services providing interoperability and scalability. *Signal Process. Image Commun.*, **22** (2), 217–234.

Kim, W.-S., Ortega, A., Lai, P. *et al.* (2010a) Depth map coding with distortion estimation of rendered view, in *Visual Information Processing and Communication* (eds A. Said and O. G. Guleryuz), Proceedings of the SPIE, Vol. **7543**, SPIE, Bellingham, WA, pp. 75430B–75430B-10.

Kim, W.-S., Ortega, A., Lee, J., and Wey, H. (2010b) 3-D video coding using depth transition data, in *Picture Coding Symposium (PCS), 2010*, IEEE, pp. 178–181.

Kolmogorov, V. (2006) Convergent tree-reweighted message passing for energy minimization. *IEEE Trans. Pattern. Anal.*, **28** (10), 1568–1583.

Kolmogorov, V. and Zabih, R. (2002) Multi-camera scene reconstruction via graph cuts, in *ECCV '02 Proceedings of the 7th European Conference on Computer Vision – Part III*, Springer-Verlag, London, pp. 82–96.

Konrad, J. and Halle, M. (2007) 3-D displays and signal processing. *IEEE Signal Proc. Mag.*, **24** (6), 97–111.

Lee, E.-K., Jung, Y.-K., and Ho, Y.-S. (2010) Three-dimensional video generation using foreground separation and disocclusion detection, in *3DTV Conference: The True Vision – Capture, Transmission and Display of 3D Video (3DTV-CON,) 2010*, IEEE.

Lee, S., Lee, S., Wey, H., and Lee, J. (2012) 3D-AVC-CE6 related results on Samsung's in-loop depth resampling, ISO/IEC JTC1/SC29/WG11, Doc. m23661, February, San Jose, CA, USA.

Lee, S.-B. and Ho, Y.-S. (2010) View consistent multiview depth estimation for three-dimensional video generation, in *3DTV Conference: The True Vision – Capture, Transmission and Display of 3D Video (3DTV-CON), 2010*, IEEE.

Martinian, E., Behrens, A., Xin, J., and Vetro, A. (2006) View synthesis for multiview video compression. Proceedings of the Picture Coding Symposium (PCS'2006), April, Beijing, China.

Merkle, P., Smolic, A., Müller, K., and Wiegand, T. (2007) Efficient prediction structures for multiview video coding. *IEEE Trans. Circ. Syst. Vid.*, **17** (11), 1461–1473.

Merkle, P., Morvan, Y., Smolic, A. *et al.* (2009) The effects of multiview depth video compression on multiview rendering. *Signal Process. Image Commun.*, **24** (1–2), 73–88.

Merkle, P., Bartnik, C., Müller, K. *et al.* (2012) 3D video: depth coding based on inter-component prediction of block partitions, in *Picture Coding Symposium (PCS), 2012*, IEEE, pp. 149–152.

Min, D., Yea, S., and Vetro, A. (2010) temporally consistent stereo matching using coherence function, in *3DTV Conference: The True Vision – Capture, Transmission and Display of 3D Video (3DTV-CON,) 2010*, IEEE.

Müller, K., Smolic, A., Dix, K. *et al.* (2008) View synthesis for advanced 3D video systems. *EURASIP J. Image Vid. Process.*, **2008**, 1–11, article ID 438148, doi: 10.1155/2008/438148.

Müller, K., Merkle, P., and Wiegand, T. (2011) 3D video representation using depth maps. *Proc. IEEE*, **99** (4), 643–656.

Oh, K.-J., Yea, S., Vetro, A., and Ho, Y.-S. (2009) Depth reconstruction filter and down/up sampling for depth coding in 3-D video, *IEEE Signal Proc. Lett.*, **16** (9), 747–750.

Ozaktas, H. M. and Onural, L. (eds) (2007) *Three-Dimensional Television: Capture, Transmission, Display*, Springer, Heidelberg.

Palaz, D., Tosic, I., and Frossard, P. (2011) Sparse stereo image coding with learned dictionaries, in *2011 IEEE International Conference on Image Processing (ICIP)*, IEEE, pp. 133–136.

Redert, A., de Beeck, M.O., Fehn, C. *et al.* (2002) ATTEST: advanced three-dimensional television system techniques, in *Proceedings. First International Symposium on 3D Data Processing Visualization and Transmission, 2002*, IEEE, pp. 313–319.

Rusert, T. (2012) 3D-CE3 summary report: in-loop depth resampling, ISO/IEC JTC1/SC29/WG11, Doc. m24823, May, Geneva, Switzerland.

Scharstein, D. and Szeliski, R. (2002) A taxonomy and evaluation of dense two-frame stereo correspondence algorithms. *Int. J. Comput. Vision*, **47** (1), 7–42.

Schwarz, H., Bartnik, C., Bosse, S. *et al.* (2012) 3D video coding using advanced prediction, depth modeling, and encoder control methods, in *Picture Coding Symposium (PCS) 2012*, IEEE.

Szeliski, R., Zabih, R., Scharstein, D. *et al.* (2006) A comparative study of energy minimization methods for Markov random fields, in *Computer Vision – ECCV 2006: 9th European Conference on Computer Vision* (eds A. Leonardis, H. Bischof, and A. Pinz), Lecture Notes in Computer Science, Vol. **3952**, Springer, Berlin, pp. 16–29.

Tanimoto, M., Fujii, T., and Suzuki, K. (2008) Improvement of depth map estimation and view synthesis, ISO/IEC JTC1/SC29/WG11, Doc. m15090, January, Antalya, Turkey.

Tech, G., Schwarz, H., Müller, K., and Wiegand, T. (2012a) 3D video coding using the synthesized view distortion change, in *Picture Coding Symposium (PCS) 2012*, IEEE.

Tech, G., Schwarz, H., Müller, K., and Wiegand, T. (2012b) Synthesized view distortion based 3D video coding for extrapolation and interpolation of views, in *2012 IEEE International Conference on Multimedia and Expo (ICME)*, IEEE Computer Society Press, Washington, DC, pp. 634–639.

Vetro, A., Tourapis, A., Müller, K., and Chen, T. (2011a) 3D-TV content storage and transmission. *IEEE Trans. Broadcast.*, **57** (2), 384–394.

Vetro, A., Wiegand, T., and Sullivan, G. J. (2011b) Overview of the stereo and multiview video coding extensions of the H.264/AVC standard. *Proc. IEEE*, **99** (4), 626–642.

Wiegand, T., Sullivan, G. J., Bjøntegaard, G., and Luthra, A. (2003) Overview of the H.264/AVC video coding standard. *IEEE Trans. Circ. Syst. Vid.*, **13** (7), 560–576.

Winken, M., Schwarz, H., and Wiegand, T. (2012) Motion vector inheritance for high efficiency 3D video plus depth coding, in *Picture Coding Symposium (PCS), 2012*, IEEE, pp. 53–56.

Würmlin, S., Lamboray, E., and Gross, M. (2004) 3D video fragments: dynamic point samples for real-time free-viewpoint video. *Comput. Graph.*, **28** (1), 3–14.

Yea, S. and Vetro, A. (2009) View synthesis prediction for multiview video coding. *Signal Process. Image Commun.*, **24** (1–2), 89–100.

9

Coding for Interactive Navigation in High-Dimensional Media Data

Ngai-Man Cheung[1] and Gene Cheung[2]

[1]*Information Systems Technology and Design Pillar, Singapore University of Technology and Design, Singapore*
[2]*Digital Content and Media Sciences Research Division, National Institute of Informatics, Japan*

9.1 Introduction

An essential aspect of an immersive experience is the ability for an observer to navigate freely through the 3D visual scenery within a remote/virtual environment as if they are physically there. The observer may interact manually via a traditional keypad (Lou *et al.*, 2005) or more naturally via a head-mounted tracking device (Kurutepe *et al.*, 2007), but in either case an immersive communication system must, in response, quickly produce media data that correspond to the observer's input; for example, if the observer tilts their head to the right, the view corresponding to the right-shifted view must be decoded and rendered for viewing in real time. If the media data representing the environment already resides at the observer's terminal prior to media interaction, then the right subset of media corresponding to the observer's input can simply be fetched from memory, decoded, and displayed. If the media data reside remotely in a server, however, then sending the entire dataset over the network before an observer starts interacting with it can be prohibitively costly in bandwidth or delay; for example, the size of a set of light-field data (Levoy and Hanrahan, 1996) – a densely sampled two-dimensional array of images taken by a large array of cameras (Wilburn *et al.*, 2002) where a desired new view is synthesized using neighboring captured views via image-based rendering (IBR) (Shum *et al.*, 2003) – has been shown to be on the order of tens of gigabytes (Levoy and Pulli, 2000), while multiview video datasets have been captured using up to 100 time-synchronized cameras (Fujii *et al.*, 2006).

Hence, a more practical communication paradigm is one where the server continuously and reactively sends appropriate media data in response to an observer's periodic requests for data subset – we call this interactive media streaming (IMS). This is in sharp contrast to noninteractive media streaming scenarios, such as terrestrial digital TV broadcast (Digital Video Broadcasting, 2012), where all available channels that are live at stream time are delivered from server to client before a client interacts with the received channels (e.g., switching TV channels, superimposing picture-in-picture with two TV channels). IMS can potentially reduce bandwidth utilization since only the media subsets corresponding to the observer's requests are transmitted. However, efficiently coding the sequence of requested media subsets prior to the streaming session – in a store-and-playback scenario – becomes a substantial technical challenge. More specifically, standard coding tools such as H.264/AVC (Wiegand *et al.*, 2003) exploit correlation among neighboring frames using closed-loop motion compensation, where a decoded frame is used as a predictor for a frame to be encoded. This creates coding dependencies among media subsets and induces a correspond- ing encoding/decoding order (frames used as predictors need to be encoded/decoded before frames they predict). If this encoding/decoding order differs from the media navigation order chosen by an observer during stream time, then a penalty in transmission rate (e.g., sending pre-encoded subsets that are not requested but are required for the decoding of desired sub- sets) will occur. This is an example of the inherent tension between media interactivity and coding efficiency; that is, providing maximum "navigation" flexibility can come at the cost of lower coding efficiency.

Over the past few years, researchers have devised novel coding structures and techniques to achieve different tradeoffs between media interactivity and coding efficiency. While, as will be shown, many different IMS applications have been studied, the solutions proposed to optimize these tradeoffs can in fact be grouped into just a few basic strategies. In this chapter, before providing a detailed overview of various proposals in the literature, we first provide a summary of these methods and how they relate to each other. This allows us to provide a unified view of these seemingly disparate problems and can serve as the basis to apply simi- lar optimization techniques to a variety of IMS applications.

The outline of the chapter is as follows. In Section 9.2 we describe at a high level the challenges in source coding for interactive navigation in high-dimensional media space and outline general approaches one can employ to address this problem. In Section 9.3 we pro- vide an overview of different IMS applications and describe their unique approaches to the IMS source coding problem. We then narrow our focus to a single IMS application, inter- active multiview video streaming (IMVS), and discuss different variations of the problem under different application scenarios in Section 9.4. Finally, we provide concluding remarks in Section 9.5.

9.2 Challenges and Approaches of Interactive Media Streaming

9.2.1 Challenges: Coding Efficiency and Navigation Flexibility

As discussed, the design of an IMS system needs to address the conflicting issues of coding efficiency and navigation flexibility. This can be illustrated by an example of streaming of multiview video. Figure 9.1 shows a multiview video dataset, which consists of multiple video sequences capturing different views of the same scene (Flierl and Girod, 2007). An interactive multiview application allows the user to switch between different views during playback. For instance, the system could support the user to switch from view 1 to view 2 at time instant 3 as shown in the figure.

Now we consider several different approaches to encode the multiview dataset for inter- active streaming. Suppose the multiview video is encoded by intra coding – that is, each

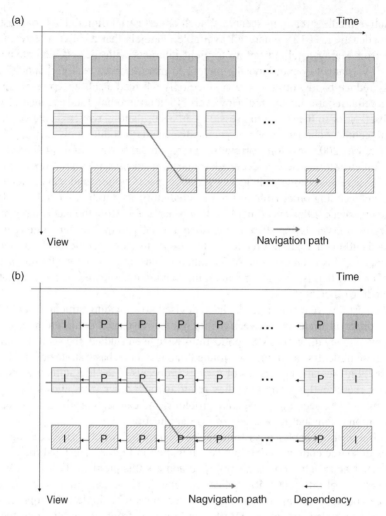

Figure 9.1 (a) Interactive streaming of multiview video. The user can navigate from one view to another view during playback. (b) Interactive multiview video streaming with simulcast coding structure.

frame is encoded independently from each other – and there is no coding dependency between different frames. In this case, the user can choose to navigate the video in any playback path, and the server simply sends the information corresponding to the frames in the playback path. While intra coding is flexible for interactive streaming, a drawback of this approach is that inter-frame redundancy is not exploited and, hence, each encoded frame would consume more bits. Therefore, intra coding provides maximum playback flexibility but suffers penalty in coding efficiency.

Alternatively, the multiview video dataset can be encoded by a simulcast approach. Each view can be encoded using the so-called IPPP coding structure, with each I-frame followed by a certain number of consecutive P-frames (Figure 9.1). The frame interval between successive I-frames corresponds to the size of the group of pictures (GOP). By exploring inter-frame redundancy, P-frames would consume smaller numbers of bits than their intra-coded counterparts, so this approach would require less server storage than the previously discussed intra coding approach. However, the switching from one view to another view can require the transmission of additional bits to reconstruct the reference frames that are needed for correct decoding. For instance, switching from view 1 to view 2 at time instant 3 requires transmission of additional information to reconstruct the frames at the time instants 0, 1, and 2 at view 2. Thus, while this simulcast approach can reduce the frame storage requirement at the server, it would require additional information to be transmitted to the clients and decoded at the client devices during view switching. Therefore, navigation flexibility comes at the cost of additional bandwidth requirement and decoding complexity.

To alleviate the bandwidth and decoding overhead, the system may opt to reduce the navigation flexibility. For instance, in the above simulcast example, the system may allow view switching only at I-frames. Doing so, coding efficiency can be maintained at the expense of reduction in playback flexibility. Alternatively, additional switching positions can be introduced by encoding certain frames with special coding techniques. For instance, some frames within a GOP can be encoded instead using H.264 SP/SI-frames (Karczewicz and Kurceren, 2003), using previous frames from the same view and from adjacent views as references. H.264 SP/SI-frames have the particular property that they can be identically reconstructed using different predictors. Thus, reference frames from the same view or from adjacent views shall lead to identical reconstruction of the current frame. This prevents occurrence and propagation of drifting errors during view switching. H.264 SP/SI-frames tend to consume more bits then P-frames but are smaller than I-frames, as they exploit inter-frame redundancy during encoding (Karczewicz and Kurceren, 2003). Thus, additional switching positions can be added to improve navigation flexibility at the expense of some penalty in server storage cost and transmission overhead.

9.2.2 Approaches to Interactive Media Streaming

Following the previous discussion, in this section we review different approaches to general interactive streaming applications.

First, media data can be encoded with an intra coding approach, where the system encodes each frame independently. When the sender receives a request from the observer corresponding to a specific media subset, the sender does not need to know in what state the decoder is in (i.e., what media data has already been transmitted and decoded at the observer). The sender can simply send the media subset that directly corresponds to the observer's request. One example is the above-mentioned intra coding of multiview video. Though simple and flexible, there is no exploitation of inter-frame correlation in intra coding, impairing the coding gain.

Alternatively, the sender can keep track of the decoder's state (what media data have been transmitted) and sends different subsets of media data according to both the observer's request and the decoder's state. The simple sub-category here is the class of techniques where the media is encoded such that there is only one unique way of decoding the requested media data subset. In this case, the sender simply sends the subset of the requested media that are missing from the receiver's cache, so that the decoder can perform the required unique decoding. An example is the above-mentioned simulcast coding of multiview video with IPPP coding of each view. As discussed, if the dependency structure has not been carefully designed, this approach may require sending of significant amount of overhead information in order to reconstruct the media data. For instance, in the simulcast example, the system has to send an overhead of three frames in order to navigate to the successive frame in the adjacent view. To alleviate transmission/decoding overhead, previous work has judiciously designed various coding structures for different applications: JPEG2000-based scheme for region-of-interest (RoI) image browsing (Taubman and Rosenbaum, 2003), rerouting for light-field streaming (Bauermann and Steinbach, 2008a,b), and RoI video streaming (Mavlankar and Girod, 2009, 2011).

In the next class of structures, there is more than one way of decoding each requested media subset, but producing only one decoded version. An example is the above-mentioned use of H.264 SP-frames for multiview video, where the SP-frames at the switching position can be decoded from the same-view reference or cross-view reference. Other examples are distributed source coding (DSC) for reversible video playback (Cheung *et al.*, 2006), SP-frames for light-field streaming (Ramanathan and Girod, 2004), DSC-based techniques for light-field streaming (Jagmohan *et al.*, 2003), and DSC-based techniques for IMVS (N.-M. Cheung *et al.*, 2009). Note that DSC can also achieve identical reconstruction from different predictors similar to H.264 SP-frames, although based on a remarkably different mechanism (Jagmohan *et al.*, 2003; Aaron *et al.*, 2004; Cheung and Ortega, 2008).

Finally, the last class of structures result in multiple decoded versions for a given observer request. Here, the multiple decoded versions lead to a multiplicative increase in the number of decoder states, which the server must keep track of. JPEG2000 + CR (conditional replenishment), JPEG2000 + MC (motion compensation) for video browsing (Devaux *et al.*, 2007; Naman and Taubman, 2007), and redundant P-frames (G. Cheung *et al.*, 2009a,b) belong to this class. Typically, some methods to reduce the number of possible decoder states, such as novel DSC implementations, must be used before the exponential growth of decoder states becomes intractable.

9.3   Example Solutions

We first provide an overview of various proposed applications in the literature for IMS (refer to Table 9.1). Essentially, proposed IMS applications can provide users with the ability to navigate media content in a flexible manner along one or more of several possible dimensions, namely, time, viewpoint, and resolution. First, users can navigate media content across time. This can be playback of media pieces in adjacent time locations (e.g., forward or backward playback of video) or of media pieces in nonconsecutive time locations (e.g., video playback of every Nth frame). Second, users can navigate media content across viewpoints. Adjacent viewpoints will likely have overlapping spatial regions whose correlation can be exploited for coding gain. Finally, users can navigate media content across resolution (zoom). This means a zoomed media piece represents a subset of spatial region of the pre-zoomed piece, but at a higher resolution. Table 9.1 shows how different proposed IMS applications provide different navigation capabilities.

We first discuss three interactive image applications with no notion of time: RoI image browsing, interactive light-field streaming, and volumetric image random access. We then discuss two single-view, single-resolution video applications with notion of time: reversible video playback and video browsing. Finally, we discuss a single-view, multiple-resolution video application (RoI video streaming) and a multi-view, single-resolution video application (IMVS).

9.3.1   Region-of-Interest (RoI) Image Browsing

RoI image browsing is an application where an observer can successively and interactively view (at different resolutions and qualities) spatial regions in a possibly very large image; for

Table 9.1 Categories of IMS applications.[a]

Application	Time	View point	Zoom	Structuring technique
RoI image browsing	0	0	1	JPEG2000 (Taubman and Rosenbaum, 2003)
Light-field streaming	0	1	0	Retracing (Bauermann and Steinbach, 2008a,b), H.264/AVC SP-frames (Ramanathan and Girod, 2004), intra DSC (Aaron et al., 2004), DSC + MC + coset (Jagmohan et al., 2003)
Volumetric image random access	0	1	0	Redundant tile (Fan and Ortega, 2009)
Video browsing	1	0	0	JPEG2000 + CR (Devaux et al., 2007), JPEG2000 + MC (Naman and Taubman, 2007)
Reversible video playback	1	0	0	DSC + MC (Cheung et al., 2006)
RoI video streaming	1	0	1	Multi-res MC (Mavlankar et al., 2007; Mavlankar and Girod, 2009, 2011)
Multiview video streaming	1	1	0	DSC + MC (N.-M Cheung et al., 2009d), retracing (G. Cheung et al., 2009a), redundant P-frames (G. Cheung et al., 2009a,b)

[a]RoI: region of interest; CR: conditional replenishment, MC: motion compensation, DSC: distributed source coding

example, a geographical map. The technical challenge for RoI image browsing is how to pre-encode the image to support such observer interactivity, while minimizing transmission rates of the requested RoIs and storage of the entire image. Taubman and Rosenbaum (2003) proposed the combined use of JPEG2000 image coding standard, for its fine-grained spatial, resolution, and quality scalability, and the JPEG2000 interactive protocol (JPIP; JPEG, 2007), for its flexibility of image random access. More specifically, given the set of subband coefficients of a discrete wavelet transform already residing at the observer's cache (JPIP supports both stateless and stateful modes of operation), the sender sends to the observer only the "missing" coefficients, those corresponding to the requested spatial location and scale of the requested RoI that are not yet available at the receiver. These coefficients are sent in an incremental rate-distortion optimal fashion, grouped in quality layers called packets (Taubman and Rosenbaum, 2003). The image can be displayed continuously at the observer as more packets are received with gradually improving quality. Given the scalability of the JPEG2000 representation, RoI interactivity can be achieved without multiple encodings of the same media data (which will be the approach selected by schemes in other applications to be discussed later).

9.3.2 Light-Field Streaming

Bauermann and Steinbach, 2008a,b considered interactive streaming of compressed image-based scene representations. Remote interactive viewing of photo-realistic three-dimensional scenes is a compelling technology with applications such as gaming and virtual reality. Image-based scene representations such as light fields provide superior quality rendering results. Moreover, they require less computation in rendering compared with the conventional geometric modeling process. However, this IBR usually requires tens of gigabytes of reference image data. Efficient compression is therefore necessary, and only relevant subsets of reference images should be streamed to the client in remote-viewing applications.

Interactive walkthrough systems using image-based scenes require random access to arbitrary images to compose arbitrary virtual views. For a compressed IBR dataset with inter-image prediction, due to coding dependency, the interactive systems generally need to transmit and decompress more pixels than actually needed for rendering. In particular, image data that the requested pixels rely on has to be transmitted and decoded as well as the requested pixels themselves. Bauermann and Steinbach (2008a,b) proposed a detailed analysis where (i) storage rate R, (ii) distortion D, (iii) transmission data rate T, and (iv) decoding complexity C are jointly considered and optimized. Notice that the transmission data rate T can be significantly larger than the storage rate R, as coding dependencies might require more information to be sent in order to decode the requested data. For the same reason, the mean number of pixels that have to be decoded to render a pixel at the client (i.e., C) can be significantly larger than one.

9.3.3   Volumetric Image Random Access

Fan and Ortega (2009) addressed the problem of random access of low-dimensional data from a high-dimensional dataset in particular for volumetric images. Volume visualization is important in many fields, such as biomedical imaging and Earth sciences. In specific visualization tasks, the researchers often would request only relevant low-dimensional data from a very large high-dimensional dataset over client–server networking systems. For instance, in some medical imaging applications, arbitrary oblique planes of the medical image volumes may be requested and extracted at the practitioner's requests, and rendered and displayed on the clients' displays. In this case, because tiles are the basic unit for compression, complete tiles have to be retrieved and transmitted. Standard tiling, however, can be inefficient in this scenario because for each retrieved cubic tile there could be only a small number of useful voxels that are lying in desired two-dimensional plane. This leads to inefficient bandwidth usage. Fan and Ortega (2009) proposed a multiple redundant tiling scheme for volumetric data, with each tiling corresponding to a different orientation. This reduces transmission of overhead data at the expense of higher server storage requirement.

9.3.4   Video Browsing

Exploiting the scalable nature of JPEG2000, proposals (Devaux *et al.*, 2007; Naman and Taubman, 2011) have also been made to use JPEG2000 for video browsing, where the streaming video can be randomly accessed with complete accessibility at the frame level: random access any frame in sequence, forward/backward playback in time, playback at K speed by decoding only every K frames, and so on. Though at encoding time each frame is encoded independently, inter-frame redundancy can, nevertheless, be exploited. Devaux *et al.* (2007) proposed to use conditional replenishment (CR), where the coefficients of a code-block of a desired frame were sent or replenished only if the corresponding code-block of the previous frame already at the decoder did not provide a good enough approximation. Along similar lines, assuming a motion model, Naman and Taubman (2011) performed motion compensation (MC) at the server, so that the transmitted motion information could be used in combination of code-blocks of previous frames to approximately reconstruct code-blocks of requested frames. The server would send new code-blocks only if the motion-compensated approximation was not good enough.

9.3.5 Reversible Video Playback

There has been a fair amount of interest to develop techniques to enable user-friendly browsing of digital video. In particular, video cassette recording (VCR) functionalities such as forward-play, backward-play, stop, pause, fast-forward, fast-backward, and random access allow users to conveniently navigate the content and are compelling for many digital video applications. For instance, surveillance applications can utilize forward/backward-play to inspect video footages in both forward and backward directions for unusual events.

Among these VCR functionalities, backward-play, or reversible playback of video, is the most difficult one to support for compressed digital video. Motion-compensated predictive coding (e.g., MPEG, H.26X) uses motion prediction and may use the reconstructed past frames as the references to encode the current frame. Consequently, the decoder needs to reconstruct the past frames first to decode the current frame. These coding dependencies pose challenges to reversible playback. To achieve reversible frame-by-frame playback, one approach would be to decode all the frames in the entire group of pictures (GOP), store all the frames in a large buffer memory, and display the decoded frames in reverse order. Clearly, this requires a significant amount of frame memory and may also incur some playback delay (Lin *et al.*, 2001; Fu *et al.*, 2006; Cheung and Ortega, 2008).

To address reversible playback, Wee and Vasudev (1999) proposed a scheme based on transcoding. The input I-P frames are converted into another I-P compressed bitstream with a reversed frame order. They proposed techniques to speed up the estimation of the reverse motion vectors using the forward motion vectors in the input I-P frames. Lin *et al.* (2001) proposed to add a reverse-encoded bitstream in the server. After encoding the frames in the forward direction, their system performs another offline encoding starting from the last frame in the reverse order. An optimization scheme has been developed to minimize the switching cost between the forward and backward bitstreams, focusing on the client decoder complexity or the network traffic. Fu *et al.* (2006) proposed to select macroblocks necessary for backward-play, manipulate them in the compressed domain, and send the processed macroblocks to the client. Motion information of the original bitstream is utilized to assist this processing. Cheung and Ortega (2008) proposed to encode the video using DSC to facilitate reversible playback. Unlike conventional motion-compensated predictive coding, DSC generates parity information to represent the current frame. The amount of parity is chosen so that the current frame is bidirectionally decodable; that is, it can be reconstructed using either the previous or the subsequent frame as side information, for forward- and backward-play, respectively.

9.3.6 Region-of-Interest (RoI) Video Streaming

Driven by advances in CCD and CMOS image sensor technologies, high-spatial-resolution digital videos are becoming widely available and more affordable. Ultra-high definition television (UHDTV) of spatial resolution 7680×4320 has been demonstrated in trade shows and soon will be available for consumers (Broadcast, 2008). Also, recent research on stitching views from multiple cameras to generate high-resolution contents has produced very encouraging results (Fehn *et al.*, 2006).

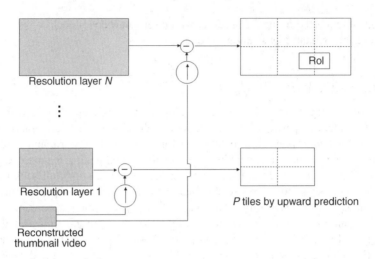

Figure 9.2 Video coding scheme for RoI streaming proposed by Mavlankar and Girod (2011).

Despite the increasing availability of high-resolution video, challenges in delivering high-resolution contents to the client are posed by the limited resolution of client displays and/or limited communication bandwidth (Mavlankar and Girod, 2011). For instance, the recent iPhone 4S has a 960×640 pixel display, and it is unlikely that there will be significant increase in resolution of some client devices given the constraint in form factors. If the user were made to watch a spatially downsampled version of the entire video scene, then the local high-resolution RoI may not be available.

To address this issue, Mavlankar and Girod (2011) proposed a video coding scheme and framework for interactive RoI streaming that provide spatial random access and pan/tilt/zoom functionality while streaming high-resolution video. Notice that the coding dependencies among successive frames in video compression make it difficult to provide spatial random access. In particular, the decoding of a block of pixels requires that other reference frame blocks used by the predictor have previously been decoded. These reference frame blocks might, however, lie outside the RoI, and might not have been transmitted or decoded earlier.

Figure 9.2 depicts the video coding scheme proposed by Mavlankar and Girod (2011). The thumbnail overview constitutes a base layer video and is coded with H.264/AVC using I, P, and B pictures. The reconstructed base layer video frames are upsampled by a suitable factor and used as a prediction signal for encoding video corresponding to the higher resolution layers. Each frame belonging to a higher resolution layer is partitioned into a grid of rectangular tiles and encoded using the upsampled base layer as predictor. Notice that in their scheme they employ only upward prediction from the upsampled base-layer thumbnail. As the system always transmits this low-resolution, low-bit-rate thumbnail to the client, their scheme enables efficient random access to an RoI at any spatial resolution.

For a given frame interval, the display of the client is rendered by transmitting the corresponding frame from the base layer and a few P tiles from exactly one higher resolution layer (Mavlankar and Girod, 2011). The system selects tiles from that resolution layer which corresponds closest to the user's current zoom factor. At the client's side, the corresponding RoI from this resolution layer is resampled to match the user's zoom factor. Only a few spatial resolution layers are required to be generated and stored at the server to support smooth zoom control. In their experiments, the spatial resolution layers stored at the server are dyadically spaced. Hence, the reconstructed thumbnail frame needs to be upsampled by powers of two horizontally and vertically to generate the corresponding prediction signals. Using the six-tap filter defined in H.264/AVC for luminance upsampling and a simple two-tap filter with equal coefficients for chrominance, thumbnail upsampling is repeated an appropriate number of times depending on the resolution layer.

Mavlankar and Girod (2011) also considered the minimization of average transmission bit rate. In their proposed scheme, the choice of tile size would affect the expected number of bits transmitted and/or decoded per pixel rendered at the client. On the one hand, a smaller tile size entails lower pixel overhead. The pixel overhead consists of pixels that have to be transmitted and/or decoded because of the coarse tile partition, but are not used to render the client's display. On the other hand, the smaller the tile size is the worse is the coding efficiency. This is because of increased amount of tile header information, lack of context continuation across tiles for context adaptive coding, and inability to exploit inter-pixel correlation across tiles. They formulated the problem and derived a framework to determine the optimal tile size.

A variation of Mavlankar and Girod's scheme that is standard compliant was later developed in the Stanford ClassX system – an online lecture streaming system using interactive RoI technology (Mavlankar *et al.*, 2010; Halawa *et al.*, 2011; Pang *et al.*, 2011). The scheme supports spatial random-access using a multiple-layer multiple-tile representation. Specifically, the video encoder (transcoder) creates multiple resolution layers from a high-resolution video input. Each layer is further subdivided into tiles. Each tile is encoded independently using H.264/AVC. This one-time encoding process generates a set of tiles at the server, with each tile corresponding to a certain resolution and spatial region. During the streaming session, relevant tiles can be served to different users depending on their individual RoIs. Thus, this coding scheme is easily scalable to large number of users. The system also generates a low-resolution thumbnail base layer showing the whole video scene. This thumbnail video is H.264/AVC encoded independently from any of the tile videos; thus, the scheme is standard compliant. When a user wants to zoom into the video, tiles in the higher resolution layers are retrieved to provide higher quality details. Compared with the conventional scheme where the original high-resolution video is directly encoded, this spatial random-access scheme allows rate and decoding complexity savings, as sending the entire high-resolution is avoided and only the tiles with the matching resolution and spatial region are transmitted. The overhead of storing multiple tiles at the server can be justified by the fact that storage cost is much cheaper than bandwidth cost.

Additional design considerations are needed to overcome the specific challenges in mobile interactive RoI video streaming (Pang *et al.*, 2011). For example, a nonoverlapping tiling scheme usually requires the client to decode and synchronize multiple tiles for rendering a particular RoI. Most mobile devices often do not have the computational capability to do so. Therefore, they employ an overlapping tiling scheme such that only one tile is needed to satisfy any RoIs. Manual pan/tilt/zoom controls are not always desirable for a viewer, especially when watching video content with high-motion and dynamic scenes as found in many sport videos. To minimize human interactions, the system can produce an automatic RoI view that only tracks the object of interest within a video using standard tracking algorithms. For example, in the ClassX mobile implementation, the encoder can generate a dedicated tile that provides the tracking view of the lecturer.

9.4 Interactive Multiview Video Streaming

We now restrict our attention to one particular type of high-dimensional media data for inter-active navigation: multiview video. While, in theory, visual observation of the same physical scene can be made from any camera viewpoint of 3D coordinate (x, y, z) and tilt angles (θ, φ, ψ) around x-, y-, and z-axes, a large majority of multiview videos are captured syn-chronously from a densely spaced one-dimensional camera array (Fujii et al., 2006). The media interactivity for the viewer is thus restricted to viewpoint movement in the x-axis: either by switching from one captured camera viewpoint to another horizontally, or by syn-thesizing novel intermediate virtual views between two captured cameras for observation. In this section, we discuss how coding structures can facilitate this type of interactivity for mul-tiview video. We first discuss in Section 9.4.1 the simple case where switching only among a finite set of captured camera viewpoints is permitted. We then discuss in Section 9.4.2 an extension where switching to intermediate virtual views, synthesized on-the-fly at the decoder, is also possible. We also consider in Section 9.4.3 the case where network transmis-sion delay is nonnegligible and must be accounted for both in the design of the coding struc-ture and in the data transmission protocol.[1]

9.4.1 Interactive Multiview Video Streaming (IMVS)

Like other interactive media navigation systems discussed earlier, we assume the intended IMVS application is a store-and-playback one.[2] In particular, the multiview video data are first pre-encoded and stored on a server in a redundant representation, so that during a subse-quent streaming session an interactive client can periodically request view-switches from the streaming server at a predefined period of T frames, and corresponding video views are streamed from the server and played back to the client uninterruptedly. Figure 9.3a shows an overview of an IMVS system, where a video server, having encoded the multiview data cap-tured by K cameras into a redundant representation, sends multiview video frames corre-sponding to interactive clients' requests for client decoding and display. Note that an alternative approach of real-time encoding a decoding path tailored for each client's unique view traversal across time is computationally prohibitive as the number of interactive clients increases.

The challenge in IMVS is to design a frame structure to pre-encode multiview video data so that streaming rate is optimally traded off with storage required to store the multiview video data, in order to support a desired level of media interaction, which in IMVS means an application-specific view-switching period T (small T leads to faster view-switches and more interactivity). As building blocks to construct an IMVS frame structure, redundant P-frames (Cheung et al., 2009a), DSC-based merge-frames (M-frames; N.-M. Cheung et al., 2009) have been proposed in the literature. Examples of these tools are shown in Figure 9.3b. Redundant P-frames encode one P-frame for each previous frame just prior to a view-switch, using the previous frame as a predictor for differential coding, resulting in multiple frame representations for a given original picture. In Figure 9.3b(i) we see that there are three P-frame $F_{i,2}$s representing the same original picture $F^o_{i,2}$ of time instant i and view 2, each using a different frame in the previous time instant – one of $F_{i-1,1}$, $F_{i-1,2}$, and $F_{i-1,3}$ – as

---

[1] An earlier and less extensive overview of IMVS was presented in Cheung et al. (2010).
[2] Coding structures can also be optimized for live IMVS scenarios (Kurutepe et al., 2007; Tekalp et al., 2007; Cheung et al., 2009c), but is outside the scope of this chapter.

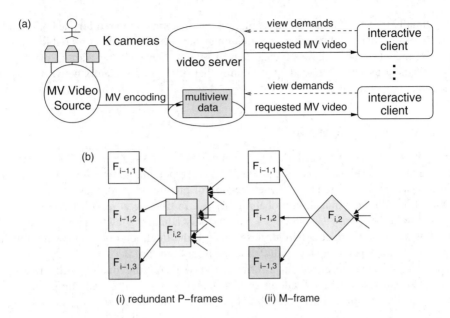

Figure 9.3 (a) IMVS system overview and (b) example frame types. Reproduced with permission of IEEE

predictor. While redundant P-frames result in low transmission bandwidth, it is obvious that using them alone would lead to exponential growth in storage as the number of view-switches across time increases.

As an alternative coding tool, an M-frame was proposed (N.-M. Cheung *et al.*, 2009), where a single version $F_{i,j}$ of the original picture $F_{i,j}^o$ can be decoded no matter from which frame a user is switching from. Petrazzuoli *et al.* (2011) proposed a distributed video framework with depth-image-based rendering (DIBR), where a Wyner–Ziv frame can be first interpolated from two neighboring key frames, and the error of the interpolated frame can then be further reduced by parity bits precomputed and stored on the server. This is essentially the same idea as N.-M. Cheung *et al.* (2009), with the additional assumption that depth information is available for view synthesis to improve the quality of the side information. In Figure 9.3b(ii), the same $F_{i,2}$ can be decoded no matter which one of $F_{i-1,1}$, $F_{i-1,2}$, and $F_{i-1,3}$ a user is switching from. One straightforward implementation of M-frame is actually an independently coded I-frame. It was shown (N.-M. Cheung *et al.*, 2009), however, that DSC-based coding tools exploiting correlation between previous frames $F_{i-1,k}$ and target frame $F_{i,j}$ can result in implementations that outperform I-frame-based implementations in both storage and transmission rate. An M-frame,[3] however, remains a fair amount larger than a corresponding P-frame.

It should be obvious that a frame structure composed of large portions of redundant P-frames relative to M-frames will result in low transmission rate but large storage, while a structure composed of small portions of redundant P-frames relative to M-frames will result in higher transmission rate but smaller storage. The optimization is then to find a structure

---

[3] A new frame type in H.264 called SP-frame (Karczewicz and Kurceren, 2003) can also be used as an M-frame. It was shown by Cheung *et al.* (2009d) that DSC frame is more efficient in most view-switching scenarios, however.

that optimally trades off expected streaming rate with storage required for the entire coding structure. To properly define expected streaming rate, we first define a probabilistic view-switching model to model how a typical viewer selects view-switches across time. We will then examine a few heuristic coding structures that offer good tradeoffs between expected streaming rate and storage.

9.4.1.1 View-Switching Models

Cheung *et al.* (2011a) proposed a simple memoryless view-switching model, where, upon watching any decoded version of the picture $F_{i,j}^o$, corresponding to time instant i and view j, an interactive client will request a coded version of picture $F_{i+1,k}^o$ of view k and next time instant $i + 1$ with view transition probability $\alpha_{i,j}(k)$. In Cheung *et al.*, (2011a), view k was restricted to be between $j - 1$ and $j + 1$, so that only adjacent view-switches are possible. This view-switching model is memoryless because the likelihood of view-switch taken at instant i does not depend on view-switches taken prior to instant i.

Xiu *et al.* (2011) proposed an alternative view-switching model with memory as follows. Suppose a client is watching view k at instant i, after watching view k' at instant $(i - 1)$. The probability that the client will switch to view l at instant $(i + 1)$ is $\Omega_{k',k}(l)$, $l \in \{k - L, \ldots, k + L\}$:

$$
\Omega_{k',k}(l) = \begin{cases} \Phi(l - (2k - k')), & \text{if } k - L < l < k + L \\ \displaystyle\sum_{n=l}^{\infty} \Phi(n - (2k - k')) & \text{if } l = k + L \\ \displaystyle\sum_{n=-\infty}^{l} \Phi(n - (2k - k')) & \text{if } l = k - L \end{cases} \tag{9.1}
$$

where $\Phi(n)$ is a symmetric *view-switching probability function* centered at zero, and L is the maximum number of views a client can skip in one view-switching instant. L essentially limits the speed at which a client can switch views. See Figure 9.4 for an example

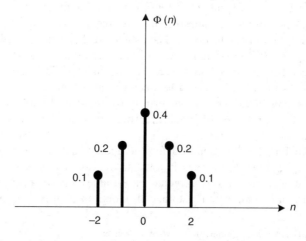

Figure 9.4   Example of $\Phi(n)$ for view-switching model with memory

of $\Phi(n)$. In words, (9.1) states that the probability $\Omega_{k',k}(l)$ that a client selects view coordinate l depends on both the current view coordinate k and previous selected coordinate k'; the probability is the highest at position $k + (k - k')$, where the client continues in view-switch direction $k - k'$. If l is at the view difference bound $k \pm L$, then the probability $\Omega_{k',k}(l)$ needs to sum over probabilities in $\Phi(n - (2k - k'))$ that fall outside the feasible views as well.

9.4.1.2 Path-Based Heuristic IMVS Structures

Instead of detailing the formal coding structure optimization in Cheung *et al.* (2011a), in this section we illustrate good heuristically built IMVS frame structures for intuition. Specifically, we outline two heuristics – a path-based heuristic and a tree-based heuristic – to construct mixtures of redundant P-frames and M-frames, trading off expected streaming rate and storage. These heuristics, though simple, perform well when the probabilities $\alpha_{i,j}(k)$ of switching to adjacent views are very small or very large. We will assume the memoryless view-switching model as discussed in Section 9.4.1.1.

We first consider the case when the probabilities $\alpha_{i,j}(k)$ of switching to adjacent views are very small; that is, an observer will very likely remain in the same view throughout all potential view-switches. To construct a good IMVS frame structure for this case, we start with the minimum storage frame structure; namely, one where an M-frame is used at each switching point. Figure 9.5a shows an example of this structure when the number of captured views is two (each view shown in a different color). Note that this structure has no redundant representation; each picture is represented by only one pre-encoded frame. Note also that this

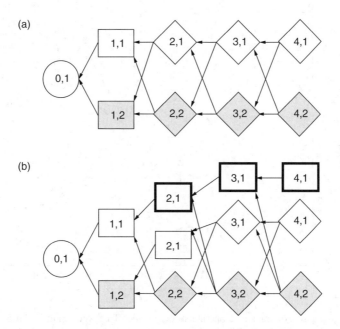

Figure 9.5 Examples of frame structures generated using path-based heuristic. Additional path in (b) is shown in thick-line rectangles: (a) minimum storage frame structure; (b) minimum storage structure plus added sub-path

structure offers the best tradeoff regardless of view-switching probabilities if storage is considered vastly more important that streaming rate.

This minimum storage frame structure comes with high transmission cost, since a transmission-expensive M-frame (relative to a P-frame) must be transmitted at each view-switch. To reduce transmission cost by incrementally increasing redundancy in the structure (thereby increasing storage), one can do the following. Because an observer is very likely to stay in a path of the same view throughout, a simple heuristic is to locate the most likely transition from a frame $F_{i,j}$ to the next frame $F_{i+1,k}$ at a switch point in the structure, and add a sub-path of P-frames for this transition till the end of the structure. In Figure 9.5b, a sub-path $\{P_{2,1}, P_{3,1}, P_{4,1}\}$ is added to the transition from $P_{1,1}$ to the same view 1. Note that, in this case, the original M-frame $M_{2,1}$ is subsequently replaced by a P-frame, because the sub-path addition has caused $M_{2,1}$'s set of previous frame predictors to reduce to a single predictor; an M-frame with a single predictor is reverted to a P-frame.

Sub-paths of P-frames as described above can be added incrementally to induce different tradeoffs between transmission rate and storage. We compare the performance of structures generated using the path-based heuristic with structures generated by a full optimization procedure described in Cheung *et al.* (2011a). The performance of each structure is shown as a data point in Figure 9.6a, representing the tradeoff between the expected transmission rate and storage required to store the structure. We see that when the switching probabilities are low, the path-based heuristic indeed produced structures (path-lo) that closely approximate the performance of the optimal (opt-lo), demonstrating that the use of the path abstraction is

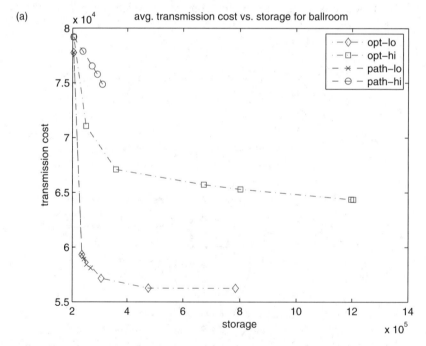

Figure 9.6 Comparison of transmission rate/storage tradeoff between path-based heuristic and full optimization, and between tree-based heuristic and full optimization, respectively: (a) rate/storage tradeoff for path-based heuristic; (b) rate/storage tradeoff for tree-based heuristic. Reproduced with permission of IEEE

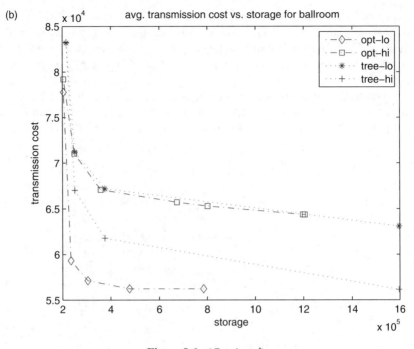

Figure 9.6 (*Continued*)

indeed appropriate. When the switching probabilities are high, however, the performance of the path-based heuristic (path-hi) and the optimal (opt-hi) are quite far apart. This motivates us to derive a different heuristic when the view-switching probabilities are high.

9.4.1.3 Tree-Based Heuristic IMVS Structures

When the probabilities $\alpha_{i,j}(k)$ of switching to adjacent views are very large – for example, when it is just as likely for an observer to stay in the same view as to switch to the adjacent views – Cheung *et al.* (2011a) constructed a different heuristic as follows. Start from the most bandwidth-efficient structure, where only P-frames are used at every switching point. This is essentially a full-tree of depth N with the lone I-frame I_{0,K_0} as the root of the tree; Figure 9.7a shows such a full-tree of depth $N = 4$. The tree has exponential $O(3^N)$ number of frames, which is not practical for large N. However, this redundant structure offers the best tradeoff when expected streaming rate is considered vastly more important than storage.

We now construct a family of structures starting from the bandwidth-efficient structure by incrementally removing frame redundancy (thereby decreasing storage) at the expense of increase in expected transmission rate. Because each view transition is equally likely, the probability of arriving at a certain P-frame of a given tree depth is exactly the same as arriving at any other P-frame of the same depth. That means the transmission costs of individual P-frames, weighted by the frame's access probability, at the same tree depth are roughly the same. On the other hand, as P-frames of one tree depth transition to P-frames of the next tree depth, the total transmission cost of P-frames of the next tree depth remains the same (still one P-frame per transition), while the storage has increased by factor of 3. That means the per-unit costs of the P-frame set at deeper tree depth are strictly more expensive than costs of the P-frame set at shallower tree depth. A reasonable heuristic then is to eliminate P-frames of tree depth $\geq d$ by replacing P-frames at tree depth d with M-frames. Figure 9.7b shows a structure where P-frames at a certain depth of the full tree in Figure 9.7a are replaced by M-frames.

We now compare the performance of structures generated using the tree-based heuristic with that of the full optimization procedure in Cheung *et al.* (2011a), shown in Figure 9.6b. We see that when the switching probabilities are high, the tree-based heuristic produced

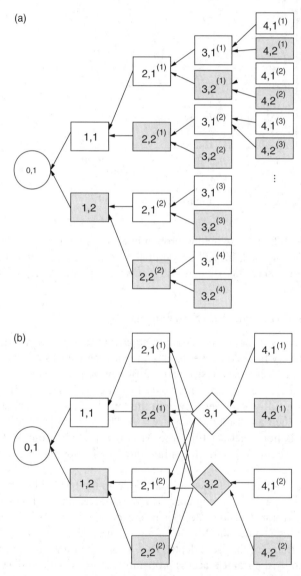

Figure 9.7 Examples of frame structures generated using path-based heuristic: (a) minimum transmission structure; (b) minimum transmission structure with limited depth. Reproduced with permission of IEEE

structures (tree-hi) that approximate that of the optimal (opt-hi), proving empirically that the use of the tree abstraction is indeed appropriate. On the other hand, when the switching probabilities are low, the performance of the tree-based heuristic (tree-lo) compared with the optimal (opt-lo) is inferior. This suggests that an optimal structure must use a combination of paths and trees corresponding to different transition probabilities to optimize the rate/storage tradeoff. This is indeed how a more involved optimization will consider optimal combination of paths and trees in constructing a well-performing frame structure. See Cheung *et al.* (2011a) for details.

9.4.2   IMVS with Free Viewpoint Navigation

One shortcoming of the IMVS system discussed in Section 9.4.1 is that the available views for a client are limited by the few discrete number of camera-captured views pre-encoded on the server, which means a view-switch can appear abrupt and unnatural to a viewer. To remedy this problem, Xiu *et al.* (2011, 2012) introduced arbitrary view switching into IMVS: in addition to camera-captured views, virtual views between captured views can also be requested by clients. Theoretically, arbitrary view switching offers viewers an infinite number of views, so that a view-switch can now take place between views as close/far as the user desires. A virtual view can be synthesized using images of the two nearest captured views via IBR (Xiu and Liang, 2009) or the recently popular DIBR (Na *et al.*, 2008). In the latter case, both texture and depth images at captured viewpoints need to be available for view synthesis. Encoding both texture and depth images at multiple camera-captured viewpoints is commonly called *video-plus-depth* or *texture-plus-depth* format (Merkle *et al.*, 2007).

To enable arbitrary view switching but to maintain a reasonable workload at the server, Xiu *et al.* (2011, 2012) proposed having the server transmit the two nearest coded views[4] to the client as references for view synthesis of the requested virtual view at the client. We call this system *IMVS with View Synthesis* (IMVS-VS). Similar to the original IMVS, the goal is to design a pre-encoded frame structure of a multiview video sequence at the server to facilitate arbitrary view switching. There are several different alternatives to enable arbitrary view switching. To avoid decoder complexity as a result of performing inpainting to fill in disoccluded pixels in real time, Maugey and Frossard (2011) proposed to pre-encode all DIBR-synthesized intermediate frame differentials at the encoder so that, at stream time, a desired intermediate view frame can be constructed by combining a DIBR-synthesized frame and decoding of pre-encoded differentials. If the number of desired intermediate views is large, this will result in a large storage requirement, however. Instead of transmitting the two nearest coded views for DIBR-based synthesis of an intermediate view, Maugey *et al.* (2012) proposed to transmit texture and depth maps of a single captured view plus *auxiliary information* (AI) – additional information to aid the decoder's inpainting algorithm so that the disoccluded regions in the synthesized view can be filled properly. (What is the most appropriate AI, however, remains an open question.)

A natural approach to enable arbitrary view switching in IMVS-VS is to use the same IMVS frame structure composed of I-frames, M-frames, and cross-time (CT)-predicted P-frames, where a P-frame is pre-encoded using a frame in a previous time instant as predictor. However, in IMVS-VS the server always transmits two frames from two neighboring views of the same instant, and those two frames typically exhibit high spatial correlation. One can

---

[4] It has been shown (Cheung *et al.*, 2011b) that using anchor views further away from the desired intermediate viewpoint for view synthesis will further degrade visual quality due to larger amount of disoccluded pixels.

hence achieve better performance by enabling within-time (WT) prediction also, where one transmitted frame can be predicted using the other transmitted frame as predictor.

9.4.2.1 Examples of IMVS-VS Structures

To illustrate the benefit of WT-predicted P-frames for IMVS-VS, Figure 9.8 shows two example structures with three coded views. The structure in Figure 9.8a has an I-frame followed by successive CT-predicted P-frames. Only two P-frames need to be transmitted for any view-switch. In Figure 9.8b, WT prediction is used in addition to CT prediction, where, after the first I-frame of view 1, the other frames of view 1 are coded as CT-predicted P-frames by using the previous frame of view 1 as predictor. Pictures of view 0 and view 2 are coded as WT-predicted P-frames by using the frame of view 1 of the same instant as predictor. This structure also only sends two P-frames for any possible view-switch. However, compared with Figure 9.8a, WT prediction can greatly reduce the number of pre-encoded frames, offering a much better storage/streaming rate tradeoff.

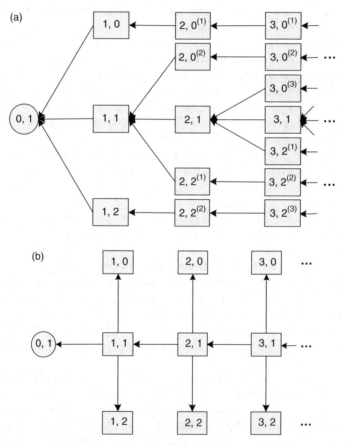

Figure 9.8 Example frame structures of IMVS-VS for three coded views. Circles and rectangles denote I- and P-frames, respectively. (i,j) denotes a frame at instant i of view j. (a) Using CT prediction only; (b) using both CT and WT predictions. Reproduced with permission of IEEE

Given the potential benefits of WT-predicted P-frames, Xiu *et al.* (2011, 2012) optimized coding structures using I-frames, M-frames, and CT- and WT-predicted P-frames for IMVS-VS, extending the structure optimization described in Cheung *et al.* (2011a). See Xiu *et al.* (2011, 2012) for details of the optimization.

9.4.3   IMVS with Fixed Round-Trip Delay

In the previous two sections we discussed coding structure optimization for IMVS (switching among discrete captured camera viewpoints) and IMVS-VS (switching to virtual intermediate views between camera viewpoints), but the overriding assumption is that a client-transmitted view-switch request can be fulfilled by the server by transmitting appropriate pre-encoded data with little or no delay. In practice, however, the streaming server can be located at a non-negligible network distance from the client, which means that the server's response to a client's view-switching request will suffer from a noticeable round-trip-time (RTT) delay. We first discuss a transmission protocol to transmit extra pre-encoded data to effect zero-delay view-switching under fixed RTT delay assumption. We then discuss a corresponding optimization to find the best coding structure under this scenario of IMVS with network delay (IMVS-ND).

9.4.3.1   Timing Events in Server–Client Communication

We first discuss timing events during server–client communication in a generic IMVS system, shown in Figure 9.9, where we assume a client can make a view-switch request every Δ frames. At time 0, the server first transmits an initial chunk of coded multiview data to the client, arriving at the client $(1/2)$RTT frame time[5] later. The client starts playback at time $(1/2)$RTT and makes their first view-switch Δ frame time later; we assume the IMVS system allows a client to request a view-switch every Δ frames. Their first view-switch decision (feedback) is transmitted immediately after the view-switch, and arrives at server at time RTT $+ \Delta$. Responding to the client's first feedback, the server immediately sends a *structure slice*, arriving at the client $(1/2)$RTT frame time later, or RTT frame time after the client transmitted their feedback. More generally then, the client sends feedbacks in interval of Δ

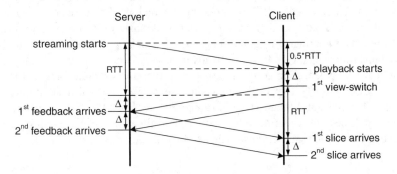

Figure 9.9   Timing diagram showing communication between streaming server and client during session start-up

---

[5] We express time in number of frames for a constant video playback speed.

frame time, and in response the server sends a structure slice for each received feedback every Δ frame time. For simplicity, we assume there are no packet losses and the RTT delay between server and client remains constant.

Notice that from the time the client started playback to the time the first structure slice is received from server, $\Delta + RTT$ frame time has elapsed. That means the initial chunk must contain enough data to enable $\delta = \text{floor}((\Delta + RTT)/\Delta) = 1 + \text{floor}(RTT/\Delta)$ view-switches before the first structure slice arrives. In other words, given initial view v^o at the start of the IMVS-ND streaming session and each view-switching instant can alter view position by ± 1, the initial chunk must contain frames spanning contiguous views $v^o - \delta$ to $v^o + \delta$. Because subsequent structure slices arrive every Δ frame time, each structure slice only needs to enable one more view-switch of one instant for the client to play back video uninterrupted and select a view without RTT delay. The view span of each structure slice, like the initial chunk, is also $2\delta + 1$.

9.4.3.2 View-Switching Example

Figure 9.10a illustrates a concrete example, where the number of views is $N = 7$, initial view is $v^o = 4$, and RTT $= \Delta - \varepsilon$ for some small positive constant $\varepsilon > 0$. Therefore, the initial chunk contains only enough multiview data to enable $\delta = 1$ view-switch, spanning view $v^o - \delta = 3$ to $v^o + \delta = 5$. If the first client selected view-switch is view 3, then the first structure slice must provide multiview data at instant 2 for possible view-switches to view 2, 3, and 4. If the first view-switch is instead view 5, then the next structure slice must contain data for possible view-switches to view 4, 5, and 6.

9.4.3.3 Frame Structure Example

Given the communication protocol described above, one can now optimize a frame structure for IMVS-ND, where the server transmits an initial chunk at the start of the streaming session, and subsequently transmits a structure slice Ξ of $2\delta + 1$ contiguous views (with center view $v(\Xi) = h$) corresponding to the client's selected view h. Figure 9.10b shows an example of a frame structure for seven captured views. We see that original image $F^o_{3,2}$ is represented by two P-frames, $P^{(1)}_{3,2}$ and $P^{(2)}_{3,2}$, each encoded using a different predictor, $P_{2,2}$ and $M_{3,3}$ respectively (indicated by edges). Corresponding to which predictor frame is available at the decoder buffer, different coded frames $F_{i,j}$s can be transmitted to enable correct (albeit slightly different) rendering of original image $F^o_{i,j}$. For example, in Figure 9.10b, the initial chunk contains frames $I_{0,4}, P_{1,3}, P_{1,4}, P_{1,5}$ to cover view-switches to view 3, 4, and 5 at instant 1. If the client's view selection at instant 1 is $h = 3$, then the structure slice that needs to be transmitted is $\Xi^{(1)}_2 = \{P_{2,2}, P_{2,3}, P_{2,4}\}$ with center view $v(\Xi^{(1)}_2) = 3$. Instead, if client remains in view $h = 4$ at instant 1, then structure slice $\Xi^{(2)}_2 = \{P_{2,3}, P_{2,4}, P_{2,5}\}$ with center view $v(\Xi^{(2)}_2) = 4$ will be sent to the decoder. Notice that different slices ($\Xi^{(1)}_2$ and $\Xi^{(2)}_2$ in this example) can contain the same frame(s) ($P_{2,3}$ and $P_{2,4}$).

9.4.3.4 Transmission Schedule

Given a frame structure T, the structure slice Ξ_i of instant i to be transmitted depends on the structure slice Ξ_{i-1} of the previous instant available in the decoder, and the client's view selection h at instant $i - \delta$. The center view $v(\Xi_i)$ of Ξ_i will necessarily be h, where

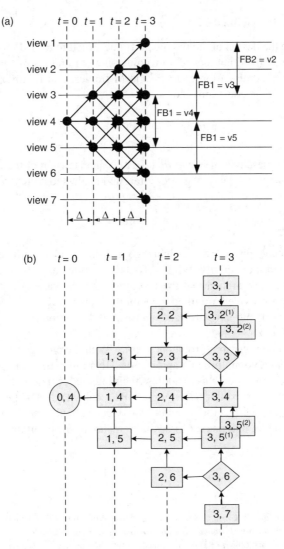

Figure 9.10 Two examples for seven total views, initial view $v^o = 4$, and RTT $= \Delta - \varepsilon$: (a) progressive view-switches; (b) redundant frame structure where I-, P-, and DSC-frames are represented by circles, squares, and diamonds, respectively. (i,j) denotes a frame at view-switching instant i of view j. A solid edge from $F_{i',j'}$ to $F_{i,j}$ means $F_{i',j'}$ is predictively coded using reference frame $F_{i,j}$

$h \in \{v(\Xi_{i-1}), v(\Xi_{i-1}) \pm 1\}$. We can formalize associations among Ξ_{i-1}, h, and Ξ_i for all i given frame structure T via a *transmission schedule* G. More precisely, G dictates which structure slice Ξ_i will be transmitted for decoding at instant i, given previous slice Ξ_{i-1} is available at the decoder and the client selects view h at instant $i - \delta$:

$$G : (\Xi_{i-1}, h) \Rightarrow \Xi_i, \qquad h \in \{v(\Xi_{i-1}) - 1, v(\Xi_{i-1}), v(\Xi_{i-1}) + 1\}$$

where the center view of Ξ_i is $v(\Xi_i) = h$.

9.4.3.5 Optimization Defined

We can now define the design of the redundant frame structure for IMVS-ND as an optimization problem: given a fixed number of viewpoints, how to find a structure T^* and associated schedule G^*, using a combination of I-, P- and M-frames, that minimizes the transmission cost $C(T)$ while a storage constraint is observed:

$$\arg \min_T C(T) \quad \text{s.t.} \quad B(T) \leq \bar{B}$$

The optimization presented in Xiu *et al.* (2012) is an extension of the one in Cheung *et al.* (2011a) to include network delay. We refer interested readers to Xiu *et al.* (2012) for details.

9.5 Conclusion

In this chapter we have discussed the challenge in IMS: to address the conflicting issues of coding efficiency and interactivity. While different solutions have been proposed for different applications, we have argued that these solutions can be categorized into a few general approaches and have provided a unified view on them. We have reviewed the solutions proposed in previous work and have also given a focused discussion on interactive streaming of multiview video.

There are many open research problems in IMS. With the emergence of many new media modalities and interactive multimedia applications, there remains an important issue of how to design efficient coding strategies to address their interactive requirements. Moreover, typical transmission networks exhibit nonnegligible RTT delay and packet losses between server and client, which greatly affect the reaction time of each client's media subset request. Overcoming both RTT delay and packet losses is tremendously important for satisfactory user experience in IMS.

References

A. Aaron, P. Ramanathan, and B. Girod (2004) Wyner–Ziv coding of light fields for random access, in *2004 IEEE 6th International Workshop on Multimedia Signal Processing*, IEEE, pp. 323–326.

I. Bauermann and E. Steinbach (2008a) RDTC optimized compression of image-based scene representation (part I): modeling and theoretical analysis. *IEEE Trans. Image Process.*, **17** (5), 709–723.

I. Bauermann and E. Steinbach (2008b) RDTC optimized compression of image-based scene representation (part II): practical coding. *IEEE Trans. Image Process.*, **17** (5), 724–736.

Broadcast (2008) Live super-HD TV to debut at IBC. http://www.broadcastnow.co.uk/news/multi-platform/news/live-super-hd-tv-to-debut-at-ibc/1425517.article#.

N.-M. Cheung and A. Ortega (2008) Compression algorithms for flexible video decoding. Proceedings Visual Communications and Image Processing (VCIP).

N.-M. Cheung, H. Wang, and A. Ortega (2006) Video compression with flexible playback order based on distributed source coding, in *Visual Communications and Image Processing 2006* (eds J. G. Apostolopoulos and A. Said), Proceedings of the SPIE, Vol. **6077**, SPIE, Bellingham, WA, p. 60770P.

G. Cheung, A. Ortega, and N.-M. Cheung (2009a) Generation of redundant coding structure for interactive multiview streaming. Proceedings of the Seventeenth International Packet Video Workshop, Seattle, WA, May.

G. Cheung, A. Ortega, and N.-M. Cheung (2009b) Bandwidth-efficient interactive multiview video streaming using redundant frame structures. Proceedings of APSIPA Annual Summit and Conference, Sapporo, Japan, October.

G. Cheung, N.-M. Cheung, and A. Ortega (2009c). Optimized frame structure using distributed source coding for interactive multiview streaming, in *ICIP'09 Proceedings of the 16th IEEE International Conference on Image Processing*, IEEE, Piscataway, NJ, pp. 1377–1380.

N.-M. Cheung, A. Ortega, and G. Cheung (2009). Distributed source coding techniques for interactive multiview video streaming, in *PCS 2009 Picture Coding Symposium*, IEEE.

G. Cheung, A. Ortega, N.-M. Cheung, and B. Girod (2010). On media data structures for interactive streaming in immersive applications, in *Visual Communications and Image Processing 2010* (eds P. Frossard, H. Li, F. Wu, B. Girod, S.Li, and G. Wei), Proceedings of the SPIE, Vol. **7744**, SPIE, Bellingham, WA, p. 77440O.

G. Cheung, A. Ortega, and N.-M. Cheung (2011a) Interactive streaming of stored multiview video using redundant frame structures. *IEEE Trans. Image Process.*, **20** (3), 744–761.

G. Cheung, V. Velisavljevic, and A. Ortega (2011b) On dependent bit allocation for multiview image coding with depth-image-based rendering. *IEEE Trans. Image Process.*, **20** (11), 3179–3194.

F.-O. Devaux, J. Meessen, C. Parisot, J. Delaigle, B. Macq, and C. D. Vleeschouwer (2007) A flexible video transmission system based on JPEG2000 conditional replenishment with multiple references, in *ICASSP 2007. IEEE International Conference on Acoustics, Speech, and Signal Processing*, IEEE, pp. I-825–I-828.

Digital Video Broadcasting (2012) Digital video broadcasting. http://www.dvb.org/.

Z. Fan and A. Ortega (2009) Overlapped tiling for fast random access of 3-D datasets. Proceedings of Data Compression Conference (DCC), March.

C. Fehn, C. Weissig, I. Feldmann, M. Mueller, P. Eisert, P. Kauff, and H. Bloss (2006) Creation of high-resolution video panoramas of sport events, in *ISM'06. Eighth IEEE International Symposium on Multimedia, 2006*, IEEE, pp. 291–298.

M. Flierl and B. Girod (2007) Multiview video compression. *IEEE Signal Proc. Mag.*, **24** (6), 66–76.

C.-H. Fu, Y.-L. Chan, and W.-C. Siu (2006) Efficient reverse-play algorithms for MPEG video with VCR support. *IEEE Trans. Circ. Syst. Vid.*, **16** (1), 19–30.

T. Fujii, K. Mori, K. Takeda, K. Mase, M. Tanimoto, and Y. Suenaga (2006) Multipoint measuring system for video and sound – 100-camera and microphone system, in *2006 IEEE International Conference on Multimedia and Expo*, IEEE, pp. 437–440.

S. Halawa, D. Pang, N.-M. Cheung, and B. Girod (2011) ClassX – an open source interactive lecture streaming system, in *MM'11 Proceedings of the ACM International Conference on Multimedia*, ACM, New York, NY, pp. 719–722.

A. Jagmohan, A. Sehgal, and N. Ahuja (2003) Compression of lightfield rendered images using coset codes, in *Conference Record of the Thirty-Seventh Asilomar Conference on Signals, Systems and Computers*, Vol. **1**, IEEE, pp. 830–834.

JPEG (2007) JPEG2000 interactive protocol (part 9 – JPIP). http://www.jpeg.org/jpeg2000/j2kpart9.html.

M. Karczewicz and R. Kurceren (2003) The SP- and SI-frames design for H.264/AVC. *IEEE Trans. Circ. Syst. Vid.*, **13** (7), 637–644.

E. Kurutepe, M. R. Civanlar, and A. M. Tekalp (2007) Client-driven selective streaming of multiview video for interactive 3DTV. *IEEE Trans. Circ. Syst. Vid.*, **17** (11, 1558–1565.

M. Levoy and P. Hanrahan (1996) Light field rendering, in *SIGGRAPH '96 Proceedings of the 23rd Annual Conference on Computer Graphics and Interactive Techniques*, ACM, New York, NY, pp. 31–42.

M. Levoy and K. Pulli (2000) The digital Michelangelo project: 3-D scanning of large statues, in *SIGGRAPH'00, Proceedings of the 27th Annual Conference on Computer Graphics and Interactive techniques*, ACM Press-/Addison-Wesley, New York, NY, pp. 131–144.

C. W. Lin, J. Zhou, J. Youn, and M. T. Sun (2001) MPEG video streaming with VCR functionality. *IEEE Trans. Circ. Syst. Vid.*, **11** (3), 415–425.

J.-G. Lou, H. Cai, and J. Li (2005) A real-time interactive multi-view video system, in *MULTIMEDIA '05 Proceedings of the 13th ACM International Conference on Multimedia*, ACM, New York, NY, pp. 161–170.

A. Mavlankar, P. Baccichet, D. Varodayan, and B. Girod (2007) Optimal slice size for streaming regions of high resolution video with virtual pan/tilt/zoom functionality. Proceedings of the European Signal Processing Conference (EUSIPCO-07), September.

A. Mavlankar and B. Girod (2009) Background extraction and long-term memory motion-compensated prediction for spatial-random-access enabled video coding, in *PCS 2009 Picture Coding Symposium, 2009*, IEEE.

A. Mavlankar, P. Agrawal, D. Pang, S. Halawa, N.-M. Cheung and B. Girod (2010) An interactive region-of-interest video streaming system for online lecture viewing, in *2010 18th International Packet Video Workshop (PV)*, IEEE, pp. 64–71.

A. Mavlankar and B. Girod (2011) Spatial-random-access-enabled video coding for interactive virtual pan/tilt/zoom functionality. *IEEE Trans. Circ. Syst. Vid.*, **21** (5), 577–588.

T. Maugey and P. Frossard (2011) Interactive multiview video system with low decoding complexity, in *2011 18th IEEE International Conference on Image Processing (ICIP)*, IEEE, pp. 589–592.

T. Maugey, P. Frossard, and G. Cheung (2012) Temporal and view constancy in an interactive multiview streaming system. Proceedings of IEEE International Conference on Image Processing, Orlando, FL, September 2012.

P. Merkle, A. Smolic, K. Mueller, and T. Wiegand (2007) Multi-view video plus depth representation and coding, in *ICIP 2007. IEEE International Conference on Image Processing, 2007*, IEEE, pp. I-201–I-204.

S.-T. Na, K.-J. Oh, C. Lee, and Y.-S. Ho (2008) Multi-view depth video coding using depth view synthesis, in *ISCAS 2008. IEEE International Symposium on Circuits and Systems*, IEEE, pp. 1400–1403.

A. T. Naman and D. Taubman (2007) A novel paradigm for optimized scalable video transmission based on JPEG2000 with motion, in *ICIP 2007. IEEE International Conference on Image Processing, 2007*, IEEE, pp. V-93–V-96.

A. Naman and D. Taubman (2011) JPEG2000-based scalable interactive video (JSIV). *IEEE Trans. Image Process.*, **20** (5), 1435–1449.

D. Pang, S. Halawa, N.-M. Cheung, and B. Girod (2011) Mobile interactive region-of-interest video streaming with crowd-driven prefetching, in *IMMPD '11 Proceedings of the 2011 International ACM Workshop on Interactive Multimedia on Mobile and Portable Devices*, ACM, New York, NY, pp. 7–12.

G. Petrazzuoli, M. Cagnazzo, F. Dufaux, and B. Pesquet-Popescu (2011) Using distributed source coding and depth image based rendering to improve interactive multiview video access, in *2011 18th IEEE International Conference on Image Processing (ICIP)*, IEEE, pp. 597–600.

P. Ramanathan and B. Girod (2004) Random access for compressed light fields using multiple representations, in *2004 IEEE 6th Workshop on Multimedia Signal Processing*, IEEE, pp. 383–386.

H.-Y. Shum, S. B. Kang, and S.-C. Chan (2003) Survey of image-based representations and compression techniques. *IEEE Trans. Circ. Syst. Vid.*, **13** (11), 1020–1037.

D. Taubman and R. Rosenbaum (2003) Rate-distortion optimized interactive browsing of JPEG2000 images, in *ICIP 2003. Proceedings. International Conference on Image Processing, 2003*, Vol. **III**, IEEE, pp. 765–768.

A. M. Tekalp, E. Kurutepe, and M. R. Civanlar (2007) 3DTV over IP: end-to-end streaming of multiview video. *IEEE Signal Proc. Mag.*, **24** (6) 77–87.

S. J. Wee and B. Vasudev (1999) Compressed-domain reverse play of MPEG video streams, in *Multimedia Systems and Applications* (eds A. G. Tescher, B. Vasudev, V. M. Bove, Jr., and B. Derryberry), Proceedings of the SPIE, Vol. **3528**, SPIE, Bellingham, WA, pp. 237–248.

T. Wiegand, G. Sullivan, G. Bjontegaard, and A. Luthra (2003) Overview of the H.264/AVC video coding standard. *IEEE Trans. Circ. Syst. Vid.*, **13** (7), 560–576.

B. Wilburn, M. Smulski, K. Lee, and M. A. Horowitz (2002) Light field video camera, in Media Processors 2002, (eds S. Panchanathan, V. M. Bove, Jr., and S. I. Sudharsanan), Proceedings of the SPIE, Vol. **4674**, SPIE, Bellingham, WA, pp. 29–36.

X. Xiu and J. Liang (2009) Projective rectification-based view interpolation for multiview video coding and free viewpoint generation, in *PCS 2009 Picture Coding Symposium, 2009*, IEEE.

X. Xiu, G. Cheung, A. Ortega, and J. Liang (2011) Optimal frame structure for interactive multiview video streaming with view synthesis capability. Proceedings of the IEEE International Workshop on Hot Topics in 3D (in conjunction with ICME 2011), Barcelona, Spain, July.

X. Xiu, G. Cheung, and J. Liang (2012) Delay-cognizant interactive multiview video with free viewpoint synthesis. *IEEE Trans. Multimed.*, **14** (4), August 2012.

10

Adaptive Streaming of Multiview Video Over P2P Networks

C. Göktuğ Gürler and A. Murat Tekalp
College of Engineering, Koç University, Turkey

10.1 Introduction

With increasing availability of high-quality three-dimensional (3D) displays and broadband communication options, 3D media is destined to move from the movie theaters to home and mobile platforms. Furthermore, stereoscopic 3DTV is advancing to multiview video (MVV) format with associated spatial audio to allow users to experience a scene from multiple viewing directions without special glasses. In Chapter 6, we introduced the emerging format in detail. In this chapter, we discuss the transmission aspect of MVV, which is distinguished from the other video formats with its variable number of views. Considering that different display technologies require different numbers of views, the bit-rate requirement to support MVV is not the same for each user. The heterogeneous bandwidth requirement is one of the key challenges in broadcasting MVV, because available standards such as digital video broadcasting (DVB) are designed to operate at a fixed bit rate and cannot perform customer-specific content delivery.

The Internet Protocol (IP), along with its underlying infrastructure, can serve as a flexible platform to deliver MVV content because, unlike DVB, it does not have fixed bit rate restriction. With services such as IPTV and WebTV, the IP network can serve as many views and quality levels as allowed by the capacity of users' connection. However, considering the high bandwidth requirement of MVV, in a classic server–client scenario, the traffic at the content provider can reach critical levels where it may be difficult to maintain scalable service with an increasing number of recipients. Moreover, even if the infrastructure allows such high-bandwidth connections, it may be possible at a high cost.

In this chapter we introduce peer-to-peer (P2P) overlay networks for the delivery of MVV to alleviate the problem of bandwidth scarcity at the sender side. Section 10.2 explains key features that can distinguish different P2P solutions and introduces the details of the BitTorrent protocol that we use as an inspiration for building a successful video streaming solution over P2P overlays. Section 10.3 focuses on methods and policies that are needed to enable

Emerging Technologies for 3D Video: Creation, Coding, Transmission and Rendering, First Edition.
Frédéric Dufaux, Béatrice Pesquet-Popescu, and Marco Cagnazzo.
© 2013 John Wiley & Sons, Ltd. Published 2013 by John Wiley & Sons, Ltd.

adaptive video streaming and explains the steps that should be taken to transform a binary-file-sharing solution to a video delivery service. The following two, Sections 10.4 and 10.5, extend the video streaming platform to support stereoscopic video and MVV. The prime subject of these sections is new adaptation strategies, which form the main difference between streaming monoscopic and 3D video with increased number of video streams. In Section 10.5 we also discuss a hybrid mechanism that can deliver multiview content over both DVB and IP channels synchronously and combines the strength of both channels, namely stability of DVB and flexibility of IP.

10.2 P2P Overlay Networks

The vast majority of services over the Internet adopt the server–client model to send or receive small amounts of data and timeliness is not a strict requirement; for example, sending an e-mail or loading a web site can take several seconds. However, as the size of transmitted data increases and the server side reaches its bandwidth limitation, it becomes difficult to maintain responsive service against an increasing number of clients. Content distribution networks, which are based on deploying multiple servers at the edges of the network, are a popular method to overcome this problem, but require additional costs. As an alternative, P2P solutions can distribute the task of forwarding data over receivers (peers) and alleviate the problem of high bandwidth requirement at the sender side, creating a more feasible solution for data delivery over IP. Today, P2P technology is widely adopted for sharing large files, and P2P traffic constitutes the largest portion of the global IP traffic.

In almost all P2P solutions that operate over IP, peers form end-to-end connections with each other and create an overlay network. A data packet that is destined to multiple peers is first received by an intermediate P2P client application over one connection and then reinjected back to the IP network to forward the data to another peer. This is because network layer multicasting (duplicating a packet that has multiple recipients at network routers) is not widely supported in the Internet. In other words, the transmission of the same data to multiple recipients is handled by the end nodes without any assistance from the network layer because the majority of routers that form the core of the Internet are not capable of forwarding a single data packet to multiple recipients that are spread across different networks.

Although overlaying is the basic principle (besides a local area network environment) there are significant differences among different P2P solutions. In this section, we focus on P2P video streaming solutions and introduce the basic properties that allow us to categorize them. We elaborate three concepts: first topology, which defines the way peers are connected to each other; second, differences between sender- versus receiver-driven streaming; and third, layered versus cross-layer design from a P2P video streaming point of view. Then, we take a look at the BitTorrent protocol, which is a successful P2P protocol that is designed to share large files over the Internet. We later discuss modifications needed in the BitTorrent protocol to enable video streaming in Section 10.3; therefore, understanding the fundamentals of the BitTorrent protocol is very critical for following the rest of this chapter.

10.2.1 Overlay Topology

The P2P network topology can be tree based, mesh based or hybrid.

Tree-based P2P solutions are efficient at delivering data from a root node to other peers that are connected to each other in a parent–child fashion. Once a peer joins a tree, data flow starts immediately with no time spent for peer search. Moreover, data are (commonly) pushed from the root to peers, allowing significantly lower latency in data dissemination when compared with mesh-based approaches, in which receivers request data as an overhead. These features make tree-based solutions suitable for time-critical applications; however, the rigid structure that is required for high efficiency also introduces some drawbacks.

The major problem with tree-based solutions is the construction and maintenance of the tree structure in the case of ungraceful peer exit events, which may cause the descendants to starve. Besides on-the-fly tree reconstruction, there are two alternative approaches to address this problem in the literature: having multiple parents for each peer (Jeon *et al.*, 2008; Fesci *et al.*, 2009) or using multiple distribution trees (Castro *et al.*, 2003; Noh *et al.*, 2009). In the first solution, each peer has backup parent(s) to request data just in case the current parent leaves the network. The second solution is based on building multiple trees so that descendants can continue to receive data from an alternative path whenever a peer leaves. We refer the reader to excellent studies on the construction of multiple trees, such as split-stream and Stanford P2P multicast, which guarantees complementary trees for complete path diversity (Baccichet *et al.*, 2007a,b).

We note that widely used open or commercial P2P applications/services do not commonly rely on purely tree-based solutions. The major reason is the difficulty in maintaining the tree formation in an optimum structure. For example, considering that a peer with more descendants should have more network capacity, peers should go up and down on the tree based on their current status of connections. However, while it is easy to detect a bottleneck connection between two nodes, it is not always trivial to understand which side is the cause of the problem. Therefore, the question of which peer should be replaced is not easy to answer. Considering that there can be malicious nodes, which disseminate false information to improve their position in the tree, constructing an efficient topology becomes a challenging task and leads to suboptimum results.

In theory, constructing a multiple tree structure may alleviate the problem by creating alternative paths for data dissemination and decrease the effect of suboptimal tree formation. However, using multiple trees has its own drawbacks. First, multiple paths decrease link utilization if redundant data are transmitted over each tree. Second, even if we assume that optimum tree problem is achieved by one of the paths; additional trees are suboptimal by definition and may become very ineffective if a weak peer becomes the root. Last, each additional tree connects more peers together, making the topology look similar to a mesh network but with static connections.

In mesh-based P2P solutions, data are distributed over an unstructured network, in which each peer can be connected to multiple other peers. The connections are established dynamically and have a full duplex nature, meaning that both parties can send/receive data with no parent–child relationship in between. The increased connectivity and dynamic nature of mesh overlays alleviate the problem of ungraceful peer exit, since a peer may switch to another peer as soon as it detects a problem. Building multiple connections for a newcomer peer takes a certain amount of time, which we call the *initiation interval*. Until this period is over, the peer may not be able to fully utilize its resources and, therefore, mesh-based solutions are more suitable for applications that may tolerate some initial delay.

10.2.2 *Sender-Driven versus Receiver-Driven P2P Video Streaming*

Video streaming applications can be classified as sender driven and receiver driven, depending on where the intelligence resides. There are some significant differences between these approaches on what information is exchanged between the peers and how the rate adaptation is performed.

In *sender-driven streaming*, the sender side has full access to transmitted content and may choose to split the bitstream into pieces as small as needed for adaptation purposes. In one extreme, video packets can go down to network abstraction layer (NAL) units like in a legacy server–client-based video streaming over user datagram protocol (UDP) and the server side can discard enhancement layer NAL units individually for rate adaptation purposes. In such a case, it is possible to think each branch in a tree as an isolated server–client-based adaptive streaming session that works with high adaptation capability.

Sender-driven streaming allows fine-grain rate adaptation; however, it is difficult to perform coordinated data delivery from multiple sources to a single receiver (as in multiple tree) because a sender side may not have up-to-date information about what the receiver side currently has received or is receiving. Therefore, there is the risk of forwarding redundant data over different paths and decrease the link utilization. At this point, multiple description coding (MDC) can be of interest as a promising technique. Using MDC, it is possible to generate self-decodable bitstreams (descriptions) with some redundancy (Goyal, 2001). Any single description allows the receiver side to decode the content at a certain quality while additional descriptions augment the quality. Therefore, the combination of multiple trees and multiple-description coding is a reasonable approach that is addressed in many studies (Padmanabhan *et al.*, 2003; Jurca *et al.*, 2007; Setton *et al.*, 2008). Unfortunately, due to the problems of both the tree-based approach and the inherent redundancy that decreases the throughput, the approach has not been appreciated by the industry.

In *receiver-driven streaming*, it is relatively easier to handle multiple sources since the peer has the instantaneous knowledge about the current state and does not reschedule the same video-segment twice unless necessary. The peer can also make intelligent decisions, such as requesting a chunk with a close deadline from a peer that has better connection and utilize its resources efficiently.

Applying MDC to a receiver-driven streaming solution does not make much sense. Considering that each description has some redundant information, the peer should prefer requesting nonredundant data instead of receiving the redundant part twice. It is also reasonable to assume that the level of overhead for requesting individual parts (nonredundant data) is less costly in terms of bitrate than receiving redundant data.

On the other hand, receiver-driven solutions cannot perform fine adaptation because the minimum transmission units cannot be too short since each unit is individually requested and also notified to neighbors upon its arrival. Therefore, very short duration transmission units increase the messaging overhead.

10.2.3 Layered versus Cross-Layer Architecture

The Internet has been designed in a layered architecture in which each layer has distinct roles and requirements. However, it is always possible to increase system performance by designing solutions that can consider layers jointly.

In a *layered approach*, the application layer relies on the assumption that the network is good enough to accommodate video streaming and does not employ feedback from the network layer. Although it may have inferior performance when compared against cross-layer design, this approach has been widely adopted owing to the ease of implementation. For example, an application developer does not adjust the rate of video transmission or repeat lost data fragments because the transmission control protocol handles these at the network layer. On the other hand, since the application does not perform rate adaptation, the viewer can experience loss of quality or service interruptions in the case of network congestion or bandwidth scarcity.

In a *cross-layer approach*, the application layer adapts the video source rate according to feedback received from the transport layer. Video rate adaptation is possible if video is encoded using scalable codecs or encoded as multiple streams at different quality and bitrates. Then, in the case of congestion, the application may discard enhancement layers (in the case of scalable encoding) or switch to a low-quality bitstream (in the case of multiple encodings) and achieve adaptive streaming. Adaptation is a critical step to achieve stable video quality in P2P video streaming applications.

10.2.4   When P2P is Useful: Regions of Operation

P2P overlays are promising to alleviate the server bottleneck problem; however, they are not magical tools that solve it under all conditions. In a successful P2P platform, there are two fundamental requirements.

The first requirement is about the bit rate of the content and peers' upload and download capacities. Assuming that the download capacity of a peer is higher than its upload capacity, there are three cases:

(i) If the content bit rate is higher than the peers' download capacity, then peers cannot receive content at an acceptable quality regardless of the P2P architecture. This is a trivial case where video streaming is not possible.

(ii) If the content bitrate is in between the average download and upload rate of the peers, then peers may receive the content at an adequate quality; however, the P2P architecture may not be self-sufficient. A streaming server or some helper nodes are needed in order to help the peers that cannot receive content from other peers due to their lack of upload capacity. In this case, the P2P overlay can only assist to decrease the bit-rate load at the server side to some extent.

(iii) If the content bit rate is less than the peers' average upload capacity, then the P2P overlay may be self-sufficient, and the server load saturates at a certain point even if increasing numbers of peers join the session.

The second requirement is the round-trip time, which is one of the key factors that affect the delivery of a video packet from one peer to another peer. As the number of peers increases, a data packet needs to be transferred over multiple hops (peers) until it reaches its final destination. If the transmission interval is longer than the buffering interval, then users may experience interruptions.

10.2.5   BitTorrent: A Platform for File Sharing

BitTorrent is a popular protocol for P2P distribution of large files over the Internet. The protocol has become so successful that it generates the majority of the IP traffic. The main idea is to divide content into *equally sized* pieces called chunks. A metadata file (.torrent) is generated that contains an IP address for a tracker server along with the SHA1 hash values for each chunk. The protocol is initiated once a peer (newcomer) connects to the tracker server and receives a sub-list of all peers (swarm) in the same session. There are two types of peers: a seeder has all the chunks and only contributes a chunk; a leecher has some chunks missing that it tries to receive. Upon connecting to a tracker server and receiving a sub-list of peers, the newcomer starts to exchange chunks with its neighbors. Once a chunk is received, its hash value is compared with the one in the metadata file. Only then are the neighboring peers notified about the availability of the chunk.

In BitTorrent, chunk exchange is managed with three fundamental policies. Using the *rarest first policy*, a peer aims to download a chunk that is least distributed in the swarm. Doing so, a peer increases the likelihood of obtaining unique data to exchange with its neighbors and also increases robustness of data delivery against peer churns by increasing the availability of rare chunks. Using the *tit-for-tat policy*, a peer ranks chunk upload requests and favors requests from a neighbor that has provided more data. A peer may choose to reject requests from other neighbors if the contribution from that neighbor is low to punish free-riders. The only exception is the *optimistical un-choking policy*, which forces a peer to upload a chunk to a *lucky* neighbor even if the neighbor has not provided any data to the peer yet in order to allow a new peer to start the session. Moreover, BitTorrent adopts *pipelining* and *end-game-mode* methods to increase the efficiency of data transmission. In pipelining, fixed-sized chunks are considered to be further divided to sub-pieces that serve as the data transmission unit. The pipelining refers to requesting multiple sub-pieces at the same time to saturate the downloading capacity, and increase link utilization and throughput, but sub-pieces of only one particular chunk can be requested in order to receive a whole chunk as soon as possible The end-game policy is to eliminate the prolonging delays in finishing the last download event due to a peer with very low upload capacity. Normally, a sub-piece is requested only once from a single peer. If a sub-piece is requested from a weak peer (that has limited upload capacity) it does not deteriorate the downloading process since the peer can always download other sub-pieces from other peers and saturate its link capacity. However, when the session is about to end, a weak peer may avoid finalizing the session. When end-game mode is enabled, the remaining sub-pieces are requested from multiple peers to avoid delays due to a weak peer (Cohen, 2003).

10.3 Monocular Video Streaming Over P2P Networks

The BitTorrent protocol has proven itself as a successful and reliable file sharing protocol (Pouwelse *et al.*, 2005). When compared with its predecessors, BitTorrent achieves robust delivery with higher throughput rates. Unfortunately, the protocol is designed for lossless data delivery with no sense of timeliness, assuming that the media will be consumed only after the whole content is received; therefore, BitTorrent is not suitable as it is for delivering time-sensitive data, such as in video streaming. Nevertheless, its robust architecture is a good starting point towards achieving a P2P video streaming platform.

There have been significant studies in the literature that modify BitTorrent for video streaming purposes. For example, BiToS (Vlavianos *et al.*, 2006) and Tribler (Pouwelse *et al.*, 2008) modify the rarest first chunk scheduling policy to deliver video chunks in a timely manner. NextShare (Eberhard *et al.*, 2010a) aims to enable adaptive streaming using scalable coding. There are also hybrid architectures that use BitTorrent to assist server–client architectures and decrease the workload at the server side (Dana *et al.*, 2005). Besides video streaming there are also studies for incentive mechanisms that are more suitable in video streaming, such as give-to-get (Mol *et al.*, 2008). All of these approaches are successful under certain conditions; however, in order to build the most efficient and robust P2P video streaming platform, the modifications should be done from a broader perspective, which includes modifying fundamental policies and methods of BitTorrent such as fixed-sized chunking, rarest-first chunk scheduling, pipelining, and end-game-mode. In the remainder of this section, we describe the modifications in detail and provide some additional concepts to create a P2P video streaming platform that is adaptive, scalable, server-load sensitive, and achieve high throughput and video quality.

10.3.1   Video Coding

Coding efficiency refers to achieving a desirable video quality at the lowest possible bit rate. Currently, H.264/AVC standard is the state of the art for the most efficient single-layer video coding. For video streaming applications over IP, the rule of thumb is to keep the overall video bit rate at about 70–75% of the available network throughput because each additional byte increases the possibility of experiencing augmented delays, higher packet loss rates, and so on.

Besides high efficiency, a secondary goal can be achieving scalability, which refers to the ability to perform graceful quality degradation by discarding some portions of the encoded bitstream. Scalability is helpful for both terminal adaptation (adapting to native resolution and/or processing power of the receiver device) and rate adaptation (adapting to the rate of available network connection). H.264/SVC provides temporal, spatial and signal-to-noise (SNR) scalability. The spatial scalability is intended for terminal adaptation and generates a nonconforming bitstream even if a single enhancement NAL unit is discarded, which makes it unsuitable for adaptive streaming over IP. The temporal scalability introduces no overhead because it is mainly achieved by discarding B-frames in hierarchical coding. There are two issues regarding temporal scalability. First, the effect of decreasing frame rate on the quality of experience is content dependent. Second, discarded B-frames are actually the ones that require the least bit rate compared with other frames. Therefore, we are most interested in SNR scalability for network rate adaptation, and refer to SVC extension of H.264/AVC (SVC) for multilayer (scalable) video coding.

10.3.2   Variable-Size Chunk Generation

Bitstream packetization refers to formation of transmission units (e.g., real-time transport protocol packets over UDP/IP) and has a critical impact on the performance of video streaming solutions over a lossy network. In a chunk-based P2P system, packetization corresponds to formation of video chunks. The main consideration is to avoid having too many chunks, which may cause significant overhead due to messaging.

In the BitTorrent protocol, all chunks have a fixed size, which is determined based on the total size of the shared content. This approach is not very suitable for video streaming because chopping the bitstream at fixed locations creates chunks that are not independently decodable. Therefore, a single chunk loss may render neighboring chunks useless due to inter-chunk dependencies. In order to achieve self-decodable video chunks, it is desirable to use group of pictures (GOP) as a chunk boundary. If the spatial resolution of video is low and GOP size is too small, it is also possible to put multiple GOPs in a chunk (Gurler *et al.*, 2012).

The chunk formation methodology from an H.264 (both AVC and SVC) encoded video bit stream is as follows. The bitstream starts with non-VCL (video coding layer) NAL units that provide metadata information such as vertical and horizontal resolution (in terms of number

NAL Units of SVC Bitstream

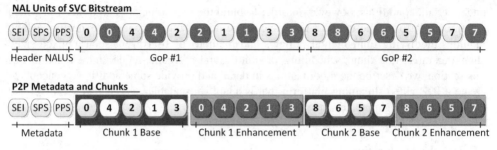

Figure 10.1 Chunk formation using one GOP per chunk

of macroblocks) which are vital for the decoding operation. The content of non-VCL NAL units can either be stored to a session metadata file that corresponds to a torrent file in Bit-Torrent or it can be repeated at the beginning of each GOP and included in every chunk. After the non-VCL NAL units there exists one base layer NAL unit and possibly multiple enhancement-layer NAL units (if SVC is utilized) for each frame. Then, two types of chunks can be generated: base-layer chunks are obtained by merging base-layer NAL units for a GOP, and are prioritized; enhancement-layer chunks are obtained by merging all enhancement-layer NAL units for a GOP, and can be discarded in the case of bandwidth scarcity as depicted in Figure 10.1, in which the rounded boxes represents the NAL units with the shaded ones as enhancement-layer NAL. In the case of single-layer coding, the content is composed of only base-layer chunks.

10.3.3   Time-Sensitive Chunk Scheduling Using Windowing

A common approach to enable timely chunk scheduling is to employ a temporal window that restricts a peer to schedule (for downloading) chunks that are in the vicinity of the current play-out time. There are various studies to achieve an optimum chunk-picking algorithm within a window that maintains steady video delivery (receiving at least the base-layer chunk) while trying to maximize the quality by receiving as many enhancement layers as possible (Eberhard *et al.*, 2010b).

A scheduling window starts from the beginning of the video and slides forward under two cases. The good case is when the peer receives all chunks in the window. Otherwise, if the deadline (represented as a vertical line) is reached, the window is forced to slide, aborting some chunks. Figure 10.2 depicts a state in which there are two layers (the enhancement-layer chunks are darker), the window size is four chunks and the ready-to-play chunk buffer duration is 2 sec. if we assume that each chunk is 1 sec. in duration. The vertical list represents the player deadline, and chunks that are before the deadline should not be scheduled.

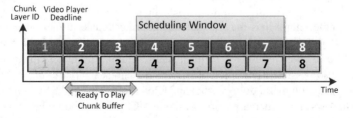

Figure 10.2 A sample downloading window

Alternatively, one may have considered sequential scheduling instead of windowing; however, there should be some randomness in the order peers request chunks. If all peers request chunks in the same order, then it is less likely that peers will have unique data to exchange among themselves. In this sense, randomization increases the P2P activity. On the other hand, prioritization may increase decoding performance because receiving an enhancement-layer chunk without receiving the corresponding chunks that are in the lower layers (e.g., base-layer chunk) is a waste of available network resource of the peer.

10.3.4 Buffer-Driven Rate Adaptation

The duration of the ready-to-play buffer is critical information that can be used in the adaptation process. In server–client architecture, it may be possible to measure the instantaneous bandwidth and perform rate adaptation accordingly. However, in the P2P case, there are multiple connections, which are not mutually independent. Therefore, it is a good practice to use the buffer duration to assess the overall downloading capability. The buffer duration is calculated by measuring the total duration of chunks starting from the end of last played chunk up to the start of downloading window, as depicted in Figure 10.2. At this point, it should be evident that using variable-length chunks has another advantage over fixed-sized chunks because the variable-length chunks have fixed durations, allowing the P2P solution to determine the duration of the buffer more easily.

The rate adaptation using buffer duration works as follows. Once a streaming session is initiated, peers request chunks of all layers to deliver at the highest possible quality. Once all the chunks are downloaded, the window slides to include new chunks for the scheduling procedure. Meanwhile, the player starts to consume received multimedia. From now on, the number of layers to be scheduled is updated each time a chunk is to be scheduled. If the buffer duration is below a certain threshold (e.g., 2 sec.), the number of downloaded streams is reduced by discarding the layer that is least important. In the opposite case, if the duration of the buffer is large and there is an unscheduled layer, then one more layer is added to the scheduling procedure. In this way a peer may regulate the quality it should receive such that if the downloading capacity is high then the video is received in high quality and vice versa.

10.3.5 Adaptive Window Size and Scheduling Restrictions

The original BitTorrent protocol follows a policy called "rarest first," which forces peers to download chunks that are distributed least in the swarm (among all peers in the same session). In other words, rarest first policy allows more chunks to be available for the peers that are likely to be requested by other peers and thus increases the P2P activity.

Windowing on the other hand, while maintaining timely chunk scheduling, significantly decreases the chance of having unique data among peers, especially in a session in which peers are synchronized to a broadcast stream. Therefore, it is best to allow the window size to be as large as possible, considering that increased buffer duration allows a longer time duration to download chunks and thus a wider window duration. Then, it is possible to estimate a window size using

$$w_s = \text{Round}\left(\frac{d_r \times b_d}{c_s}\right) \tag{10.1}$$

where w_s (number of chunks) is the window size, c_s (bytes) is the average chunk size, d_r (byte/s) is the download rate, and b_d (s) is the buffer duration.

Equation 10.1 states that the next window size is found by dividing the total number of bytes to be downloaded within the next window (equal to average download rate multiplied by the buffer duration) by the average chunk size (provided in the metadata). The result is rounded down to obtain an integer value (Gurler *et al.*, 2012).

In addition to extending the window size, restricting a peer from scheduling chunks from content originator when the buffer duration is beyond a certain value is important for decreasing the workload of the server. Without this feature, a peer with high connection speed keeps growing its buffer and if it is ahead of other peers in the session, then it continues to request more chunks from the server since these chunks are not available any other peer. However, with this feature enabled, a peer does not spend precious bandwidth of the server for a chunk that is not time critical, leaving more resources to handle time-critical peers. Therefore, this policy aims at load balancing for the upload connection of the server among peers.

10.3.6 Multiple Requests from Multiple Peers of a Single Chunk

The BitTorrent protocol adopts the end-game-mode method, which allows a peer to request the same sub-piece from multiple peers multiple times when the downloading session is about to end and only a few numbers of pieces left to fully receive the content. The main idea behind the method is to avoid delays due to an unfinished sub-piece that is downloaded from a peer with limited upload capacity (weak peer). To this end, requesting the same sub-piece from multiple peers increases the chance of receiving the remaining sub-pieces in a short time and avoids prolonging delays. However, one must be aware that this policy is enabled only at the very end of the downloading session because even a weak peer may continue to contribute to chunk retrieval at a slow pace in the middle of a session and does not prolong the total duration of content retrieval.

In video streaming, the timeliness concern is not restricted to the end of the download session and there is always the chance of receiving a chunk after its play-out deadline has been surpassed due to a weak peer. However, unlike the BitTorrent case, it is not a good practice to request every chunk from multiple peers in order to increase the chance of receiving a chunk on time as this approach can lead to a significant amount of redundant data and decrease link utilization. Instead, a peer should be aware of the state of the network and behave adaptively. First, the peer should always trace the rate of data transmission with its neighbors and mark the peers as *risky* if chunks are received at a pace that slower than real time. (This is the case in which duration of receiving a chunk is longer than the duration of its content.) Afterwards, if the available-buffer duration is large it is safe to request chunks from *risky* peers. This is analogous to BitTorrent's regular downloading policy where end-game-mode is off and content is requested from any neighbor even if the rate of transmission is slow because there is no timeliness concern. In the other extreme, if the available buffer duration is low the P2P engine should avoid requesting chunks from risky peers. It is also possible that if the state of the link to a neighbor has recently worsened and the last chunk is about to be delayed then such chunks can be requested from trusted peers/seeders.

10.4 Stereoscopic Video Streaming over P2P Networks

10.4.1 Stereoscopic Video over Digital TV

Stereoscopic 3D broadcast has already begun over digital TV channels in frame-compatible format, which is based on pixel subsampling to keep the frame size and rate the same as that of monocular video. For instance, side-by-side format applies horizontal subsampling to the left and right frames and reduces the resolution by 50%. Then, the subsampled frames are stitched together to form a single frame and encoded as a monocular video frame using standard video codecs such as H.264/AVC. Although the spatial resolution is same as a monocular video, the bit rate of the encoded frame-compatible stereoscopic video can be higher than its monoscopic counterpart because of the increased prediction error at the border of two views and increased complexity due to spatial scaling.

Stereoscopic frame-compatible 3D broadcast is a simple solution because it only requires a 3D-ready display without any infrastructural change. However, it has a number of drawbacks that cannot be resolved in a frame-compatible transmission mechanism. First, the perceived visual quality can be inferior to that of monocular video because subsampled stereo views have decreased horizontal spatial resolution. Switching to full-resolution stereo format without any infrastructural modification is not possible because of the increased (almost doubled) bit-rate requirement. Although the extra bandwidth requirement may be kept around 50% using multiview video coding (MVC), using an advanced codec would still require modifications at both the sender (encoder) and receiver (decoder) sides. The second drawback of using frame-compatible video broadcast is the inability to change the viewing point in stereoscopic 3D video, restricting a user to watch a scene from a fixed point of view. On the other hand, it could have been possible to change the viewing point if different stereo-pairs are streamed to an end user based on their preferences. Naturally, legacy video broadcast solutions cannot provide user-specific content and restricted to the same video for all users. Considering this limitation of legacy broadcast systems, the flexibility of IP is obvious because IP can deliver any content (two-dimensional (2D) video, stereoscopic 3D video, or MVV) with custom use specifications such as resolution, quality, viewing angle, and the number of views.

10.4.2 Rate Adaptation in Stereo Streaming: Asymmetric Coding

When advancing from streaming monoscopic video to streaming stereoscopic 3D video, there are two important challenges besides handling the increased number of views and increased bit-rate requirements. First, an efficient video coding scheme must be developed that will minimize the redundancy between stereo-view pairs and decrease the overall bit rate. Next, intelligent adaptation methods must be defined that will allow source rate scaling with minimum noticeable visual artifacts.

Both of these challenges are dominated by features of human visual perception. For example, in the monoscopic video case, researchers have revealed that the human visual system is more sensitive to brightness (luminance) than to the color information (chrominance). Based on this fact, the standard digital video frame format (YCrCb, 4:2:0) has been formed in a way that the brightness component (Y) is represented using more bits (twice the number) to achieve finer quantization when compared against color components (UV). In other words, asymmetric rate allocation is performed between luminance and chrominance to achieve the best perceived quality.

It is also possible to apply the same approach for the stereoscopic video case. Studies of the human visual system on the perception of stereoscopic 3D content have shown that the human visual system can perceive high frequencies in 3D stereo video, even if that information is present in only one of the views. Stelmach *et al.* (2000) compared spatial and temporal asymmetry among views and concluded that applying spatial asymmetry between right and left stereo-pairs using low-pass filtering causes only minor artifacts that may be neglected by the observers. On the other hand, the artifacts due to applying temporal asymmetry are more noticeable and also content dependent. Therefore, temporal asymmetry should be avoided. Based on this finding, some researchers believe that the best perceived 3D stereo video quality may be achieved by using asymmetric video coding in which the reference and auxiliary views are coded at unequal quality levels; for example, the peak signal-to-noise (PSNR) value. However, the question of what should be the level of the asymmetry between the PSNR values is an important question, which we elaborate in the following.

Determining the Level of Asymmetry: There exists a PSNR threshold such that when the view whose PSNR is reduced is encoded above this PSNR, the perceived 3D stereo video quality is dominated by the higher quality view and visual artifacts are perceptually unnoticeable. Moreover, the threshold value may be dependent on the 3D display technology. To test this proposition and determine the so-called just-noticeable asymmetry level in terms of a PSNR threshold, Saygili *et al.* (2011) have generated various stereo video pairs at different unequal quality levels using different test sequences and conducted subjective tests using different stereoscopic displays. Their subjective test results indicate that if the PSNR value of the low-quality pair of the stereoscopic views is above ∼32 dB thresholds then assessors may neglect the quality degradation in one view and the overall perception of the content is at the quality of the high-quality pair. Interestingly, this threshold does not vary among different test sequences covering different types of contents, and the effect of different display technologies is only minor. It has also been noted that ∼32 dB threshold is a roughly the boundary for the blockiness artifacts due to video coding and suggests that such artifacts may not be concealed by the human visual system.

Asymmetric Video Coding: There are multiple possible video coding schemes for achieving asymmetric rate allocation and there are two major criteria to evaluate each method. First, the encoding scheme should have high rate-distortion performance, achieving high PSNR value for a given bit rate. The second criterion should have a wide range of adaptation capability to enable smooth streaming in different network environments.

Nonscalable codecs such as H.264/AVC and its MVC extension are superior to the scalable video coding approach in terms of rate distortion performance. MVC can be considered as the benchmark because it introduces inter-view dependencies and decrease the overall redundancy. Unfortunately, neither H.264/AVC nor its multiview extension provides graceful quality degradation, leaving the SVC as the only option for enabling adaptive streaming.

When SVC is utilized, asymmetric coding can be achieved in two ways. The trivial method is to encode both views using SVC, as depicted in Figure 10.3. In this method, the base layer is encoded around 32 dB and the enhancement layer is encoded at maximum

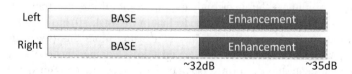

Figure 10.3 Asymmetric coding using SVC for both views

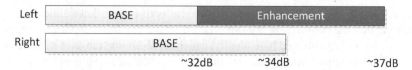

Figure 10.4 Asymmetric coding using SVC only for one view

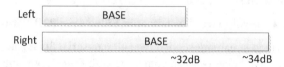

Figure 10.5 Asymmetric coding using SVC only for one view (rate scaled)

quality for a given bit rate. When the available channel capacity is not enough to transmit the content at full quality, only one of the enhancement layers is discarded first, to enforce asymmetry among quality of stereo pairs. In this approach, since both views are coded using SVC, there exists considerable encoding overhead.

Alternatively, it is possible to use SVC only for one view, as depicted in Figure 10.4. In this approach the base layer is again encoded at around 32 dB. The second view is encoded above that value using a nonscalable codec, saving some bit rate that can be dedicated to increase the quality level of the enhancement layer for the first view. In this scheme, asymmetric coding is exploited at both extremes in terms of available bit rate. At one extreme, the available bandwidth is above the maximum rate of the video. According to the asymmetric coding theory, the first view (high quality) dominates the perceived quality in such a condition. At the other extreme, the enhancement layer is discarded and the nonscalable bitstream becomes the high-quality pair of the asymmetry, as depicted in Figure 10.5.

Table 10.1 provides sample results that can help us to compare the above-mentioned scalable asymmetric video coding schemes for three different 3D sequences. First, we compare

Table 10.1 Comparison of scalable asymmetric video coding schemes

		One view scalable					Both views scalable				
		Left (SVC)		Right (H.264)		Total	Left (SVC)		Right (SVC)		Total
		Bit rate	PSNR	Bit rate	PSNR		Bit rate	PSNR	Bit rate	PSNR	
Adile	Full quality	898	38.5	435	34.3	**1333**	649	36.1	650	36.1	**1299**
	Base quality	357	32.8	435	34.3	792	358	32.7	359	32.2	717
	Difference	541	5.7			541	291	3.4	291	3.93	582
Flower pot	Full QUALITY	1640	36.5	714	33.8	**2354**	1152	35.1	1115	35.3	**2267**
	Base quality	556	31.0	714	33.8	1270	515	31.3	494	31.3	1009
	Difference	1084	4.57			1084	637	3.8	621	4.07	1258
Train	Full quality	2458	37.1	1012	33.8	**3470**	1805	35.9	1743	35.8	**3548**
	Base quality	732	31.5	1012	33.8	1744	736	31.2	708	31.5	1444
	Difference	1726	5.6			1726	1069	4.7	1035	4.34	2104

the total size of the encoded streams and we see that both options have almost the same bit rate, showing that the comparison is a fair one. The same condition applies for the total difference between full quality and base quality, which represent the range of scalability. While these conditions are equal, we can see that the one view scalable coding option yields higher PSNR values, indicating its superiority.

10.4.3   Use Cases: Stereoscopic Video Streaming over P2P Network

10.4.3.1   Use Case 1: Asymmetric Stereoscopic Video over P2P Networks

A stereoscopic 3D video streaming service over IP can serve 3D content at full resolution, yielding superior perceived quality when compared with frame-compatible services. In such a service, one of the views may be encoded using SVC, whereas the other one using H.264/AVC as described in Section 10.4.2, creating three different types of chunks; right-view chunks, base-layer chunks of left view, enhancement-layer chunks of left view. The first two streams (right view and base layer of left view) have equal importance and should be prioritized against the enhancement layer of the left view. Once the chunks and the corresponding metadata file are created, a P2P session can be initiated to start sending and receiving video chunks.

10.4.3.2   Use Case 2: Full-Resolution Stereoscopic Video Delivery over Synchronized DVB and IP Channels

A more interesting application is to synchronize the video delivery over IP with the broadcast over DVB channel. The main motivation for this approach is to utilize the robustness of DVB, which reliably provides stereo views at reduced resolution while the unreliable IP channel provides the auxiliary content that contains the difference of spatially reduced frames from their full resolution counterparts.

Figure 10.6 summarizes such a service in which the difference signal is computed, encoded, chunked and then forwarded to peers over the IP channel. Meanwhile, the DVB broadcast provides the low-resolution stereo pairs, which are *upsampled* upon reception. Then, if the IP content is available, the difference is added to achieve full-resolution stereoscopic 3D. One key issue is the feedback from the DVB receiver that will force the P2P client to download corresponding video chunks and maintain the synchronicity between the two channels.

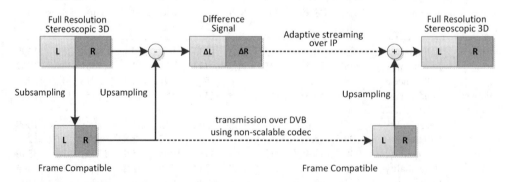

Figure 10.6   Synchronized full-resolution stereoscopic video delivery

10.5   MVV Streaming over P2P Networks

10.5.1   MVV Streaming over IP

Today, there are no known streaming services that provide MVV content to home users. As explained briefly in Section 10.1, the fundamental reasons for this can be listed as: (i) lack of specifications for MVV, such as resolution and number of views, making it difficult to create universal content that is suitable for all multiview displays; (ii) heterogeneous bandwidth requirement of different multiview displays, making it infeasible to perform transmission over fixed bit-rate channels; (iii) high volume of multiview contents, rendering frame-compatible format useless and forcing alternative methods to decrease the bit-rate requirement; (iv) lack of availability for MVV content and, consequently, the lack of public interest and awareness in MVV; (v) low visual quality of today's MVV displays and concerns regarding image quality, brightness, sharpness, and eye discomfort.

There are also a series of proposed solutions to overcome these problems. The problem of heterogeneous specifications for different MVV displays can be alleviated by an application at the receiver side that can generate artificial views by interpolating the available information. Then, it would be possible to convert any received signal into what is needed by the display. Depth-image-based rendering (DIBR) techniques would be a valuable tool for this purpose. In DIBR, a single channel frame that indicates the pixel depths is transmitted along with the corresponding view. Then, it becomes possible to use simple geometry to estimate the position of pixels from different directions, making it possible to render frames that will represent the scene from a different viewing angle. Moreover, depth maps have less high-frequency components (no textures) and can be encoded with high efficiency, easing the transmission of depth maps. Chapter 7 presents the details of this approach.

It is known that IP can serve different users at different rates and that, therefore, IP is a good candidate to overcome the diverse bandwidth requirements. Moreover, with the increased access options to the Internet, IP is available everywhere. Therefore, using IP decreases the deployment cost of such a streaming service, making it accessible by more people.

The high volume of the content can be addressed by two means. First is to use advanced video coding techniques such as MVC to decrease the redundancy by introducing inter-view dependences. Second is to use DIBR techniques that allow the receiver side to render multiple artificial views using low-cost depth maps instead of transmitting every view and decrease the overall bit-rate requirement. Nevertheless, the required bit rate for a large number of views can be at a critical level, making it difficult to serve by IP. This is the fundamental reason for utilizing the P2P network and distributing the burden of content transmission over multiple recipients.

The last two issues remain as a challenge, but it is natural to assume that the display quality will be improved over time. Once high-quality displays are available, it is also natural to assume that content generators would want to use MVV experience to create an immersive experience for their customers. Thus, we can trust that, over time, MVV contents will be available and that will draw the public attention and drag them to looking into MVV services.

10.5.2   Rate Adaptation for MVV: View Scaling

Although there are no publicly available MVV streaming services yet, there are studies in the literature and research projects that aim to deliver MVV content. The European project called DIOMEDES aims to establish a hybrid MVV distribution service that operates over both DVB and IP channel and utilizes P2P networks to minimize the server load (Aydogdu *et al.*, 2011). Similarly, there are other studies that suggest using IP to transmit, and it is evident that to establish a successful service one must investigate the rate adaptation options specific to MVV.

In monoscopic video, adaptation is simple and achieved by discarding the highest level of enhancement layer. In stereoscopic 3D, the adaptation process can be a bit more complex and involves considering the asymmetric rate allocation to exploit the features human visual systems. Although it is possible to think that view adaptation (switching to 2D) is another option in stereoscopic 3D, it has been revealed that this method leads to significant viewing discomfort because switching to 2D effectively is similar to closing one eye while watching. For MVV, however, even if we drop one view, the user can still watch the content in 3D, especially if the missing view is interpolated using depth maps. Moreover, it is also possible that dropping a particular view has no effects over the perceived quality if it is not in the user's field of view.

This fact reveals a key concept in rate adaptation for MVV streaming. Unlike a monoscopic video streaming, the prioritization among different layers is not obvious in MVV. The enhancement layer of one view may be more important than the base layer of another if it is not visible by the user. Since the client application cannot compute which views are currently visible such information should be provided to the adaptation engine by external means, such as a head-tracking system. Only then is it possible to perform efficient rate adaptation while transmitting the content over IP.

10.5.3   Use Cases: MVV Streaming over P2P Network

10.5.3.1   Use Case 1: Free-View TV for a Single Viewer using Head Tracking and Stereoscopic Display

As long as the view pairs on the screen can change according to a user's viewing position, a stereoscopic display can provide free-view functionality since the user will perceive only two views even in a multiview display at a given time. Based on this fact, it is possible to extend the stereoscopic video streaming scenario and introduce free-view functionalities by adding a head-tracking device, but such a service can only support a single person due to the single viewing point restriction of the stereoscopic 3D display.

At the server side, the adjacent views are encoded at asymmetric quality levels to exploit unequal rate allocation and achieve the best perceived quality. The tracking device detects the user's head position and informs the P2P client about the selected views. Based on this information, the P2P engine schedules corresponding video chunks. When the user requests another stereo pair, the P2P client notifies this information to neighboring peers and to the tracker server. In this way, a neighbor who does not share a common view can remove the peer from their list. Meanwhile, the tracker server also updates the peers' downloading state in order to match the ones that have a common interest.

10.5.3.2   Use Case 2: MVV Streaming for a Light-Field Display

As the second use case scenario, we present the setup in project DIOMEDES, which is a P2P video streaming service for the delivery of MVV content that is accompanied by object-based audio. In this setup, a seeder server provides the MVV chunks that are composed of base and enhancement-layer NAL units and the corresponding depth maps. Two baseline camera views and 5.1 audio are forwarded over a DVB channel that is synchronized with the remaining views that are streamed over IP. In order to drive the light-field display, all views must be received (unlike the stereoscopic case). Therefore, there exists a minimum bit-rate requirement to operate the setup. When it is necessary to perform rate adaptation, first the object-based audio is discarded, leaving only 5.1 audio. If further bit-rate reduction is imminent, then views (with their associated depth maps) are discarded one by one, starting from the enhancement layer. Then the intelligent rendering software mechanism tries to render the scene from the remaining depth and color views and conceal the missing views.

References

Aydogdu, K., Dogan, E., Gokmen, H.H. *et al.*, (2011) DIOMEDES: content aware and adaptive delivery of 3D media over P2P/IP and DVB-T2. NEM Summit, Implementing Future Media Internet, Turin, Italy, http://www.diomedes-project.eu/publications/NEMSummit_2011.pdf.

Baccichet, P., Noh, J., Setton, E., and Girod, B. (2007a) Content-aware P2P video streaming with low latency. Proceedings of the IEEE International Conference on Multimedia and Expo (ICME), Beijing, China.

Baccichet, P.P., Schierl, T., Wiegand, T., and Girod, B. (2007b) Low-delay peer-to-peer streaming using scalable video coding, in *Packet Video 2007*, IEEE, pp. 173–181.

Castro, M., Druschel, P., Kermarrec, A.-M. *et al.*, (2003) Split-stream: high-bandwidth content distribution in cooperative environments. Proceedings of the 9th ACM Symposium on Operating Systems Principles, New York, USA, pp. 298–313.

Cohen, B. (2003) Incentives build robustness in BitTorrent. Workshop on Economics of P2P Systems.

Dana, D., Li, D., Harrison, D., and Chuah, C.N. (2005) Bass: BitTorrent assisted streaming system for video-on-demand. Proceedings of the International Workshop on Multimedia Signal Processing (MMSP), pp. 1–4.

Eberhard, M., Mignanti, S., Petrocco, R., and Vehkaperä, J. (2010a) An architecture for distributing scalable content over peer-to-peer networks. Second International Conferences on Advances in Multimedia (MMEDIA), pp. 1–6.

Eberhard, M., Szkaliczki, T., Hellwagner, H. *et al.*, (2010b) Knapsack problem-based piecepicking algorithms for layered content in peer-to-peer networks, in *AVSTP2P'10 Proceedings of the 2010 ACM Workshop on Advanced Video Streaming Techniques for Peer-to-Peer Networks and Social Networking*, ACM, New York, NY, pp. 71–76.

Fesci, M., Tunali, E.T., and Tekalp, A.M. (2009) Bandwidth-aware multiple multicast tree formation for P2P scalable video streaming using hierarchical clusters, in *ICIP'09 Proceedings of the 16th IEEE International Conference on Image Processing*, IEEE Press, Piscataway, NJ, pp. 945–948.

Goyal, V.K. (2001) Multiple description coding: compression meets the network. *IEEE Signal Proc. Mag.*, **18** (5), 74–93.

Gurler, C.G., Savas, S., and Tekalp, A.M. (2012) A variable chunk size and adaptive scheduling window for P2P streaming of scalable video. Proceedings of the IEEE International Conference on Image Processing (ICIP), Orlando, Florida.

Jeon, J.H., Son, S.C., and Nam, J.S. (2008) Overlay multicast tree recovery scheme using a proactive approach. *Comput. Commun.*, **31** (14), 3163–3168.

Jurca, D., Chakareski, J., Wagner, J.-P., and Frossard, P. (2007) Enabling adaptive video streaming in P2P systems. *IEEE Commun. Mag.*, **45** (6), 108–114.

Mol, J.J.D., Pouwelse, J.A., Meulpolder, M. *et al.* (2008) Give-to-get: free-riding-resilient video-on-demand in P2P systems, in *Multimedia Computing and Networking 2008* (eds R. Rejaie and R. Zimmermann), Proceedings of the SPIE, Vol. **6818**, SPIE, Bellingham, WA, p. 681804.

Noh, J., Baccichet, P., Hartung, F., *et al.* (2009) Stanford peer-to-peer multicast (SPPM) overview and recent extensions. *PCS'09 Proceedings of the 27th Conference on Picture Coding Symposium*, IEEE Press, Piscataway, NJ, pp. 517–520.

Padmanabhan, V. N., Wang, H.J., and Chou, P.A. (2003) Resilient peer-to-peer streaming, in *Proceedings. 11th IEEE International Conference on Network Protocols, 2003*, IEEE, pp. 16–27.

Pouwelse, J., Garbacki, P., Epema, D., and Sips, H. (2005) The BitTorrent P2P file-sharing system: measurements and analysis, in *IPTPS'05 Proceedings of the 4th International Workshop on Peer-to-Peer Systems*, Springer-Verlag, Berlin, pp. 205–216.

Pouwelse, J., Garbacki, P., Wang, J. *et al.* (2008) Tribler: a social-based peer-to-peer system. *Concurr. Comput.: Pract. E.*, **20** (2), 127–138.

Saygili, G., Gurler, C. G., and Tekalp, A. M. (2011) Evaluation of asymmetric stereo video coding and rate scaling for adaptive 3D video streaming. *IEEE Trans. Broadcast.*, **57** (2), 593–601.

Setton, E., Baccichet, P., and Girod, B. (2008) Peer-to-peer live multicast: a video perspective *Proc. IEEE*, **96** (1), 25–38.

Stelmach, L., Wa, J.T., Meegan, D., and Vincent, A. (2000) Stereo image quality: effects of mixed spatio-temporal resolution. *IEEE Trans. Circ. Syst. Vid.*, **10** (2), 188–193.

Vlavianos, A., Iliofotou, M., and Faloutsos, M. (2006) BiToS: enhancing BitTorrent for supporting streaming applications. Global Internet Workshop in conjunction with IEEE INFOCOM, pp. 1–6.

Part Three

Rendering and Synthesis

11

Image Domain Warping for Stereoscopic 3D Applications

Oliver Wang,[1] Manuel Lang,[1] Nikolce Stefanoski,[1] Alexander Sorkine-Hornung,[1] Olga Sorkine-Hornung,[2] Aljoscha Smolic,[1] and Markus Gross[1]

[1]*Disney Research Zurich, Switzerland*
[2]*ETH Zurich, Switzerland*

11.1 Introduction

Post-production techniques for stereoscopic video are a key piece in the widespread adoption of three-dimensional (3D) technology. Many of the post-production practices used in two-dimensional (2D) cinematography cannot be directly applied to 3D; depth perspectives make it harder to "fake" scene layouts, and changes that occur in one view must be correctly reproduced in the second. Without the ability to correct for errors in stereo cinematography, easily edit film content, or adjust the 3D viewing volume after filming, stereo production will remain tedious and costly, and realizable only by highly trained stereographers and big-budget productions.

In this chapter, we will discuss a new class of methods called image domain warping (IDW) that present simple but robust solutions to several classic stereoscopic post-production problems.

One of the most important issues with producing stereoscopic content is related to the so-called accommodation/vergence conflict. In our optical system, perception of depth is greatly influenced by two parameters: retinal disparity (controlled by vergence) and focus (controlled by accommodation). A central challenge in stereoscopic movie production is that current display technologies only have indirect control over vergence, which is achieved by presenting a pair of slightly different images to the left and right eyes (the difference between these images is called image-disparity, which in the remainder of the chapter we will refer to as disparity, but should not be confused with actual retinal disparity). Accommodation cannot be controlled at all; we have to focus our eyes on the screen surface, even if objects are positioned in three dimensions in front or behind the screen surface (see Chapter 15 for more information).

Emerging Technologies for 3D Video: Creation, Coding, Transmission and Rendering, First Edition.
Frédéric Dufaux, Béatrice Pesquet-Popescu, and Marco Cagnazzo.
© 2013 John Wiley & Sons, Ltd. Published 2013 by John Wiley & Sons, Ltd.

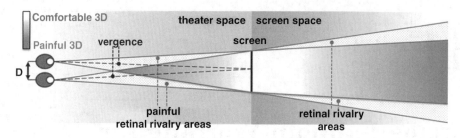

Figure 11.1 A diagram showing the comfort zone where viewers can converge on 3D points without strain. This zone is positioned relative to the location of the screen, and extends a short distance on either side. The limits of this comfort zone require stereoscopic cinematographers to map the arbitrary depth range of the world into this compressed zone around the screen

Several problems arise from this conflict. For one, if we focus on one object, things that are out of focus cannot be fused by our visual system owing to too large retinal disparities and will appear as double images (Howard and Rogers, 2002). There is, therefore, a restricted disparity range around the *horopter*, called Panum's area, which permits proper stereo vision and depth perception. In addition, these accommodation/vergence conflicts lead to considerable viewing problems ranging from distortions of the 3D structure of a scene to visual fatigue (Hoffman *et al.*, 2008).

In production, the admissible disparity range is commonly referred to as the *comfort zone* (see Figure 11.1). One of the most fundamental tasks of stereographers is to map the arbitrary depth range of the real world into this comfort zone. This is done by careful modification of the stereo camera baseline and convergence settings during filming. However, using only these limited degrees of freedom greatly restricts the creativity of cinematographers and can be difficult to accomplish when filming live or unpredictable content. One of the motivations of stereo post-production is to allow modification of this disparity range to fit the comfort zone *after* filming has occurred. Other tasks requiring stereo processing involve remapping depth ranges to compensate for different viewing conditions, user preference, or artistic depth expression.

All of the above-mentioned stereo editing tasks can be thought of as instances of *virtual view synthesis*, where editing is achieved by synthesizing novel views at different positions along the camera's initial baselines.

11.2 Background

Virtual view synthesis techniques require knowing the 3D scene geometry to correctly render new views. However, because obtaining 3D scene models is often impractical for real-world video footage, the most common method for representing the required geometry is in the form of depth maps, which describe the distance from each pixel to the scene. When combined with knowledge about camera position, orientation, and focal length, this can be used to compute a 3D point location for each pixel. A commonly used technique for creating images from image plus depth-map pairs (Figure 11.2) is called depth-image-based rendering (DIBR) (Chang *et al.*, 1999; Zitnick *et al.*, 2004).

For example, let us say that we desired to take a stereoscopic video, and compress the disparity range by reducing the camera baseline by a factor of two (a common stereoscopic

Figure 11.2 An image with corresponding depth map

processing operation). The traditional solution using DIBR would be to project pixels from the left and right views into 3D world coordinates. Then, re-project this scene into two new virtual camera positions along the baseline to create the final images. While simple and physically correct, like most problems, the devil is in the details, and DIBR is a notoriously difficult problem to get right.

This is due to a couple of significant challenges. For one, owing to its design, DIBR is highly dependent on pixel-perfect depth maps. Creating these is a very difficult problem. While techniques such as pre-filtering depth maps can be used to reduce artifacts, automatic methods for computing depth maps are not robust enough to be used for most applications. Furthermore, the type of artifact that depth-map errors generates is immediately visible, as high-frequency stray pixels do not correspond to any believable real-world situations.

In this chapter, we discuss an alternate approach called IDW. IDW attempts to solve the same problem as DIBR, but does so with an approach that is more robust to deficiencies of the depth map. This robustness is achieved at the expense of physical realism and, as such, introduces a different class of artifacts. However, it has been found that the lower frequency lapses in realism associated with IDW are often less detectable to viewers than the most common artifacts that arise from DIBR (MPEG, 2012). Furthermore, by operating strictly in the image domain, no calibration information of the cameras is needed, making the method more applicable to production workflows.

11.3 Image Domain Warping

We define a warp as a function that deforms space (Figure 11.3):

$$w : \mathbb{R}^2 \to \mathbb{R}^2 \tag{11.1}$$

In most of the examples that we discuss here, w is both continuous and C1-smooth, which prevents tearing artifacts in the warped images and, as we will discuss later, avoids disocclusion inpainting.

Image warps have a long history of use in computer vision and graphics-based problems. One area that has seen a large amount of development in image warping is in the problem of aspect-ratio retargeting, for which we refer to the recent survey paper of Rubinstein *et al.* (2010). In these cases, the image warp is used to specify a mapping from an image at one aspect ratio to an image at another aspect ratio.

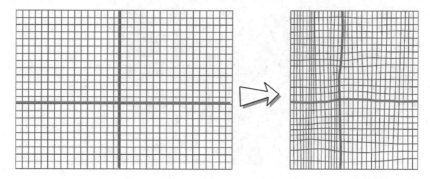

Figure 11.3 A visualization of a warping function that deforms an input image. A grid is placed over an image and deformed. Image content is then rendered into each of these deformed grid cells to produce the final warped image

The goal of our stereo editing tasks is to compute some warping of the initial stereo images that produces an image pair that exhibits different *stereo* properties. To do this, we formulate a quadratic error term that combines point-based constraints that enforce specific stereo properties at feature locations with a saliency weighted smoothness constraint that forces distortions into less important areas. A linear solve can then be used to compute new warped grid cells. The details of this algorithm will be discussed in Section 11.6.

11.4   Stereo Mapping

The first IDW-based stereo post-production task that we will discuss is stereo mapping. It is often desirable to fix stereo disparities after filming, due to conflicts that arise due to our fundamental perceptual and technical limitations. However, once these conflicts are sufficiently minimized, our depth perception is quite robust, even if the resulting 3D scene is not geometrically consistent. This is due to the additional depth information from cues such as relative size and order of objects in a scene, and motion parallax (Siegel and Nagata, 2000). In this section, we exploit these properties and present an algorithm for automatic approximate disparity remapping based on stereoscopic warping (Lang *et al.*, 2010).

11.4.1   Problems in Stereoscopic Viewing

As motivated in Section 11.1, stereoscopic 3D (S3D) production and display is an extremely complex field, involving a broad range of research and experience on human visual perception (Howard and Rogers, 2002; also see Chapter 18: 3D Media and the Human Visual System), display technology (Hoffman *et al.*, 2008), and industrial best practice (Mendiburu, 2009). We will begin by discussing some common problems related to disparity perception, proposing a set of disparity mapping *operators* which formalize these ideas, and describing the specific warping that we use to achieve these disparity edits.

11.4.2   Disparity Range

The maximal and minimal disparity an output device can create is limited. These disparity range limitations are the most obvious issue in stereoscopic perception and production. If the disparity exceeds one of the limits then the scene will violate the stereo comfort zone and stereo fatigue will occur. Concrete applications for disparity range mapping are the adaptation of stereoscopic content to display devices of different size, the uniform compression of scene depth for a more comfortable viewing experience, or moving scene content forward or backward by adding a constant offset to the disparity range. In practice, such a mapping can be achieved by modifying the camera baseline (the interaxial distance) during filming and by shifting the relative position of the left and right view after filming to control the absolute disparity offset.

Another reason that ranges often must be remapped is objects that are floating in front of the screen and intersect with the image borders will cause retinal rivalry. In post-production this can be corrected using the floating window technique, which is a virtual shift of the screen plane towards the viewer (Mendiburu, 2009). In general, however, such adaptations have to be performed by expensive and cumbersome manual per-frame editing, since the camera baseline of recorded footage cannot be easily modified.

11.4.3 Disparity Sensitivity

Our ability to discriminate different depths decreases with increased viewing distance. One result from perceptual research is that the stereoacuity is inversely proportional to the square of the viewing distance (Howard and Rogers, 2002). This means that our depth perception is generally more sensitive and accurate with respect to nearby objects, while for distant objects other depth cues such as occlusion or motion parallax are more important (Banks *et al.*, 2004).

This effect can be exploited in stereoscopic movie production by compressing the disparity values of distant objects. For example, a disadvantage of the previously mentioned linear range adaptation is that strong disparity range reduction leads to apparent flattening of objects in the foreground. Using the insights about stereoacuity, the decreased sensitivity to larger depths can be used to apply *nonlinear* remapping instead, resulting in less flattening of foreground objects. Effectively, this corresponds to a compression of the depth space at larger distances. This idea can be extended to composite nonlinear mapping, where the disparity range of single objects is stretched, while the space in between the objects is compressed. Such nonlinear operations which exploit the limitations in sensitivity of our visual system have been successfully employed in related areas such as media retargeting. But so far, they are difficult to apply to stereoscopic footage of live action, since this would require an adaptive modification of the camera baseline. In production, the only way to achieve such effects is complex multi-rigging by capturing a scene with camera rigs of varying baseline and manual composition in post-production.

11.4.4 Disparity Velocity

The last important area is the temporal aspect of disparity. For real-world scenes without conflicting stereo cues, it has been shown that our visual system can rapidly perceive and process stereoscopic information. The reaction time, however, can increase considerably for conflicting or ambiguous cues, such as inconsistent vergence and accommodation. Moreover, there is an upper limit to the temporal modulation frequency of disparity (Howard and Rogers, 2002).

These temporal properties have considerable importance in the production of stereoscopic content. In the real world we are used to disparities varying smoothly over time. In stereoscopic movies, however, transitions and scene cuts are required. Owing to the above-mentioned limitations, such strong discontinuities are perceptually uncomfortable and might again result in the inability to perceive depth (Mendiburu, 2009). Therefore, stereoscopic film makers often employ a continuous modification and adaption of the depth range at scene cuts in order to provide smooth disparity velocities, so that the salient scene elements are at similar depths over the transition. Additionally, such depth discontinuities can be exploited explicitly as a storytelling element or visual effect and are an important tool used to evoke emotional response.

11.4.5 Summary

In summary, we have discussed a few central aspects of disparity mapping, which we utilize to design the disparity mapping operators:

- **Disparity Range:** Mapping of the global range of disparities; for example, for display adaptation.
- **Disparity Sensitivity:** Disparity mapping for global or locally adaptive depth compression and expansion.
- **Disparity Velocity:** Temporal interpolation or "smoothing" between different disparity ranges at scene transitions.

11.4.6 Disparity Mapping Operators

Without loss of generality, we assume that the input footage is recorded with a stereo camera rig and is approximately rectified. For a digital stereo image pair (I_1, I_r), let $\vec{x}_1 \in \mathbb{R}^2$ be a pixel position in the left image I_1.

We define the disparity $d(\vec{x}_1) \in \mathbb{R}$ as the distance (measured in pixels) to the corresponding pixel in I_r (and vice versa); or, in other words, $d(\vec{x}_1) = \vec{x}_1 - \vec{x}_r \rightarrow$. The range of disparities between the two images is an interval $\Omega = [d_{min}, d_{max}] \subset \mathbb{R}$. The disparity mapping operators are now defined as functions $\phi : \Omega \rightarrow \Omega'$ which implement the rules and guidelines described in Section 11.4.5 by mapping an original range Ω to a new range Ω'.

11.4.7 Linear Operator

Globally linear adaptation of a disparity $d \in \Omega$ to a target range $\Omega' = [d'_{min}, d'_{max}]$ can be obtained by a mapping function:

$$\phi_1(d) = \frac{d'_{max} - d'_{min}}{d_{max} - d_{min}} (d - d_{min}) + d'_{min} \tag{11.2}$$

By changing the interval width of Ω', the depth range can be scaled and offset such that it matches the overall available depth budget of the comfort zone.

11.4.8 Nonlinear Operator

Global nonlinear disparity compression can be achieved by any nonlinear function; for example:

$$\phi_n(d) = \log(1 + sd) \tag{11.3}$$

with a suitable scale factor s.

For more complex, locally adaptive nonlinear editing, the overall mapping function can be composed from basic operators. For example, given a set of different target ranges $\Omega_1, \ldots, \Omega_n$ and corresponding functions $\phi_0, \ldots, \phi_n$, the target operator would be

$$\phi_a(d) = \begin{cases} \phi_0(d), & d \in \Omega_0 \\ \ldots & \ldots \\ \phi_n(d), & d \in \Omega_n \end{cases} \tag{11.4}$$

An elegant approach to generate such complex nonlinear functions in a depth authoring system is to use the histogram of disparity values and identify dominant depth regions, or to analyze the visual saliency of scene content in image space. From these importance values, which essentially correspond to the first derivative ϕ'_a, the actual disparity operator can be generated as the integral

$$\phi_a(d) = \int_0^d \phi'_a(x)\, dx \tag{11.5}$$

11.4.9   Temporal Operator

Temporal adaptation and smoothing, as is required for smooth scene transitions or visual effects, can be defined by weighted the interpolation of two or more of the previously introduced operators; for example:

$$\phi_t(d, t) = \sum_i w_i(t)\phi_i(d) \tag{11.6}$$

where $w_i(t)$ is a suitable weighting function.

11.5   Warp-Based Disparity Mapping

In Section 11.4 we showed how disparity is perceived and derived a set of handy disparity mapping operators. In this section we will now discuss how those disparity mapping operators can be applied as a stereo post-processing technique.

11.5.1   Data Extraction

To apply the disparity mapping operators, we first detected a sparse set of disparity features F in the input images (Figure 11.4). These sparse disparities are estimated in an automatic, accurate, and robust way. While any number of methods can be used, two methods were applied in the results shown in this chapter. The first method (Zilly *et al.*, 2011) computes descriptors and finds matches between features in both input images. Descriptor-based methods are characterized by their high robustness and accuracy, but can lead to clustering of features in a few image regions. For this reason, additional features and disparities were added with a second method such that they cover a wider array of scene points. These weaker features are then matched using a Lucas–Kanade method. In the end, disparity outliers are detected and removed using the RANSAC (random sample consensus) algorithm.

Figure 11.4 Sparse feature correspondences (see Plate 2 for the colored figure)

Additionally, the image saliency maps (Figure 11.5) are extracted and used to prevent artifacts from being introduced into salient regions during warping. This is achieved by using a saliency map, which is automatically estimated from the input images (Guo *et al.*, 2008). The saliency map indicates the level of visual significance for each image pixel. Information about image saliency is then explicitly used during the warp calculation.

11.5.2 Warp Calculation

The warp calculator computes now two warps which are needed to synthesize a new output stereoscopic image pair. Each warp is computed as the result of an energy minimization problem. The energy functional $E(w)$ is defined, which provides in its minimum a warp w that leads to the desired change of disparities as required for the virtual view synthesis, with the minimal distortion in salient regions. The energy functional consists of four terms which are related to a particular type of constraint, as described below. Each term is weighted with a parameter λ:

$$E(w) = \lambda_d E_d(w) + \lambda_s E_s(w) + \lambda_t E_t(w) + \lambda_e E_e(w) \tag{11.7}$$

Disparity Constraints. Disparity constraints are derived from the extracted disparities $(\vec{x}_l, \vec{x}_r) \in F$ and the desired disparity mapping operator. For each correspondence pair we require

$$w_l(\vec{x}_l) - w_r(\vec{x}_r) - \phi(\vec{x}_l - \vec{x}_r) = 0 \tag{11.8}$$

Figure 11.5 Saliency maps

meaning that the disparity of a warped correspondence pair $(w_l(\vec{x}_1), w_r(\vec{x}_1))$ should be identical to applying the disparity mapping operator ϕ to the original disparity $d(\vec{x}_1)$. Since the above constraints only prescribe relative positions, we require a small set of absolute position constraints (for features $\mathbb{F}^* \subset \mathbb{F}$) which fix the global location of the warped images. The warped positions are defined by the average previous position and the mapped disparity:

$$w_l^*(\vec{x}_1) := \frac{\vec{x}_1 + \vec{x}_r}{2} + \frac{\phi(\vec{x}_1 - \vec{x}_r)}{2}$$

$$w_r^*(\vec{x}_r) := \frac{\vec{x}_1 + \vec{x}_r}{2} - \frac{\phi(\vec{x}_1 - \vec{x}_r)}{2}$$

(11.9)

This defines the basic stereoscopic warping constraints so that the warped images match the target disparity range Ω'. The energy term for minimization of the disparity constrains is therefore

$$E_d(w) = \sum_{(\vec{x}_1, \vec{x}_r) \in \mathbb{F}} \left\| w_l(\vec{x}_1) - w_r(\vec{x}_r) - \phi(\vec{x}_1 - \vec{x}_r) \right\|^2$$

$$+ \sum_{(\vec{x}_1, \vec{x}_r) \in \mathbb{F}^*} \left\| w_l(\vec{x}_1) - w_l^*(\vec{x}_1) \right\|^2 + \left\| w_r(\vec{x}_r) - w_r^*(\vec{x}_r) \right\|^2$$

(11.10)

Smoothness Constraints. This term measures the distortion of quad faces of the warping grid. It penalizes local warp deformations by increasing the cost if quad edges change their angle or length:

$$E_s(w) = \sum_{(a,b,c,d)} S[a, b, c, d]\{ \left\| w(b) - w(a) - b + a \right\|^2 + \left\| w(c) - w(b) - c + b \right\|^2$$

$$+ \left\| w(d) - w(c) - d + c \right\|^2 + \left\| w(a) - w(d) - a + d \right\|^2 \}$$

(11.11)

The cost for each quad with vertices (a, b, c, d) is weighted with the average saliency $S[a, b, c, d]$ of this quad. Consequently, the warp is stiffer in salient regions and allows stronger distortions in less salient regions.

Temporal Constraints. Temporal constraints are applied to minimize temporal artifacts. If w^t denotes the warp at time instant t, then the energy term for one warp to be minimized is

$$E_t(w^t) = \sum_v \left\| w^t(v) - w^{t-1}(v) \right\|^2$$

(11.12)

Vertical Edge Constraints. Straight vertical image edges are particularly important for the stereo fusion. To prevent vertical image edges from bending, the following constraints are introduced. Assuming that a vertical edge e is described by a set of 2D positions on that edge $e = \{p_1, \ldots, p_n\}$, the energy term to be minimized for all edges is

$$E_e(w) = \sum_e \sum_{p,q \in e} \left\| w_x(p) - w_x(q) \right\|^2$$

(11.13)

where $w_x(p)$ represents the x-coordinate of $w(p) = (w_x(p), w_y(p))$.

Figure 11.6 Left: stereo correspondences and the disparity histogram for the cow example. Right: close-ups of the warped stereo pair showing the deformed isolines with respect to the input views (see Plate 3 for the colored figure)

Energy Minimization. After specifying the four terms of the energy functional

$$E(w) = \lambda_d E_d(w) + \lambda_s E_s(w) + \lambda_t E_t(w) + \lambda_e E_e(w) \qquad (11.14)$$

the warp F is computed by finding the minimum of the functional by solving a sparse least-squares system. The degree in which the postulated constraints are fulfilled is controlled by the weights λ_d, λ_s, λ_t, and λ_e. These weights are sequence independent. In each time instant, warp calculation computes two warps, one for each input image. Finally, each computed warp is filtered using a bilateral filter. Bilateral filtering mainly removes noise from straight edges in the synthesized images. In the case of input images with full HD resolution, experiments showed that solving for warps with a resolution of 180×100 is sufficient in terms of synthesis quality.

Solving the least-squares system yields the coordinates of the two warp grids (shown in Figure 11.6). These warp grids can now be used to synthesize the output image. This can for example be done efficiently with bilinear interpolation with existing graphics hardware (*uv* texture mapping).

11.5.3 Applications

The stereoscopic warping framework presented can be applied to solve different practical problems. We discuss three production scenarios in this section: nonlinear and linear editing for post-production, automatic disparity correction, and display adaptation.

The first major application area for disparity mapping operators and warping is post-production of stereoscopic content. Using the proposed methodology and algorithms, depth composition can be modified as desired using different types of nonlinear and linear disparity operators as artistic tools. Figure 11.7 shows a nonlinear disparity mapping operation (note the change of disparity on the cow's nose versus the cow in the background).

Figure 11.8 shows an automatic disparity correction; the car comes out of the screen, causing a window violation on the left-hand side of the screen. This can be automatically corrected by remapping the scene depth using IDW.

Figure 11.7 Nonlinear disparity remapping. The disparity range of the original (left) is quite large leading to diplopia on large screens. Our nonlinearly remapped image (right) pushes the cow behind the screen, while leaving the background depth untouched (see Plate 4 for the colored figure)

Figure 11.8 Automatic correction of disparity. The original stereo pairs are shown on the left; our result is on the right. The cropped racing car captured with strong negative disparities and a large overall scene depth results in the so-called framing problem. With a simple linear disparity scaling the car is pushed behind the screen without increasing the background disparity (see Plate 5 for the colored figure), © KUK Filmproduction GmbH, All rights reserved

The third application area of stereo mapping technology that we illustrate here is 3D display adaptation. The background is that the 3D content that was optimized for certain viewing conditions consisting of display size and viewing distance will look different for other viewing conditions. In order to keep a good 3D impression and to preserve the artistic intention, disparity adaptation is necessary when reformatting content, for example, from cinema to TV or even to a handheld device (Figure 11.9).

Figure 11.9 Display adaptation into both directions. The middle image is the original stereo pair, while the left image features a linear reduction of the disparity range to 50% and the right image an increase to 200%, © www.vidimensio.com (see Plate 6 for the colored figure)

11.6   Automatic Stereo to Multiview Conversion

S3D content production workflows, distribution channels, data compression techniques, and display technologies are constantly improving and being adapted to enable a higher quality experience, for example, in the 3D cinema or on S3D displays in the home. However, the necessity to wear glasses is often regarded as a main obstacle of today's mainstream 3D display systems. Multi-view autostereoscopic displays (MADs) allow glasses-free stereo viewing and support motion parallax viewing in a limited range. In contrast to stereo displays, MADs require not two, but multiple different views as input. Research communities and standardization bodies investigate novel 3D video formats, which on the one hand are well compressible and on the other hand enable an efficient generation of novel views as required by MADs (Farre *et al.*, 2011).

11.6.1   Automatic Stereo to Multiview Conversion

Figure 11.10 shows a block diagram of the view synthesizer, where two-view video is converted to N-view video by view synthesis ($N > 2$). The view synthesis algorithm generates N output images for each time instance. This happens in the following four steps:

- **Data Extraction:** image saliency and vertical edges are extracted from each of the input images. In addition, a sparse set of features between the two images is extracted.
- **Warp Calculation:** based on the previously extracted data, two image warps are calculated.
- **Warp Interpolation/Extrapolation:** based on the previously calculated warps, a set of $2N$ image warps is computed by interpolation or extrapolation. The positions of the N views, as specified by a particular multiview display, are provided here as input to the warp interpolator/extrapolator in order to indicate which warps to compute.
- **IDW:** based on the previously interpolated/extrapolated image warps and the two input images, N output view images are synthesized by IDW.

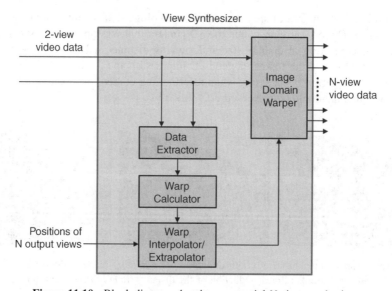

Figure 11.10   Block diagram showing a potential N-view synthesizer

11.6.2 Position Constraints

A warp formulation similar to the one described in Section 11.5 can be used following the same basic steps: computing features between the initial stereo pair and enforcing saliency-weighted smoothness. However, our position constraints are modified. Instead of remapping input disparities by some function ϕ, we instead want to create virtual views evenly spaced along the initial baseline. We do this by replacing the disparity constraints with a set of *position constraints* that define a warp mapping each of the initial images into a virtual camera position at the *center* of the initial baseline. This can be expressed by the following energy term:

$$E_p(w) = \sum_{(\vec{x}_l, \vec{x}_r) \in F} \left\| w(\vec{x}_l) - \vec{x}_l - \frac{\vec{x}_l - \vec{x}_r}{2} \right\|^2 \tag{11.15}$$

which is part of the energy functional

$$E(w) = \lambda_p E_p(w) + \lambda_s E_s(w) + \lambda_t E_t(w) + \lambda_e E_e(w) \tag{11.16}$$

similar to the functional shown in Section 11.5. The computed warp can then be interpolated to compute any number of in-between views or extrapolated to compute views outside of this range.

11.6.3 Warp Interpolation and Extrapolation

Warp interpolation and extrapolation are used to generate N views as required by the display system at arbitrary locations while only having to compute the actual warps once. Warps are extrapolated or interpolated according to

$$w_{out} = a w_{in} + (1 - a)q \tag{11.17}$$

where w_{in} is one of the warps calculated during warp calculation, q is a uniform warp, and a is a weight which is computed based on the desired output view position.

Depending on the position of the view to be synthesized, two warps are computed in the warp interpolator/extrapolator, w_{left} and w_{right}, which are used to warp the left and right input image, respectively. A new image I_{synth} is then synthesized according to

$$I_{synth} = I_{mask} \circ \Psi(I_{left}, w_{left}) + (1 - I_{mask}) \circ \Psi(I_{right}, w_{right}) \tag{11.18}$$

where I_{mask} is a binary mask or an alpha mask with values in $[0, 1]$, operator $\circ$ represents a component-wise multiplication, and operator

$$\Psi(I, w)[i,j] := I(w^{-1}[i,j]) \tag{11.19}$$

warps image I with a 2D image warp w (Wolberg, 1990).

Figure 11.11 shows eight views that were synthesized from a stereo image pair. The third and the sixth images are the original input, while the first two and the last two views are obtained by extrapolation and the fourth and fifth views are obtained by interpolation.

Figure 11.11 Eight synthesized views from most left to most right view (see Plate 7 for the colored figure)

11.6.4 Three-Dimensional Video Transmission Systems for Multiview Displays

In March 2011, the Moving Pictures Experts Group (MPEG) issued a Call for Proposals (CfP) (MPEG, 2011) with the goal to identify (i) a 3D video format, (ii) a corresponding efficient compression technology, and (iii) a view synthesis technology which enables an efficient synthesis of new views based on the proposed 3D video format. In the scope of this CfP, the view synthesis method presented in Section 11.6.3 was proposed as part of the transmission system shown in Figure 11.12. Multiview video coding was performed with a novel coding method presented in Schwartz *et al.* (2012). See Chapter 8 for more information on video formats and coding.

The proposed transmission system is compatible with existing stereo or multiview video formats. This is enabled by using the view synthesis method presented in this section, since it uses only M-view video data as input. In particular, no auxiliary data like depth maps are needed. A large subjective study coordinated by MPEG proved that multiview video coding in combination with IDW-based view synthesis leads to high-quality view synthesis results (Figure 11.13). Consequently, the proposal (Stefanoski *et al.*, 2011) was considered as one of the four winning proposals in MPEG (2012).

In order to reduce the expected computational complexity at the receiver side, a modified transmission system was recently proposed and investigated which transmits multiview video plus warp data (Stefanoski *et al.*, 2012). The modified system is shown in Figure 11.14. Thus, it is proposed to shift the warp extraction and warp calculation part to the sending side and, in addition to the multiview data, to efficiently compress and transmit the warp calculation result; that is, a restricted set of warps. The two modules shifted to the encoder represent the computationally most expensive parts of the view synthesizer. Thus, the view synthesis at

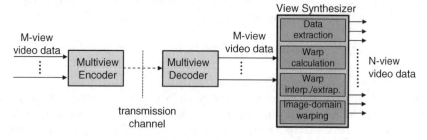

Figure 11.12 Diagram of transmission system

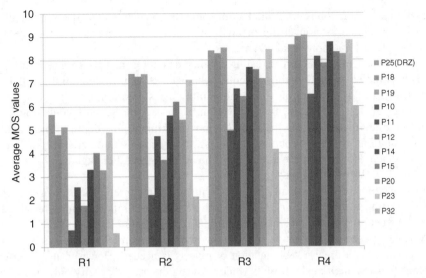

Figure 11.13 Subjective results for the different MPEG proposals in the multiview display test scenario. P25 indicates the proposal shown in Figure 11.12. Here, R1 to R4 indicate increasing target bit rates and a mean opinion score (MOS) value of above 8 indicates transparent visual quality

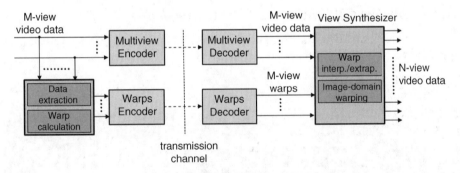

Figure 11.14 Modified transmission system

the receiver side is now a process which requires significantly less computational complexity than the system presented in Figure 11.12.

The modified system shown in Figure 11.14 leads to a significant reduction of the complexity at the receiving side and leads to an increase of the view synthesis speed up to a factor of 8.7. In addition, evaluation experiments showed that compressed warps represent a practically negligible portion of about 3.7% on average of the overall (video + warp) bit rate (Stefanoski *et al.*, 2012).

11.7 IDW for User-Driven 2D–3D Conversion

In the previous sections of this chapter, we have discussed using IDW to modify existing stereo footage and to create multiple views from a stereo pair. In both of these cases, initial feature correspondences between left and right views can be used to drive the warping.

However, this information is not available for some stereo applications; one such application is 2D–3D conversion, or converting monoscopic footage into stereoscopic footage. Please refer to Chapter 3 for a more detailed analysis.

The 2D–3D conversion requires creating a virtual camera from a position near the original view. This virtual view synthesis again requires knowing 3D information of the initial scene; unfortunately, in this application we cannot simply use stereo relations between a pair of images to estimate scene depth. Instead, we must guess at the scene layout from a single image. Fully automatic 2D–3D methods attempt to estimate this depth information from scene assumptions (such as ground/sky planes), local image characteristics (such as contrast), or motion. However, these assumptions are all not general enough to allow for robust conversion of real-world footage.

In this chapter, we will instead specifically discuss a technique for user-driven 2D–3D conversion using IDW (Wang *et al.*, 2012), where the depth information is provided by a user, and the goal is to make this interaction as pain free as possible.

The naive approach would be to simply draw the depth map. However, this is a difficult task in practice, as a precise segmentation of objects is needed. However, using IDW, it is possible to generate high-quality results while allowing the user to specify only sparse disparity cues, rather than accurate 3D geometry. This information is easily provided in the form of grayscale scribbles (Figure 11.15). The IDW algorithm can then convert this sparse input into a desired disparity range to generate a warped stereo pair with appropriate parallax.

11.7.1 Technical Challenges of 2D–3D Conversion

It is not possible to simply use the technique described in the above sections verbatim, as a couple of novel technical challenges present themselves in this application.

Figure 11.15 This shows the sparse input given by a user in the form of grayscale scribbles, as well as a 3D representation and a resulting disparity map (see Plate 8 for the colored figure)

The first of these is the resolution of the warp. The requirements for creating disparity from a single image are such that neighboring pixels on image borders must be moved to very different final positions. In order to accomplish this, the warp must be able to introduce higher frequency distortions into the image than would be required when simply modifying existing disparities. As such, for this application, large grid cells for warping are insufficient, and the solve must be done on a per-pixel basis. This requires more memory to compute, but allows the warp to make more detailed modifications to the scene.

Second, in order to create a convincing 2D–3D conversion, sharp discontinuities in the depth must exist around object borders. When editing stereo footage, such discontinuities already exist and need only to be increased or decreased. A single image, however, contains no disocclusion information at depth boundaries. If we use the continuous image warping method described above, we essentially reconstruct a wavy 3D surface on which the image is projected. This does not yield very convincing results, as our eyes are very sensitive to exactly these sharp jumps in depth.

To allow for the creation of disocclusions, we make a *piecewise-continuous* depth assumption, meaning that we expect the scene depth to vary smoothly, *except* at discontinuity locations. In order to incorporate this into our warp, we use a similar formulation as before, but allow some of the neighbor smoothness weights to be set to zero, allowing the warp to separate at these locations without penalty:

$$E_s(w) = \sum_{(a,b,c,d)} S[a,b,c,d]\{\delta(b,a)\|w(b) - w(a) - b + a\|^2 + \delta(c,b)\|w(c) - w(b) - c + b\|^2$$

$$+ \delta(d,c)\|w(d) - w(c) - d + c\|^2 + \delta(a,d)\|w(a) - w(d) - a + d\|^2\}$$

$$(11.20)$$

This can be expressed by the following modified smoothness term shown in the previous equation. Basically, we see that the same smoothness as before is augmented by a delta function, such that $\delta(i,j) = 0$ when there is an edge between pixels i and j and $\delta(i,j) = 1$ otherwise.

Ideally, we would like these discontinuities to exist at all depth edges. However, determining depth edges from an image is a very challenging task; image edges can correspond to either texture or depth edges, and depth edges can exist without image edges. Therefore, instead of trying to find depth edges, an oversegmentation of image edges is computed, containing both depth and texture edges (Figure 11.16). This is justified by two observations. First, depth edges that do not have a color difference are not important locations to create

Figure 11.16 This figure shows an example of an image edge oversegmentation, followed by a visualization of the synthesized disparity after the warping. Even though the warp is cut at many more locations than correspond to depth edges, sharp discontinuities in the disparity exist only at the important depth edges, due to the locations of user strokes

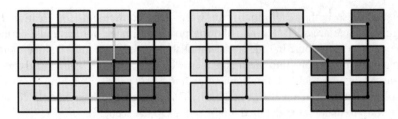

Figure 11.17 This figure shows how discontinuities are introduced in the warp. Edges that are colored gray have weight equal to zero. By cutting the smoothness term at these edges, the two regions separate while preserving their internal smoothness

disocclusions, as no 3D fusion will be possible in these regions. Second, the user input provides additional information, such that even if a cut in the warp exists it will only be used if this cut occurs near competing strokes. In other words, misclassified texture edges will not adversely effect the warp unless determined so by corresponding user strokes. Furthermore, by overselecting edges, we reduce the chance that visible depth edges will be missed, at the expense of allowing the warp to cut at some texture edges.

Allowing the warp to be cut in these edge locations creates additional difficulties; our warps can now contain both overlaps and tears (Figure 11.17). Overlaps can be simply rendered in simple depth order, but tears introduce disoccluded regions where no known information exists. Filling these disocclusions is a difficult problem, and many methods have been proposed to do this, most commonly inpainting or texture synthesis (Wei *et al.*, 2009).

However, the IDW framework can be used to provide a novel and robust method for fixing these disoccluded regions. This is done by conceptually performing a warp-based stretching on the background region, pulling it into the disocclusion. This approach sacrifices some realism by creating a wider region behind the disocclusion than existed, but by using visual saliency to drive the stretching (similar to the aspect ratio retargeting problem), these deformations can be hidden in nearly all situations. Because we are working in a stereo framework, the final disparity must be considered as well, and is included into the constraints.

This stretching is implemented as a second solve after the first, again by using a modified formulation. The first solve is performed with the cut at discontinuity edges as described above, yielding the optimal foreground positions. The second solve then fixes the location of foreground edge pixels (shown by the filled circles in Figure 11.18) by removing them from the system of equations and reintroduces the smoothness term in the cut $\forall_{ij} : \delta(i,j) = 1$. This has the effect of pulling the background region into the disocclusion.

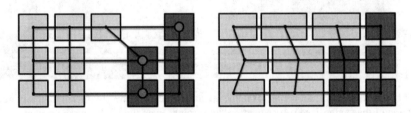

Figure 11.18 The disocclusion filling step. Foreground border pixels are fixed in location, and the edges are reconnected, causing the background region to fill in the hole

Figure 11.19 Before and after inpainting. The background is stretched in a content-and-disparity aware sense to fill in the disocclusions, which are visible here as solid regions by depth boundaries (see Plate 9 for the colored figure)

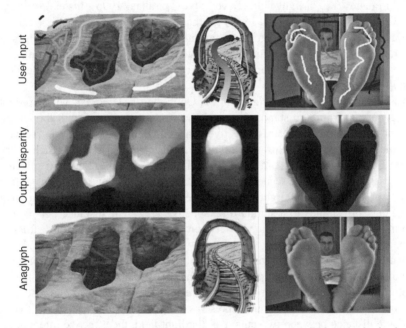

Figure 11.20 Some results from this approach, showing some user scribbles, synthesized per-pixel disparities, and grayscale anaglyph output image pairs (see Plate 10 for the colored figure)

Using this combination of approaches and minor modifications to the warping formulation in the previous sections, IDW can be used to generate a stereoscopic pair from single input view. Figure 11.19 and Figure 11.20 show some results from this approach.

11.8 Multi-Perspective Stereoscopy from Light Fields

The previous sections have shown how image warping can provide a very effective and practical solution to the problems of optimizing stereo input, stereo to multiview conversion, and monoscopic image to stereo conversion. The key insight of those methods is that changes in disparity can be realized by selectively deforming (stretching and compressing) image regions. Consequently, the resulting output images do not correspond to a standard single perspective anymore. The deformation effectively causes local changes in perspective; hence, one can interpret the results of those stereoscopic image warping algorithms as *multi-perspective* images (Yu *et al.*, 2010). As mentioned in Section 11.4.3, in movie production combining different perspectives in such a manner is known as the concept of multi-rigging. By compositing the captured views of multiple stereo camera rigs with different interaxial distances, multi-perspective stereo images are generated in order, for example, to make the stereoscopic experience more comfortable for the viewer (Mendiburu, 2009). The warping techniques described so far are the computational equivalent of creating such images from just a single standard perspective stereo pair.

For scenarios with more than two input views, such as when capturing a scene with multiple cameras or when rendering computer-animated movies, a complementary approach for generating optimized stereoscopic output exists. Given a sufficiently dense sampling of a scene with images (i.e., a light field; Levoy and Hanrahan, 1996), instead of warping the input views we can *select* the set of light rays that produces the stereoscopic output with the desired properties. Fortunately, core concepts such as the earlier introduced disparity mapping operators remain valid. We simply replace the synthesis part for generating the stereoscopic image pairs with an algorithm that is more suitable for handling multiple input views (Kim *et al.*, 2011).

The basic idea is illustrated in Figure 11.21. On the left a stereo pair with fixed interaxial distance generated from two standard perspective images is shown, with the left and right views color-coded in red and cyan, respectively. In a 3D light field of the same scene created, for example, with a whole linear array of cameras instead of just a pair, each of these two views corresponds to a *planar cut* through the light-field volume. Each cut selects a set of light rays passing through the scene and generates a corresponding output image from those light rays. For a planar cut, all light rays have been captured by one of the input images, and hence the cut corresponds to a standard perspective view. The distance between two cuts represents the fixed interaxial distance between the two respective cameras that captured those images.

For generating stereoscopic output images with per-pixel control over the interaxial distance, the core idea is to compute *curved cuts* instead of planar ones through the 3D light-field volume. As before, each cut surface represents one of the two stereoscopic output images, and the distance between cuts represents interaxial distance. However, thanks to the now variable distance between two cuts over the light field, the effective interaxial distance can be controlled individually for every output pixel; spatially close cuts select light rays from a pair of cameras with a small interaxial distance, whereas distant cuts select light rays from a pair of input images farther apart. Hence, one gains true per-pixel control over the generated output disparities.

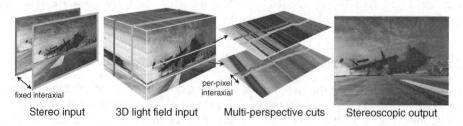

Figure 11.21 Basic concept behind the generation of multi-perspective stereoscopic output images from a light field, © Disney (see Plate 11 for the colored figure)

Figure 11.22 Per-pixel disparities, © Disney (see Plate 12 for the coloured figure)

As an example, Figure 11.22 shows a stereoscopic output image. The interaxial distance is shown per pixel as a different grayscale value.

In a nutshell, the algorithm works as follows. From the set of input images the user selects a reference view and defines the desired output disparities; for example, by employing one of the previously introduced disparity operators or manually prescribing output disparity constraints. From a 3D representation of the scene and the goal disparities, an energy function can be defined over the 3D light-field volume. The minimizer of that energy function which, for example, can be computed using graph min-cuts (Boykov and Kolmogorov, 2004), is a cut surface that, together with the given reference view, represents a stereoscopic output pair with the desired disparity constraints. An example is shown in Figure 11.23, where a window violation due to excessive negative disparities is automatically resolved by suppressing disparity gradients that are too strong.

This basic algorithm can be extended and generalized to incorporate temporal coherence of the cuts to process video, generate n-view stereo output using multiple cuts instead of standard stereo in order to reduce ghosting artifacts on automultiscopic displays (see Figure 11.24), deferred light-field generation for computer-animated movies, and to handle different input than linear arrays of cameras (Kim *et al.*, 2011).

Figure 11.23 The standard stereo pair on the left features strong negative disparities, resulting in a stereoscopic window violation. With the light field cuts too high, disparity gradients can be automatically compressed, yielding the solution on the right. The bottom images show the corresponding cuts through a 2D slice of the light field, © Disney (see Plate 13 for the colored figure)

Figure 11.24 *N*-view automultiscopic displays support only a limited disparity range, leading to ghosting artifacts (on the left) if exceeded by the content. With the light field cuts, multiple output images can be computed for such displays (bottom image) that adhere to those disparity constraints and, hence, can be displayed without artifacts (see Plate 14 for the colored figure)

Since each output image is assembled using light rays captured from various different perspectives, the above approach generates multi-perspective stereoscopic output images. An advantage of that over the previously described warping-based methods is that occlusion effects are properly reproduced and that warping artifacts such as noticeable image deformation are avoided, since the output images are generated from actually captured light rays rather than stretching or compressing the input images. However, this comes at the expense of requiring a sufficiently dense light field generated from tens of images rather than just a stereo pair, and the need for corresponding 3D scene information such as depth maps compared with sparse features and saliency maps required for the warping-based approaches. Therefore, the two approaches represent a quite complementary set of tools for generating optimized stereoscopic output.

11.9 Conclusions and Outlook

S3D is an accepted technology in a variety of professional and consumer applications today. The technology is mature enough, and the art of content creation is understood well enough, to provide users and audiences with a pleasant and expressive experience. However, there are still open questions and enough room for improvement regarding technology from capture to display. Autostereoscopy is still in its infancy, but is promising solutions to a variety of drawbacks that still remain in the traditional stereo viewing experience.

Processing of the captured video and other sensor data is one of the main focus areas of research for 3D content creation, transmission, and display (Smolic *et al.*, 2011). In that context, many of the classical 3D video processing approaches rely on some kind of 3D scene geometry representation, which has to be generated in some way. This can, for instance, be with a 3D model or depth estimation (DE). Using depth estimates in conjunction with DIBR is currently the most popular approach to manipulate and synthesize views. However, unreliability, inaccuracy, and computational complexity of DE still prohibit adoption of such approaches or limit application to user-assisted, semi-automatic systems.

On the other hand, many application scenarios basically only require manipulation of images or synthesis of virtual views close to available camera views. In order to do that, it is in many cases not necessary to go over 3D geometry representations like dense depth maps, which are difficult to acquire. Instead, the desired changes to available images can be formulated mathematically directly in image space. The solutions to those constraint systems of equations can be computed as warping functions, which will implement the postulated manipulations of images.

These principles of IDW for stereoscopic video processing have been demonstrated for a variety of applications such as post-production, automatic error correction, display adaptation, stereo to multiview conversion, or 2D–3D conversion. Solutions are available for automatic, accurate, and reliable 3D video processing avoiding the burden of 3D geometry reconstruction and related rendering. Interactive systems for intuitive 2D–3D conversion have also been demonstrated based on IDW. In consequence, IDW is a powerful approach to 3D video processing, which often overcomes limitations of DE + DIBR. Improvements of algorithms and components, extensions to additional applications, and integration into products and services can be expected in the future.

Initial software implementations of IDW algorithms have proven efficiency compared with DE + DIBR counterparts. Recent advances indicate even more efficient algorithms, avoiding complex energy minimizations and solvers (Griesen *et al.*, 2012; Lang *et al.*, 2012). Such approaches are also highly interesting for efficient hardware implementation, especially compared with DE + DIBR algorithms. Finally, boundaries between DE + DIBR and IDW may weaken over time and technology may converge, towards the goal to enable advanced 3D video functionalities (Smolic *et al.*, 2011).

Acknowledgments

Images for Figure 11.2 are used with the permission of Tanimoto Lab at Nagoya University and are available at http://www.tanimoto.nuee.nagoya-u.ac.jpp.

References

Banks, M., Gepshtein, S., and Landy, M. (2004) Why is spatial stereoresolution so low? *J. Neurosci.*, **24**, 2077–2089.

Boykov, Y. and Kolmogorov, V. (2004) An experimental comparison of min-cut/max-flow algorithms for energy minimization in vision. *IEEE PAMI*, **26** (9), 1124–1137.

Chang, C.F., Bishop, G., and Lastra, A. (1999) LDI tree: a hierarchical representation for image-based rendering, in *SIGGRAPH '99, Proceedings of the 26th Annual Conference on Computer Graphics and Interactive Techniques*, ACM Press/Addison-Wesley, New York, NY, pp. 291–298.

Farre, M., Wang, O., Lang, M. *et al.* (2011) Automatic content creation for multiview autostereoscopic displays using image domain warping. Hot3D Workshop.

Griesen, P., Heinzle, S., and Smolic, A. (2012) Hardware-efficient real-time HD video retargeting. High Performance Graphics (2012), June.

Guo, C., Ma, Q., and Zhang, L. (2008) Spatio-temporal saliency detection using phase spectrum of quaternion Fourier transform, in *CVPR 2008. CVPR 2008. IEEE Conference on Computer Vision and Pattern Recognition, 2008*, IEEE.

Hoffman, D., Girschik, M., Akeley, K., and Banks, M.S. (2008) Vergence–accommodation conflicts hinder visual performance and cause visual fatigue. *J. Vision*, **8** (3), 1–30.

Howard, I.P. and Rogers, B.J. (2002) *Seeing in Depth*, Oxford University Press, New York, NY.

Kim, C., Hornung, A., Heinzle, S. *et al.* (2011) Multi-perspective stereoscopy from light fields. *ACM Trans. Graph.*, **30** (6), 190.

Lang, M., Hornung, A., Wang, O. *et al.* (2010) Nonlinear disparity mapping for stereoscopic 3D, in *SIGGRAPH '10 ACM SIGGRAPH 2010 Papers*, ACM, New York, NY, article no. 75.

Lang, M., Wang, O., Aydin, T. *et al.* (2012) Practical temporal consistency for image-based graphics applications. *ACM Trans. Graph.*, **31** (4), article no. 34.

Levoy, M. and Hanrahan, P. (1996) Light field rendering, in *SIGGRAPH '96 Proceedings of the 23rd Annual Conference on Computer Graphics and Interactive Techniques*, ACM, New York, NY, pp. 31–42.

Mendiburu, B. (2009) *3D Movie Making: Stereoscopic Digital Cinema from Script to Screen*, Focal Press.

MPEG (2011) Call for proposals on 3D video coding technology, MPEG N12036, March.

MPEG (2012) Report of subjective test results from the call for proposals on 3D video coding, MPEG N12036.

Rubinstein, M., Gutierrez, D., Olga, S., and Shamir, A. (2010) A comparative study of image retargeting. *ACM Trans. Graph.*, **29** (6), article no. 160.

Schwartz, H., Bartnik, C., Bosse, S. *et al.* (2012) 3D video coding using advanced prediction, depth modeling, and encoder control methods, in *Picture Coding Symposium (PCS), 2012*, IEEE.

Siegel, M. and Nagata, S. (2000) Just enough reality: comfortable 3-D viewing via microstereopsis. *IEEE T. Circ. Syst. Vid. Technol.* **10** (3), 387–396.

Smolic, A., Kauff, P., Knorr, S. *et al.* (2011) Three-dimensional video postproduction and processing. *Proc. IEEE*, **99** (4), 607–625.

Stefanoski, N., Espinosa, P., Wang, O. *et al.* (2011) Description of 3D video coding technology proposal by Disney Research Zurich and Fraunhofer HHI. MPEG, Doc. M22668.

Stefanoski, N., Lang, M., and Smolic, A. (2012) Image quality vs rate optimized coding of warps for view synthesis in 3D video applications. International Conference on Image Processing, October.

Wang, O., Lang, M., Frei, M. *et al.* (2012) StereoBrush: interactive 2D to 3D conversion using discontinuous warps, SBIM.

Wei, L and LeFebvre, L-Y and Kwatra, V and Turk, G (2009) State of the art in example-based texture synthesis, EG-STAR.

Wolberg, G. (1990) *Digital Image Warping*. IEEE Computer Society Press, Los Alamitos, CA.

Yu, J., McMillan, L., and Sturm, P. (2010) Multi-perspective modelling, rendering and imaging. *Comput. Graph. Forum* **29** (1): 227–246.

Zilly, F., Riechert, C., Eisert, P., and Kauff, P. (2011) Semantic kernels binarized – a feature descriptor for fast and robust matching, in *2011 Conference for Visual Media Production (CVMP)*, IEEE, pp. 39–48.

Zitnick, L., Bing-Kang, S., Uyttendaele, M. *et al.* (2004) High-quality video view interpolation using a layered representation. *ACM Trans. Graph.*, **23** (3), 600–608.

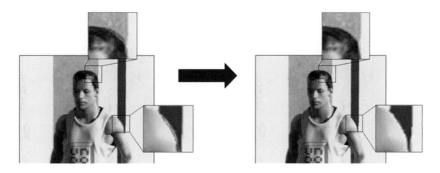

(Plate 1) Figure 8.9 Comparison of linear interpolation for depth up-sampling (left) versus a nonlinear filtering approach (right). Original images provided courtesy of Nokia

(Plate 2) Figure 11.4 Sparse feature correspondences

(Plate 3) Figure 11.6 Left: stereo correspondences and the disparity histogram for the cow example. Right: close-ups of the warped stereo pair showing the deformed isolines with respect to the input views

Emerging Technologies for 3D Video: Creation, Coding, Transmission and Rendering, First Edition. Frédéric Dufaux, Béatrice Pesquet-Popescu, and Marco Cagnazzo.

(Plate 4) Figure 11.7 Nonlinear disparity remapping. The disparity range of the original (left) is quite large leading to diplopia on large screens. Our nonlinearly remapped image (right) pushes the cow behind the screen, while leaving the background depth untouched

(Plate 5) Figure 11.8 Automatic correction of disparity. The original stereo pairs are shown on the left; our result is on the right. The cropped racing car captured with strong negative disparities and a large overall scene depth results in the so-called framing problem. With a simple linear disparity scaling the car is pushed behind the screen without increasing the background disparity, © KUK Filmproduction GmbH, All rights reserved

(Plate 6) Figure 11.9 Display adaptation into both directions. The middle image is the original stereo pair, while the left image features a linear reduction of the disparity range to 50% and the right image an increase to 200%, © www.vidimensio.com

(Plate 7) Figure 11.11 Eight synthesized views from most left to most right view

(Plate 8) Figure 11.15 This shows the sparse input given by a user in the form of grayscale scribbles, as well as a 3D representation and a resulting disparity map

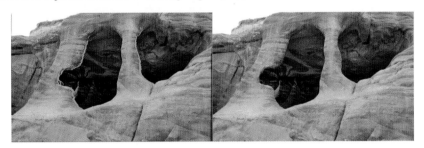

(Plate 9) Figure 11.19 Before and after inpainting. The background is stretched in a content-and-disparity aware sense to fill in the disocclusions, which are visible here as solid regions by depth boundaries

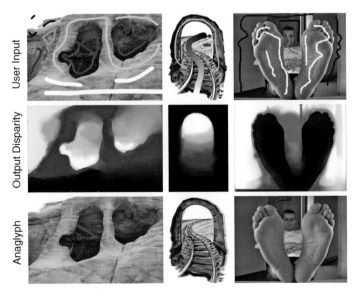

(Plate 10) Figure 11.20 Some results from this approach, showing some user scribbles, synthesized per-pixel disparities, and grayscale anaglyph output image pairs

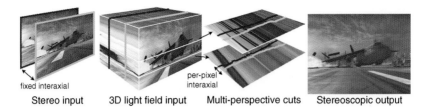

fixed interaxial   per-pixel interaxial

Stereo input   3D light field input   Multi-perspective cuts   Stereoscopic output

(Plate 11) Figure 11.21   Basic concept behind the generation of multi-perspective stereoscopic output images from a light field, © Disney

(Plate 12) Figure 11.22   Per-pixel disparities, © Disney

(Plate 13) Figure 11.23   The standard stereo pair on the left features strong negative disparities, resulting in a stereoscopic window violation. With the light field cuts too high, disparity gradients can be automatically compressed, yielding the solution on the right. The bottom images show the corresponding cuts through a 2D slice of the light field, © Disney

(Plate 14) Figure 11.24   *N*-view automultiscopic displays support only a limited disparity range, leading to ghosting artifacts (on the left) if exceeded by the content. With the light field cuts, multiple output images can be computed for such displays (bottom image) that adhere to those disparity constraints and, hence, can be displayed without artifacts

(a)   (b)   (c)

(Plate 15) Figure 14.1 (a) Sample 2D query image Q (kitchen); (b) NN #1 (c) NN #2. The corresponding depth fields are shown below the images. All image + depth pairs are from NYU Kinect dataset (Silberman and Fergus, 2011). From Konrad and Halle (2012b), reproduced with permission of IEEE © 2012

(a)   (b)   (c)

(Plate 16) Figure 14.2 Depth field z fused from $k = 45$ NNs. (a) Median fused depth z before smoothing; (b) fused depth after smoothing via cross-bilateral filtering $\hat{z}$; (c) $\hat{Q}_R$ right image rendered using the 2D query Q (Figure 14.1) and the smoothed depth field $\hat{z}$. From Konrad *et al.* (2012b), reproduced with permission of IEEE © 2012

(a)

(b)

(c)

(Plate 17) Figure 14.3 Anaglyph images generated using (a) ground-truth depth, (b) depth estimated by the proposed algorithm, and (c) depth estimated by the Make3D algorithm (Saxena *et al.*, 2009). From Konrad *et al.* (2012b), reproduced with permission of IEEE © 2012

(a) (b) (c)

(Plate 18) Figure 14.6   Example of color balancing in a stereo pair captured by identical cameras but with different exposure parameters: (a) original left view I^l; (b) original right view I^r; and (c) right view $\hat{I}^r$ after linear transformation (14.10)

(Plate 19) Figure 16.15   Example of a reconstructed image. From Arai *et al.* (2010b), reproduced with permission from IET

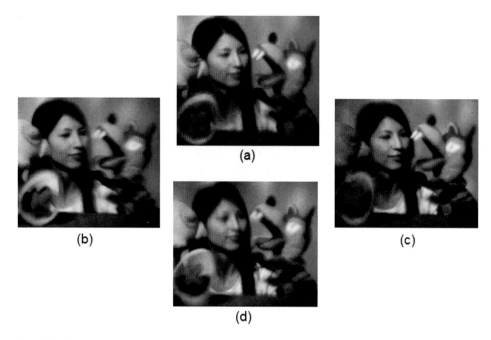

(a)

(b)

(c)

(d)

(Plate 20) Figure 16.16 Partial enlargement of the reconstructed image: (a) upper viewpoint; (b) left viewpoint; (c) right viewpoint; (d) lower viewpoint. From Arai *et al.* (2010b), reproduced with permission from IET

(a)

(b)

(Plate 21) Figure 16.17 Reconstructed image projected onto the diffuser plate. (a) Diffuser plate placed on the lens array. (b) Diffuser plate placed in front of the lens array. From Arai *et al.* (2010b), reproduced with permission from IET

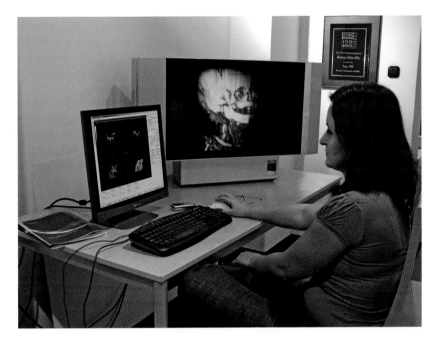

(Plate 22) Figure 17.6 Desktop light-field display showing volumetric data

(Plate 23) Figure 17.7 Large-scale light-field display showing pre-rendered 3D data

(a)

(b)

(c)

(d)

(e)

(f)

(g)

(Plate 24) Figure 20.2 (a-f) the six pictures acquired by the Ladybug camera (after lens distortion). (g) stitching to create an omnidirectional image

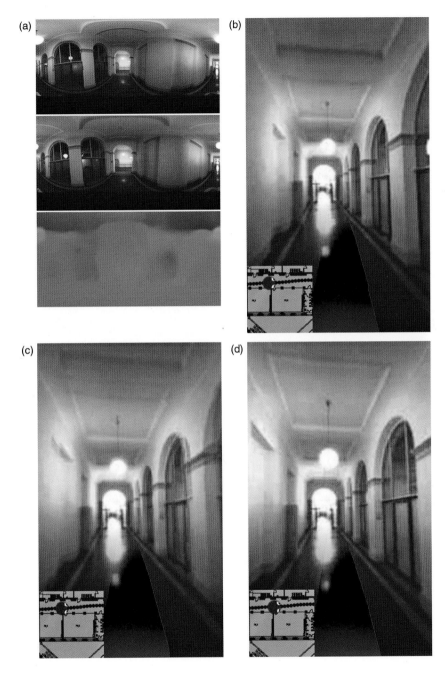

(Plate 25) Figure 20.5 View interpolation for smooth visualization. (a) Two spatially adjacent panorama views (top) and the corresponding optical flow (bottom). (c) Interpolated view between the two existing images (b) and (d) issued from the two spatially adjacent panorama views

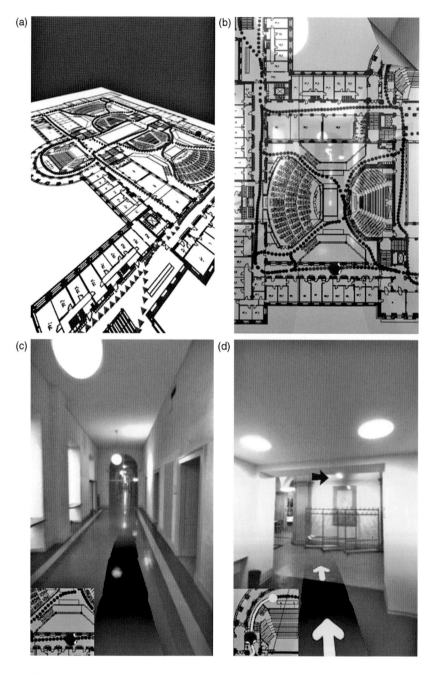

(Plate 26) Figure 20.6 Visualization application to interactively explore the virtual tour. (a, b) A mini-map shows the user's current position and viewing direction (large dot with a small arrow) on the floor plan as well as the mapped locations (small dots). (c, d) Display of the image corresponding to the user's current position and viewing direction from the panoramic view together with exploration information. The wide lines show a possible exploration direction. In guiding scenario (d), the lower arrows indicate the shortest path to a location of interest and the top arrow indicates turns at bifurcations

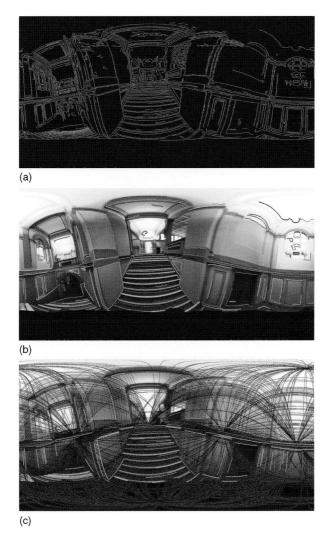

(a)

(b)

(c)

(Plate 27) Figure 20.8 The three main steps of line extraction: (a) edge detection (edge pixels have been enlarged for a better display); (b) edge chaining; (c) line detection by split-and-merge

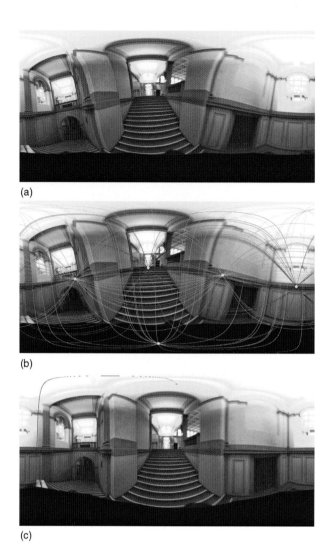

(a)

(b)

(c)

(Plate 28) Figure 20.9   Vertical rectification: (a) original image with a tilted camera in a stairway; (b) automatic VP extraction by Bazin *et al.* (2012); (c) rectified image where the camera becomes aligned with gravity and the vertical lines of the world become vertical straight lines

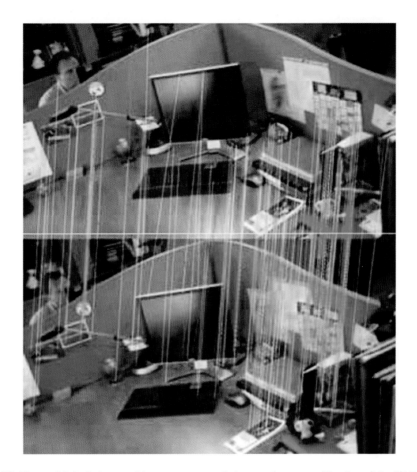

(Plate 29) Figure 22.6 Point matching on stereoscopic images for auto-calibration of the ST-Ericsson platform

(Plate 30) Figure 23.6 Five-view interleaving of an interactive menu

(Plate 31) Figure 23.15   3D carousel menu (2D flat objects composition)

(Plate 32) Figure 23.16   Natural 2D + depth picture

(Plate 33) Figure 23.17   Composition of 2D menu and synthetic 2D + depth model

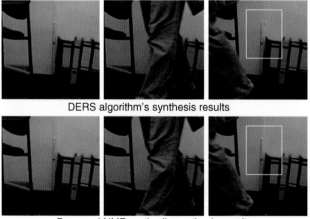

(Plate 34) Figure 24.3   Successive synthesis results of the DERS algorithm and the proposed NMR method

12

Image-Based Rendering and the Sampling of the Plenoptic Function

Christopher Gilliam, Mike Brookes, and Pier Luigi Dragotti

Department of Electrical and Electronic Engineering, Imperial College London, UK

12.1 Introduction

Image-based rendering (IBR) is an effective technique for rendering novel views of a scene from a set of multi-view images. Instead of generating novel views by projecting three-dimensional (3D) scene models and their textures, new views are rendered by interpolating nearby images. The advantage of this method is that it produces convincing photorealistic results, as it combines real images, without requiring a detailed 3D model of the scene. A drawback, however, is that a very large number of images are needed to compensate for the lack of geometric information. Therefore, an important goal in IBR is to reduce the size of the multi-view image set whilst still achieving good-quality rendering.

The central concept behind IBR is that a scene can be represented as a collection of light rays emanating from the scene. The multi-view image set, therefore, captures this representation of the scene as each image records the intensity of a set of light rays travelling from the scene to the camera. The light rays in question are parameterized using a seven-dimensional (7D) function introduced by Adelson and Bergen (1991): the *plenoptic function*. It describes the intensity of the light ray passing through the camera centre at a 3D spatial location (x, y, z) with an angular viewing direction $(\vartheta_x, \vartheta_y)$ for a wavelength μ at a time τ. In many situations, however, it is more convenient to parameterize the viewing direction in pixel coordinates (v, w).

IBR can be seen, therefore, as the problem of sampling and reconstructing the plenoptic function. That is, a finite set of images samples the continuous plenoptic function and the rendering of a new view is a reconstruction from the samples (Zhang and Chen, 2004). If the plenoptic function is incorrectly sampled, then blurring or ghosting artefacts appear in the rendered views (Chai *et al.*, 2000). As a result, challenges in plenoptic sampling include minimizing the number of images required for artefact-free rendering and deciding the best position of the cameras to maximize the rendering quality.

Emerging Technologies for 3D Video: Creation, Coding, Transmission and Rendering, First Edition.
Frédéric Dufaux, Béatrice Pesquet-Popescu, and Marco Cagnazzo.
© 2013 John Wiley & Sons, Ltd. Published 2013 by John Wiley & Sons, Ltd.

Bearing this in mind, this chapter presents the state of the art in plenoptic sampling theory. The chapter is outlined as follows. Section 12.2 covers the variety of plenoptic parameterizations that occur by restricting aspects of the scene and viewing position. In particular, the section focuses on two common plenoptic representations: the light field and the surface light field. Section 12.3 covers the sampling of the plenoptic function assuming a uniform distribution of the cameras. In this context, the sampling analysis is approached in a Fourier framework. Following this, Section 12.4 presents the state of the art in adaptive sampling of the plenoptic function, known as adaptive plenoptic sampling. Finally, Section 12.5 concludes the chapter and presents an outlook on possible future directions in IBR and plenoptic sampling. For other surveys on the topic we refer to Shum *et al.* (2007) and Zhang and Chen (2004).

12.2   Parameterization of the Plenoptic Function

The high dimensionality of the plenoptic function makes both theoretical analysis and practical application a challenging problem. However, the dimensionality of the plenoptic function can be reduced by restricting certain aspects of the scene and sensing set-up. With these restrictions, we can re-parameterize the 7D plenoptic function into more tractable representations. In Zhang and Chen (2004) the plenoptic representations are categorized based on a combination of six assumptions used to generate them. The proposed six assumptions can be divided into those that restrict the scene and those that progressively restrict the viewing position; that is, from three dimensions to a two-dimensional (2D) surface, then a one-dimensional (1D) path and finally a fixed position.

For example, the six-dimensional *surface plenoptic function* (Zhang and Chen, 2003b) removes one dimension by assuming the radiance of a light ray is constant along its path through empty space. In McMillan and Bishop (1995) a five-dimensional parameterization is constructed by removing both the time and wavelength parameters. The time is removed by assuming a static scene and then the wavelength is removed by partitioning the spectrum into three bands (red, green and blue). By using these assumptions, and restricting the viewing position to a surface, Shum and He (1999) constructed the *concentric mosaic* representation. In this representation the scene is captured by a single camera mounted on the end of a rotating beam. As a result, the intensity of a light ray is described using just three parameters: the 2D pixel location and the beam's angle of rotation.

12.2.1   Light Field and Surface Light Field Parameterization

In this chapter we will focus on two popular plenoptic representations: the *light field*[1] (Levoy and Hanrahan, 1996) and the *surface light field* (Miller *et al.*, 1998; Wood *et al.*, 2000). Both parameterizations assume that the scene is static and the radiance of a light ray is constant along its path through space. With this assumption, the spatial location of the cameras can be simplified to a 2D surface. As a result, both the light field and the surface light field are four-dimensional (4D) plenoptic representations. They differ, however, in their method of defining a light ray using these four dimensions.

Assuming a pinhole camera model, the light field parameterization defines each light ray by its intersection with two parallel planes:[2] the camera plane (t, u) and the image plane

---

[1] This representation is similar to the *lumigraph* introduced in Gortler *et al.* (1996) and the *ray-space* in Fujii *et al.* (1996).

[2] This implicitly assumes that the scene can be bounded within a box.

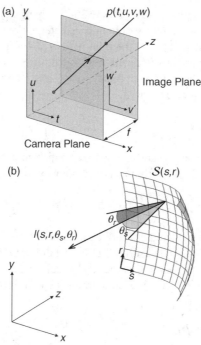

Figure 12.1 (a) The 4D light field $p(t, u, v, w)$ in which a light ray is defined by its location on the camera plane (t, u) and its corresponding pixel location on the image plane $(v, w) = (v' - t, w' - u)$. The distance between the two planes is the focal length f. (b) The surface plenoptic function $l(s, r, \theta_s, \theta_r)$, which is the intensity of a light ray emitted from a point (s, r) on the scene surface S at a viewing direction (θ_s, θ_r)

(v, w). The separation between the two planes is equal to the focal length f. Therefore, the intensity of the light ray at camera location (t, u) and pixel location (v, w) is

$$p_4 = p(t, u, v, w) \tag{12.1}$$

see Figure 12.1a for a diagram. Wilburn *et al.* (2002) proposed a time-dependent version of the light field in order to handle dynamic scenes. This new representation was termed *light field video*. Another variant is the *spherical light field* proposed by Ihm *et al.* (1997). In this case the scene is bounded within a sphere and each light ray is defined by its intersection with two concentric spheres. Notice, however, that in this framework, and its variations, the light ray is defined with respect to the receiving camera position.

In contrast, the light ray in the surface light field is defined relative to its point of origin on the scene surface S. This surface is parameterized using two curvilinear surface coordinates (s, r) such that a point on the surface is defined as $S(s, r) = [x(s, r), y(s, r), z(s, r)]^T$, where $[x, y, z]^T$ are the point's Cartesian coordinates (Nguyen and Do, 2009). The direction the light ray leaves the surface is defined by the viewing angle (θ_s, θ_r), where θ_s and θ_r are measured relative to the z-axis. Therefore, the intensity of a light ray emitted from a point (s, r) on the scene surface at a viewing direction (θ_s, θ_r) is

$$l_4 = l(s, r, \theta_s, \theta_r) \tag{12.2}$$

A diagram of this framework is shown in Figure 12.1b.

On a final note, several authors (Chai *et al.*, 2000; Zhang and Chen, 2003b; Do, *et al.*, 2012) further reduce the dimensionality of both parameterizations by considering only a horizontal slice of the scene. In the case of the light field, u and w are fixed; this corresponds to the situation where the camera positions are constrained to a line parallel to the x-axis and only one scan-line is considered in each image. Therefore, the intensity of the light ray at camera location t and pixel location v is

$$p_2 = p(t, v) \tag{12.3}$$

For the surface light field, r and θ_r are fixed corresponding to a 1D surface, $S(s)$. Therefore, the intensity of the light ray emitted from a surface point $[x(s), z(s)]^{\mathrm{T}}$ at viewing angle θ_s is

$$l_2 = l(s, \theta_s) \tag{12.4}$$

Diagrams of the 2D light field and 2D surface light field are shown in Figure 12.2. For the remainder of this chapter we shall use the 2D light field and 2D surface light field when examining the sampling of the plenoptic function. Accordingly, in the 2D surface light field, we drop the s subscript from the viewing angle, referring to it only as θ.

12.2.2 Epipolar Plane Image

As defined in Section 12.2.1, the 2D light field representation explicitly defines the intensity of a light ray captured at a coordinate (t, v). By considering all possible (t, v) coordinates, then a visual representation of the 2D light field can be constructed. This visual

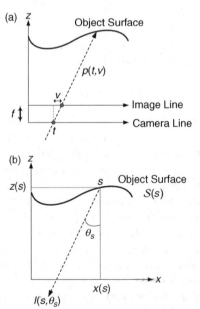

Figure 12.2 (a) The 2D light field $p(t, v)$ in which a light ray is defined by its intersection with the camera line at a location t and the corresponding pixel location v on the image line. (b) The 2D surface plenoptic function $l(s, \theta_s)$ in which a light ray is defined by its point of origin s on the object surface S at a viewing angle θ_s

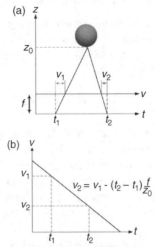

Figure 12.3 Diagram showing (a) the 2D parameterization of the light field and (b) its EPI representation. A point in (a) translates to a line in the EPI with a slope inversely proportional to the depth of the point

representation is known as the *epipolar plane image* (EPI) (Bolles *et al.*, 1987) or *EPI-volume* if we are considering 2D images. It highlights how the inherent structure in the linear camera path leads to structure in the light field. For example, a point at a depth z_0 (see Figure 12.3a) is mapped to a line in the EPI with a slope that is inversely proportional to z_0 (see Figure 12.3b). This structure leads to the following important characteristic: lines with steeper slopes will always occlude lines with shallower slopes in the EPI domain since a point close to the camera will occlude those that are more distant. Feldmann *et al.* (2003) extended this visual tool to nonlinear camera paths and termed it *image cube trajectories.*

This concept of mapping a point in space to a line in the EPI can be generalized to higher dimensional structures. For instance, all the points at a certain depth will have the same gradient or trajectory in the EPI. Thus, a region in space, made from neighbouring points, will result in a collection of trajectories in the EPI-volume. Using this observation, Criminisi *et al.* (2005) decomposed the scene into layers and grouped the resulting trajectories, from each layer, into volumes called *EPI-tubes.* More generally, for higher dimensional plenoptic representations, these volumes or hypervolumes are termed *plenoptic manifolds* (Berent and Dragotti, 2007).

12.3   Uniform Sampling in a Fourier Framework

When cameras are uniformly spaced, at a spacing Δt, it is natural to analyse plenoptic sampling within a classical Fourier framework. In such a framework, uniform sampling leads to spectral replication in frequency and the minimum sampling requirement – the Nyquist rate – is such that the replicas do not overlap. If these replicas overlap, then aliasing will occur, which manifests itself as artefacts in the rendering process (Chai *et al.*, 2000). Therefore, the spectral support of the plenoptic function is examined in order to determine the minimum spatial sampling density, which is inversely proportional to the maximum spacing between cameras. With this in mind, we are interested in the properties of the plenoptic spectrum, the Fourier transform of the plenoptic function.

12.3.1 Spectral Analysis of the Plenoptic Function

The first spectral analysis of the plenoptic function was performed in Chai *et al.* (2000). Their approach involved using the inherent structure in the EPI to map each image to a reference position. Assuming a Lambertian scene[3] with no occlusion, an arbitrary image is mapped to the reference at $t = 0$ as follows:

$$p(t, v) = p\left(0, v - \frac{ft}{z}\right) \qquad (12.5)$$

Therefore, by analysing points in the scene at different depths, Chai *et al.* (2000) derived approximate bounds on the plenoptic spectrum dependent on the maximum and minimum depths of the scene. This mapping procedure was also used in Chen and Schonfeld (2009) to examine the effects on the plenoptic spectrum when the camera path is varied. However, this analysis implicitly assumes the scene depth is approximately piecewise constant; thus, it does not take into account the effects of depth variation in the scene, which is instead considered in Do *et al.* (2012).

Specifically, Do *et al.* (2012) and Zhang and Chen (2003b) derived spectral properties for a broader range of scenes by exploiting the equivalence between the plenoptic function and the surface light field. This equivalence is formalized by modelling the scene with a functional surface. In this framework, the depth of the scene surface, relative to the real world coordinate x, is defined by the function $z(x)$ and its texture is modelled as a bandlimited signal $g(s)$, where s is the curvilinear coordinate on the surface. Now, assuming the camera line t coincides with the x coordinate system, the authors link a light ray arriving at (t, v) to its point of origin on the surface at $(x, z(x))$ using the following geometric relationship:

$$t = x - z(x)\tan(\theta) = x - z(x)\frac{v}{f} \qquad (12.6)$$

where f is the focal length and θ is the viewing angle. An illustration of this relationship is shown in Figure 12.4. Provided this geometric relationship is a one-to-one mapping, then the spatial position $(x, z(x))$ specifies a single curvilinear position s, which allows the plenoptic function to be mapped to the surface light field and vice versa. The provision of a one-to-one mapping is enforced in Do *et al.* (2012) by excluding scenes with occlusions. Therefore, $z(x)$ is constrained such that

$$|z'(x)| < \frac{f}{v_{\mathrm{m}}} \qquad (12.7)$$

where $z'(x)$ is the first derivative of z with respect to x, and v_{m} is the maximum value of v for a camera with a finite field of view; hence $v \in [-v_{\mathrm{m}}, v_{\mathrm{m}}]$. Although this constraint is not directly enforced in Zhang and Chen (2003b), a one-to-one relationship is achieved by selecting the closest point to the scene that satisfies (12.6).

Formally, the mapping between the plenoptic function and the surface light field is as follows. Equation 12.6 allows the mapping of the plenoptic function $p(t, v)$ to $l_x(x, v_{\mathrm{d}})$ the intensity of a light ray emitted from the spatial position $(x, z(x))$ at a viewing direction defined by $v_{\mathrm{d}} = v/f$; that is:

$$l_x(x, v_{\mathrm{d}}) = p(x - z(x)v_{\mathrm{d}}, f v_{\mathrm{d}}) \qquad (12.8)$$

---

[3] The Lambertian assumption means that the intensity of a light ray leaving a point on the scene surface is independent of the angle the light ray leaves the surface. In other words, the point looks the same from any viewing angle.

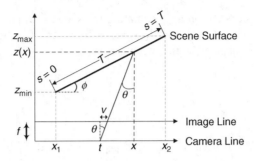

Figure 12.4 Scene model of a slanted plane showing the intersection of a light ray (t, v) with the scene surface at $(x, z(x))$, where $z(x) \in [z_{min}, z_{max}]$ is the depth of the surface. Note that f is the focal length of the cameras, θ is the viewing angle and ϕ is the angle of slant for the plane

The surface light field $l(s, \theta)$ is then obtained by mapping the spatial position x to the curvilinear coordinate s and the viewing direction v_d to the viewing angle θ; hence

$$l(s, \theta) = l_x(x(s), v_d(\theta)) \tag{12.9}$$

The importance of this mapping is that spectral properties of the plenoptic function can be derived by assuming properties of the surface light field without explicitly defining the scene's geometry. Therefore, Do *et al.* (2012) derive the plenoptic spectrum in terms of $l_x(x, v_d)$ and determine spectral properties based on the behaviour of $l_x(x, v_d)$.

The plenoptic spectrum in question is obtained as follows. Starting from its definition

$$\begin{aligned} P(\omega_t, \omega_v) &= \mathcal{F}_{t,v}\{p(t, v)\} \\ &= \int_{-\infty}^{\infty} \int_{-\infty}^{\infty} p(t, v)\, e^{-j(\omega_t t + \omega_v v)}\, dt\, dv \end{aligned} \tag{12.10}$$

the integration variables are changed using (12.6) and $v_d = v/f$, which results in a Jacobian of $(1 - z'(x)v_d)f$. Consequently, the following is obtained:

$$P(\omega_t, \omega_v) = \int_{-\infty}^{\infty} \int_{-\infty}^{\infty} l_x(x, v_d)\, e^{-j[\omega_t(x - z(x)v_d) + \omega_v f v_d]}(1 - z'(x)v_d)f\, dx\, dv_d \tag{12.11}$$

At this point two identities are introduced, the first is $h(x, v_d) = l_x(x, v_d)(1 - z'(x)v_d)$ and the second is $L_x(x, \omega_v) = \mathcal{F}_{v_d}\{l_x(x, v_d)\}$. Using these identities, the integral in v_d becomes

$$H(x, \omega_v) = \mathcal{F}_{v_d}\{h(x, v_d)\} = \int_{-\infty}^{\infty} h(x, v_d)\, e^{-j\omega_v v_d}\, dv_d$$

$$= L_x(x, \omega_v) - jz'(x)\frac{\partial L_x(x, \omega_v)}{\partial \omega_v} \tag{12.12}$$

Finally, inserting the above into (12.11), Do *et al.* (2012) obtained a general equation for the plenoptic spectrum that is independent of the scene's geometry:

$$P(\omega_t, \omega_v) = f \int_{-\infty}^{\infty} H(x, \omega_v f - z(x)\omega_t)\, e^{-j\omega_t x}\, dx \tag{12.13}$$

The first point to take from this equation is the dependency of the plenoptic spectrum on the slope of the surface $z'(x)$. This dependency was not apparent when using the mapping procedure in (12.5) as it relied on approximating the scene with piecewise constant depth segments. Second, by using the fact that $l_x(x, v_d) = l_x(x)$ for a Lambertian scene, Do *et al.* (2012) showed that in frequency the following is true:

$$L_x(x, \omega_v) = 0, \quad \text{if} \quad \omega_v \neq 0 \qquad (12.14)$$

which leads to

$$P(\omega_t, \omega_v) = 0, \quad \text{if} \quad \omega_v f - z(x)\omega_t \neq 0 \qquad (12.15)$$

Therefore, as $z(x) \in [z_{\min}, z_{\max}]$, the plenoptic spectrum is precisely bounded by lines relating to the maximum and minimum depths of the scene; see Figure 12.5a. Moreover, they formalized bounds for non-Lambertian scenes by assuming $l_x(x, v_d)$ is bandlimited in v_d to

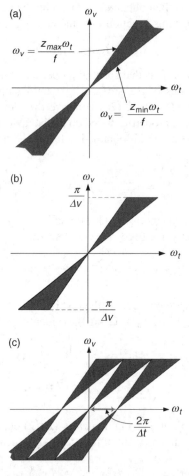

Figure 12.5 Diagrams of the plenoptic spectrum: (a) the plenoptic spectrum bounded between $\omega_v = \omega_t z_{\min}/f$ and $\omega_v = \omega_t z_{\max}/f$. (b) The 'bow-tie'-shaped plenoptic spectrum caused by the pixel resolution Δv inducing lowpass filtering in ω_v. (c) The optimal packing for the sampled plenoptic spectrum, where Δt is the camera spacing

B_L. Consequently, relaxing the Lambertian assumption on the scene results in an extended region around the plenoptic spectrum, a fact also highlighted in Zhang and Chen (2003b).

Assuming a Lambertian scene, the plenoptic spectrum shown in Figure 12.5a is bandlimited if lowpass filtering is applied in ω_v. Such lowpass filtering occurs owing to the finite pixel resolution of the cameras Δv. Therefore, the plenoptic spectrum is bandlimited in ω_v at $\pi/\Delta v$, which results in the 'bow-tie'-shaped spectrum shown in Figure 12.5b. Based on this shape, the optimal packing of the replicated spectra at critical sampling is shown in Figure 12.5c. To achieve this packing, without overlap occurring, Chai *et al.* (2000) derived the following maximum camera spacing:

$$\Delta t = \frac{2\pi}{\Omega_v f\left(z_{min}^{-1} - z_{max}^{-1}\right)} \tag{12.16}$$

where Ω_v is the maximum frequency in ω_v (in the worst case it is equal to $\pi/\Delta v$).

12.3.2   The Plenoptic Spectrum under Realistic Conditions

In Do *et al.* (2012) the derivation of the plenoptic spectrum, stated in (12.13), requires the scene surface be restricted by a no-occlusion constraint given by (12.7). In turn this no-occlusion constraint depends on the camera's field of view; if the field of view is finite, then the scene depth can vary provided its gradient obeys the constraint. However, if the camera's field of view is not restricted, then the scene surface must be flat (relative to the camera line). This observation implies that the spectrum in (12.13) is only valid for scenes without depth variation as the spectral analysis does not limit the camera's field of view.

In view of this, Gilliam *et al.* (2010) re-examined the spectral analysis of the plenoptic function when incorporating two realistic conditions: finite scene width (FSW) and cameras with finite field of view (FFoV). Assuming a Lambertian scene with no occlusion, the imposition of their realistic conditions on (12.11) results in

$$P(\omega_t, \omega_v) = \int_{x_1}^{x_2} l_x(x)\, e^{-j\omega_t x} \int_{-v_m/f}^{v_m/f} (1 - z'(x)v_d) f\, e^{-j[\omega_v f - z(x)\omega_t]v_d}\, dv_d\, dx \tag{12.17}$$

where $x \in [x_1, x_2]$ is the FSW condition and $v_d \in [-v_m/f, v_m/f]$ is the FFoV condition. Solving the integral in v_d, the plenoptic spectrum becomes

$$\begin{aligned}
P(\omega_t, \omega_v) = 2v_m \Bigg[&\int_{x_1}^{x_2} l_x(x)\, \text{sinc}\left(\omega_v v_m - z(x)\frac{\omega_t v_m}{f}\right) e^{-j\omega_t x} dx \\
&- j\frac{z'(x)v_m}{f} \int_{x_1}^{x_2} l_x(x)\, \text{sinc}'\left(\omega_v v_m - z(x)\frac{\omega_t v_m}{f}\right) e^{-j\omega_t x} dx \Bigg]
\end{aligned} \tag{12.18}$$

where $\text{sinc}(h) = \sin(h)/h$ and sinc' is the first derivative of the sinc function with respect to its argument. Thus, for scenes with depth variation, this equation indicates that the realistic conditions cause significant spectral spreading in the plenoptic spectrum.

The effect of this spectral spreading is explored in Gilliam *et al.* (2010) by examining the plenoptic spectrum for a simple geometric scene, a slanted plane with bandlimited texture.

The geometry for such a scene is described using the following set of equations:

$$
G_s = \begin{cases} z(x) = (x - x_1)\tan(\phi) + z_1 \\ x(s) = s\cos(\phi) + x_1 \\ g(s) \xrightarrow{\mathcal{F}} G(\omega) \text{ and } G(\omega) = 0 \text{ for } |\omega| > \omega_s \end{cases} \tag{12.19}
$$

where ω_s is the maximum frequency of the texture signal and ϕ is the angle between the plane and the x-axis. The spatial coordinate (x_1, z_1) indicates the starting point of the plane – in other words, the origin for the curvilinear coordinate s – and (x_2, z_2) indicates the end point of the plane. At (x_2, z_2), the curvilinear point is equal to the width of the plane; that is, $s \in [0, T]$, which leads to the following relationship:

$$
T = \frac{x_2 - x_1}{\cos(\phi)} = \frac{z_2 - z_1}{\sin(\phi)} \tag{12.20}
$$

An illustration of this geometric description is shown in Figure 12.4 in which $z_{\min} = z_1$ and $z_{\max} = z_2$.

Using this scene geometry and assuming a complex exponential texture, an exact closed-form expression for the plenoptic spectrum of a slanted plane is derived. The expression is obtained by applying the description of the scene geometry to (12.18), to give

$$
\begin{aligned}
P_S(\omega_t, \omega_v) = 2v_{\mathrm{m}} \cos(\phi)\, e^{-j\omega_t x_1} & \left[\int_0^T g(s)\mathrm{sinc}(\omega_{\mathrm{I}}(s))\, e^{-j\omega_t \cos(\phi)s}\, ds \right. \\
& \left. - j\frac{v_{\mathrm{m}}\tan(\phi)}{f} \int_0^T g(s)\mathrm{sinc}'(\omega_{\mathrm{I}}(s))\, e^{-j\omega_t \cos(\phi)s}\, ds \right]
\end{aligned} \tag{12.21}
$$

where

$$
\omega_{\mathrm{I}}(s) = \omega_v v_{\mathrm{m}} - (s\sin(\phi) + z_1)\frac{v_{\mathrm{m}}}{f}\omega_t
$$

The remaining integrals are then solved, assuming $g(s) = e^{j\omega_s s}$, using the following identity from Abramowitz and Stegun (1964):

$$
\int_0^{jh} \frac{1 - e^{-w}}{w}\, dw = \mathrm{E}_1(jh) + \ln(jh) + \gamma
$$

where $h \in \mathbb{R}$, E_1 is the exponential integral and γ is Euler's constant. Therefore, the expression for the plenoptic spectrum of a slanted plane with complex exponential texture is

$$
\begin{aligned}
P_S(\omega_t, \omega_v) = e^{-j\omega_t x_1} & \left\{ \frac{j2v_{\mathrm{m}}}{\omega_t} \left[\mathrm{sinc}(a)\, e^{-jT(\omega_t \cos(\phi) - \omega_s)} - \mathrm{sinc}(b) \right] \right. \\
& \left. + \frac{j\omega_s f}{\sin(\phi)\omega_t^2} [\zeta\{jb(c-1)\} - \zeta\{ja(c-1)\} - \zeta\{jb(c+1)\} + \zeta\{ja(c+1)\}]\, e^{jbc} \right\}
\end{aligned} \tag{12.22}
$$

if $\omega_t \neq 0$, else

$$P_S(0, \omega_v) = 2v_m T \operatorname{sinc}\left(\frac{\omega_s T}{2}\right)\left[\cos(\phi)\operatorname{sinc}(\omega_v v_m) - j\frac{\sin(\phi)v_m}{f}\operatorname{sinc}'(\omega_v v_m)\right]e^{j\omega_s(T/2)}$$

The parameters a, b and c are defined as

$$a = \omega_v v_m - \omega_t\frac{z_2 v_m}{f}, \quad b = \omega_v v_m - \omega_t\frac{z_1 v_m}{f} \quad \text{and} \quad c = \frac{\omega_s f - f\omega_t\cos(\phi)}{\sin(\phi)\omega_t v_m}$$

and the function ζ is

$$\zeta\{jh\} = \begin{cases} E_1(jh) + \ln|h| + j\dfrac{\pi}{2} + \gamma & \text{if} \quad h > 0 \\[2mm] E_1^*(j|h|) + \ln|h| - j\dfrac{\pi}{2} + \gamma & \text{if} \quad h < 0 \\[2mm] 0 & \text{if} \quad h = 0 \end{cases}$$

where $E_1^*(jh)$ is the complex conjugate of $E_1(jh)$. Figure 12.6a shows an example of a plenoptic spectrum for a slanted plane with a cosine texture.

This analysis shows that the realistic conditions of FSW and FFoV result in a band-unlimited plenoptic spectrum in both ω_t and ω_v. Moreover, the plenoptic spectrum defined in (12.22) is no longer bounded by lines relating to the maximum and minimum depths of the scene even in the case of a Lambertian scene. As a result Gilliam *et al.* (2010, 2011) approximated the bandwidth of the plenoptic spectrum using its essential bandwidth, a region in frequency containing 90% of the signal's energy. Based on the structure of (12.22), they defined the essential bandwidth corresponding to a slanted plane in terms of a four-parameter model. These four parameters are: A, the width of the region in ω_t; z_{opt}/f, the slant of the model; Ω_t, the maximum frequency in ω_t; and Ω_v, the maximum frequency in ω_v. This

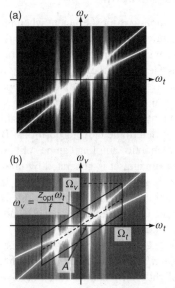

Figure 12.6 (a) The plenoptic spectrum for a slanted plane with a texture signal $g(s) = \cos(\omega_s s)$. (b) The same spectrum as in (a) with the parameterization of the essential bandwidth superimposed

model is superimposed on the actual plenoptic spectrum in Figure 12.6b. In Gilliam, *et al.* (2011) the parameters Ω_t and Ω_v are defined as

$$\Omega_t = \frac{f\omega_s}{f\cos(\phi) - v_{\mathrm{m}}|\sin(\phi)|} + \frac{2\pi}{T} \tag{12.23}$$

and

$$\Omega_v = \frac{z_{\max}\Omega_t}{f} + \frac{\pi}{v_{\mathrm{m}}} \tag{12.24}$$

respectively. From these fixed points, the other parameters are

$$z_{\mathrm{opt}} = \frac{z_{\max} + z_{\min}}{2} \tag{12.25}$$

and

$$A = \frac{T|\sin(\phi)|\Omega_t}{z_{\mathrm{opt}}} + \frac{2\pi f}{z_{\mathrm{opt}} v_{\mathrm{m}}} \tag{12.26}$$

Therefore, by assuming the plenoptic spectrum is approximately bandlimited to its essential bandwidth, the following maximum camera spacing for a slanted plane is derived:

$$\Delta t = \frac{2\pi}{A} = \frac{2\pi z_{\mathrm{opt}} v_{\mathrm{m}}}{v_{\mathrm{m}}\Omega_t T|\sin(\phi)| + 2\pi f} \tag{12.27}$$

When compared with (12.16) the above equation results in the cameras being positioned farther apart. This less conservative sampling is possible as (12.22) gives greater insight into the structure of the plenoptic spectrum.

12.4 Adaptive Plenoptic Sampling

In general, uniform plenoptic sampling is most efficient when the scene in question is relatively constant (in either depth or texture); however, in practice, this is rarely the case. In order to avoid undersampling, the uniform camera spacing, in (12.16) or (12.27), becomes conservative and is determined by the largest depth and the largest texture variation in the scene. The solution is to allow irregular or adaptive camera placement that depends on the scene, termed free-form sampling in Zhang and Chen (2006). The difficulty with sampling the plenoptic function in such a way is that there are many possible camera configurations to choose from. Therefore, more constraints are required to determine camera positions. Based on the heuristics used to position the cameras, adaptive plenoptic sampling can be split into three categories.

In the first category, an initial image set is generated by uniformly oversampling the scene. This image set is then minimized based on some quantitative error metric. For this reason, this category is termed sample reduction (SR). An example of SR is presented in Fleishman *et al.* (2000). In their approach the scene is modelled using a mesh and each image is ranked based on the proportion of scene elements covered. The reduced image set is then generated by keeping the images with the highest rank (i.e. those that cover the highest proportion of scene elements). In a similar fashion, Namboori *et al.* (2004) filtered the initial image set based on an analysis of the scene geometry. Another, slightly different, approach involves mapping the plenoptic function to a different sampling matrix using multidimensional lattice theory (Zhang and Chen, 2006). Thus, redundancy in the image set is removed by nonrectangular downsampling of the plenoptic function. However, the main limitation of SR is that it initially requires a large number of images to ensure the scene is oversampled and then discards those that are deemed redundant.

A valid alternative is active incremental capturing (AIC). In contrast to SR, the scene is initially undersampled uniformly and intermediate samples are added to reduce a local reconstruction error. This is repeated until the average local reconstruction error is below a threshold. Figure 12.7 illustrates how new samples are introduced in a concentric mosaic representation (Zhang and Chen, 2003a) and in a 4D light-field representation (Zhang and Chen, 2006). A systematic framework for AIC was presented in Zhang and Chen (2003a), where they defined the position-interval error (PIE) as a measure of the average reconstruction error for any pair of samples. Thus, the goal is to have a uniform PIE for each pair of samples. A good estimate of the PIE is the local colour consistency (Zhang and Chen, 2003a). Another example of AIC is in Schirmacher *et al.* (1999), in which an adaptive mesh is used to define the camera positions. New camera positions are introduced on the edges of the mesh in order to reduce the estimated reconstruction error between the nearest source images.

The final category in adaptive plenoptic sampling is active rearranged capturing (ARC; Zhang and Chen, 2006). It involves repositioning a limited set of cameras in order to improve rendering quality on the fly. The main advantage of repositioning the cameras on the fly is that ARC can sample and render dynamic scenes. This is in contrast to both SR and AIC methods, which require a two-stage sampling process: a uniform stage followed by a selective stage. Consequently, both SR and AIC are restricted to static scenes. An example of ARC is presented in Zhang and Chen (2007), where N_c cameras are used to render N_p views, assuming $N_p > N_c$. The new camera positions are determined by minimizing the sum of the squared rendering errors. This minimization is solved iteratively using the local colour consistency as an estimate of the rendering error. The cameras are initialized on a uniform grid and then progressively moved to new positions based on the minimization, hence allowing the system to be applied to dynamic scenes.

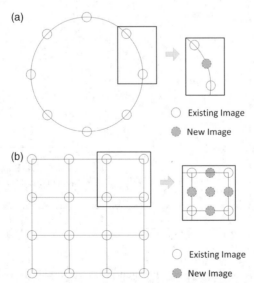

Figure 12.7 Diagram illustrating how new images are introduced when using AIC to nonuniformly sample the scene. In (a) a single image is introduced in a concentric mosaics representation (Zhang and Chen, 2003a) and in (b) multiple images are introduced in a 4D light-field representation (Zhang and Chen, 2006). A new image is inserted between an image pair based on the reconstruction error

12.4.1 Adaptive Sampling Based on Plenoptic Spectral Analysis

Adaptive plenoptic sampling is re-examined in Gilliam *et al.* (2011) with the aim of developing a framework that combines the theoretical results from plenoptic spectral analysis with adaptive camera placement. They proposed an ARC algorithm to adaptively position a finite number of cameras depending on the local geometry of the scene and the total number of cameras available. It is used to adaptively sample the plenoptic function corresponding to a scene with a smoothly varying surface and bandlimited texture. The framework of this algorithm is based on modelling the local geometric complexity of a scene with a sequence of slanted planes and then positioning cameras using the plenoptic spectral analysis of a slanted plane from Gilliam *et al.* (2010). The adaptive element of this framework comes from how the scene surface is modelled using the sequence of slanted planes.

To determine the model of the surface, Gilliam *et al.* (2011) defined a distortion metric for the reconstruction of the plenoptic function. This metric is the combination of the geometric error, caused by approximating the surface with a sequence of slanted planes, and the aliasing error, caused by undersampling the scene. The aliasing error is approximated based on the slanted plane model of the surface. Therefore, the distortion decreases as the model of the surface becomes more accurate but increases if too few samples are available (as planes within the model are undersampled). Consequently, the optimum model of the surface that minimizes this distortion depends on the number of cameras available; hence, the camera positions themselves also depend on this. Taking inspiration from rate distortion, this problem is posed as a Lagrangian minimization and a binary tree framework is used to determine the optimum model of the surface given the finite number of cameras available.

An example of this adaptive camera placement is shown in Figure 12.8. The scene in question is a piecewise quadratic surface with bandlimited texture. The figure shows the camera locations, along the x-axis, and the approximation of the surface when there are 15,

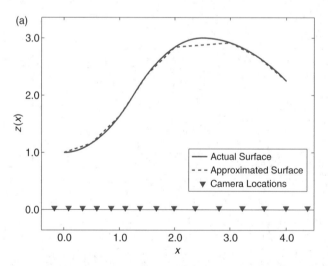

Figure 12.8 An example of the adaptive algorithm presented in Gilliam *et al.* (2011) positioning a varying number of cameras for a piecewise quadratic surface with bandlimited texture. The number of cameras varies from 15 in (a), to 25 in (b), then 35 in (c) and, finally, 45 in (d). The resulting positions of the cameras are indicated along the x-axis. Each graph also shows the approximation of the surface generated by the algorithm. The number of planes in each approximation is (a) 7, (b) 9, (c) 13 and (d) 16

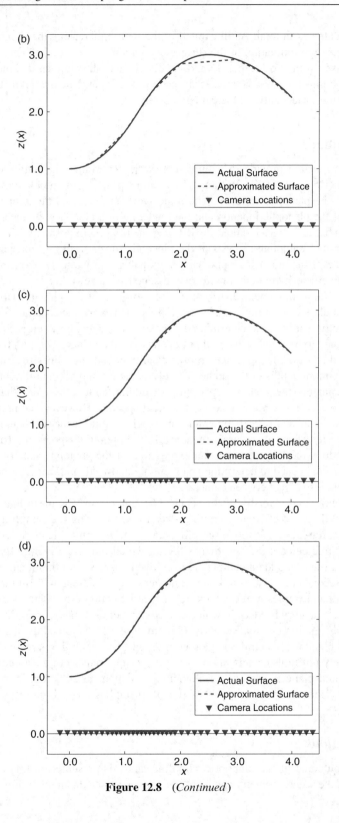

Figure 12.8 (*Continued*)

25, 35 or 45 cameras in total. As illustrated in the figure, increasing the number of samples results in a finer approximation of the surface. The number of slanted planes in the approximation increases from 7 to 11, then 13 and, finally, to 16 as the number of cameras increases. The figure also illustrates how a large local scene variation, and the overall distance the cameras are from the surface, affects the camera density.

12.5 Summary

The plenoptic function provides a natural framework for examining multi-view acquisition and rendering (Do *et al.*, 2012). Within it, IBR can be posed as the problem of sampling and reconstructing the plenoptic function. The multi-view image set, in this scenario, represents the samples of the plenoptic function and the rendering of a new view its reconstruction. The minimum number of images required in IBR, and their optimum positioning, can therefore be determined through sampling analysis of the plenoptic function. In this chapter we have presented the state of the art in plenoptic sampling analysis, in particular focusing on the optimum sampling efficiency in both a uniform and adaptive framework.

Assuming a uniform camera distribution, the sampling of the plenoptic function can be examined in a classical Fourier framework. In this framework the spectral properties of the plenoptic spectrum are used to determine the maximum spacing between each camera. If the scene is assumed to be Lambertian without occlusions, then Chai *et al.* (2000) and Do *et al.* (2012) showed that the plenoptic spectrum is exactly bounded by lines relating to the minimum and maximum depths of the scene. Therefore, taking into account the finite pixel resolution of the cameras, the plenoptic spectrum is bandlimited to a 'bow-tie'-shaped spectrum, which leads to a corresponding maximum camera spacing. However, Gilliam *et al.* (2010) showed that the inclusion of realistic conditions, such as finite scene width and cameras with finite field of view, lead to spectral spreading in the plenoptic spectrum. To examine this spectral spreading, they derived an exact expression for the plenoptic spectrum of a slanted plane. This was then used to determine an essential bandwidth and, hence, a new expression for the maximum spacing between cameras.

The complexity and diversity of the scenes encountered in IBR mean that uniform sampling is rarely the most efficient sampling strategy. Optimizing the sampling efficiency in IBR, therefore, requires an adaptive sampling approach. Adaptive plenoptic sampling can be divided into three categories. The first is SR, in which the scene is initially oversampled uniformly and then redundant samples are removed based on a minimization criterion. In contrast, the second category initially undersamples the scene in a uniform manner and then adds intermediate samples to reduce the local reconstruction error. This is known as AIC. Finally, the last category is ARC, in which a fixed number of cameras are positioned to maximize the reconstruction quality. Recently, Gilliam *et al.* (2011) joined the theoretical results derived from plenoptic spectral analysis with an adaptive sampling strategy. Their framework adaptively positions cameras based on the spectral analysis of a slanted plane. Therefore, the slanted plane is used as an elementary element to construct more complicated scenes and the cameras are positioned based on the sampling analysis corresponding to it.

12.5.1 Outlook

An implicit outcome of the plenoptic sampling analysis presented in this chapter is that knowledge of the scene geometry can be used to either reduce the number of images required in IBR or enhance the rendering quality. The interplay between amount of geometric information available and the number of cameras required for IBR has been examined in Chai *et al.* (2000). As a result, a new breed of IBR techniques has recently come to the fore that use per-pixel depth information, as well as pixel colour information, to render new views. This type of approach was termed depth IBR (DIBR) in Fehn (2004).

Currently, the required depth information is generated from passive cameras using 3D reconstruction techniques. Unfortunately, such techniques are computationally intensive and prone to inaccuracies (Do *et al.*, 2011). However, advances in sensing technologies, for example Microsoft's Xbox Kinect, may allow the large-scale deployment of 3D cameras using active depth-sensing systems (Kolb *et al.*, 2010). The ability of such cameras to accurately, and reliably, estimate the depth of a scene makes possible the augmentation of traditional colour images with depth images. Therefore, it paves the way for multi-view systems that contain both colour and depth cameras.

Such technological advancement allows the prospect of multi-view depth imaging as a research topic in its own right. For example, Gilliam *et al.* (2012) performed an initial spectral analysis on multi-view depth images. However, the following open questions remain: How many depth cameras are necessary to infer the scene geometry? How do the number of depth cameras relate to the number of colour cameras and vice versa? Can an excess of one be used to compensate a reduction of the other?

References

Abramowitz, M. and Stegun, I.A. (eds) (1964) *Handbook of Mathematical Functions with Formulas, Graphs, and Mathematical Tables*. Dover.

Adelson, E.H. and Bergen, J.R. (1991) *The Plenoptic Function and the Elements of Early Vision. Computational Models of Visual Processing*. MIT Press, Cambridge, MA, pp. 3–20.

Berent, J. and Dragotti, P.L. (2007) Plenoptic manifolds. *IEEE Signal Process. Mag.*, **24** (6), 34–44.

Bolles, R., Baker, H. and Marimont, D. (1987) Epipolar-plane image analysis: an approach to determining structure from motion. *Int. J. Comput. Vis.*, **1** (1), 7–55.

Chai, J.X., Chan, S.C., Shum, H.Y. and Tong, X. (2000) Plenoptic sampling, in *SIGGRAPH '00 Proceedings of the 27th Annual Conference on Computer Graphics and Interactive Techniques*, ACM Press/Addison-Wesley, New York, NY, pp. 307–318.

Chen, C. and Schonfeld, D. (2009) Geometrical plenoptic sampling, in *ICIP'09 Proceedings of the 16th IEEE International Conference on Image Processing*, IEEE Press, Piscataway, NJ, pp. 3769–3772.

Criminisi, A., Kang, S.B., Swaminathan, R. *et al.* (2005) Extracting layers and analyzing their specular properties using epipolar-plane-image analysis. *Comput. Vis. Image Und.* **97** (1), 51–85.

Do, M.N., Nguyen, Q.H., Nguyen, H.T. *et al.* (2011) Immersive visual communication. *IEEE Signal Process. Mag.*, **28** (1), 58–66.

Do, M.N., Marchand-Maillet, D. and Vetterli, M. (2012) On the bandwidth of the plenoptic function. *IEEE Trans. Image Process.*, **21** (2), 708–717.

Fehn, C. (2004) Depth-image-based rendering (DIBR), compression, and transmission for a new approach on 3D-TV, in *Stereoscopic Displays and Virtual Reality Systems XI* (eds A.J. Woods, J.O. Merritt, S.A. Benton and M.T. Bolas), Proceedings of the SPIE, Vol. **5291**, SPIE, Bellingham, WA, pp. 93–104.

Feldmann, I., Eisert, P. and Kauff, P. (2003) Extension of epipolar image analysis to circular camera movements, in *Proceedings. 2003 International Conference on Image Processing, 2003. ICIP 2003*, Vol. **3**, IEEE, pp. 697–700.

Fleishman, S., Cohen-Or, D. and Lischinski, D. (2000) Automatic camera placement for image-based modeling. *Comput. Graph. Forum*, **19** (2), 101–110.

Fujii, T., Kimoto, T. and Tanimoto, M. (1996) Ray space coding for 3D visual communication. Proceedings of Picture Coding Symposium, pp. 447–451.

Gilliam, C., Dragotti, P.L. and Brookes, M. (2010) A closed-form expression for the bandwidth of the plenoptic function under finite field of view constraints, in *2010 17th IEEE International Conference on Image Processing (ICIP)*, IEEE, pp. 3965–3968.

Gilliam, C., Dragotti, P.L. and Brookes, M. (2011) Adaptive plenoptic sampling, in *2011 18th IEEE International Conference on Image Processing (ICIP)*, IEEE, pp. 2581–2584.

Gilliam, C., Brookes, M. and Dragotti, P.L. (2012) Image based rendering with depth cameras: How many are needed? in *2012 IEEE International Conference on Acoustics, Speech and Signal Processing (ICASSP)*, IEEE, pp. 5437–5440.

Gortler, S., Grzeszczuk, R., Szeliski, R. and Cohen, M. (1996) The lumigraph, in *SIGGRAPH '96 Proceedings of the 23rd Annual Conference on Computer Graphics and Interactive Techniques*, ACM, New York, NY, pp. 43–54.

Ihm, I., Park, S. and Lee, R.K. (1997) Rendering of spherical light fields, in *PG '97 Proceedings of the 5th Pacific Conference on Computer Graphics and Applications*, IEEE Computer Society, Washington, DC, pp. 59–68.

Kolb, A., Barth, E., Koch, R. and Larsen, R. (2010) Time-of-flight sensors in computer graphics. *Comput. Graph. Forum* **29** (1), 141–159.

Levoy, M. and Hanrahan, P. (1996) Light field rendering, in *SIGGRAPH '96 Proceedings of the 23rd Annual Conference on Computer Graphics and Interactive Techniques*, ACM, New York, NY, pp. 31–40.

McMillan, L. and Bishop, G. (1995) Plenoptic modelling: an image-based rendering system, in *SIGGRAPH '95 Proceedings of the 22nd Annual Conference on Computer Graphics and Interactive Techniques*, ACM, New York, NY, pp. 39–46.

Miller, G., Rubin, S. and Ponceleon, D. (1998) Lazy decompression of surface light fields for precomputed global illumination, in *Rendering Techniques '98* (eds G. Drettakis and N. Max), Springer, pp. 281–292.

Namboori, R., Teh, H.C. and Huang, Z. (2004) An adaptive sampling method for layered depth image, in *CGI '04 Proceedings of the Computer Graphics International*, IEEE Computer Society, Washington, DC, pp. 206–213.

Nguyen, H.T. and Do, M.N. (2009) Error analysis for image-based rendering with depth information. *IEEE Trans. Image Process.*, **18** (4), 703–716.

Schirmacher, H., Heidrich, W. and Seidel, H.P. (1999) Adaptive acquisition of lumigraphs from synthetic scenes. *Comput. Graph. Forum*, **18** (3), 151–160.

Shum, H.Y. and He, L.W. (1999) Rendering with concentric mosaics, in *SIGGRAPH '99 Proceedings of the 26th Annual Conference on Computer Graphics and Interactive Techniques*, ACM Press/Addison-Wesley, New York, NY, pp. 299–306.

Shum, H.Y., Chan, S.C. and Kang, S.B. (2007) *Image-Based Rendering*, Springer.

Wilburn, B., Smulski, M., Lee, K. and Horowitz, M. (2002) The light field video camera, in *Media Processors 2002* (eds S. Panchanathan, V.M. Bove, Jr, and S.I. Sudharsanan), Proceedings of the SPIE, Vol. **4674**, SPIE, Bellingham, WA, pp. 29–36.

Wood, D., Azuma, D., Aldinger, K. *et al.* (2000) Surface light fields for 3D photography, in *SIGGRAPH '00 Proceedings of the 27th Annual Conference on Computer Graphics and Interactive Techniques*, ACM Press/Addison-Wesley, New York, NY, pp. 287–296.

Zhang, C. and Chen, T. (2003a) Non-uniform sampling of image-based rendering data with the position-interval error (PIE) function, in *Visual Communications and Image Processing 2003* (eds T. Ebrahimi and T. Sikora), Proceedings of the SPIE, Vol. **5150**, SPIE, Bellingham, WA, pp. 1347–1358.

Zhang, C. and Chen, T. (2003b) Spectral analysis for sampling image-based rendering data. *IEEE Trans. Circ. Syst. Vid. Technol.*, **13** (11), 1038–1050.

Zhang, C. and Chen, T. (2004) A survey on image-based rendering - representation, sampling and compression. *Signal Process. Image Commun.*, **19** (1), 1–28.

Zhang, C. and Chen, T. (2006) *Light Field Sampling, Synthesis Lectures on Image, Video, and Multimedia Processing*, Morgan & Claypool Publishers.

Zhang, C. and Chen, T. (2007) Active rearranged capturing of image-based rendering scenes – theory and practice. *IEEE Trans. Multimedia*, **9** (3), 520–531.

13

A Framework for Image-Based Stereoscopic View Synthesis from Asynchronous Multiview Data

Felix Klose, Christian Lipski, and Marcus Magnor
Institut für Computergraphik, TU Braunschweig, Germany

13.1 The Virtual Video Camera

In this chapter we would like to explore the possibilities to re-render a scene captured by multiple cameras. As we will show, it is possible to synthesize new views of the scene from previously unrecorded virtual camera positions. Let us start off with a very brief overview.

The framework we describe relies purely on *image-based* operations. The final image is generated by applying 2D transformations to recorded input images. At the core of our pipeline are dense image correspondences. We use a multi-image warping and blending scheme to create virtual in-between views of our captured scene. The pipeline generally does not distinguish between images that have been recorded at different points in time and those captured with a changed viewpoint. As a result, the framework can naturally handle *asynchronously* recorded video footage. We show that the renderer is capable of generating *stereoscopic* three-dimensional (3D) content without additional preprocessing.

The objective common to all free-viewpoint navigation systems is to render photo-realistic vistas of real-world, dynamic scenes from arbitrary perspective (within some specified range), given a number of simultaneously recorded video streams. Most systems exploit epipolar geometry based on either dense depth/disparity maps (Goldlücke *et al.*, 2002; Naemura *et al.*, 2002; Zitnick *et al.*, 2004) or complete 3D geometry models (Matusik *et al.*, 2000; Carranza *et al.*, 2003; Vedula *et al.*, 2005; Starck and Hilon, 2007; de Auiar *et al.*, 2008). In order to estimate depth or reconstruct 3D geometry of dynamic scenes, the input multi-video data must be precisely calibrated as well as captured synchronously. This dependence on synchronized footage can limit practical applicability: high-end or even custom-built acquisition hardware must be employed, and the recording setup indispensably includes some sort of camera interconnections (cables, WiFi). The cost, time, and effort involved in recording synchronized multi-video data prevent widespread use of free-viewpoint navigation methods.

Our approach accepts unsynchronized, uncalibrated multi-video footage as input. It is motivated by the pioneering work on view interpolation by Chen and Williams (1993). We pick up on the idea of interpolating different image acquisition attributes in higher dimensional space and suitably extend it to be applicable to view interpolation in the spatial and temporal domains. By putting the temporal dimension on a par with the spatial dimensions, a uniform framework is available to continuously interpolate virtual video camera positions across space and time.

In our approach, we make use of dense image correspondences (Stich *et al.*, 2011), which take the place of depth/disparity or 3D geometry. We show how dense correspondences allow extending applicability to scenes whose object surfaces are highly variable in appearance (e.g., due to specular highlights) or hard to reconstruct for other reasons. Perceptually plausible image correspondence fields can often still be established where ground-truth geometry (or geometry-based correspondences) cannot (Beier and Nelly, 1992). Dense image correspondences can also be established along the temporal dimension to enable interpolation in time.

Our virtual video camera system features continuous viewpoint navigation in space and time for general, real-world scenes from a manageable number of easy-to-acquire, unsynchronized and uncalibrated video recordings. To facilitate intuitive navigation in space and time, we later describe the construction of a 3D navigation space. We show that special care has to be taken to create the notion of a static virtual camera if the temporal dimension is involved.

We now continue with a description of the image formation and view synthesis pipeline (Lipski *et al.*, 2010c); see Figure 13.1. To be able to perform this synthesis the input data have to be preprocessed to calculate the dense correspondence fields needed. The algorithm we developed for automatic estimation of these correspondences is then described in Section 13.2 (Lipski *et al.*, 2010a). When using the virtual video camera in a production environment it is necessary to be able to correct the image correspondences in failure cases. A custom-developed computer-aided approach to correspondence editing is shown in Section 13.3 (Klose *et al.*, 2011). In Section 13.4 we show the extension of the virtual video camera for stereoscopic 3D content creation (Lispki *et al.*, 2011).

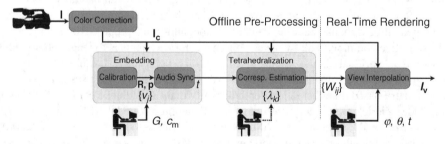

Figure 13.1 Virtual video camera processing pipeline: multi-video data I is color-corrected first (I_c). To embed the video frames into navigation space $\{v_i\}$, the user specifies the master camera c_m and a common ground plane G. Extrinsic camera parameters ($\mathbf{R}, \mathbf{p}$) and the time offsets t are automatically estimated. Adjacency of video frames induces a tetrahedralization $\{\lambda_k\}$ of the navigation space, and dense correspondences $\{W_{ij}\}$ are estimated along the edges of the tetrahedralization. We allow for manual correction of spurious correspondence fields. After these offline processing steps, the navigation space can be interactively explored (viewing directions φ, θ, time t) by real-time rendering the virtual view I_v

13.1.1   Navigation Space Embedding

Our goal is to explore the captured scene in an intuitive way and render a (virtual) view I_v of it, corresponding to a combination of viewing direction and time. To this end, we choose to define a 3D navigation space $\mathbf{N}$ that represents spatial camera coordinates as well as the temporal dimension. In their seminal paper, Chen and Williams (1993) proposed to interpolate the camera rotation $\mathbf{R}$ and position $\mathbf{p}$ directly in six-dimensional hyperspace. While this is perfectly feasible in theory, it has several major drawbacks in practice: it neither allows for intuitive exploration of the scene by a user, nor is it practical to handle the amount of emerging data needed for interpolation in this high-dimensional space. Additionally, cameras would have to be arranged in a way that they span an actual volume in Euclidean space. With this requirement, it would be hard to devise an arrangement of cameras where they do not occlude each other's view of the scene. The crucial design decision in our system is hence to map the extrinsic camera parameters to a lower dimensional space that allows intuitive navigation. The temporal dimension already defines one axis of the navigation space $\mathbf{N}$, leaving two dimensions for parameterizing the camera orientation and position. Practical parameterizations that allow for realistic view interpolation are only possible if the cameras' optical axes cross at some point (possibly at infinity).

A natural choice for such an embedding is a spherical parameterization of the camera setup. While, for example, a cylindrical embedding or an embedding in a plane is also feasible, a spherical embedding allows for all reasonable physical camera setups, ranging from cameras arranged in a one-dimensional (1D) arc, over cameras placed in a spherical setup to linear camera arrays with parallel optical axes in the limit. Regarding existing data sets, it is obvious that spherical/arc-shaped setups are the predominant multiview capture scenarios (e.g., Zitnick *et al.*, 2004; de Auiar *et al.*, 2008). Even in unordered image collections, it can be observed that a majority of camera positions is distributed in arcs around certain points of interest (Snavely *et al.*, 2006). Other placements, such as panoramic views, are also possible, but would suffer from the small overlap of the image regions of cameras. For all the results presented in this chapter, we hence employ a spherical model with optical axes centered at a common point. Camera setups such as Bézier splines or patches are also feasible and our approach may adapt to any setup as long as a sensible parameterization can be obtained. As the extension of our approach to these settings is straightforward, we will not discuss it in detail.

Cameras are placed on the surface of a virtual sphere; their orientations are defined by azimuth φ and elevation θ. Together with the temporal dimension t, φ and θ span our 3D navigation space $\mathbf{N}$. If cameras are arranged in an arc or curve around the scene, θ is fixed, reducing $\mathbf{N}$ to two dimensions. As this simplification is trivial, we will only cover the 3D case in our discussion. In contrast to the conventional partition of space suggested by Chen and Williams (1993), we thus restrict the movement of the virtual camera to a subspace of lesser dimensionality (2D approximate spherical surface or 1D approximated arc). Although this might appear as a drawback at first sight, several advantages arise from this crucial design decision; see Figure 13.2:

1. The amount of correspondence fields needed for image interpolation is reduced significantly, making both preprocessing and rendering faster.
2. An unrestricted partition of Euclidean space leads to degenerated tetrahedra. Particularly when cameras are arranged along a line or arc, adjacencies between remote cameras are established for which no reliable correspondence information can be obtained.
3. Our parameterization of the camera arrangement provides an intuitive navigation around the scene.

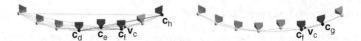

Figure 13.2 Two possibilities to partition a half arc camera setup. Both images depict the camera setup seen from above. When partitioning the Euclidean space directly (left), several problems arise. Although an actual volume is spanned by the cameras, many tetrahedra are degenerated. If the virtual camera v_c is reconstructed, captured images from cameras c_d, c_e, c_f and c_h as well as the wide-baseline correspondence fields between them are required. Using our navigation space embedding (right), interpolation only takes place between neighboring cameras. Less data has to be processed and correspondence estimation is much easier. Additionally, spatial navigation is intuitively simplified. Instead of a 3D position, only the position on the one-dimensional arc has to be specified. The small spatial error (distance between v_c and line segment between c_f and c_g) is negligible even for ad-hoc setups

To define our navigation space **N**, we assume that we know ground-truth extrinsic camera parameters rotation **R** and position **p** for every camera, as well as a few point correspondences with their 3D world. For a specific virtual image I_v, we want to interpolate the image at a given point in navigation space defined by the two spatial parameters φ and θ as well as recording time t. To serve as sampling points, the camera configuration of our recorded multi-video input in Euclidean world space is embedded into navigation space **N**:

$$\Psi : (\mathbf{R}, \mathbf{p}, t) \rightarrow (\varphi, \theta, t) \tag{13.1}$$

In our system, Ψ is simply a transformation from Cartesian coordinates to spherical coordinates, where the sphere center $\mathbf{p}_s$ and the radius of the sphere r_s are computed from the cameras' extrinsic parameters **R** and **p** in a least-squares sense. The embedding is uniquely defined by specifying a ground plane in the scene and by labeling one of the cameras as the master camera c_m; see Figure 13.3.

It is evident that the recording hull spanned by all camera positions is, in general, only a crude approximation to a spherical surface. Fortunately, we found that nonspherical camera arrangements do not cause any visually noticeable effect during rendering.

13.1.2 Space–Time Tetrahedralization

In order to interpolate viewpoints, Chen and Williams (1993) proposed to generate some arbitrary graph to partition the space such that every possible viewpoint lies in a d-simplex

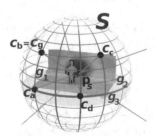

Figure 13.3 Navigation space: we define a sphere **S** around the scene. For the center $\mathbf{p}_s$ of **S**, we least-squares-determine the point closest to the optical axes of all cameras. The user selects three points g_1, g_2, g_3 to define the ground plane. We take the normal of the plane as the up vector of the scene, and thus as the rotation axis of the sphere. The embedding is uniquely defined by labeling one of the cameras as the master camera $\mathbf{c}_m$

and can be expressed as a linear combination of the $d + 1$ vertices of the enclosing simplex. When including the temporal dimension, however, an arbitrary graph leads to rendering artifacts due to inconsistent mappings between Euclidean space and navigation space.

Our navigation space consists of two view-directional dimensions and the temporal dimension. We subdivide the navigation space into tetrahedral λ such that each embedded video frame represents one vertex (Lipski *et al.*, 2010c). We interpolate any virtual camera view from the enclosing tetrahedron $\lambda = \{v_i\}$, $i = 1 \ldots 4$. As already mentioned, arbitrary partitions are feasible in static scenes. However, our spherical approximation and/or calibration imprecision will lead to interpolation errors in practice if the temporal dimension is included.

To prove this point, let us assume that we have captured a dynamic scene with four static, unsynchronized cameras c_a, c_b, c_c, and c_d. As we stated before, the view-directional subspace relates to the spherical camera placement. To be more exact, it represents an approximation of a spherical surface; examples for possible approximations are given in Figure 13.4, upper right and lower right. We will show that although the chosen spherical approximation is arbitrary, it is crucial to enforce consistency of the approximation when the temporal dimension is introduced, even if ground-truth camera parameters $\mathbf{R}$ and $\mathbf{p}$ are available.

Imagine now that the user intends to re-render the scene from a static virtual camera. The position of this camera is somewhere between the four original cameras, so that the new view has to be interpolated. Because the camera is static in navigation space, φ and θ stay fixed while t is variable such that the virtual camera moves along a line l in navigation space (Figure 13.5). Assume that the navigation space has been partitioned by applying a standard Delaunay tetrahedralization (Edelsbrunner and Shah, 1995) to the set of navigation space vertices v_a^i, v_b^i, v_c^i, v_d^i, where i indexes the temporal dimension. Now consider the camera path l. In line section l_1, images associated with camera c_a, c_b, and c_c are used for interpolation. The virtual camera thus represents a point that lies on plane Δ_3 in the original Euclidean space (Figure 13.4, lower right). In line section l_2, however, the virtual camera is interpolated from all four cameras; that is, it moves within the volume spanned by the four different cameras in Euclidean world space (Figure 13.5a). This violates the notion of a static virtual camera. In line segment l_3 the virtual camera is represented by a point in Euclidean world space situated on plane Δ_1 (Figure 13.4, upper right). Since Δ_1 and Δ_3 are not coplanar, the position of the virtual camera in Euclidean space is different in both line sections, again violating the static camera assumption.

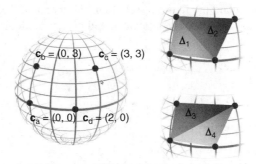

Figure 13.4 Surface tessellation: for the spherical coordinates of four cameras $c_a, \ldots, c_d$ two possible tessellations $\{\Delta_1, \Delta_2\}$ and $\{\Delta_3, \Delta_4\}$ exist. During space–time interpolation, the configuration of the initial tessellation may not change to avoid interpolation discontinuities

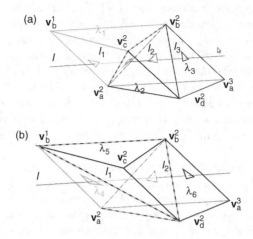

Figure 13.5 (a) Unconstrained Delaunay tetrahedralization. (b) Constrained Delaunay tetrahedralization. Unconstrained vs. constrained tessellation of the navigation space: with an unconstrained tetrahedralization (a), tetrahedra λ_1, λ_2 and λ_3 are created. A static virtual camera moves along line l and passes through all three tetrahedra. When comparing the line sections l_1, l_2, and l_3, the point in navigation space relates to different points in Euclidean space; thus, the virtual camera seems to move. By introducing boundary faces, such errors can be avoided (b). The virtual camera passes solely through tetrahedra (λ_4, λ_6) which are spanned by the same three cameras and is not influenced by $\mathbf{v}_c^2$ (tetrahedron λ_5 remains obsolete). Therefore, the virtual camera remains static in Euclidean space, as intended

In navigation-space interpolation, this error can be avoided if any two points $\mathbf{v}_t = (\varphi, \theta, t)$ and $\mathbf{v}_{t+\delta} = (\varphi, \theta, t + \delta)$ re-project to the same point in Euclidean space; that is, if the virtual camera is static in Euclidean space. Instead of building an arbitrary graph structure on the navigation space vertices, we apply a combination of constrained and unconstrained Delaunay tessellations to navigation space vertices.

To obtain a temporally consistent tetrahedralization, we start with an unconstrained Delaunay tessellation on the sphere; that is, we neglect the temporal dimension for now. This initial tessellation must be preserved; that is, the spatial ordering of cameras must stay the same for all recorded frames of the video streams. Referring back to the example given above, we arrive at the tessellation shown in Figure 13.6 (left). The edges of this tessellation serve as boundaries that help maintain the initial tessellation. We then add the temporal dimension. The initial tessellation of the sphere is duplicated and displaced along the temporal axis. In

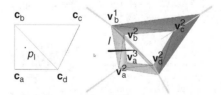

Figure 13.6 Boundary faces: a 2D unconstrained Delaunay tessellation reveals the boundary edges between cameras (left). In 3D navigation space (right), boundary faces are inserted for each boundary edge. The virtual static camera p_l that moves on a line l in navigation space always lies in tetrahedra constructed from $\mathbf{c}_a$, $\mathbf{c}_b$, and $\mathbf{c}_d$, as postulated in Figure 13.4

order to maintain the structure of the tessellation computed so far, we add boundary faces for each edge of the initial tessellation; for example, we add a face $(\mathbf{v}_b^1, \mathbf{v}_b^2, \mathbf{v}_d^2)$ in the example above (Figure 13.6, right).

A constrained Delaunay tessellation can then be computed on this 3D point cloud in navigation space (Si and Gaertner, 2005). In our example, the added boundary faces separate the data in such a way that every tetrahedron corresponds to one of the two planes in Euclidean space; that is, either to λ_1 or λ_2 in Figure 13.4, but never to λ_3 or λ_4. All tetrahedra also comprise four images that were captured by exactly three cameras. If each image vertex in a tetrahedron was captured by a different camera, the situation shown in line section l_2 (Figure 13.5a) would arise.

The undesired flipping along the temporal dimension also occurs when working with dynamic cameras. Here, we assume that the cameras move but their relative positions do not change, that is, we do not allow cameras to switch position. In this case, the same reasoning as presented above for static cameras applies. However, now the notion of a static virtual camera is defined with respect to the moving input cameras. By doing so, sudden changes of the spherical surface approximation can be prevented, just as in the case of stationary recordings. The only limitation is that when cameras move in Euclidean space, the approximated spherical surface also changes. In these cases our approximation error – that is, the distance of the rendered virtual camera to our idealized spherical camera arrangement – does not remain fixed. In general, this is only noticeable with fast-moving cameras and can usually be neglected for small camera movements. The same applies to imprecise camera calibration. Small errors in $\mathbf{R}$ and $\mathbf{p}$ are inevitable, but usually do not manifest in visible artifacts.

13.1.3 Processing Pipeline

Based on the above-described navigation space tetrahedralization, we propose a fully functional processing pipeline for free-viewpoint navigation from unsynchronized multi-video footage. Our processing pipeline makes use of known techniques and suitably adapts them to solve the problems encountered. We now go through the required preprocessing steps that prepare recorded material for input into our system.

Color correction. To correct for color balance differences among cameras, we use the master camera c_m defined in the embedding stage and apply the color correction approach presented by Snavely *et al.* (2006) to all video frames.

Camera calibration. We need to determine $\mathbf{R}$ and $\mathbf{p}$ of the camera setup to define the mapping Ψ from world space coordinates to navigation space $\mathbf{N}$ (Section 13.1.1). Recent structure-from-motion algorithms for unordered image collections (Goesele *et al.*, 2007; Schwartz and Klein, 2009; Lipski *et al.*, 2010b) solve this problem robustly and can also provide a set of sparse world space points needed for constructing the common ground plane for our navigation space. We found that this algorithm yields robust results also for dynamic scenes. For dynamic camera scenarios (i.e., handheld, moving cameras), $\mathbf{R}$ and $\mathbf{p}$ have to be computed for every frame of each camera.

Temporal registration. The mapping Ψ additionally needs the exact recording time t of each camera. We estimate the sub-frame temporal offset by recording a dual-tone sequence during acquisition and analyzing the audio tracks afterwards (Hasler *et al.*, 2009). If recording a separate audio track is not feasible, pure post-processing approaches (Meyer *et al.*, 2008) can be employed instead. Please note that, after temporal registration, the recorded images are still not synchronized. Since the camcorder recordings were triggered manually, a subframe offset still prevents exploiting the epipolar constraint in the presence of moving objects.

Dense correspondence field estimation. In order to interpolate between two images I_i, I_j, we need bidirectional dense correspondence maps $\mathbf{W}_{ij}$ for each tetrahedral edge in navigation space $\mathbf{N}$. We automatically compute the correspondences based on dense scale-invariant feature transform (SIFT) feature matching and a belief propagation scheme; more detail is given in Section 13.2. In regions where the automatic correspondence estimation fails or the rendering algorithm produces visible artifacts, utilities exist that enable user-guided correspondence correction (Section 13.3).

13.1.4 Rendering

Having subdivided the navigation space $\mathbf{N}$ into tetrahedra, each point $\mathbf{v}$ is defined by the vertices of the enclosing tetrahedron $\lambda = \{\mathbf{v}_i\}$, $i = 1 \ldots 4$. Its position can be uniquely expressed as $\mathbf{v} = \sum_{i=1}^{4} \mu_i \mathbf{v}_i$, where μ_i are the barycentric coordinates of $\mathbf{v}$. Each of the four vertices $\mathbf{v}_i$ of the tetrahedron corresponds to a recorded image I_i. Each of the 12 edges e_{ij} corresponds to a correspondence map $\mathbf{W}_{ij}$ that defines a translation of a pixel location $\mathbf{x}$ on the image plane. We are now able to synthesize a novel image $I_\mathbf{v}$ for every point $\mathbf{v}$ inside the recording hull of the navigation space $\mathbf{N}$ by multi-image interpolation:

$$I_\mathbf{v} = \sum_{i=1}^{4} \mu_i \tilde{I}_i \tag{13.2}$$

where

$$\tilde{I}\left(\Pi_i(\mathbf{x}) + \sum_{j=1,\ldots,4 j \neq i} \mu_j \Pi_j(\mathbf{W}_{ij}(\mathbf{x})) \right) = I_i(\mathbf{x}) \tag{13.3}$$

are the forward-warped images (Mark *et al.*, 1997). $\{\Pi_i\}$ defines a set of re-projection matrices Π_i that map each image I_i onto the image plane of $I_\mathbf{v}$, as proposed by Seitz and Dyer (1996). Those matrices can be easily derived from camera calibration. Since the virtual image $I_\mathbf{v}$ is always oriented towards the center of the scene, this re-projection corrects the skew of optical axes potentially introduced by our loose camera setup and also accounts for jittering introduced by dynamic cameras. Image re-projection is done on the GPU without image data resampling.

The spatial relationship between corresponding pixels is encoded in the correspondence maps when combined with the camera calibration matrices. By assuming linear object motion we can determine where an object is located in two different views at the same point in time. For every pixel, we determine its position in a neighboring view. If the two cameras did not capture the images synchronously, we determine the actual position of the object by assuming linear motion. We least-squares determine the 3D location of this point in Euclidean space. We re-project this point into the original view and store the depth value. We can use the depth maps obtained during the following image interpolation phase to resolve occlusion ambiguities.

Hole filling is done during the blending stage where image information from the other three warped images is used. As a last step, the borders of the rendered images are cropped (10% of the image in each dimension), since no reliable correspondence information is available for these regions.

13.1.5   Application

The virtual video camera can be used for the creation of virtual camera movements during post-production. Arbitrary panning, time-freeze, and slow-motion shots can be defined and rendered. We implemented a graphical user interface that allows intuitive exploration of our navigation space and allows the user to compose arbitrary space–time camera movements interactively. The user interface makes use of common 3D animation concepts. A visualization of the navigation space is presented, and the current virtual camera view is rendered in real time (Figure 13.7). The user can continuously change viewing direction and time by simple click-and-drag movements within the rendering window. Camera paths are defined by placing, editing, or deleting control points in navigation space at the bottom of the screen. The virtual video camera path through space–time is interpolated by Catmull–Rom splines.

To evaluate the applicability to a real-world production scenario, the virtual video camera has been used in a stereoscopic music video production (Lispki *et al.*, 2011). Section 13.4 shows the final renderings and the production pipeline used.

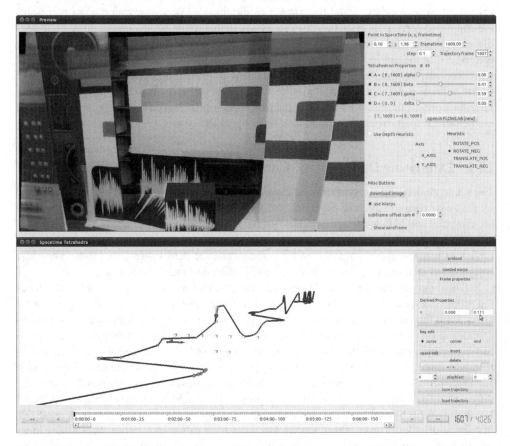

Figure 13.7   Virtual video editing: to interactively create arbitrary virtual video camera sequences, the user moves the camera by click-and-drag movements in the rendering window (top). The spline curve representing the space–time camera path is automatically updated and visualized in the navigation space window (bottom)

13.1.6 Limitations

Similar to depth/disparity-based free viewpoint navigation systems (Zitnick *et al.*, 2004), our virtual viewpoint is spatially restricted: we can viewpoint-navigate on the hull spanned by all camera recording positions, looking at the scene from different directions, but we cannot, for example, move into the scene or fly through the scene.

Output rendering quality obviously depends on the visual plausibility of the correspondence fields. While we found the automatic, pair-wise correspondence estimation algorithm (Lipski *et al.*, 2010a) to yield convincing and robust results overall, we explicitly allow for human interaction in addition to correct for remaining spurious correspondences. While for special scenarios other specifically tailored matching algorithms might yield better results, we incorporated a matching algorithm that provides robust results for most scenes. For the computation of a correspondence field, our implementation took about 10 min per image pair, depending on image resolution and scene content. Correspondence correction typically takes about 1 min per video frame pair. Small uncorrected inaccuracies in automatically computed fields manifest in cross-fading/ghosting artifacts visible in some cases. In our system, occlusion and disocclusion are handled by estimating approximate depth maps using the calculated correspondences and the camera calibration. We hence rely on the correctness of the computed correspondence fields. In cases where this rough occlusion handling fails, visible rendering artifacts occur at occlusion border. For high-quality rendering results, we observed that the angle between adjacent cameras should not exceed 10°, independent of scene content. For greater distances, missing scene information appears to become too large to still achieve convincing interpolation results. The same is true for scene motion that is too fast. As a rule of thumb, we found that, across space or time, scene correspondences should not be farther apart than approximately 10% of linear image size.

13.2 Estimating Dense Image Correspondences

The scene model used for synthesizing the virtual free-viewpoint camera positions is defined by pair-wise dense image correspondences. As described in Section 13.1.4, multiple images are morphed and blended to create a new virtual camera position. For a given camera trajectory, the number of required correspondence fields is quite large and the quality of the final rendering results depends heavily on the correspondence field validity. A robust automatic computation method is needed to make the use feasible in longer sequences.

Establishing the needed dense image correspondences between images is still a challenging problem, especially when the input images feature long-range motion and large occluded areas. With the increasing availability of high-resolution content, the requirements for correspondence estimation between images are further increased. High-resolution images often exhibit many ambiguous details, where their low-resolution predecessors only show uniformly colored areas; thus, the need for smarter and more robust matching techniques arises. Liu *et al.* (2008) recently proposed a dense matching approach for images possibly showing different scene content. We describe an approach for establishing pixel correspondences between two high-resolution images and pick up on their idea to incorporate dense SIFT feature descriptors (Lowe, 2004), yet we use them for a different purpose. While they identify visually similar regions in low-resolution images, we use them as a descriptor for fine detail in high-resolution images. Our approach provides a versatile tool for various tasks in video post-production. Examples are image morphing, optical flow estimation, stereo rectification, disparity/depth reconstruction, and baseline adjustment.

In order to match fine structural detail in two images, we compute a SIFT descriptor for each pixel in the original high-resolution images. To avoid ambiguous descriptors and to speed up computation, we downsample each image by selecting the most representative SIFT descriptor for each $n \times n$ grid cell (typically $n = 4$). An initial lower resolution correspondence map is then computed on the resulting downsampled versions of both images. The 131-dimensional descriptor of each pixel is a combination of the mean color (3D) and the representative SIFT descriptor of this cell (128-dimensional). The L_1-norm of this vector describes dissimilarity between two pixels and allows for much clearer distinction between non-corresponding pixels when compared with using just the pixel color as in many other approaches.

While the original belief propagation implementation by Felzenszwalb and Huttenlocher (2006) might not retain crisp borders due to the grid-based message passing scheme, we employ a non-gridlike regularization technique as proposed by Smith *et al.* (2009). As memory consumption of belief propagation on this scale is still too high for long-range correspondence estimation, we use a simple minima-preserving data term compression. During belief propagation, a symmetry term ensures consistent results. Occluded regions are identified and inpainted: assuming that each occluded area is surrounded by two independently moving regions, we use geodesic matting (Bai and Sapiro, 2009) to propagate correspondence information. The resulting image correspondence map is upsampled to its original size and refined locally.

13.2.1   Belief Propagation for Image Correspondences

Belief propagation was introduced to computer vision by Felzenszwalb and Huttenlocher (2006). Since then, it has received considerable attention by the community. It was used by Liu *et al.* (2008) in combination with SIFT features for dense correspondence estimation between similar images potentially showing different scene content. However, these correspondences were only established for thumbnail-sized images and did not suffice for the tasks our approach can cope with (i.e., occlusion handling, symmetric correspondences, and high-resolution data).

Optical flow algorithms are closely linked to our approach; a survey of state-of-the-art algorithms has been conducted by Baker *et al.* (2007). The key difference in concept is that optical flow algorithms typically derive continuous flow vectors instead of discrete pixel correspondences. Additionally, optical flow computation assumes color/brightness constancy between images.

Today, commercial tools for stereo footage post-production have reached a mature stage of development (e.g., Ocula; Wilkes, 2009). Our image correspondence algorithm could be easily integrated into any stereoscopic post-production pipeline, since it works in an unsupervised fashion, only requires image pairs as input, and can be applied to various post-production tasks, including image rectification and disparity estimation.

Belief propagation estimates discrete labels for every vertex in a given graph; that is, for every pixel in a given image. Although we do not achieve sub-pixel accuracy with belief propagation, its robustness makes it an appealing option for discrete energy minimization problems. In a nutshell, establishing pixel correspondences between two images with belief propagation works as follows. Matching costs for every possible pixel match in a given search window are computed for each pixel. Typically, the L_1 norm of the color vectors serves as a basic example for this matching cost. Neighboring pixels iteratively exchange their (normalized) matching costs for the correspondences. This message-passing process regularizes the image correspondences and finally converges to a point where consensus about final pixel correspondences is reached. As a result, a discrete correspondence vector $w(\mathbf{p}) = (u(\mathbf{p}), v(\mathbf{p}))$ is assigned to every pixel location $\mathbf{p} = (x, y)$ that encodes the correspondence to pixel location $\mathbf{p}' = (x + u(\mathbf{p}), y + v(\mathbf{p}))$ in the second image. For a thorough introduction we refer the reader to Felzenszwalb and Huttenlocher (2006).

Assuming that the search space is the whole image, computational complexity and memory usage are as high as $O(L^4)$, where L is the image width. By decoupling u and v (the horizontal and vertical component of the correspondence vector), the complexity for message passing can be reduced to $O(L^3)$, as proposed by Liu et al. (2008). Still, the evaluation of the matching costs runs in $O(L^4)$.

We formulate the correspondence estimation as an energy minimization problem; our energy functional is based on the one proposed by Liu et al. (2008):

$$E(\mathbf{x}) = \sum_p \|d_1(\mathbf{p}) - d_2(\mathbf{p} + w)\|_1 + \sum_{(\mathbf{p},\mathbf{q})\in\varepsilon} \min(\alpha|u(\mathbf{p}) - u(\mathbf{q})|, d)$$
$$+ \sum_{(\mathbf{p},\mathbf{q})\in\varepsilon} \min(\alpha|v(\mathbf{p}) - v(\mathbf{q})|, d) \qquad (13.4)$$

where $w(\mathbf{p}) = (u(\mathbf{p}), v(\mathbf{p}))$ is the correspondence vector at pixel location $\mathbf{p} = (x, y)$. In contrast to the original SIFT flow implementation, $d_i(\mathbf{p}) = c_i(\mathbf{p}) + s_i(\mathbf{p})$ is a 131-dimensional descriptor vector, containing both color information $c_i(\mathbf{p}) \in \mathbb{R}^3$ and the SIFT descriptor $s_i(\mathbf{p}) \in \mathbb{R}^{128}$. Each descriptor entry has a value between 0 and 255, we set $\alpha = 160$ and $d = $ image width $\times 5$. In addition, the pixel neighborhood ε is not simply defined by the image lattice, as we will show in Section 13.2.4. In contrast to SIFT flow, we do not penalize large motion vectors, since we explicitly try to reconstruct such scenarios.

As Liu et al. (2008) demonstrated, this energy functional can be minimized with efficient belief propagation. Before we start with a detailed description of our pipeline, we would like to present our extension of the energy functional to symmetric correspondence maps.

13.2.2 A Symmetric Extension

Since we want to enforce symmetry between bidirectional correspondence maps, we introduce a symmetry term similar to the one proposed by Alvarez et al. (2007).

To our energy functional we add a symmetry term:

$$E(\mathbf{x}) = \sum_p \|d_1(\mathbf{p}) - d_2(\mathbf{p} + w)\|_1 + \sum_{(\mathbf{p},\mathbf{q})\in\varepsilon} \min(\alpha|u(\mathbf{p}) - u(\mathbf{q})|, d)$$
$$+ \sum_{(\mathbf{p},\mathbf{q})\in\varepsilon} \min(\alpha|v(\mathbf{p}) - v(\mathbf{q})|, d) + \sum_{\mathbf{p}} \min(\alpha\|w_{12}(\mathbf{p}) + w_{21}(\mathbf{p} + w_{12}(\mathbf{p}))\|_2, d) \qquad (13.5)$$

Please note that now two correspondence maps coexist: w_{12} and w_{21}. They are jointly estimated and evaluated after each belief propagation iteration. It proved to be sensible to assign the same weighting and truncation values α and d to the symmetry term that are also used for message propagation.

Our correspondence estimation consists of six consecutive steps. The first three steps (Sections 13.2.9, 13.2.10 and 13.2.5) are preprocessing steps for the actual belief propagation optimization. After an initial low-resolution solution has been computed, possibly occluded parts are inpainted, as described in Section 13.2.6. As a last step, the correspondence map is upsampled and locally refined as described in Section 13.2.7.

13.2.3 SIFT Descriptor Downsampling

Liu *et al.* (2008) designed their SIFT flow with the goal in mind to match images that may only be remotely similar, which comes close to the original intention to find only a few dominant features (Lowe, 2004). Our goal, on the other side, is to match very similar images. We are faced with the challenge to discard possible matching ambiguities that occur when only color information is used as a dissimilarity measure between pixels. We use SIFT features to capture detail information about the scene; hence, we generate one feature for every pixel of the full-resolution images and search at the bottom layer of the SIFT scale-space pyramid. In order to capture only the most prominent details, a single representative feature $s_i(\mathbf{p}_g)$ is kept for every $n \times n$ grid g of pixel locations.

The search for a representative descriptor is inspired by the work of Frey and Dueck (2007), who used their affinity propagation technique to search for clusters in data and simultaneously identify representatives for these clusters. Since we have a predefined arrangement of clusters – that is, we want to roughly preserve the $n \times n$ pixel block structure – we used our fixed clusters and only adopt their suggestion that a cluster's representative should be the one most similar to all other members of the cluster.

Hence, the representative descriptor for each pixel block is the one in the $n \times n$ pixel cell that has the lowest cumulative L_1 distance to all other descriptors.

A downsampled representation of the image is computed, where every grid cell is represented by a single descriptor. This descriptor consists of the mean color values of the cell and the representative SIFT descriptor.

13.2.4 Construction of Message-Passing Graph

The fact that image regions of similar color often share common properties (e.g., similar motion) is often exploited in regularization techniques. Typically, this is achieved by applying an anisotropic regularization; that is, neighboring pixels with different colors exert less influence on each other than pixels with a similar color. This technique has two drawbacks. First, regularization is decelerated. Second, the grid-aligned regularization still manifests in jaggy borders around correspondence discontinuity edges. Recently, Smith *et al.* (2009) proposed the construction of a non-grid-like regularization scheme. While they applied this technique to stereo computation with a variational approach, we adapt their idea to our belief propagation approach; see Figure 13.8. We build an initial graph where each vertex represents a pixel location of the image. Edges connect pixels that have a certain maximal distance. Typically, we set this maximal distance to 20 pixels, since pixels further apart are rarely selected as neighbors.

Each edge is assigned a weight that corresponds to the L_1 norm of the color and position of the connected pixels. As in Smith *et al.* (2009), a minimum spanning tree is calculated using Kruskal's algorithm (Kruskal, 1956). The edges of the spanning tree are stored and removed from the overall graph; the procedure is repeated once. Afterwards, the mean valence of a vertex is 4. We further add the four original neighbors from the image grid to the neighborhood ε of a pixel location.

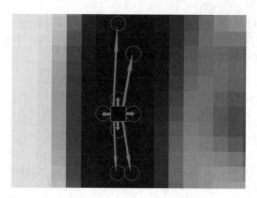

Figure 13.8 In our belief propagation scheme a single pixel (square) exchanges messages with its spatial neighbors as well as pixels of similar color (circles). The underlying graph structure is obtained by computing minimal spanning trees

13.2.5 Data Term Compression

One bottleneck in belief propagation with SIFT features is the computation of matching costs $\|s_1(\mathbf{p}) - s_2(\mathbf{p} + \mathbf{w})\|_1$ between pixel locations $s_1(\mathbf{p})$ and $s_2(\mathbf{p} + \mathbf{w})$. Liu *et al.* (2008) precomputed the matching costs before message passing. The alternative is to reevaluate matching costs on demand, which happens at least once during each iteration. This results in $262(= 131 \times 2)$ memory lookups per pixel comparison. Since storing data terms is not an option with our high-resolution data and the on-the-fly evaluation leads to run times of several days, we design a simple data term compression technique. We precompute all possible matching costs for a single pixel $s_1(\mathbf{p})$ and its potential matching candidates; see Figure 13.9. Since it is quite likely that a pixel will be finally matched to a candidate with a low dissimilarity (i.e., low matching cost), we employ a minima-preserving compression technique that loses detail in areas where high matching scores prevail. For each $m \times m$ grid cell of the

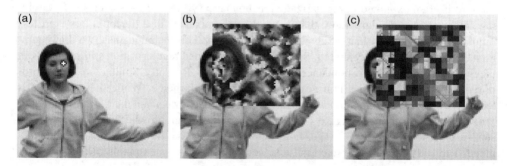

Figure 13.9 Data term compression. For each pixel in a source image (a), matching costs have to be evaluated for belief propagation. One common approach is to precompute matching costs in a predefined window (b). However, this leads to very high memory load. Our approach uses a simple minima-preserving compression of these matching cost windows (c). The minima of each pixel block, each column, row, and the diagonal lines of the search window are stored. During decompression, the maximum of these values determines the matching cost for a given location. While regions with high matching costs are not recovered in detail, local minima are preserved with high accuracy

original data term, the minimum is stored. In addition, for each row and column of the matching window, the respective minimum is stored. The same applies to the minima along the two diagonal directions.

During decompression, the maximum of these minima is evaluated, resulting in five memory lookups (the minimal grid cell value, the minimal row and column values, and the minimal values of the two diagonals). When setting $m = 4$, at a data term window size of typically 160×160 pixels, the memory usage per term is reduced from $160 \times 160 = 25\,600$ float values to $40 \times 40 + 4 \times 160 = 2240$ float values.

13.2.6   Occlusion Removal

It can be observed that the introduction of a symmetry term leads to quasi-symmetric warps in nonoccluded areas. Hence, we use the symmetry $\|w_{12}(\mathbf{p}) + w_{21}(\mathbf{p} + w_{12}(\mathbf{p}))\|_2$ of two opposing correspondence maps w_{12} and w_{21} as a measure of occlusion.

For each of these two simultaneously estimated maps, asymmetric correspondence regions are identified and treated independently. First, all occluded regions are filled with correspondence information values using diffusion. All pixels which would lie outside the actual image boundaries according to their correspondence vector are discarded as boundary occlusions and their diffused values are kept.

Assuming that each of the remaining occlusion regions is confined by a foreground and a background region that move incoherently, we perform a k-means ($k = 2$) clustering of the border region outside each occluded area; all pixels in these border regions are clustered according to their two correspondence vector components $u(\mathbf{p})$ and $v(\mathbf{p})$. The resulting pixel sets serve as input data for a binary geodesic matting (Bai and Sapiro, 2009) that assigns each pixel in the occluded area a foreground or background label. After the labeling is computed, the median value of the n nearest neighbors of the foreground or backround region is assigned. Typically, we set $n = 20$ to get smooth results.

13.2.7   Upsampling and Refinement

The low-resolution correspondence is upsampled as follows. On the high-resolution map, each pixel that was chosen as the SIFT descriptor representative is assigned the values that result from the low-resolution belief propagation (scaled by factor n). For all remaining pixels, the value of the nearest representative pixel in gradient space is assigned. These assigned correspondence values serve as a prior for a local refinement. Like on the low-resolution level, symmetric belief propagation is used to obtain the final per-pixel correspondence. The crucial difference is that the search window is set to a very small size (typically $(n \times 2 + 1) \times (n \times 2 + 1)$ pixel) and that it is located around a correspondence prior $\mathbf{p} + w_{12}$ and not around the pixel location $\mathbf{p}$ itself.

13.2.8   Limitations

The most severe limitation is the long run time of our algorithm. Until now, the implementation has been completely CPU based. This is also due to the fact of high memory consumption. Since we compute two symmetric correspondence maps in parallel, memory consumption is as high as 4 GB for large displacements (e.g., 400 pixels displacement is not uncommon at full HD resolution). In order to move the computation to the GPU, an even more compact data representation than our currently employed compression scheme has to be developed. Obtaining a suitable search window size is another problem, since the maximum displacement has to be known prior to the correspondence estimation. If the window size is set too small, then some correspondences will not be established correctly. If it is set too high, then the overall run-time will be excessively high.

13.3 High-Quality Correspondence Edit

As discussed in the previous sections, the rendering quality of the virtual video camera free-viewpoint rendering system depends on the image correspondences.

For all image correspondence algorithms, pathological cases exist where the underlying model assumptions are violated and the results degrade dramatically. The most prominent examples include glares on nondiffuse surfaces and poorly textured regions. Other scenarios include long-range correspondences, illumination changes, complex scenes, and occlusions, where any contemporary algorithm has failure cases. Purely automatic solutions will, therefore, not be able to produce results that deliver convincing results when used for image interpolation. On the other hand, user-driven editing tools often require a lot of manual interaction. With our interactive tool called Flowlab, user corrections can be applied to pre-computed correspondences, aided by automatic methods which refine the user input. Provided with an initial solution, the user selects and corrects erroneous regions of the correspondence maps. In order to keep user interaction at a minimum level, we employ state-of-the-art techniques that assist in segmenting distinct regions and locally refining their correspondence values.

Up to the point of writing this chapter, precious few editing tools exist for correspondences, because correction is mostly performed in the image domain after image interpolation has been applied. A pioneering approach by Beier and Nelly (1992) employs sparse correspondences in the form of lines at salient edges in human faces. The commercial movie production tool Ocula (Wilkes, 2009) provides an editing function for stereo disparity mapping, which considers correspondences only for the stereo case.

Flowlab is novel in that it is the first general tool to edit the relationship between images, rather than the images themselves. It is general in that it is not directly linked to a specific algorithm. This contrasts with previous ad-hoc approaches to manual correspondence correction which have been tied to the correction of one specific approach (Baker *et al.*, 2007).

13.3.1 Editing Operations

For the general Flowlab workflow the user passes two images I_1 and I_2 as input to the application. The user can assess the quality of initial correspondences by rendering in-between images. The user may apply editing operations on mismatched areas and may switch between rendering and editing until all visible errors are corrected. The overall interface is designed to provide a very efficient and fast workflow. Flowlab supports three basic editing operations: removal of small outliers, direct region mapping, and optical flow-assisted region mapping.

Small outliers can be eliminated with median filtering. The user specifies a polygonal region, and a median filter is applied within the selected region. The correspondences are filtered with a fixed window size. This removes very high frequency noise, which can appear around object boundaries.

Region mapping is performed by first selecting a polygon in the source image, defined by an arbitrary number of control points. These can be moved to better approximate, for example, a distinct object. A zoom function can be used to improve the matching accuracy in complex regions.

The selected region is then mapped onto the destination image. A bounding box with four control points allows the projective transformation of the selection. The user-specified transformation can either be applied directly or used as a prior for a state-of-the-art optical flow alignment step.

13.3.2 Applications

Considering image interpolation or image-based rendering techniques, the quality of correspondences can be rated by viewing the interpolation results. If the results are not visually convincing, Flowlab can then be used to correct the erroneous regions within the underlying correspondences. We found that this makes any later correction pass on the output images a lot faster or even unnecessary. Furthermore, if the correspondences are used for multiple output frames (e.g., in a slow-motion scenario), the effort for correction is reduced even more. Only one correspondence pair instead of each frame has to be modified.

Ultimately, it is up to the user to decide whether an artifact is best corrected in the correspondence or a later stage. While Flowlab was primarily designed with movie production in mind, there are other possible applications. For example, ground-truth generation of correspondence maps is not a straightforward task owing to the lack in tools.

13.4 Extending to the Third Dimension

After discussing the rendering process for monoscopic video content and the estimation of the required image correspondence scene model, we now investigate the possibility to automatically generate stereoscopic 3D content using the virtual video camera system.

The popularity of stereoscopic 3D content in movies and television creates a huge demand for new content. Three major methodologies for creating stereoscopic material are currently employed: purely digital content creation for animation films and games, content filmed with specialized stereoscopic cameras, and the artificial enhancement of monocular content.

For computer animated films, the creation of a stereoscopic second view is straightforward. Since the first camera is purely virtual, creating a second virtual camera poses no problem. The major drawback is that the creation of naturalistic real-world images is extremely complex and time consuming.

The enhancement of monocular recordings suffers from a similar problem. Although the recorded footage in this case has the desired natural look, the creation of a proxy geometry or a scene model can be tedious. The depth map or proxy geometry used to synthesize a second viewpoint has to be created by skilled artists (for example View-D from Prime Focus). The complexity of the model creation directly depends on the complexity of the recorded scene.

While directly recording with a stereoscopic camera rig eliminates the need to create an additional scene model, it requires highly specialized and, therefore, expensive stereo camera hardware. Leaving aside monetary constraints, the on-set handling of the stereoscopic cameras poses a challenge. The view and baseline selection, for example, requires careful planning to give the viewer a pleasing stereoscopic experience. Changing the parameters in post-production is difficult or even impossible.

We examine the stereoscopic content creation from a purely image-based-rendering per-spective. Using only recorded images, a naturalistic image impression can be achieved with-out the need for manual modeling. Using multiview datasets of only a few cameras, it becomes possible to interpolate arbitrary views on the input camera manifold. Thereby, it is not only possible to create stereoscopic 3D views of an arbitrary scene, but also the control over stereo parameters such as baseline and convergence plane is kept during post-production. Since the virtual video camera image-based stereoscopic free-viewpoint video framework is capable of time and space interpolation, it combines the natural image impression from direct stereoscopic recording with the full viewpoint and time control of digitally created scenes.

In order to demonstrate the level of maturity of our approach, we incorporated our free-viewpoint framework into an actual post-production pipeline of a full-length music video; see Figure 13.10. The goal of this project is to test how well our approach integrates into existing post-production pipelines and to explore how movie makers can benefit from image-based free-viewpoint technology.

13.4.1 Direct Stereoscopic Virtual View Synthesis

The virtual video camera is capable of rendering any viewpoint within the camera manifold. Therefore, the easiest way of creating S3D content is using the camera position as the left

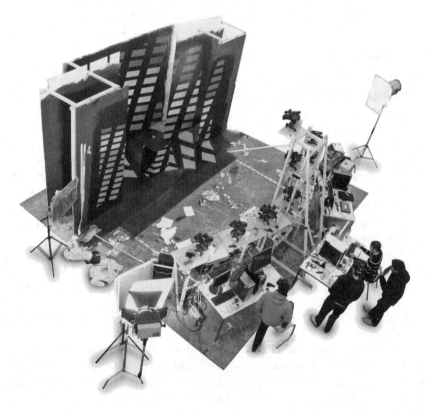

Figure 13.10 "Who Cares?" set. Eleven HD camcorders captured various graffiti motifs. With our approach we are able to synthesize novel spatial and temporal in-between viewpoints

eye view $I_v^L(\varphi, \theta, t)$ and render the right eye view by horizontally offsetting the camera position.

The view for the right eye $I_v^R(\varphi + \Delta, \theta, t)$ is synthesized, similar to Equations 13.1 and 13.2 by offsetting the camera position along the φ-direction. A common rule for stereoscopic capture is that the maximal angle of divergence between the stereo camera axes should not exceed $1.5°$. Otherwise, the eyes are forced to diverge to bring distant objects in alignment, which usually causes discomfort. By the nature of construction, our approach renders converging stereo pairs and angles of divergence between 0.5 and $1.5°$ to give the most pleasing stereoscopic results.

13.4.2 Depth-Image-Based Rendering

An alternative approach is to apply depth-image-based rendering (DIBR) as an intermediate step. Only a center camera view is rendered and in addition a per-image depth map is obtained. We render the depth maps for novel views, by interpolating Euclidean 3D position of pixels instead of RGB information. The location of each pixel is computed on the fly in the vertex shader using the depth map and the camera matrix. The final depth value is obtained by reprojecting the 3D position into the coordinate space of the synthetic virtual camera. The color images and depth maps are used to render both a left- and a right-eye view. This is done by using the naive background extrapolation DIBR techniques (Schmeing and Jiang, 2011).

13.4.3 Comparison

Since both approaches seem appealing at first sight, we would like to compare the individual benefits. The most striking advantage of the first approach is that it does not require any additional building blocks in the production pipeline. Basically, the scene is rendered twice with slightly modified view parameters. Since there is no second rendering pass (as there is in the depth-image-based method), no additional resampling or occlusion handling issues arise.

The most valuable advantage of the second approach is that no restriction is made on the camera placement. When using the direct approach, the two virtual views must fit into the navigation space boundaries, prohibiting a vertical placement of camera. In contrast, the DIBR method allows stereoscopic rendering even if the cameras are aligned in a vertical arc; see Figure 13.11. In addition, the whole navigation space can be used: while the DIBR method allows a placement of the center camera to the far left and far right of the navigation space, the direct method is restricted by the placement of the left- or right-eye view; see Figure 13.11. Another benefit becomes obvious when the linear approximation of object movement is considered. Nonlinear object motion will lead to unwanted vertical disparity when the direct method is applied. Although nonlinear motion might lead to errors in the depth estimation, the DIBR scheme prevents vertical disparity at the rendering stage. The additional re-rendering of the material (color images and depth maps are turned into a left-eye and a right-eye view) might seem as a disadvantage at first. This notion changes if the material is in some way edited between these two rendering passes. When any kind of image/video editing operation is applied to the center image (e.g., correction of rendering artifacts), it is automatically propagated to the left/right-eye view. In the case of the direct method, any post-production operations would have to be applied individually (and possibly incoherently) to both views.

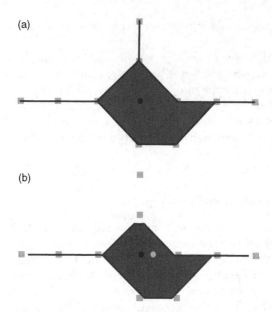

Figure 13.11 Navigation space boundaries, original camera positions are shown in gray: (a) when a single image (circle) is rendered, the full volume (dark gray) can be accessed. Please note that for the horizontal and vertical arcs (left, top, right) only purely horizontal or vertical camera movement is feasible. Otherwise, error-prone long-range image correspondences would have to be used. (b) When an actual camera pair (two circles) is rendered, the navigation space volume is effectively eroded, since there has to be a minimum horizontal distance between the both views. This effect prohibits stereoscopic view interpolation if cameras are placed along a vertical arc (b, top)

13.4.4 Concluding with the "Who Cares?" Post-Production Pipeline

Finally, we show the actual production pipeline; see Figure 13.12 of the music video project, which featured the virtual video camera.

In the "Who Cares?" project, we process HD input material featuring two timelines (live foreground action and background time-lapse) captured with nonsynchronized camcorders. To our knowledge it is the first project that makes use of this input material and allows horizontal, vertical, and temporal image interpolation. The basic idea of the video is that two DJs appear on a stage and perform a live act with their audio controller. During the course of the music video, the background is painted over with various graffiti motifs (e.g., giant stereo speakers, equalizer bars, or a night skyline). Although it would be possible to create and animate these graffiti motifs with traditional CGI tools, the artists decided to use an actual graffiti time-lapse to maintain a credible "low-fi" look. We embedded our stereoscopic free-viewpoint video into a traditional 2D post-production timeline. By collaborating with independent film makers that were familiar with traditional 2D content post-production, we successfully integrated our system into a tradition compositing/post-production pipeline.

The production project shows that the Virtual Video Camera described in this chapter is able to produce high-quality results, suitable for use in video production scenarios.

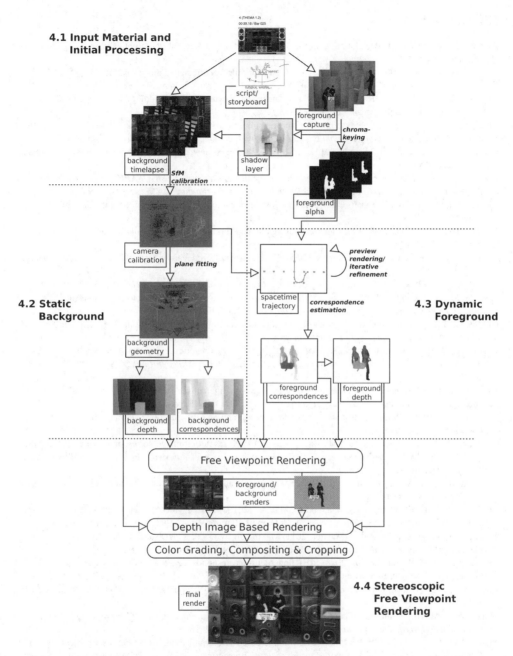

Figure 13.12 An overview of the production pipeline. After an initial processing stage (top), foreground and background footage are processed independently. While background depth and correspondence maps are derived from a static geometric model (left), foreground data are computed using state-of-the-art correspondence estimation algorithms (right). The processed footage is passed to a free-viewpoint renderer before the actual left and right eye views are rendered using DIBR (bottom)

References

Alvarez, L., R. Deriche, T. Papadopoulo, and J. Sanchez (2007) Symmetrical dense optical flow estimation with occlusion detection. *Int. J. Comput. Vision*, **75** (3), 371–385.

Bai, X., and G. Sapiro (2009) Geodesic matting: a framework for fast interactive image and video segmentation and matting. *Int. J. Comput. Vision*, **82** (2), 113–132.

Baker, S., D. Scharstein, J.P. Roth *et al.* (2007) A database and evaluation methodology for optical flow, in *IEEE 11th International Conference on Computer Vision, 2007. ICCV 2007*, IEEE, pp. 1–8.

Beier, T. and S. Nelly (1992) Feature-based image metamorphosis, in *SIGGRAPH '92 Proceedings of the 19th Annual Conference on Computer Graphics and Interactive Techniques*, ACM, New York, NY, pp. 35–42.

Carranza, J., C. Theobalt, M. Magnor, and H. P. Seidel (2003) Free-viewpoint video of human actors. *ACM Trans. Graph.*, **22** (3), 569–577.

Chen, S. E. and L. Williams (1993) View interpolation for image synthesis, in *SIGGRAPH '93 Proceedings of the 20th Annual Conference on Computer Graphics and Interactive Techniques*, ACM, New York, NY, pp. 279–288.

De Auiar, E., C. Stoll, C. Theobalt *et al.* (2008) Performance capture from sparse multi-view video. *ACM Trans. Graph.*, **27** (3), 1–10.

Edelsbrunner, H., and N. Shah (1995) Incremental topological flipping works for regular triangulations. *Algorithmica*, **15** (3) 223–241.

Felzenszwalb, P. F. and D. P. Huttenlocher (2006) Efficient belief propagation for early vision. *Int. J. Comput. Vision*, **70**, 41–54.

Frey, B. J. and D. Dueck (2007) Clustering by passing messages between data points. *Science*, **315**, 972–976.

Goesele, M., N. Snavely, B. Curless *et al.* (2007) Multiview stereo for community photo collections, in *IEEE 11th International Conference on Computer Vision, 2007. ICCV 2007*, IEEE, pp. 1–8.

Goldlücke, B., M. Magnor, and B. Wilburn (2002) Hardware-accelerated dynamic light field rendering, in *Proceedings of the Vision, Modeling, and Visualization 2002 (VMV 2002)*, Aka GmbH, pp. 455–462.

Hasler, N., B. Rosenhahn, T. Thormählen *et al.* (2009) Markerless motion capture with unsynchronized moving cameras, in *IEEE Conference on Computer Vision and Pattern Recognition, 2009. CVPR 2009*, IEEE, pp. 224–231.

Klose, F., K. Ruhl, C. Lipski, and M. Magnor (2011) Flowlab – an interactive tool for editing dense image correspondences, in *2011 Conference for Visual Media Production (CVMP)*, IEEE, pp. 1–8.

Kruskal, J. B. (1956) On the shortest spanning subtree of a graph and the traveling salesman problem. *Proc. Am. Math. Soc.*, **7** (1), 48–50.

Lipski, C., C. Linz, T. Neumann, and M. Magnor (2010a) High resolution image correspondences for video postproduction, in *CVMP '10 Proceedings of the 2010 Conference on Visual Media Production*, IEEE Computer Society, Washington, DC, pp. 33–39.

Lipski, C., D. Bose, M. Eisemann *et al.* (2010b) Sparse bundle adjustment speedup strategies. WSCG Short Papers Post-Conference Proceedings.

Lipski, C., C. Linz, K. Berger, A. Sellent, and M. Marcus, (2010c) Virtual Video Camera: Image-Based Viewpoint Navigation Through Space and Time, in Computer Graphics Forum, Volume 29 (2010), number 8, pp. 2555–2568.

Lispki, C., F. Klose, K. Ruhl, and M. Magnor (2011) Making of "Who Cares" HD stereoscopic free viewpoint music video, in *2011 Conference on Visual Media Production (CVMP)*, IEEE, pp. 1–10.

Liu, C., J. Yuen, A. Torralba *et al.* (2008) SIFT flow: dense correspondence across different scenes, in *ECCV '08 Proceedings of the 10th European Conference on Computer Vision: Part III*, Springer Verlag, Berlin, pp. 28–42.

Lowe, D. G. (2004) Distinctive image features from scale invariant keypoints. *Int. J. Comput. Vision*, **60** (2), 91–110.

Mark, W., L. McMillan, and G. Bishop (1997) Post-rendering 3D warping, in *I3D '97 Proceedings of the 1997 Symposium on Interactive 3D Graphics*, ACM, New York, NY, pp. 7–16.

Matusik, W., C. Buehler, R. Raskar *et al.* (2000) Image-based visual hulls, in *SIGGRAPH '00 Proceedings of the 27th annual conference on Computer graphics and interactive techniques*, ACM Press/Addison-Wesley, New York, NY, pp. 369–374.

Meyer, B., T. Stich, M. Magnor, and M. Pollefeys (2008) Subframe temporal alignment of non-stationary cameras. Proceedings of British Machine Vision Conference BMVC '08.

Naemura, T., J. Tago, and H. Harashima (2002) Real-time video-based modeling and rendering of 3D scenes. *IEEE Comput. Graph. Appl.*, **22** (2), 66–73.

Schmeing, M. and X. Jiang (2011) Time-consistency of disocclusion filling algorithms in depth image based rendering, in *3DTV Conference: The True Vision – Capture, Transmition and Display of 3D Video (3DTV-CON), 2011*, IEEE, pp. 1–4.

Schwartz, C. and R. Klein (2009) Improving initial estimations for structure from motion methods. The 13th Central European Seminar on Computer Graphics (CESCG 2009).

Seitz, S. M. and C. R. Dyer (1996) View morphing, in *SIGGRAPH '96 Proceedings of the 23rd Annual Conference on Computer graphics and Interactive Techniques*, ACM, New York, NY, pp. 21–30.

Si, H. and K. Gaertner (2005) Meshing piecewise linear complexes by constrained delaunay tetrahedralizations, in *Proceedings of the 14th International Meshing Roundtable*, Springer, Berlin, pp. 147–163.

Smith, B. M., L. Zhang, and J. Hailin (2009) Stereo matching with nonparametric smothness priors in feature space, in *IEEE Conference on Computer Vision and Pattern Recognition, 2009. CVPR 2009*, IEEE, pp. 485–492.

Snavely, N., S. M. Seitz, and R. Szeliski (2006) Photo tourism: exploring photo collections in 3D. *ACM Trans. Graph.*, **25** (3), 835–846.

Starck, J. and A. Hilon (2007) Surface capture for performance based animation. *IEEE Comput. Graph.*, **27** (3), 21–31.

Stich, T., C. Linz, C. Wallraven *et al.* (2011) Perception-motivated interpolation of image sequences. *ACM Trans. Appl. Percept. (TAP)*, **8** (2), article no. 11.

Vedula, S., S. Baker, and T. Kanade (2005) Image based spatio-temporal modeling and view interpolation of dynamic events. *ACM Trans. Graph.*, **24** (2), 240–261.

Wilkes, L. (2009) The role of Ocula in stereo post production. Whitepaper, The Foundry.

Zitnick, C, S. B. Kang, M. Uyttendaele *et al.* (2004) High-quality video view interpolation using a layered representation. *ACM Trans. Graph.*, **23** (3), 600–608.

Part Four

Display Technologies

14

Signal Processing for 3D Displays

Janusz Konrad

Department of Electrical and Computer Engineering, Boston University, USA

14.1 Introduction

A virtual reproduction of the three-dimensional (3D) world has been attempted in various forms for over a century. Today's resurgence of 3D entertainment, both in the cinema and at home, is largely due to technological advances in 3D displays, standardization of 3D data storage and transmission, and improved 3D content availability. However, the use of 3D displays is not limited to entertainment; they are also found today in scientific visualization, image-guided surgery, remote robot guidance, battlefield reconnaissance, and so on.

Despite these advances, challenges remain in the quest for an effortless "being there" experience. The main challenge is perhaps in the development of next-generation display technologies aiming at a high-quality, high-resolution 3D experience without glasses. Another challenge is in the area of 3D content compression, but significant progress has been made on this front by the adoption of the multiview coding extension of the Advanced Video Coding standard. Here, we focus on a different challenge, namely the development of effective and efficient signal processing methods in order to overcome 3D displays' deficiencies and enhance the viewers' 3D experience. In this context, signal processing is the glue that bonds together the content, transmission, and displays while accounting for viewer experience.

In this chapter, we discuss the role of signal processing in content generation for both stereoscopic and multiview 3D displays, and also in the mitigation of various 3D display deficiencies (e.g., crosstalk, aliasing, color mismatch). We describe examples of signal processing techniques that help improve the quality of 3D images and assure the ultimate 3D experience for the user. Owing to space constraints, only a high-level description of various algorithms is provided while leaving out substantial mathematical details. An extensive list of references is provided for readers seeking such details.

Emerging Technologies for 3D Video: Creation, Coding, Transmission and Rendering, First Edition.
Frédéric Dufaux, Béatrice Pesquet-Popescu, and Marco Cagnazzo.
© 2013 John Wiley & Sons, Ltd. Published 2013 by John Wiley & Sons, Ltd.

14.2   3D Content Generation

The ongoing 3D entertainment revolution will become unsustainable unless rich 3D content is available. While 3D content generation at home is a reality today (3D still cameras and camcorders), this is insufficient for ubiquitous 3D entertainment. Although, today, 3D TV programming is on the rise, the overall commercial 3D content production lags behind the 3D hardware availability. This content gap can be filled by converting two-dimensional (2D) video to 3D stereoscopic video, and many methods have been proposed to date. The most successful approaches are interactive; that is, they involve human operators (Guttmann *et al.*, 2009; Angot *et al.*, 2010; Phan *et al.*, 2011; Liao *et al.*, 2012). Such methods are a good choice for editing a few shots, but are time consuming and costly when conversion of long video sequences is needed.

An alternative to operator-assisted methods are fully automatic 2D-to-3D conversion methods. The main issue in such methods is the construction of a viable depth or disparity field for a given 2D input image. To date, several electronics manufacturers have developed real-time 2D-to-3D converters that rely on certain heuristics about typical scene scenarios, for example: faster or larger objects are closer to the viewer, higher frequency of texture belongs to objects located further away, and so on. Although impressive at first, upon more careful inspection most such methods are inferior to the operator-supervised methods. In general, it is very difficult, if not impossible, to construct a deterministic scene model that covers all possible 3D scene scenarios.

Many attempts have been made in computer vision to automatically compute depth from a single image – for example, depth-from-defocus (Subbarao and Surya, 1994), structure-from-motion (Szeliski and Torr, 1998), shape-from-shading (Zhang *et al.*, 1999) – but no reliable solutions have been found to date except for some highly constrained cases. Recently, a radically different path has been taken to attack the problem: application of machine-learning techniques based on image parsing (Saxena *et al.*, 2009; Liu *et al.*, 2010). However, they currently work only on specific types of images using carefully selected training data. In the final step of 2D-to-3D conversion, the depth obtained is used jointly with the monocular image to produce a stereopair by means of so-called depth-based rendering.

Stereoscopic content from a 3D camera or from a 2D-to-3D conversion process can be displayed directly on a stereoscopic 3D display. However, such content is insufficient for a multiview 3D display and an additional step, often called intermediate view interpolation, is needed to generate additional views.

Below, we describe a recently developed automatic 2D-to-3D image conversion method that "learns" depth from on-line 3D examples and also a simple real-time method for view interpolation that computes a multiview image from one stereopair. The reader is referred to Chapter 3 in this book for an extensive discussion of 2D-to-3D image conversion methods.

14.2.1   Automatic 2D-to-3D Image Conversion

The 2D-to-3D conversion method described below was inspired by our work on saliency detection in images (Wang *et al.*, 2011). Rather than deterministically specifying a 3D scene model for the input 2D image, we "learn" the model, and more specifically we "learn" either the disparity or depth, from a large collection of 3D images, such as YouTube 3D or the NYU Kinect database (Silberman and Fergus, 2011). While 3D content on YouTube is encoded into stereopairs (pairs of corresponding frames in left and right videos), from which disparity can be computed, the NYU Kinect database is composed of still images and corresponding depth fields. Therefore, the learning process is applied to disparities, if stereoscopic 3D images are available, or depth fields, when image + depth 3D pairs are provided.

Our method is based on the observation that among millions of 3D images on-line there likely exist many with similar content to that of a 2D query that we wish to convert to 3D. Furthermore, our method is based on the assumption that two images that are photometrically similar are likely to have similar 3D structure (depth). This is not unreasonable since photometric properties are usually correlated with 3D content (depth, disparity). For example, edges in a depth map almost always coincide with photometric edges. We rely on the above observation and assumption to "learn" the depth of the 2D query from a dictionary of 3D images and render a stereopair using the query and computed depth in the steps outlined below.

Let Q be the 2D query image, for example the left image of a stereopair whose right image Q_R is being sought. We assume that a database $\mathcal{I} = \{(I^1, z^1), (I^2, z^2), \ldots\}$ of image+depth pairs (I^k, z^k) is available. In this chapter, we will not consider a stereoscopic 3D database, such as YouTube 3D, since the image+depth and stereoscopic pair representations are largely equivalent and can be computed from each other with high accuracy. For example, disparity d can be computed from a stereoscopic image pair and then, under the assumption that this pair was captured by parallel cameras, the depth can be computed as $d = Bf/z$, where B is the baseline and f is the focal length of the cameras. Therefore, we consider here only the case of image+depth database $\mathcal{I}$ to find a depth estimate $\hat{z}$ and a right-image estimate $\hat{Q}_R$ for a query image Q.

14.2.1.1 *k*NN Search

There exist two types of images in a large 3D image repository: those that are relevant for determining the depth of a 2D query image and those that are not. Images that are photometrically dissimilar from the 2D query need to be rejected because they are not useful for estimating depth (as per our earlier assumption). Even if we miss some depth-relevant images, we are effectively limiting the number of irrelevant images that could potentially be harmful to the 2D-to-3D conversion process. The selection of a smaller subset of images provides the additional benefit of reduced computational complexity for very large 3D repositories.

One possible method for selecting a useful subset of depth-relevant images from a large database is to select only the k images that are closest to the input, where closeness is measured by some distance function capturing global image properties such as color, texture, edges, and so on. We originally used the weighted Hamming distance between binary hashes of features (Konrad *et al.*, 2012a), but owing to its high computational complexity we recently switched to the Euclidean norm of the difference between histograms of oriented gradients (HOGs) (Konrad *et al.*, 2012b). Each HOG consists of 144 real values (4×4 grid of blocks with nine gradient direction bins) that can be efficiently computed.

We search for matches to our 2D query among all 3D images in the database $\mathcal{I}$. The search returns an ordered list of image + depth pairs, from the most to the least photometrically similar to the 2D query. We discard all but the top k matches (k nearest neighbors (NNs) or kNNs) from this list and store indices of these top matches in a set $\mathcal{K}$. Figure 14.1 shows a 2D query and its top two NNs along with the corresponding depth fields. Note that the NNs have similarly oriented walls and furniture located in similar locations to those in the query.

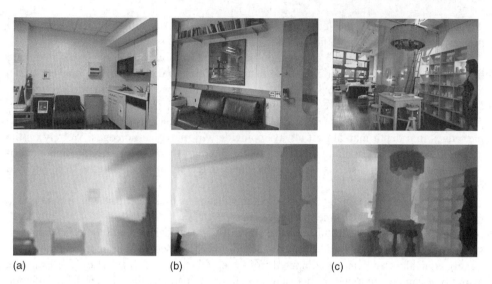

(a)   (b)   (c)

Figure 14.1   (a) Sample 2D query image Q (kitchen); (b) NN #1 (c) NN #2. The corresponding depth fields are shown below the images. All image + depth pairs are from NYU Kinect dataset (Silberman and Fergus, 2011). From Konrad and Halle (2012b), reproduced with permission of IEEE © 2012 (see Plate 15 for the colored figure)

14.2.1.2 Depth Fusion

In general, none of the NN image + depth pairs (I^i, z^i), $i \in \mathcal{K}$, will perfectly match the query Q (Figure 14.1). However, the location of the main objects (e.g., furniture) and the appearance of the background are usually fairly consistent with those in the query. If a similar object (e.g., table) appears at a similar location in several kNN images, it is likely that a similar object also appears in the query and the depth field being sought should reflect this. We compute this depth field by fusing all kNN depths at each spatial location x using the median operator:

$$z[x] = \text{median}\{z^i[x], \quad \forall i \in \mathcal{K}\} \tag{14.1}$$

The median depth field z obtained from $k = 45$ NNs of the query Q from Figure 14.1 is shown in Figure 14.2a. Although coarse and overly smooth, it provides a globally

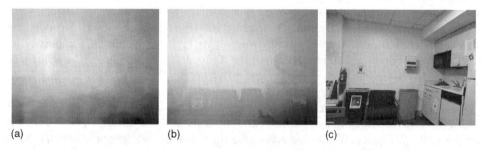

(a)   (b)   (c)

Figure 14.2   Depth field z fused from $k = 45$ NNs. (a) Median fused depth z before smoothing; (b) fused depth after smoothing via cross-bilateral filtering $\hat{z}$; (c) $\hat{Q}_R$ right image rendered using the 2D query Q (Figure 14.1) and the smoothed depth field $\hat{z}$. From Konrad *et al.* (2012b), reproduced with permission of IEEE © 2012 (see Plate 16 for the colored figure)

correct assignment of distances to various areas of the scene (bottom closer, middle further away).

14.2.1.3 Cross-Bilateral Depth Filtering

Although the median-based fusion helps make the depth more consistent globally, the resulting depth is overly smooth and locally inconsistent with the query image due to:

1. misalignment of edges between the fused depth field and query image;
2. lack of fused depth edges where sharp object boundaries occur;
3. lack of fused depth smoothness where smooth depth changes are expected.

In order to correct these inconsistencies, we use a variant of bilateral filtering, an edge-preserving image smoothing method based on anisotropic diffusion controlled by local image content (Durand and Dorsey, 2002). However, we apply the diffusion to the fused depth field z and use the 2D query image Q to control the degree of depth smoothing. This is often referred to as cross-bilateral filtering and has two goals: alignment of depth edges with those of the query image Q and local noise/granularity suppression in the fused depth z. It is applied to the whole depth field z as follows:

$$\hat{z}[x] = \frac{1}{\gamma[x]} \sum_y z[y] h_{\sigma_s}(x - y) h_{\sigma_e}(Q[x] - Q[y])$$

$$\gamma[x] = \sum_y h_{\sigma_s}(x - y) h_{\sigma_e}(Q[x] - Q[y]) \tag{14.2}$$

where $\hat{z}$ is the filtered depth field and $h_\sigma(x) = \exp(-\|x\|^2/2\sigma^2)/2\pi\sigma^2$ is a Gaussian weighting function. Note that the directional smoothing of z is controlled by the query image via the weight $h_{\sigma_e}(Q[x] - Q[y])$. For large discontinuities in Q, the weight $h_{\sigma_e}(Q[x] - Q[y])$ is small and thus the contribution of $z[y]$ to the output is small. However, when $Q[y]$ is similar to $Q[x]$ then $h_{\sigma_e}(Q[x] - Q[y])$ is relatively large and the contribution of $z[y]$ to the output is larger. In essence, depth smoothing is happening along, but not across, edges in the query image.

Figure 14.2 shows the fused depth before cross-bilateral filtering (z) and after ($\hat{z}$). The filtered depth preserves the global properties captured by the unfiltered depth field z and remains smooth within objects and in the background. At the same time, it keeps edges sharp and aligned with the query image structure.

14.2.1.4 Stereo Rendering

To compute a right image estimate $\hat{Q}_R$ from the 2D query Q, a disparity field is needed. Assuming that the fictitious image pair $(Q, \hat{Q}_R)$ was captured by parallel cameras with baseline B and focal length f, we compute this disparity from the estimated depth $\hat{z}$ as $d[x, y] = Bf/\hat{z}[x]$, where $x = [x, y]^T$. Then, we forward-project the 2D query Q to produce the right image:

$$\hat{Q}_R[x + d[x, y], y] = Q[x, y] \tag{14.3}$$

while rounding the location coordinates $(x + d[x, y], y)$ to the nearest sampling grid point. We handle occlusions (assignment of multiple values) by depth ordering: if

$(x_i + d[x_i, y_i], y_i) = (x_j + d[x_j, y_i], y_i)$ for some i, j, we assign to the location $(x_i + d[x_i, y_i], y_i)$ in $\hat{Q}_R$ an RGB value from that location (x_i, y_i) in Q whose disparity $d[x_i, y_i]$ is the largest. In newly exposed areas – that is, for x_j such that no x_i satisfies $(x_j, y_i) = (x_i + d[x_i, y_i], y_i)$ – we apply simple inpainting based on solving partial differential equations ("inpaint_nans" from *MatlabCentral*).

Figure 14.2 shows the final right image for the query from Figure 14.1. Overall, the image appears to be consistent with the query (left image), although small distortions are visible (e.g., fridge handle). Anaglyph 3D images, which combine the corresponding left and right images into one image to be viewed through red–cyan glasses, are shown in Figure 14.3. The right image in each anaglyph was rendered using either the ground-truth depth field (Figure 14.3a) or a depth field learned from $k = 45$ NNs using the proposed method (Figure 14.3b), or a depth field obtained using the Make3D algorithm (Saxena *et al.*, 2009), considered to be state of the art in automatic 2D-to-3D conversion methods today (Figure 14.3c). While the anaglyph image obtained using the proposed method has small distortions (e.g., fridge door), the one obtained using Make3D has severe geometric distortions on the back wall and fridge door. This visual improvement has been confirmed numerically by computing a correlation coefficient between the estimated and ground-truth depths (Konrad *et al.*, 2012b). While the Make3D depth attained an average correlation of 0.45 with respect to the ground truth, our method attained 0.71, both computed across all 1449 images of the NYU Kinect dataset. Admittedly, the Make3D algorithm that we used was optimized for outdoor scenes and we had no means of retraining the algorithm for indoor scenes present in the NYU database. However, even if Make3D can be adapted to perform better, there is a huge computational gap between the two methods. While the Make3D-based approach took 12 h to compute depth and then render all 1449 right images, the proposed method took only 5 s (including the kNN search). For more experimental results, including the use of stereoscopic rather than image + depth databases, the reader is encouraged to consult our earlier work (Konrad *et al.*, 2012a,b) and our web site (http://vip.bu.edu/projects/3-d/2d-to-3d).

14.2.2 Real-Time Intermediate View Interpolation

In order that a 3D display can deliver a realistic, natural experience to viewers, it must render the correct perspective for each viewer position or, in other words, permit motion-parallax. Clearly, many views must be displayed. While one could use many cameras, one for each view, the hardware cost and complexity plus the transmission cost would be prohibitive. An alternative is to use a small number of cameras and render the missing (virtual) views based on the available ones. One class of such methods falls into the category of *image-based rendering*; first, a 3D model is generated based on the available views and then used to render the missing views (Kang *et al.*, 2007). However, such methods do not work well for complex 3D scenes and are computationally demanding.

Although a multi-camera setup is not practical, we note that if applied to lenticular or parallax-barrier multiview screens such a setup would need to be constrained in terms of camera baseline (three to five interocular distances across all cameras) and camera convergence angle (no more than a few degrees between any two cameras). A larger baseline would increase disparity and, subsequently, perceived crosstalk in high-contrast images, while an increased convergence angle would additionally lead to keystoning, causing perceptual rivalry. The small baseline and convergence angle facilitate various simplifying assumptions that allow virtual view computation without explicit 3D modeling and lead to methods that

(a)

(b)

(c)

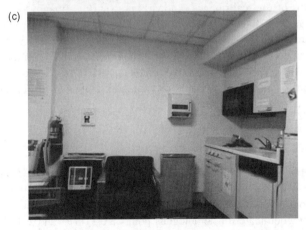

Figure 14.3 Anaglyph images generated using (a) ground-truth depth, (b) depth estimated by the proposed algorithm, and (c) depth estimated by the Make3D algorithm (Saxena *et al.*, 2009). From Konrad *et al.* (2012b), reproduced with permission of IEEE © 2012 (see Plate 17 for the colored figure)

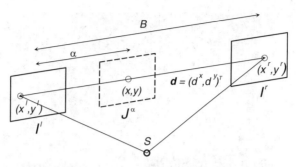

Figure 14.4 Illustration of intermediate-view interpolation for a parallel camera setup. Images I^l and I^r are assumed to have been captured by left and right cameras, respectively. (x^l, y^l) and (x^r, y^r) are homologous points (projections of 3D point S) in the corresponding images. For a virtual camera located at position α, (x, y) would be the coordinates of homologous point in virtual view J^α (dashed lines)

are commonly known as *intermediate view interpolation* or *intermediate view reconstruction* (Izquierdo, 1997; Mansouri and Konrad, 2000; Ince and Konrad, 2008b).

Thus, we assume parallel camera geometry; virtual cameras are assumed to be located on a line passing through the optical centers of real cameras and optical axes of all cameras are parallel (schematically shown for two cameras in Figure 14.4). We also assume that the stereo camera baseline B is normalized to unity and that all real and virtual images are defined on 2D lattice Λ. The goal now is as follows: given the left image I^l and right image I^r compute an intermediate image J^α that would have been captured by a virtual camera placed at the relative distance α from the left camera. This necessitates two steps: correspondence estimation between real views and interpolation of virtual-view luminance/color based on the real views. These two steps can be performed in two different ways, as explained below.

The central role in correspondence estimation is played by a transformation between the coordinate systems of the real images (known) and a virtual image (to be computed). This transformation can be defined either as a *forward projection* that maps real-image coordinates to virtual-image coordinates or *backward projection* that maps virtual-image coordinates to real-image coordinates. Typically, this transformation takes on the form of a field of disparity vectors where each disparity vector connects two homologous points, that is, points that are projections of the same 3D point onto imaging planes of the left and right cameras, as depicted in Figure 14.4.

14.2.2.1 Forward Projection

Let d^l be a disparity field defined on lattice Λ of real image I^l and let d^r be a disparity field defined on lattice Λ of real image I^r. For a disparity vector $d^l[x]$ anchored in I^l, the coordinates of homologous points are x in I^l and $x + d^l[x]$ in I^r, whereas for a disparity vector $d^r[x]$ anchored in I^r, the coordinates of homologous points are x in I^r and $x + d^r[x]$ in I^l. Ideally, homologous points should have identical brightness values and thus the following should hold:

$$I^l[x] = I^r[x + d^l[x]], \qquad I^r[x] = I^l[x + d^r[x]], \qquad \forall x \in \Lambda \qquad (14.4)$$

In practice, however, these two constraints only hold approximately, and various estimation methods are used to compute d^l and d^r, such as block matching (Izquierdo, 1997), optical flow (Ince and Konrad, 2008a), dynamic programming (Kang *et al.*, 2007), and graph cuts (Kolmogorov and Zabih, 2002).

Assuming that brightness constancy holds along the whole disparity vector, not only at the endpoints, the following is also true:

$$J^{\alpha}[x + \alpha d^{l}[x]] = I^{l}[x], \qquad J^{\alpha}[x + (1 - \alpha)d^{r}[x]] = I^{r}[x], \qquad \forall x \in \Lambda$$

because J^{α} is at the distance of α from I^{l} and at the distance of $(1 - \alpha)$ from I^{r}. Therefore, the reconstruction of intermediate-view intensities $J^{\alpha}[x + \alpha d^{l}[x]]$ and $J^{\alpha}[x + (1 - \alpha)d^{r}[x]]$ can be as simple as a substitution with $I^{l}[x]$ and $I^{r}[x]$, respectively. However, in general, $x + \alpha d^{l}[x] \notin \Lambda$ and $x + (1 - \alpha)d^{r}[x] \notin \Lambda$; that is, the homologous points in virtual view J^{α} do not belong to lattice Λ. In fact, owing to the space-variant nature of disparities, the above locations are usually irregularly spaced, whereas the goal here is to compute $J^{\alpha}[x]$ at regular locations $x \in \Lambda$. One option is to round the coordinates $x + \alpha d^{l}[x]$ and $x + (1 - \alpha)d^{r}[x]$ so that both belong to Λ. For the commonly used orthonormal lattices, this means forcing $\alpha d^{l}[x]$ and $(1 - \alpha)d^{r}[x]$ to be full-pixel vectors; that is, rounding coordinates to the nearest integer (McVeigh *et al.*, 1996; Scharstein, 1996) and then dealing with multiply occupied pixels and unassigned pixels by means of depth ordering and inpainting, respectively. An alternative is to use more advanced approaches, such as those based on splines (Ince *et al.*, 2007) in order to perform irregular-to-regular image interpolation. While simple rounding suffers from objectionable reconstruction errors, advanced spline-based methods produce high-quality reconstructions but, unfortunately, require significant computational effort.

14.2.2.2 Backward Projection

Let the disparity field d^{J} be defined on lattice Λ in the virtual view J^{α} and bidirectionally point toward the real images I^{l} and I^{r} (Mancini and Konrad, 1998; Zhai *et al.*, 2005). With this definition, disparity vectors pass through pixel positions of the virtual view (i.e., vectors are pivoted at $x = [x, y]^{T}$ in the virtual view) and the constant-brightness assumption now becomes

$$J^{\alpha}[x] = I^{l}[x - \alpha d^{J}[x]] = I^{r}[x + (1 - \alpha)d^{J}[x]], \qquad \forall x \in \Lambda \qquad (14.5)$$

Compared with equations (14.4), each pixel in J^{α} is now guaranteed to be assigned a disparity vector and, therefore, two intensities (from I^{l} and I^{r}) are associated with it. Although usually $x - \alpha d^{J}[x] \notin \Lambda$ and $x + (1 - \alpha)d^{J}[x] \notin \Lambda$, intensities at these points can be easily calculated from I^{l} and I^{r} using spatial (regular-to-irregular) interpolation.

In order to compute J^{α} at distance α from I^{l}, a disparity field pivoted at α is needed. Although this necessitates disparity estimation for each α (using similar methods to those mentioned earlier), it also simplifies the final computation of J^{α} since view rendering becomes a by-product of disparity estimation; once d^{J} that satisfies (14.5) has been found, either left or right luminance/color can be used for the intermediate-view texture. However, owing to noise, illumination effects and various distortions (e.g., aliasing), the matching pixels, even if truly homologous, usually are not identical. This is even more true when correspondence is performed at block level. Therefore, an even better reconstruction is accomplished when linear interpolation is applied to both intensities (Mancini and Konrad, 1998):

$$J^{\alpha}[x] = (1 - \alpha)I^{l}[x - \alpha d^{J}[x]] + \alpha I^{r}[x + (1 - \alpha)d^{J}[x]], \qquad \forall x \in \Lambda \qquad (14.6)$$

Clearly, all intermediate-view pixels are assigned an intensity and no additional post-processing is needed. We note that backward projection with linear intensity interpolation is a simple and effective approach to small-baseline view interpolation that can be implemented in real time. However, its performance deteriorates as the camera baseline increases since the occlusion areas are mishandled.

The quality of interpolated views strongly depends on the accuracy of estimated disparities, in particular at object boundaries. One of the more popular methods for disparity estimation derives from the optical flow formulation (Horn and Schunck, 1981). In the case of disparities pivoted in J^α, a variant of this method attempts to satisfy the constant-brightness assumption (14.5) while isotropically regularizing d:

$$\arg\min_{d} \iint_{x\in\Omega_J} (I^l[x - \alpha d[x]] - I^r[x + (1 - \alpha)d[x]])^2 + \lambda(||\nabla d^x||^2 + ||\nabla d^y||^2)\, dx \quad (14.7)$$

where Ω_J is the domain of J^α, $d[x] = [d^x[x]d^y[x]]^T$, and ∇ is the gradient operator. Owing to regularization along and across object boundaries, the resulting disparity fields tend to be blurry at object boundaries where one would expect crisp edges.

This can be corrected by anisotropic regularization that preserves disparity edges better than isotropic diffusion (Alvarez et al., 2002), but an image gradient is needed to guide the diffusion process. Since in estimation based on backward projection (14.7) the disparity is defined on the sampling grid of the unknown image J^α, no such gradient is available. However, it turns out that simple linear interpolation (14.6) produces intermediate views with sufficiently reliable edge information to guide the anisotropic diffusion.

Therefore, one can use a coarse intermediate image J_c, computed by linear interpolation (14.6) with isotropically diffused disparities (14.7), in order to guide the edge-preserving regularization as follows:

$$\arg\min_{d} \iint_{x\in\Omega_J} (I^l[x - \alpha d[x]] - I^r[x + (1 - \alpha)d[x]])^2 + \lambda(F_x(d^x, J_c) + F_x(d^y, J_c))\, dx$$

$$(14.8)$$

Above, $F_x(\cdot)$ assures anisotropic regularization (Perona and Malik, 1990) and is defined as follows:

$$F_x(d, J_c) = \nabla^T d[x] \begin{bmatrix} g(|J_c^x[x]|) & 0 \\ 0 & g(|J_c^y[x]|) \end{bmatrix} \nabla d[x] \quad (14.9)$$

where $g(\cdot)$ is a monotonically decreasing function, and J_c^x and J_c^y are horizontal and vertical derivatives of J_c. If $|J_c^x[x]| = |J_c^y[x]|$ then isotropic smoothing takes place and (14.8) simplifies to (14.7) except for a different λ. However, if, for example, $|J_c^x[x]| \gg |J_c^y[x]|$, then stronger smoothing takes place vertically and a vertical edge is preserved.

Figure 14.5 compares intermediate-view interpolation results for one-dimensional disparities (parallel cameras) estimated using isotropic regularization (14.7) and edge-preserving regularization (14.8). Note the less-diffused disparity for anisotropic regularization leading to a lower interpolation error (about 1 dB higher PSNR). Both interpolated views, however, are of excellent quality with only some localized errors at image periphery.

Intermediate-view interpolation algorithms can replace cumbersome camera arrays, thus providing input for multiview displays. They can be also used as an efficient prediction tool in multiview compression. Yet another interesting application is parallax adjustment. Among viewers there exists a very significant variation in sensitivity

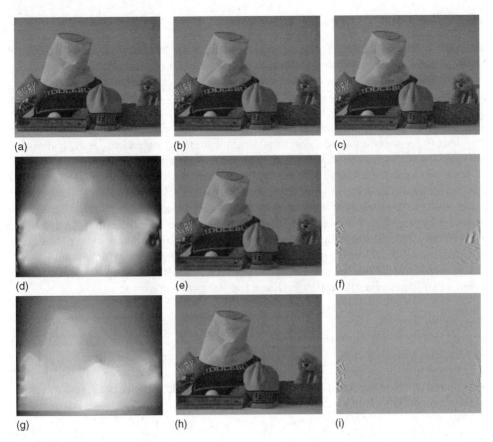

Figure 14.5 Comparison of view interpolation methods for *Middl* sequence from the Middlebury dataset: (a) left image I^l, (b) true intermediate image J, (c) right image I^r, (d) disparity estimated using isotropic regularization (14.7), (e) corresponding linearly interpolated intermediate view J (14.6), (f) interpolation error between images in (e) and (b), peak signal-to-noise ratio PSNR = 33.72 dB, (g) disparity estimated using anisotropic (edge-preserving) regularization (14.8), (h) corresponding linearly interpolated intermediate view J (14.6), and (i) interpolation error between images in (h) and (b), PSNR = 34.74 dB

to stereoscopic cues (Tam and Stelmach, 1995): while some viewers have no problem with the fusion of a particular stereo image, others may be unable to do so or may feel eye strain. The problem may be particularly acute in the presence of excessive 3D cues in the viewed images. Although viewers may experience no fusion problems with properly acquired stereoscopic images (i.e., with moderate parallax), significant difficulties may arise when parallax is excessive. For images acquired by a trained operator of a stereoscopic camera, this should occur infrequently. However, if the images are acquired under conditions optimized for one viewing environment (e.g., TV screen) and then are presented in another environment (e.g., large-screen cinema), viewer discomfort may occur. In order to minimize viewer discomfort, the amount of parallax can be adjusted by means of intermediate-view interpolation (i.e., by reducing the virtual baseline). This amounts to equipping a future 3D screen with a "3D-ness" or "depthness" knob similar to the "brightness" and "contrast" adjustments used today.

14.2.3 Brightness and Color Balancing in Stereopairs

3D cinematographers are very careful to use closely matched stereo cameras in terms of focus, luminance, color, and so on. While the human visual system can deal with small discrepancies in focus, luminance, or color, if the views are severely mismatched the viewer will experience visual discomfort. Furthermore, such a mismatch may adversely affect correspondence estimation, with severe consequences for intermediate-view interpolation (Section 14.2.2) and 3D data compression. For example, color or luminance bias in one of the images may lead to incorrect correspondences and, thus, distorted intermediate views, and also to increased transmission rates of 3D data as compression algorithms are sensitive to such drifts.

Below, we describe a simple method that helps to deal with global luminance and color mismatch issues (Franich and ter Horst, 1996). The underlying assumption is that the mismatch is due to unequal camera parameters (lighting conditions are considered the same for the two viewpoints) and can be modeled by a simple, linear transformation applied to the whole image. The transformation parameters are found by requiring that sample means and variances of color components be identical in both images after the transformation.

The method can be implemented as follows. Given an RGB stereopair $\{I_i^l, I_i^r\}$, $i = 1, 2, 3$, suppose that I_i^l undergoes the following linear transformation at each x:

$$\hat{I}_i^l[x] = a_i I_i^l[x] + b_i, \qquad i = 1, 2, 3 \tag{14.10}$$

were a_i and b_i are such parameters that

$$\hat{\mu}_i^l = \mu_i^r, (\hat{\sigma}_i^l)^2 = (\sigma_i^r)^2, \qquad i = 1, 2, 3 \tag{14.11}$$

with $\hat{\mu}_i^l$ and μ_i^r being the means and $(\hat{\sigma}_i^l)^2$ and $(\sigma_i^r)^2$ being the variances of $\hat{I}_i^l$ and I_i^r, respectively. From (14.10) it follows that

$$\hat{\mu}_i^l = a_i \mu_i^l + b_i, \qquad (\hat{\sigma}_i^l)^2 = a_i^2 (\sigma_i^l)^2 \tag{14.12}$$

Solving (14.11) and (14.12) for parameters a_i and b_i one obtains

$$a_i = \frac{\sigma_i^r}{\sigma_i^l}, \qquad b_i = \mu_i^r - \frac{\sigma_i^r}{\sigma_i^l} \mu_i^l$$

Clearly, balancing the color between two mismatched views amounts to the calculation of means and variances of two images (to obtain the transformation parameters), and then one multiplication (gain a_i) and one addition (offset b_i) per color component per pixel. Although this model is quite crude, it is surprisingly effective when viewpoint mismatch is caused by camera miscalibration. Figure 14.6 shows an example of the application of this approach; the

(a)   (b)   (c)

Figure 14.6 Example of color balancing in a stereo pair captured by identical cameras but with different exposure parameters: (a) original left view I^l; (b) original right view I^r; and (c) right view $\hat{I}^r$ after linear transformation (14.10) (see Plate 18 for the colored figure)

original images I^l and I^r have been captured by identical cameras but with different exposure and shutter-speed settings. Clearly, the original right view is much darker and has a slight blue–greenish tint compared with the left view. After the transformation, the colors and luminance level are well matched between the views.

14.3 Dealing with 3D Display Hardware

Once two or more views of a 3D image/video are available, they need to be delivered to viewers' eyes using one of several light-directing mechanisms, such as polarization, temporal multiplexing (shuttering), spectral multiplexing (anaglyph, Infitec), parallax barrier, lenticules, and so on. Each of these technologies has its strengths and weaknesses. Below, we discuss two such weaknesses, namely the perceived left/right image crosstalk in displays using polarized and shuttered glasses, and the spatial aliasing in lenticular and parallax-barrier eyewear-free displays.

14.3.1 Ghosting Suppression for Polarized and Shuttered Stereoscopic 3D Displays

Many 3D display technologies suffer from optical crosstalk that is perceived by a viewer as double edges or "ghosts" at high-contrast features misaligned between the left and right images due to disparity. Systems using polarized glasses suffer from crosstalk caused by the imperfect light extinction in the glasses. In systems using projection, an additional crosstalk may result from light depolarization on the projection screen. In liquid-crystal shutters (LCS), in addition to the imperfect light extinction in the opaque state (light leakage), LCS timing errors (opening/closing too early or too late) may also cause crosstalk. This crosstalk may be further magnified by screen characteristics, such as the green phosphor persistence of CRT monitors/projectors, phosphor afterglow in plasma panels, as well as the pixel response rate and update method in LCD panels. In systems using a single DLP projector with LCS glasses, only extinction characteristics and timing errors of the shutters play a role. The reader is referred to the literature for a more detailed discussion of the sources of crosstalk in 3D displays, and in particular to an excellent review by Woods (2010). Although improvements may be possible in each system by a careful selection of components, manipulation of image contrast/brightness, and adjustment of disparity magnitude, the degree of potential improvements is quite limited.

A significant reduction of the perceived crosstalk, even its complete cancellation under certain conditions, is possible by employing signal processing. The basic idea is to create an "anti-crosstalk"; that is, to pre-distort images so that upon display the ghosting is largely suppressed (Lipscomb and Wooten, 1994). We have developed a crosstalk suppression algorithm that is computationally efficient and accounts for specific screen and LCS characteristics (Konrad *et al.*, 2000). The algorithm is based on a simple crosstalk model:

$$J_i^l = I_i^l + \phi_i(I_i^r, I_i^l), \qquad J_i^r = I_i^r + \phi_i(I_i^l, I_i^r), \qquad i = 1, 2, 3 \tag{14.13}$$

where J_i^l and J_i^r, $i = 1, 2, 3$, are RGB components of images perceived by the left and right eyes, respectively, I_i^l and I_i^r are the corresponding RGB components driving the monitor, and ϕ_i are *crosstalk functions* for the three color channels. The crosstalk functions ϕ_i

quantify the amount of crosstalk seen by an eye in terms of the unintended *and* intended stimuli. These functions depend on the particular combination of glasses and screen used, and need to be carefully quantified; example functions for a specific CRT monitor and LCS glasses can be found in Konrad *et al.* (2000). Note, that the above crosstalk model ignores the point spread functions of the monitor and glasses.

If the mapping (14.13), which transforms (I_i^l, I_i^r) into (J_i^l, J_i^r) for $i = 1, 2, 3$, is denoted by T with the domain $D(T)$ and range $R(T)$, then the task is to find the inverse mapping T^{-1} that transforms (I_i^l, I_i^r), images we would like to see, into crosstalk-biased images (G_i^l, G_i^r) that drive the monitor; that is, find (G^l, G^r) satisfying

$$I_i^l = G_i^l + \phi_i(G_i^r, G_i^l), \qquad I_i^r = G_i^r + \phi_i(G_i^l, G_i^r), \qquad i = 1, 2, 3 \qquad (14.14)$$

For given crosstalk functions ϕ_i, this mapping can be computed off-line and stored in a 400 kB look-up table for 8-bit color components ($256 \times 256 \times 3 \times 2$) and 6.3 MB table for 10-bit components.

Since $R(T)$ is only a subset of $[0, 255] \times [0, 255]$ for $D(T) = [0, 255] \times [0, 255]$, the inverse mapping T^{-1} operating on $[0, 255] \times [0, 255]$ may result in negative tristimulus values, which cannot be displayed in practice. The algorithm is trying to "carve out" intensity notches (Figure 14.7) that will get filled with the unintended light (from the

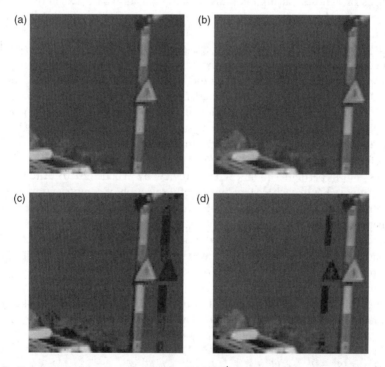

Figure 14.7 Example of crosstalk compensation: (a) left I^l and (b) right I^r original images (luminance only); (c–d) the same images (c) G^l and (d) G^r after crosstalk compensation using the algorithm described in Konrad *et al.* (2000) with luminance clipping below the amplitude of 30. From Konrad *et al.* (2000), reproduced with permission of IEEE © 2000

other view), but in dark parts of an image this is not possible. Two solutions preventing negative tristimulus values are linear mapping of RGB components, at the cost of a reduced image contrast, and black-level saturation, which leads to a loss of detail in dark image areas ("black crush"), both applied prior to T^{-1}. Since neither solution is acceptable, it is preferred to seek a compromise between the degree of crosstalk reduction and image quality degradation (Konrad *et al.*, 2000). This method has been successfully used in LCS-based systems, but it is equally applicable to polarization-based systems using CRT monitors, CRT projectors, or DLP projectors. Also, the recent 3D displays using LCD/plasma panels and LCS glasses may benefit from this approach. However, this approach is not applicable to lenticular and parallax-barrier autostereoscopic screens since the nature of crosstalk is more complicated due to the number of views involved.

An example of the application of the above algorithm is shown in Figure 14.7. Note the reduced image intensity (dark patches) in areas corresponding to bright features present in the other image; the unintended light due to optical crosstalk will fill those "holes" during display, leading to crosstalk compensation.

14.3.2   *Aliasing Suppression for Multiview Eyewear-Free 3D Displays*

When preparing a 3D image for display on a lenticular or parallax-barrier 3D screen, data from several views must be multiplexed together. If the resolution of each view is identical to that of the underlying pixel-addressable panel, each view must be first spatially subsampled and then all the extracted samples must be interleaved together to form a standard 2D image. It is only when viewing this image through a lenticular sheet or parallax barrier that a viewer sees different pixels by each eye thus invoking the perception of depth. Since each pixel of a typical panel used consists of three sub-pixels (RGB) to represent color, the subsampling occurs on the R, G, or B sub-pixel raster in each view. For example, the locations of R sub-pixels associated with one view only are shown in Figure 14.8 as circles. Ideally, all these sub-pixels would be seen by one eye only. Note that these sub-pixels may be irregularly spaced owing to the fact that the subsampling pattern depends on the properties of the lenticular sheet or parallax barrier (lenticule/slit pitch, angle, etc.) as well as those of the screen raster (sub-pixel pitch, geometry, etc.). In consequence, the single-view sub-pixel locations cannot be described by a 2D lattice. A similar observation can be made for green and blue sub-pixels. Thus, it is not obvious how to quantify and counteract the aliasing resulting from such an irregular subsampling. Below, we describe two approaches that we have developed to address this issue.

14.3.2.1   Spatial-Domain Approach

Let the irregular subsampling layout for red sub-pixels of one view, depicted by circles in Figure 14.8, be described by a set $\mathcal{V} \subset R^2$; that is, $x \in \mathcal{V}$ denotes the location of a circle. The goal is to approximate $\mathcal{V}$ by a lattice or union of cosets, denoted $\Psi \subset R^2$, in the spatial domain (Konrad and Agniel, 2006). If Γ is the orthonormal lattice of the screen raster for red sub-pixels (denoted by dots in Figure 18.8), then $\mathcal{V} \subset \Gamma \subset R^2$ by definition. The task is to find a sampling structure Ψ that best approximates the set $\mathcal{V}$ under a certain metric. One possibility is to minimize the distance between sets Ψ and $\mathcal{V}$. Let $d(x, \mathcal{A})$ be the distance from point x to a discrete set of points $\mathcal{A}$; that is, $d(x, \mathcal{A}) = \min_{y \in \mathcal{A}} ||x - y||$. Then, one can

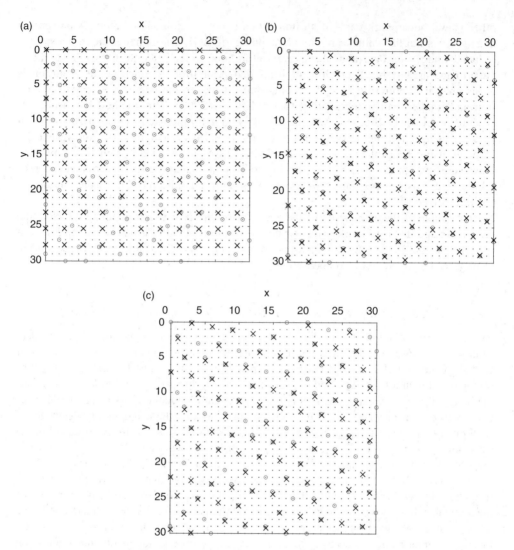

Figure 14.8 Approximation of single-view red (R) sub-pixels using (a) orthogonal lattice, (b) nonorthogonal lattice, and (c) union of 21 cosets. Dots denote the orthonormal screen raster of red sub-pixels (Γ), circles (o) denote sub-pixels activated when rendering one view ($\mathcal{V}$), while crosses ($\times$) show locations from model Ψ. From Konrad and Halle (2007), reproduced with permission of IEEE © 2007

find an optimal approximation by minimizing the mutual distance between point sets Ψ and $\mathcal{V}$ as follows:

$$\min_{\boldsymbol{\eta}} \xi_{\mathcal{V}} \sum_{x \in \mathcal{V}} d(\boldsymbol{x}, \Psi) + \xi_{\Psi} \sum_{x \in \Psi} d(\boldsymbol{x}, \mathcal{V}) \qquad (14.15)$$

where $\boldsymbol{\eta}$ is a vector of parameters describing the sampling structure Ψ. For a 2D lattice, $\boldsymbol{\eta}$ is a 3-vector, and thus the minimization (14.15) can be accomplished by a hierarchical exhaustive search over a discrete state space (Konrad and Agniel, 2006). As the weights $\xi_{\mathcal{V}}$ and ξ_{Ψ}

are adjusted, different solutions result. If $\xi_\Psi = 0$, a very dense (in the limit, infinitely dense) Ψ would result, while for $\xi_\mathcal{V} = 0$ a single-point $\Psi \subset \mathcal{V}$ would be found optimal. Instead of a combination of both distances one could use either of them under a constrained minimization (e.g., a constraint on the density of Ψ). A similar procedure should be applied to green (G) and blue (B) sub-pixels to assure aliasing suppression in all three color channels. It can also be extended to a union of K cosets instead of a single lattice, but the number of parameters in vector $\boldsymbol{\eta}$ would grow from 3 to $3 + 2(K - 1)$ (Konrad and Agniel, 2006).

Applied to a number of subsampling layouts $\mathcal{V}$, the above method has been shown effective in identifying regular approximations of $\mathcal{V}$, from orthogonal-lattice approximations (quite inaccurate), through non-orthogonal-lattice approximations (more accurate), to union-of-cosets approximations (most accurate), as shown in Figure 14.8. Note a progressively improving alignment between single-view irregular points in $\mathcal{V}$ (circles) and regular points in Ψ (crosses). Having identified an approximation of $\mathcal{V}$ by a regular sampling structure, it is relatively straightforward to find the passband of suitable low-pass anti-alias pre-filters that must precede view multiplexing (Konrad and Agniel, 2006).

14.3.2.2 Frequency-Domain Approach

In an alternative approach, the irregularity of $\mathcal{V}$ is approximated in the frequency domain (Jain and Konrad, 2007). The main idea is based on the observation that in one dimension a sequence of unit impulses (Kronecker deltas) $g[n]$ has the discrete-time Fourier transform (DTFT) in the form of a periodic impulse train (sum of frequency-shifted Dirac delta functions). Therefore, the subsampling of a signal $x[n]$ implemented by multiplying it with the sequence $g[n]$ results in the convolution of their DTFTs: $\mathcal{F}\{x[n]\} * \mathcal{F}\{g[n]\}$. Clearly, the spectrum $\mathcal{F}\{x[n]\}$ is going to be replicated at locations of the periodic impulse train $\mathcal{F}\{g[n]\}$. A similar relationship holds in two dimensions with respect to bi-sequences and 2D periodic impulse trains.

The above relationships also hold if $g[n, m]$ is a bi-sequence of *irregularly spaced* unit impulses; that is, defined on $\mathcal{V}$. Then, $\mathcal{F}\{g[n, m]\}$ is a 2D train of Dirac delta functions defined on a reciprocal lattice Λ^* of the least dense lattice Λ such that $\mathcal{V} \subset \Lambda$ (Jain, 2006). Additionally, impulses in this train are scaled differently at different frequency locations (Jain, 2006). The consequence of this is that the signal spectrum is replicated with different gains at different frequency locations, and replications with smaller gain factors may have less of an impact on aliasing than those with large gain factors. Although to completely prevent aliasing one could consider the worst-case scenario – that is, limiting the signal spectrum so that the closest off-origin spectral replications do not "leak" into the baseband Voronoi cell – such a design would be overly conservative. Alternatively, a controlled degree of aliasing may be permitted, but then either the actual signal spectrum or its model must be known. We have obtained good results for a separable Markov-1 image model (Jain and Konrad, 2007); an anti-alias filter's passband shape was found by identifying frequencies at which the ratio of the baseband spectrum magnitude to that of its closest spectral replication (with suitable gain) is at least one; that is, aliasing energy at a specific frequency does not exceed the baseband energy of the original signal. Of course, this energy ratio can be adjusted to be more conservative.

Figure 14.9 shows contour plots of the magnitude response of anti-alias filters based on the spatial- and frequency-domain approximations. Note a rotated hexagonal passband for

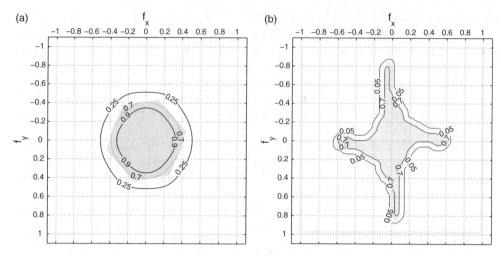

Figure 14.9 Contour plots of the magnitude response of an anti-alias filter obtained by: (a) approximating pixel layout $\mathcal{V}$ by a nonorthogonal lattice Λ in the spatial domain; (b) approximating $\mathcal{V}$ in the frequency domain by means of spectrum modeling with a Markov-1 image model for each view. The desired magnitude response passband is shaded. Both frequency axes are normalized to the Nyquist frequency. From Konrad and Halle (2007), reproduced with permission of IEEE © 2007

the model based on spatial-domain approximation using a nonorthogonal lattice and a diamond-shaped passband, with horizontal and vertical extensions, for the frequency-domain approximation. Applied in practice, a pre-filtering based on either specification results in effective suppression of aliasing artifacts, although pre-filtering based on the specifications from Figure 14.9b preserves horizontal and vertical detail of 3D images better.

The benefit of an approximation in the frequency domain is that anti-alias filtering can be adapted to specific spectra, whether of a particular image or model. In principle, given sufficient computing power (already available in today's CPUs and graphics cards), one can imagine anti-alias filtering adapting to each displayed image in real time.

14.4 Conclusions

The recent resurgence of 3D in movies and its slow introduction into homes are both largely due to the ongoing digital content revolution. Today, 3D cinematography primarily uses digital cameras, the post-production is performed on computers, and the cinemas are equipped with digital projectors. At home, the content is delivered by means of digital transmission or recorded on digital media (e.g., Blu-ray) and almost all 3D displays are pixel addressable.

The digital bitstream is one of the most important foundations for the resurgence of 3D as it facilitates 3D data compression, transmission, and storage. Furthermore, it facilitates signal processing of 3D content to help cure many deficiencies of 3D displays. In this chapter, we have provided several examples of a successful application of signal processing to content generation and curing the ills of today's 3D hardware. Signal processing is likely to continue playing an even more important role in the future as 3D displays become more advanced and their deficiencies evolve; algorithms can be easily adapted to such changes, even in real time.

The future enabled by the combination of 3D signal processing, digital content, and new display technologies looks promising. Will consumers adopt this technology and replace the ubiquitous 2D screens with 3D counterparts? Only the time will show.

Acknowledgments

I would like to acknowledge the contributions to this chapter by my past students, especially Anthony Mancini, Bertrand Lacotte, Peter McNerney, Philippe Agniel, Ashish Jain, Serdar Ince, Geoffrey Brown, and Meng Wang, as well as by Professor Prakash Ishwar of Boston University who helped develop some of the 2D-to-3D image conversion algorithms. I would also like to acknowledge the support from the National Science Foundation under grants ECS-0219224 and ECS-0905541.

References

Alvarez, L., Deriche, R., Sánchez, J., and Weickert, J. (2002) Dense disparity map estimation respecting image discontinuities: a PDE and scale-space based approach. *J. Vis. Commun. Image Represent.*, **13**, 3–21.

Angot, L.J., Huang, W.-J., and Liu, K.-C. (2010) A 2D to 3D video and image conversion technique based on a bilateral filter, in *Three-Dimensional Image Processing (3DIP) and Applications* (ed. A.M. Baskurt), Proceedings of the SPIE, Vol. **7526**, SPIE, Bellingham, WA, p. 75260D.

Durand, F. and Dorsey, J. (2002) Fast bilateral filtering for the display of high-dynamic-range images. *ACM Trans. Graph.*, **21**, 257–266.

Franich, R. and ter Horst, R. (1996) Balance compensation for stereoscopic image sequences, ISO/IEC JTC1/SC29/WG11 – MPEG96.

Guttmann, M., Wolf, L., and Cohen-Or, D. (2009) Semi-automatic stereo extraction from video footage, in *2009 IEEE 12th International Conference on Computer Vision*, IEEE, pp. 136–142.

Horn, B. and Schunck, B. (1981) Determining optical flow. *Artif. Intell.*, **17**, 185–203.

Ince, S. and Konrad, J. (2008a) Occlusion-aware optical flow estimation. *IEEE Trans. Image Process.*, **17** (8), 1443–1451.

Ince, S. and Konrad, J. (2008b) Occlusion-aware view interpolation. *EURASIP J. Image Video Process.*, **2008**, 803231.

Ince, S., Konrad, J., and Vázquez, C. (2007) Spline-based intermediate view reconstruction, in *Stereoscopic Displays and Virtual Reality Systems XIV* (eds A.J. Woods, N.A. Dodgson, J.O. Merritt *et al.*), Proceedings of the SPIE, Vol. **6490**, SPIE, Bellingham, WA, pp. 0F.1–0F.12.

Izquierdo, E. (1997) Stereo matching for enhanced telepresence in three-dimensional videocommunications. *IEEE Trans. Circ. Syst. Video Technol.*, **7** (4), 629–643.

Jain, A. (2006) Crosstalk-aware design of anti-alias filters for 3-D automultiscopic displays, Master's thesis, Boston University.

Jain, A. and Konrad, J. (2007) Crosstalk in automultiscopic 3-D displays: blessing in disguise?, in *Stereoscopic Displays and Virtual Reality Systems XIV* (eds A.J. Woods, N.A. Dodgson, J.O. Merritt *et al.*), Proceedings of the SPIE, Vol. **6490**, SPIE, Bellingham, WA, pp. 12.1–12.12.

Kang, S., Li, Y., and Tong, X. (2007) *Image Based Rendering*, Now Publishers.

Kolmogorov, V. and Zabih, R. (2002) Multi-camera scene reconstruction via graph cuts, in *ECCV' 02 Proceedings of the 7th European Conference on Computer Vision – Part III*, Springer Verlag, London, pp. 82–96.

Konrad, J. and Agniel, P. (2006) Subsampling models and anti-alias filters for 3-D automultiscopic displays. *IEEE Trans. Image Process.*, **15** (1), 128–140.

Konrad, J. and Halle, M. (2007) 3-D displays and signal processing: an answer to 3-D ills? *IEEE Signal Proc. Mag.*, **24** (6), 97–111.

Konrad, J., Lacotte, B., and Dubois, E. (2000) Cancellation of image crosstalk in time-sequential displays of stereoscopic video. *IEEE Trans. Image Process.*, **9** (5), 897–908.

Konrad, J., Brown, G., Wang, M. *et al.* (2012a) Automatic 2D-to-3D image conversion using 3D examples from the internet, *Stereosc. Disp. Appl., Proceedings of the SPIE*, Vol. **8288**, SPIE, Bellingham, WA.

Konrad, J., Wang, M., and Ishwar, P. (2012b) 2D-to-3D image conversion by learning depth from examples, in *2012 IEEE Computer Society Conference on Computer Vision and Pattern Recognition Workshops (CVPRW)*, IEEE Computer Society Press, Washington, DC, pp. 16–22.

Liao, M., Gao, J., Yang, R., and Gong, M. (2012) Video stereolization: combining motion analysis with user interaction. *IEEE Trans. Visual. Comput. Graph.*, **18** (7), 1079–1088.

Lipscomb, J. and Wooten, W. (1994) Reducing crosstalk between stereoscopic views, in *Stereoscopic Displays and Virtual Reality Systems* (eds S.S. Fisher, J.O. Merritt, and M.T. Bolas), Proceedings of the SPIE, Vol. **2177**, SPIE, Bellingham, WA, pp. 92–96.

Liu, B., Gould, S., and Koller, D. (2010) Single image depth estimation from predicted semantic labels, in *2010 IEEE Conference on Computer Vision and Pattern Recognition*, IEEE, pp. 1253–1260.

Mancini, A. and Konrad, J. (1998) Robust quadtree-based disparity estimation for the reconstruction of intermediate stereoscopic images, in *Stereoscopic Displays and Virtual Reality Systems V* (eds M.T. Bolas, S.S. Fisher, and J.O. Merritt), Proceedings of the SPIE, Vol. **3295**, SPIE, Bellingham, WA, pp. 53–64.

Mansouri, A.R. and Konrad, J. (2000) Bayesian winner-take-all reconstruction of intermediate views from stereoscopic images. *IEEE Trans. Image Process.*, **9** (10), 1710–1722.

McVeigh, J., Siegel, M., and Jordan, A. (1996) Intermediate view synthesis considering occluded and ambiguously referenced image regions. *Signal Process. Image Commun.*, **9**, 21–28.

Perona, P. and Malik, J. (1990) Scale-space and edge detection using anisotropic diffusion. *IEEE Trans. Pattern Anal. Mach. Intell.*, **12** (7), 629–639.

Phan, R., Rzeszutek, R., and Androutsos, D. (2011) Semi-automatic 2D to 3D image conversion using scale-space random walks and a graph cuts based depth prior, in *2011 18th IEEE International Conference on Image Processing (ICIP)*, IEEE, pp. 865–868.

Saxena, A., Sun, M., and Ng, A. (2009) Make3D: learning 3D scene structure from a single still image. *IEEE Trans. Pattern Anal. Mach. Intell.*, **31** (5), 824–840.

Scharstein, D. (1996) Stereo vision for view synthesis, in *Proceedings. 1996 IEEE Computer Society Conference on Computer Vision and Pattern Recognition*, IEEE Computer Society Press, Washington, DC, pp. 852–858.

Silberman, N. and Fergus, R. (2011) Indoor scene segmentation using a structured light sensor, in *2011 IEEE International Conference on Computer Vision Workshops (ICCV Workshops)*, IEEE, pp. 601–608.

Subbarao, M. and Surya, G. (1994) Depth from defocus: a spatial domain approach. *Int. J. Comput. Vision*, **13**, 271–294.

Szeliski, R. and Torr, P.H.S. (1998) Geometrically constrained structure from motion: points on planes, in *3D Structure from Multiple Images of Large-Scale Environments* (eds R. Koch and L. Van Gool), Lecture Notes in Computer Science, Vol. **1506**, Springer Verlag, Berlin, pp. 171–186.

Tam, W. and Stelmach, L. (1995) Stereo depth perception in a sample of young television viewers. International Workshop on Stereoscopic and 3D Imaging, Santorini, Greece, pp. 149–156.

Wang, M., Konrad, J., Ishwar, P. *et al.* (2011) Image saliency: from intrinsic to extrinsic context, in *2011 IEEE Conference on Computer Vision and Pattern Recognition (CVPR)*, IEEE, pp. 417–424.

Woods, A. (2010) Understanding crosstalk in stereoscopic displays. Three-Dimensional Systems and Applications Conference, Tokyo, Japan.

Zhai, J., Yu, K., Li, J., and Li, S. (2005) A low complexity motion compensated frame interpolation method, in *IEEE International Symposium on Circuits and Systems (ISCAS)*, Vol. **5**, IEEE, Piscataway, NJ, pp. 4927–4930.

Zhang, R., Tsai, P.S., Cryer, J., and Shah, M. (1999) Shape-from-shading: a survey. *IEEE Trans. Pattern Anal. Mach. Intell.*, **21** (8), 690–706.

15

3D Display Technologies

Thierry Borel and Didier Doyen
Technicolor, France

15.1 Introduction

The availability of new promising digital display technologies has been one of the important triggers of the revival of three-dimensional (3D) stereo in cinemas in the mid 2000s after several unsuccessful attempts in the second half of the twentieth century. After the success of some major 3D movies like *Avatar* from James Cameron at the end of 2009, the display manufacturing industry has started to develop 3DTV sets to deploy the 3D experience to the homes. Nevertheless, the success is not yet there. Some of the reasons are linked to the quality of experience; some others are linked to the economics (Borel, 2012).

In this chapter we present in detail the main 3D display technologies available for cinemas, for large-display TV sets, and for mobile terminals. Some use glasses and some do not. We also propose a perspective of evolution for the near and long term.

15.2 Three-Dimensional Display Technologies in Cinemas

Recent 3D display technologies appeared first on cinema screens before their appearance in homes. The most popular ones in the USA are based on light polarization concepts, but some other interesting techniques based on shutter glasses or color interference filters have also been implemented in movie theaters (Mendiburu, 2009), for instance in Europe and in Asia. All these technologies encompass various advantages and drawbacks. None of them is perfect but, thanks to digital technologies, the performances achieved are nevertheless much higher than what was available in the 1950s or 1980s; this is one of the reasons why it has been successful this time.

The following sections are going to describe in more detail the basic principles and the relative benefits of each of those approaches. The anaglyphic technologies are not going to be covered as they have been abandoned very rapidly in cinemas since they were not able to render good color and generated fatigue.

Emerging Technologies for 3D Video: Creation, Coding, Transmission and Rendering, First Edition.
Frédéric Dufaux, Béatrice Pesquet-Popescu, and Marco Cagnazzo.
© 2013 John Wiley & Sons, Ltd. Published 2013 by John Wiley & Sons, Ltd.

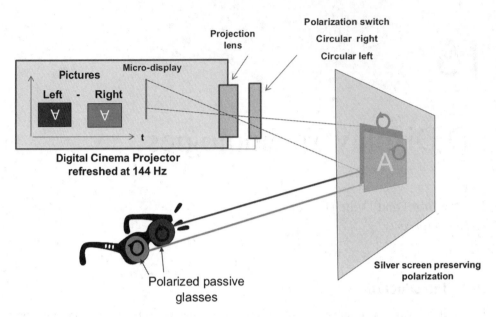

Figure 15.1   Digital cinema projector with polarized glasses

15.2.1   Three-Dimensional Cinema Projectors Based on Light Polarization

At the early stage of 3D cinema projection, two film projectors were used in concert: one equipped with a vertical linear polarization filter and the second with a horizontal polarization filter. A polarization-preserving screen, also called silver screen, was then necessary to scatter the light back towards the audience. The spectators were wearing glasses with linearly polarized filters to direct the left content towards the left eye and the right content towards the right eye. Since two mechanical projectors were used, it was really difficult to align and synchronize them perfectly over time. Headaches, nausea and discomfort were frequent. This technique is no longer used.

In order to avoid previous issues, most of today's available technologies use a unique digital projector. It is made of a lighting system based on a xenon lamp ($\sim$3 kW), which illuminates three micro-displays, one for each of the red, green, and blue primaries. An optical system recombines the three pictures into one light beam and sends them through a single projection lens onto a large cinema screen. The micro-display technology is generally based on digital micro-mirror devices (Digital Light Processing system[1]). We can also find optical architectures using transparent liquid-crystal micro-panels or reflective liquid-crystal on silicon micro-displays.

Figure 15.1 shows a very widespread 3D projection system, which is installed in 80% of 3D cinemas.[2]

The left and right images are alternatively displayed on the micro-displays. The refresh rate must be high enough to be imperceptible to the spectators' eyes. The display mode is called "triple flash," as each left and right image is refreshed three times faster than the usual

---

[1] DLP systems have been developed by Texas Instruments.

[2] RealD system; www.reald.com.

24 images/second rate used in two-dimensional (2D) movie acquisition. Consequently, the working frequency of the projector is

$$F = 3 \times 24 \times 2 = 144\,\text{Hz} \tag{15.1}$$

An additional optical component called a polarization switch is placed in front of the projection lens. It aims at changing the way light is polarized from circular left to circular right in perfect synchronization with the display of respectively the left and right images. The main reason why a circular polarization technique is used here is that it allows the spectator to move their head around the sight axis without creating crosstalk, or a double image, unlike with previously used linear polarization systems.

Once correctly polarized, the pictures are projected onto a silver screen, whose specifically designed material has the property of preserving the polarization state after reflection and scattering. The simple passive glasses worn by the spectator encompass a circular left polarization filter on the left eye and a circular right polarization filter on the right eye. Then, each left and right image is directed to the correct eye and the 3D stereo effect can occur.

The key element of the chain is the polarization switch. It is made of one liquid-crystal cell, featuring one big pixel only, sandwiched between a linear polarizer at the entrance and a quarter-wave retarder at the exit. Depending on the voltage applied across the cell, the light is linearly polarized at $\pm 45°$ with respect to the retarder optical axis (see Figure 15.2). This optical material has the following property: at the far side of the plate, the horizontal wave is exactly a quarter of a wavelength delayed relative to the vertical wave, resulting in a circular polarization, whose rotation direction depends on the entrance linear polarization direction. Fortunately, the glasses do not need an active element. They are just made of an optical retarder film (transforming the incoming circular polarized light back to a linear one) and a linear polarization analyzer. These two elements can easily be made out of plastic, the reason why it is cheap.

The technology described above uses a digital projector and a digital server, which are significant investments to be made, especially by smaller theaters equipped with classical

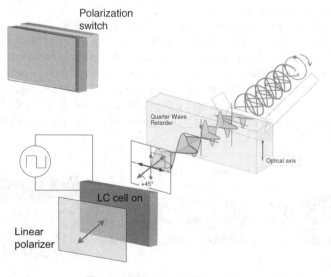

Figure 15.2   Polarization switch

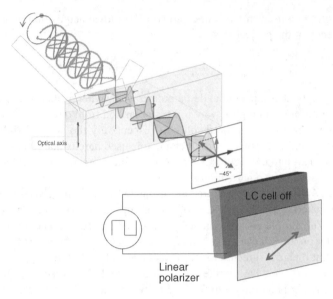

Figure 15.2 (*Continued*)

film projectors. This is the reason why an original approach has been developed by Technicolor, which allows 35 mm film projectors to display 3D movies without having to purchase expensive equipment, and without the drawbacks of historical systems based on two film projectors working in parallel. This cheaper and innovative alternative is shown on Figure 15.3. It simply consists in printing the film in a top/bottom format, the sub-images, left and right, are vertically compressed by a factor of 2 and a new double projection lens replaces the classical one. This double projection lens includes circular polarizing elements (circular left for the left image and circular right for the right image) and projects the two images on the screen with a pixel accurate precision. A silver screen is, of course, needed to preserve polarization and the same glasses as in Figure 15.1 can be used.

Generally speaking, the main advantage of the polarized light approaches is that glasses are cheap and the spectator can take them back home till the next screening or even use them in the home if they own a 3DTV set working with passive glasses. Nevertheless, some drawbacks exist. The cinema owner has to replace the classical white projection screen by a silver screen. In addition to the created over cost, silver screen brightness is not homogeneous when displaying a 2D content. A hot spot exist in the center of the screen, which makes the 2D experience more or less disappointing, depending on your position in the cinema theater.

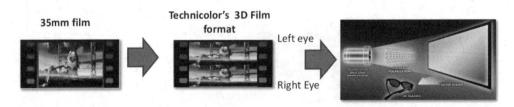

Figure 15.3   Film cinema projector with polarized glasses

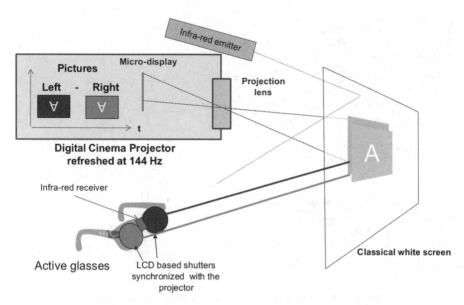

Figure 15.4 Digital cinema projector with shutter glasses

Another major drawback is the very low light efficiency of the system, as the light is passing through two polarization filters, 70% of the light is absorbed by those filters. Furthermore, as each eye sees a black picture 50% of the time, in the case of digital projection, or a half picture in the case of the 35 mm film projector, only 15% of the emitted light reaches the eyes of the spectators.

15.2.2 Three-Dimensional Cinema Projectors Based on Shutters

In order to avoid the silver screen drawbacks, some cinema owners prefer using a shutter glasses concept as described in Figure 15.4. Here, a digital projector is necessary as well but a classical white screen can be used. The images are also displayed at a rate of 144 frames per second (triple flash mode) and, instead of a polarization switching cell, an infrared (IR) emitter is used to synchronize the active shutter glasses worn by the spectators with the left–right image sequence delivered by the projector.

The active glasses encompass two liquid-crystal cells acting as shutters, which are alternatively turned on and off to select the right image for the right eye and the left image for the left eye. Some models also use radio frequencies for synchronization. As the refresh rate is high enough, the users do not perceive flickering effects when watching a movie.

As said, the main advantage of this solution is that it does not necessitate the use of a silver screen. The same projector and screen can be favorably used for both 2D and 3D experiences. Nevertheless, the active glasses are much more expensive than the passive glasses and management cost is not negligible either. They need to be collected at the end of the show, cleaned up and batteries have to be regularly loaded. They are also bulkier and less comfortable to wear. This is why some theater owner, who had in a first step chosen this solution, turned back to polarized systems to simplify the glasses management.

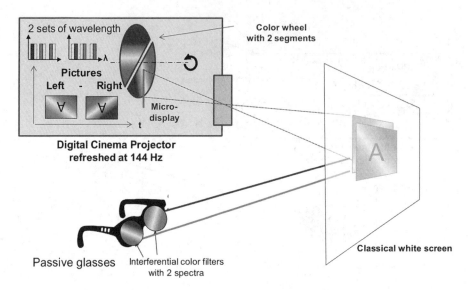

Figure 15.5   Digital cinema projector with interference filters

15.2.3   Three-Dimensional Cinema Projectors Based on Interference Filters

The last projection technique we describe in this section tries to gain all the advantages of previous systems, namely both white screen and passive glasses. This concept is described on Figure 15.5 and is based on the "metamerism" phenomenon, which considers that, for a human eye, a given color stimulus can be provided by various light spectra (Jorke, 2009).

The 3D cinema projectors based on this concept include a two-segment color wheel, whose rotation speed is synchronized with the refresh frequency of the imagers. The luminous wavelengths provided by the lamp are differently filtered by each segment of the wheel. Two independent sets of primary colors are then generated: (R1, G1, B1) and (R2, G2, B2); these two spectra have no common parts. During the left image period, (R1, G1, B1) illuminates the imagers and during the right image period (R2, G2, B2) is in charge.

On the spectator passive glasses are positioned interferential color filters corresponding to the left and right spectra described above, which steer the content to the correct eyes.

As said, this system is the only one combining both a standard white screen and passive comfortable glasses. Nevertheless, the specific interference color filters used here are pretty sophisticated as they have to be very selective in terms of wavelength; today, they can only be manufactured on glass substrates. This is why, although they are cheaper than active glasses, they are still more expensive than polarized glasses on plastic substrate. Furthermore, this kind of concept cannot be used in the home as all TV sets in the world use one set of primaries only. Consequently, glasses have to be collected at the end of the show and regularly cleaned. But the most serious drawback of this technology is that the cinema owner needs to upgrade the hardware of the existing digital projector to integrate the color wheel. This is why, despite its many advantages, this technology is not widespread in the world.

15.3   Large 3D Display Technologies in the Home

The production of more and more 3D stereo movies has encouraged the industry to put in place a larger ecosystem targeting the distribution of 3D to the home. New standards have been developed (3D Blu-ray, HDMI 1.4, DVB-3DTV, etc.) to package or transmit 3D media, major broadcasters have started to produce various kinds of 3D stereo content (sports, concerts, operas, documentaries, etc.) and of course the display and TV manufacturing industry has provided 3DTV sets to the consumer market. This business is still in its infancy, but display technologies are already available and the chicken and egg cycle is broken. In addition, consumers are now able to generate their own 3D content since mobile devices are available featuring two cameras, and 3D camcorders are also available.

We will now review the main display technologies that have been, are, and will be available to consumers to enjoy the 3DTV experience, both on large displays and mobile devices. Each of these approaches has diverse advantages and drawbacks.

15.3.1   Based on Anaglyph Glasses

The oldest 3D stereo visualization technology is called anaglyph. It was previously used in the early years of 3D cinema but, as it is based on the color coding of the left and right views, it can also be used in the home either on a standard 2D TV set or a computer monitor provided the content has been adapted to the colored glasses. Three variants exist (Figure 15.6):

Figure 15.6   Three variants of anaglyph glasses

- red–cyan
- yellow–blue
- green–purple.

The red–cyan glasses were, historically, the first anaglyph glasses to be used in the past. The yellow–blue concept needs to have a neutral density filter on top of the yellow filter to balance the light energies on each eye. This was developed to make the content visible in 2D for people not wearing glasses. But the color skewing is strong and this is pretty uncomfortable to watch. Finally, the green–purple approach is supposed to have a better balance in terms of light energy for each eye and, consequently, it is a bit more comfortable to watch than the two previous ones.

The main advantage of the anaglyph approach is that it is very inexpensive since the stereo image can be carried through any RGB video distribution channel and be displayed on any RGB display; the glasses are very cheap too. Nevertheless, the color performances are pretty poor: we can see color skewing, unsaturated colors, and the concept is pretty intolerant to color compression. It is very hard to watch a complete 2 h movie with an anaglyph system, but it can be useful for very short clips on the Internet without having to buy 3D display devices.

15.3.2 Based on Shutter Glasses

Stereoscopic 3DTV based on shutter glasses requires delivering sequentially the video dedicated to each eye in synchronization with active glasses. Flat-panel display technologies have in the last 10 years improved their capability a lot to render HD content at a high frame rate. Plasma and liquid-crystal display (LCD) technologies are both "display and hold" technologies. Unlike the old cathode ray tube technology, these displays emit light all along the frame. This is a drawback to render correctly a moving object. At a 50/60 Hz frame rate, blurring effects will be noticeable. It was then a priority for TV manufacturers to increase this display frame rate up to 100/120 Hz (reducing also large-area flicker) and then far beyond recently. Now with a refresh rate of 200/240 Hz, stereoscopic 3DTV based on shutter glasses has become feasible.

Shutter-glasses-based displays have the advantage to render the full resolution of the video for both eyes. Addressing left and right eyes with different contents is done in the temporal dimension, not in the spatial one. This will induce a lower brightness but not a lower resolution.

One of the main hurdles for TV manufacturer has been to avoid as much as possible the cross-talk effect when providing stereoscopic displays. Cross-talk is the leakage of video information into the wrong eye (e.g., left video perceived by the right eye) (Woods, 2011). Since such displays are addressed line by line, the left eye should not receive any light before the last line of the left video has been written (same principle for the right eye). This implies the introduction of a dark time in between left- and right-addressed video.

For light-emitting diode (LED)–LCD TVs, this dark time can today be realized thanks to the optimized synchronization of:

- the shutter opening time
- the LED backlight management
- the LCD panel addressing scheme.

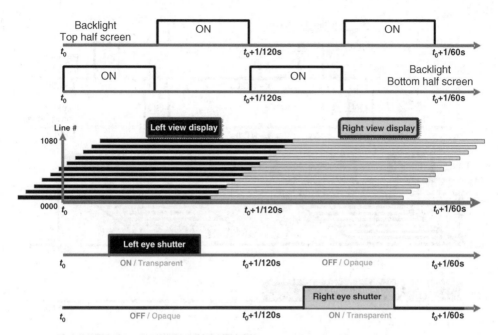

Figure 15.7 A first synchronization for LED–LCD TV using active glasses

Figure 15.7 and 15.8 illustrate two different optimizations of this same problem (Blondé *et al.*, 2011). In Figure 15.7 the shutter is only open during a quarter of the time. The backlight switch ON period is different for the top and the bottom parts of the panel to synchronize with the line addressing scheme. In Figure 15.8 the shutter is open half of the time but there is a dark time period between left and right addressing times. This dark time is also synchronized with the backlight.

Although specific care has been taken, the response time of current liquid-crystal technologies cannot ensure a cross-talk-free display. Even if the dark time is quite large, there will be in any case a leakage of one video into the wrong eye (Blondé *et al.*, 2012). The phenomenon is indeed complex to quantify since it is not constant all over the panel. The addressing scheme induces a slight timing difference between top and bottom lines in term of emitting light when shutter glasses are open the same way.

As a consequence, the resulting light efficiency of such systems is quite poor. We could expect only a bit more than 10% of light efficiency if we consider what really reaches each eye.

For a plasma display panel (PDP), this synchronization is a bit simpler. The addressing scheme is a sequence of addressing and display periods called a sub-field. There are about 10 of these sub-fields per frame. A 3D addressing scheme will be a split of these 10 sub-fields into two groups of five sub-fields (five for the left and five for the right eye). The shutter opening time should occur during the first addressing period, just before a display period. The number of sub-fields is adapted to the stereoscopic context, but the global addressing is not modified compared with a 2D one.

Nevertheless, the cross-talk is also a concern for such technology since no PDP manufacturer is able to provide a product without phosphor lag. The response time of the blue phosphor is in general better than the red and green ones. Phosphor lag is in the range of several

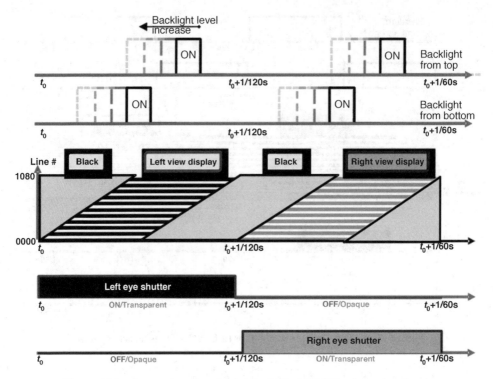

Figure 15.8 A second synchronization for LED–LCD TV using active glasses

milliseconds. It becomes difficult to find an optimized opening window of the shutter without integrating a part of the wrong eye information.

Figure 15.9 illustrates an example of the timing difference between 2D and 3D addressing schemes. The number of sub-fields per eye is half compared with the full 2D scheme. The phosphor lag is also noticeable since the light emitted does not go back to zero in between display periods. There is still light after the IR emitting time, which means potentially on the wrong eye when the shutter will be ON.

To ensure a correct synchronization between shutter glasses and display, a wireless emitter is used, mainly based on IR transmission. Although today there are many different protocols to communicate with glasses (Woods, 2012), there is a clear wish from TV manufacturers to propose universal glasses that will accept any protocol in the future. The end user should then be able to get their own active glasses. They will be compatible with any TV based on shutter glasses.

15.3.3 Based on Polarized Glasses

Stereoscopic 3DTV based on polarized glasses can deliver simultaneously the video dedicated to each eye on the same picture. Left and right information is interleaved on a line basis. Odd lines will render left video and even lines the right one. There is no need in this case to have a high refresh rate system; 100/120 Hz displays will be sufficient. The differentiation between left and right video is obtained thanks to a combination of two polarization filters. The first one is put in front of the panel and the other one is located on the glasses.

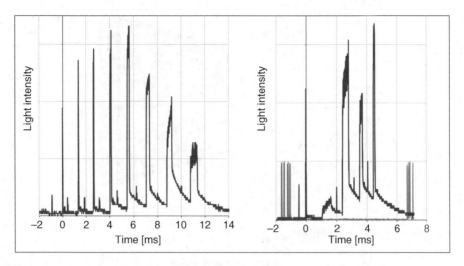

Figure 15.9 Plasma phosphor lag. The chronograms show a PDP in 2D mode on the left and 3D mode on the right. On the right side, the two serials of pulses either side of the main curve represent the IR signal emitted in 3D mode for electrical driving of the active glasses

The polarizer used for such displays could be a linear or circular polarizer. A linear polarizer will let light waves go through in only one direction. To obtain a circular polarizer, a quarter wave plate is placed in front of the linear filter. This circular polarizer will be preferred for 3DTV since it does not create any cross-talk when the user is moving their head. This is also what has been chosen in cinemas.

Polarized glasses solutions have several advantages over shutter glasses:

- low-cost solution, since glasses could be plastic;
- lightweight and easy to use solution, since they do not require any battery (passive solution);
- compatible solution, since the same circular polarization is used by every TV manufacturer (compatibility also with RealD cinema solution).

On the other hand, polarized systems have some drawbacks:

- They require the positioning of a polarization filter in front of the panel with a high level of precision. This induces an added cost for TV manufacturer and it also reduces the brightness of the TV even in 2D mode.
- The vertical resolution is reduced by half, interleaving left and right video information in a single HD frame.

As for active shutter glasses, polarized systems are not cross-talk free. There are two main sources for cross-talk on a polarized system. The first is due to an imperfect combination of filters: part of the left video signal reaches the right eye although filters should have blocked it. The second source of cross-talk is linked to a wrong observer positioning in front of the panel (Figure 15.10). Physically, the polarized filter is not exactly on top of the LCD cell;

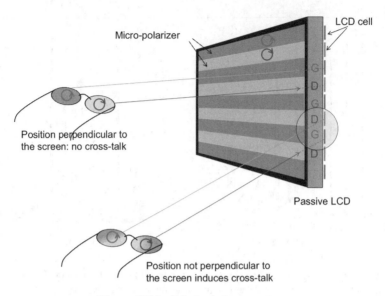

Figure 15.10   Polarized glasses system

there is a small distance in between. If the observer is not exactly placed perpendicular to the panel (but maybe few degrees off this perpendicular), they will mix a part of the left video with the right one. The positioning of the TV in the living room will then be quite important.

As for the active glasses system, a lot of light is lost to ensure a correct eye separation. Filter designs are tuned to minimize the cross-talk at the cost of a reduced remaining light. Again, the final light efficiency is generally below 20% after cascading the different filters.

15.3.4   Without Glasses

Movie spectators can stand wearing glasses in a theater during a 2 h film. However, even though manufacturers have made progress in terms of comfort, quality, and cost, some users are still reluctant to wear stereo glasses at home when watching their favorite programs, especially for people already wearing prescription glasses. As it seems to be one of the reasons of the slow deployment of 3D into the home, the display industry started some years ago to study technologies that would enable enjoying 3D content without the need of glasses. These devices are called 3D auto-stereoscopic displays (AS-3D). Some prototypes have already been demonstrated, but clearly the picture qualities, the width of the viewing spots, and the cost have not reached the expected level required by the consumer market yet.

Lens arrays and parallax barriers are the two main technologies used today to display 3D without glasses. The latter is only used for mobile applications and is addressed in Section 15.4.1. In this section, we are going to focus on large-area glasses-less technologies, namely lens arrays.

Large-area 3D glasses-less displays are made of arrays of microlenses glued on top of a classical LCD panel (Figure 15.11). The role of the lens is to focus the light provided by a set of pixels underneath towards specific directions. In the example below, for clarity of the schematic, we only depicted a three-view display, whereas most of the existing prototypes feature eight or nine views (some prototypes have been demonstrated with 28 or 46 views).

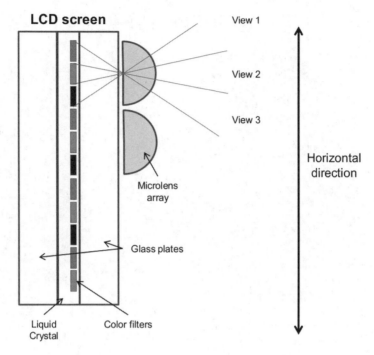

Figure 15.11   Micro-structure of a lens array panel

In this example, the width of the lens is equal to the width of three sub-pixels, the light of each sub-pixel being distributed in a small section of the horizontal space and scattered on a large angle in the vertical direction (like in 2D).

One of the first conclusions that appears here is that the resolution is divided by the number of views. Consequently, in order to build up an equivalent 3D HDTV set without glasses, the resolution of the LCD panel would need to be much higher than HD. Nevertheless, this simple mathematical analysis is not fully correct. Even though each eye sees a resolution divided by the number of views, the left and right pictures are different and are changing when the spectator is turning around the scene. The magic consequence of this is that the brain is able to reconstruct or interpolate part of the missing details of the scene, and the perceived resolution is higher than the mathematical one. To illustrate this, some companies have demonstrated Quad HD prototypes (3840×2160) with eight views, which looked like 3D HD.

In addition to the loss of resolution, the major drawback of this technology is that the user has to sit or stand in a so-called sweet spot, namely specific small areas where the 3D effect is visible. Beyond those spots, the images cannot be analyzed by the brain. As can be noticed in Figure 15.12, the two eyes must be positioned at a specific viewing distance and in a specific cone so that each eye can see one single view coming from all areas of the display. In the example below, the right eye watches the scene through viewing angle 1 and the left eye through viewing angle 2; if the head moves slightly towards the left, the 3D effect is still fine as the right eye is now seeing the second view and the left eye the third view. Going further to the left, the two eyes are now standing in two different zones, corresponding to a kind of anti-stereo zone, which is very uncomfortable to watch. Subsequently, the higher the number

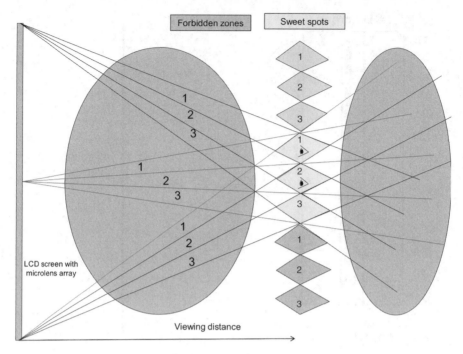

Figure 15.12 Sweet spots

of views, the bigger the sweet spot and the more comfortable it is to watch but the lower is the resolution. Designing glasses-less displays is then a matter of compromise between the picture quality in terms of resolution and the size of the viewing zones.

Another important aspect of AS-3D is that the pitch of the microlens array (the distance between two consecutive lenses) is not a multiple of the pixel pitch. In our example, the inter-lens distance is a bit smaller than the length of three sub-pixels. This condition is necessary to ensure a convergence of all rays of the panel towards the user, as shown in Figure 15.12. Pixels in the center of the screen have to emit light perpendicular to the screen, and pixels on the right or left edges have to emit light with a different angle corresponding to the position of the user. This necessitates a very precise lens array design and a very accurate positioning of it versus the pixel array, as is shown in Figure 15.13. This difficulty has consequences in terms of manufacturing cost, especially if we consider that, to spread the loss of resolution over the horizontal and vertical directions, some displays feature slightly slanted microlens arrays, which are even trickier to position.

In conclusion, AS-3D displays based on lens arrays are brighter than parallax barrier concepts (see Section 15.4.1) as they do not block part of the light coming out of the panel and they so are satisfactory for large panels. Today, they are mainly used for digital signage applications in shopping malls, museums, movie theater lobbies where the high price and narrow sweet spot are less critical than in the consumer business environment. It will still take some years before this kind of technology can reach the home. Manufacturers will first have to increase the panel resolution to enhance the number of views and consequently the size of the sweet spot. In the meantime, as 3D producers are not going to generate 3D content with multiple cameras before a much longer time, 3D stereo to multi-view algorithms will have to improve. A lot of companies are working on this topic

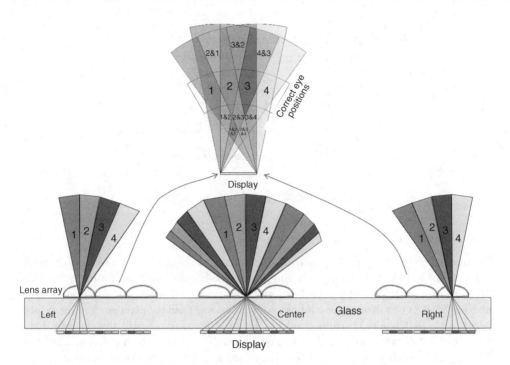

Figure 15.13   Edge and center cross-section of an AS-3D display

using, for instance, disparity map estimation between two views and view interpolation to compute the intermediary views. Nevertheless, as the two eyes can only see two adjacent intermediary views at the same time, the 3D effect is reduced versus classical stereoscopy. It is then necessary to extrapolate views beyond the incoming stereo base, which is then complicated since occlusion areas appear and automatic region filling are not performing well enough today to reach the desired picture quality. Ideally, specific content for AS-3D display should be generated in post-production and distributed to the home. This implies modifications of distribution standards to tag this particular content and make it backward compatible with stereo displays. Producers and distributors will only invest if displays are available at the right price for the consumer, which again will not appear for some years from now.

15.4 Mobile 3D Display Technologies

Nowadays, mobile displays are ubiquitous. Phones, tablets, laptops, and camcorders are all using mainly liquid-crystal small displays between 2 and 15 inches. Organic light-emitting diode (OLED) technology is starting to appear in the mobile phone sector and is likely going to be increasingly competitive for mobile applications in the coming years.

Unlike large-area 3DTV sets, it does not make much sense to use classical glasses for 3D mobile applications. Either auto-stereoscopic concepts or goggles using microdisplays can be used. As already mentioned in Section 15.3.4, the main AS-3D technology used today is based on parallax barriers, but some other approaches have also been studied or deployed.

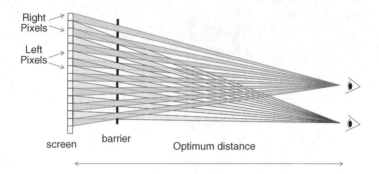

Figure 15.14   Parallax barrier principle

15.4.1   Based on Parallax Barriers

The basic principle of parallax barriers is depicted in Figure 15.14. Similar to microlens arrays, it aims at steering the light coming out of pixels containing the left (or right) video information to the left (or right) eye. For a given optimum distance of visualization, knowing the mean inter-ocular distance of a human being (∼65 mm) and the pixel pitch of the display, it is possible to define the barriers thanks to a simple geometry construction as indicated in the figure. The distance between the LCD panel and the barriers is then calculated and fixed for a specific viewing distance and the size of the apertures depends on the display pixel pitch and the position on the panel. Sweet spots are also inevitable as for microlens arrays (see Figure 15.12), but as the user is holding the device it is somewhat easier to find the correct position than in the case of a large-area 3DTV set. This is the reason why it is not mandatory to have a large number of views to be comfortable; two to four views are suffi-cient for most applications.

The main drawback of this solution is that the brightness is divided by the number of views, which is the reason why it is not used for large-area displays with eight views for instance. A TV set with eight times less light than a 2D display is not acceptable from a consumer market perspective. Another issue is that if the user is slightly moving away from the sweet spot, then he can see cross-talk and even anti-stereo content (the left information is sent to the right eye and vice versa), which is painful to watch. This is the reason why some techniques have been developed using the onboard webcam existing on some devices for video phone applications to detect and track the eye position of the user in such a way that the left/right video information is always directed towards the corresponding eye.

Nevertheless, barriers have some interesting advantages. They are uncomplicated to man-ufacture: a simple plastic film with black stripes is satisfactory. On more sophisticated mobile displays like tablets, barriers are electronically generated thanks to a second LCD placed on top of the first one. It is then possible to provide horizontal or vertical black stripes and feature a 3D mobile, which works in both portrait and landscape modes, or in full-reso-lution 2D by simply turning the bars off. The barrier design is easier to optimize since there is only one user to satisfy. In addition, it is adaptable to future technologies like OLEDs.

15.4.2   Based on Lighting Switch

In the previous sections, several concepts have been presented that consist of steering the left and right video contents to the corresponding eyes. An alternative, demonstrated by 3M,

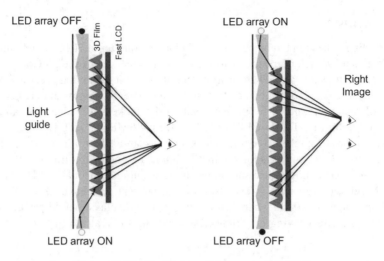

Figure 15.15 Backlight switching principle (3M)

proposes steering the illumination beams instead, as shown in Figure 15.15. It uses a specific 3D film and light guide and two arrays of backlight LEDs placed on the right and left edges of the panel; each LED array is alternatively switched on and off at a high frequency (120 or 240 Hz) and the optical system is designed in such a way that light coming out of one LED array illuminates a half portion of the horizontal space; when the eyes of the user are placed at the frontier between these two portions, 3D stereoscopy can be observed.

This approach, of course, works for a single user only, namely for mobile applications and can just display two views. It is by essence working on transmission displays (LCD) and is not applicable on self-emitting technologies like OLEDs.

15.5 Long-Term Perspectives

Glasses-less technologies are based on multiple views of a scene, but the sweet spot is pretty limited. To enlarge the viewing angle, an ideal approach would be to produce holograms in real time at a refresh frequency compatible with TV. Holograms are a mastered technology for still pictures, but the way to produce them is still complicated and cannot be easily transferred to the video world. Many research institutes are working on this and some prototypes exist for very professional or military applications, but huge and expensive machinery is necessary to achieve good performance. Furthermore, the amount of data to process is gigantic, around 100 000 times HDTV. A German company (SeeReal) has developed a concept that allows a reduction in the complexity of the system by using head trackers. Doing so, it is only necessary to generate sub-holograms in the direction of the user's eyes, thus reducing the data processing. But prototypes are still in their infancy.

Some other approaches have been demonstrated by companies like Holografika in Hungary or NICT (National Institute of Information and Communications Technology) in Japan (Agocs *et al.*, 2006; Häussler, 2008). The advantage is that it is no longer a multi-view display but a light-field generator. It allows the reproduction of each point of the 3D space using a bunch of small HD projectors and a specific screen. This technology is described in detail in Chapter 17.

15.6 Conclusion

3DTV is starting to be adopted by consumers. In the first quarter of 2012, more than 7 million sets had been sold, which corresponds to a growth of 245% versus the previous year.[3] Most of new HDTV flat panels are 3D compatible, but it does not mean that users are buying the necessary glasses. The lack of good content is still an issue, but significant efforts have been and are still being conducted by professionals. Film makers, broadcasters, and industry organizations (Blu-ray, DVB, CEA, ATSC, MPEG, SMPTE, HDMI, 3D@Home) are working to improve quality and provide standards and education so that 3D can be produced, transmitted, and displayed in the home in good conditions. In 5–10 years from now, AS-3D technologies should be mature enough to enable 3D visualization without glasses on large multi-view displays, but content will have to be adapted. And we can hope than in 20–30 years, full holographic displays would be affordable to envisage the ultimate viewing experience, reproducing the real world!

References

Agocs, T., Balogh, T., Forgacs, T. *et al.*, (2006) A large scale interactive holographic display, in *Virtual Reality Conference, 2006*, IEEE, p. 311.

Blondé, L., Sacré, J.-J., Doyen, D. *et al.* (2011) Diversity and coherence of 3D crosstalk measurements. *SID Symp. Dig. Tech.* **42**(1), 804–807.

Blondé, L., Doyen, D., Thébault, C. *et al.* (2012) Towards adapting current 3DTV for an improved 3D experience, in *Stereoscopic Displays and Applications XXIII* (eds A.J. Woods, N.S. Holliman, and G.E. Favalora), Proceedings of the SPIE, Vol. 8288, SPIE, Bellingham, WA, p. 82881Q.

Borel, T. (2012) Cinéma, la conquête de la 3D. *Pour la Science*, (416), 56–63.

Häussler, R., Schwerdtner, A., and Leister, N. (2008) Large holographic displays as an alternative to stereoscopic displays, in *Stereoscopic Displays and Applications XIX* (eds A.J. Woods, N.S. Holliman, and J.O. Merritt), Proceedings of the SPIE, Vol. 6803, SPIE, Bellingham, WA, p. 68030M.

Jorke, H., Simon, A., and Fritz, M. (2009) Advanced stereo projection using interference filters. *J. Soc. Inf. Displ.*, **17**, 407–410.

Mendiburu, B. (2009) *3D Movie Making: Stereoscopic Digital Cinema from Script to Screen*, Focal Press.

Woods, A. (2011) How are crosstalk and ghosting defined in the stereoscopic literature?, in *Stereoscopic Displays and Applications XXII* (eds A.J. Woods, N.S. Holliman, and N.A. Dodgson), Proceedings of the SPIE, Vol. 7863, SPIE, Bellingham, WA, p. 78630Z.

Woods, A. and Helliwell, J. (2012) Investigating the cross-compatibility of IR-controlled active shutter glasses, in *Stereoscopic Displays and Applications XXIII* (eds A.J. Woods, N.S. Holliman, and G.E. Favalora), Proceedings of the SPIE, Vol. 8288, SPIE, Bellingham, WA, p. 82881C.

---

[3] *Source:* Display Search.

16

Integral Imaging

Jun Arai

NHK (Japan Broadcasting Corporation), Japan

16.1 Introduction

Integral imaging is a three-dimensional (3D) photography technique that is based on integral photography (IP), by which information on 3D space is acquired and represented. The forms of representation include spatial images, 3D space slice images (tomography), the trajectory of a target object in 3D space, and a two-dimensional (2D) motion picture seen from an arbitrary viewpoint in 3D space. Here, a spatial image refers to an optically generated 3D real image or virtual image. When a spatial image is displayed, an observer perceives it in the same way as perceiving an actual object. In this chapter, we mainly describe the technology for displaying 3D space as a spatial image by integral imaging.

It may be that the first attempt to produce an image that could be viewed with stereoscopic vision was made sometime around 1600, using a technique for drawing two small images for viewing side by side with the left and right eyes (Norling, 1953). Currently, the binocular 3D video is increasingly being shown in movie theaters or via broadcasting services which uses the same basic principle as the stereogram to create the feeling of three-dimensionality. Binocular 3D video can only present one pair of left–right images, but spatial images differ in that respect.

Integral photography (IP) was invented as a 3D photographic technique in 1908 by the French physicist M.G. Lippmann (Lippmann, 1908). IP enables the viewer to observe a spatial image from an arbitrary viewpoint and experience the sense of three dimensions without requiring special glasses. The parallax panoramagram and holographic techniques also enable 3D viewing from arbitrary positions without special 3D glasses, but they were proposed after the invention of IP. Accordingly, IP was the first available technique to record 3D space for viewing the content as a spatial image.

Integral imaging, which is the basic principle of IP, requires high precision in the fabrication of optical devices and high-resolution video technology. There have been attempts to overcome the technical difficulties and construct imaging and display systems for spatial images. Before 1970, 3D still photography had been studied from the viewpoints of optical

Emerging Technologies for 3D Video: Creation, Coding, Transmission and Rendering, First Edition.
Frédéric Dufaux, Béatrice Pesquet-Popescu, and Marco Cagnazzo.
© 2013 John Wiley & Sons, Ltd. Published 2013 by John Wiley & Sons, Ltd.

and photographic technology. Since the 1970s, advances in cameras and spatial light modulation devices have brought the shift towards the study of real-time systems for imaging and displaying 3D space from the viewpoints of optics and electrical engineering.

Section 16.2 describes the technical development of 3D photography from the invention by Lippmann up to 1970. Section 16.3 describes a system that is capable of real-time capture and display of 3D space. Section 16.4 explains the quality of the spatial image generated, the geometrical relationship with actual space, resolution, and the viewing area. Section 16.5 describes the recent efforts to improve the quality of the spatial image and application of the technology to 3D measurement. Section 16.6 concludes and describes the future work.

16.2 Integral Photography

16.2.1 Principle

IP is a technique for 3D photography that was proposed by M.G. Lippmann at the 1908 conference of the Société française de Physique. The images are formed with a dry plate on which an array of many microlenses is fabricated (Figure 16.1). The result is that one minute image of the subject is formed on the dry plate for each lens in the array. We refer to those small images as elemental images. The dry plate is developed and then fixed. If the fixed dry plate is illuminated from the left in the figure, a spatial image at the subject position (position A or B in Figure 16.1) is formed from the reversibility property of light. That spatial image can be viewed from the right side of the dry plate.

However, the configuration illustrated in Figure 16.1 produces a pseudoscopic image in which the depth is reversed in the observed image. Here, we explain the pseudoscopic image with reference to Figure 16.1. When capturing the image, the subjects A and B are seen from the left side with respect to the lens array, so subject A is closer to the viewpoint than B is. When viewing the spatial image, on the other hand, it is viewed from the right side of the lens array, so subject B is now in front of A in the viewed image. An image that exhibits this reversing of the depth relationship of objects in capturing and viewing is called a pseudoscopic image. A method of forming a correct positive spatial image by reversing both the pseudoscopic and negative images is described in Lippmann (1908). The reversing operation is achieved by placing a dry plate with lens array immediately in front of the acquired

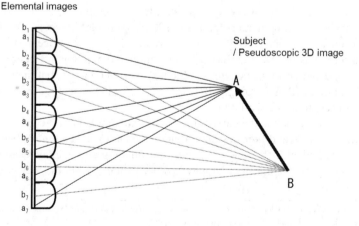

Figure 16.1 IP imaging and pseudoscopic negative 3D image

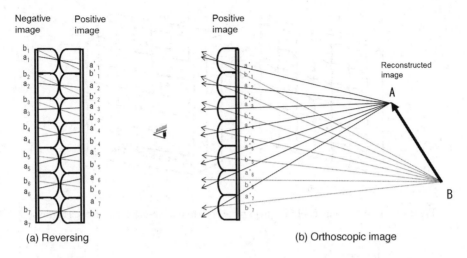

<table>
</table>

(a) Reversing	(b) Orthoscopic image

Figure 16.2 Orthoscopic generation and display of a positive 3D image by IP

negative image to acquire a new image (Figure 16.2a). If the illumination is from the right side of the new positive image, which is to say the right side of the dry plate in Figure 16.2b, the light rays from the elemental images come from the same directions as the light rays from subjects A and B. Viewing the lens array and the elemental image array of the positive image from the left side of the dry plate in Figure 16.2b, which is to say the same direction from which the subject is seen when capturing, the virtual spatial image can be seen in the same position as the subject. As we can see from Figure 16.2b, the result is an orthoscopic 3D image.

16.2.2   Integral Photography with a Concave Lens Array

A method for obtaining an orthoscopic 3D image that does not require development in the reversal process was proposed by H.E. Ives in 1939 (Ives, 1939). That method used a concave lens array, as shown in Figure 16.3. The subject is placed in front of the concave lens array. For the concave lens array, structures such as the one illustrated in Figure 16.4 have

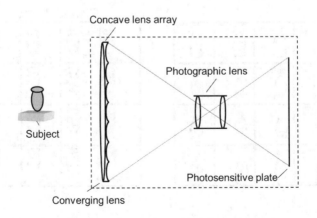

Figure 16.3   IP imaging using the concave lens array

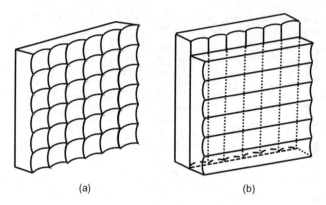

(a)   (b)

Figure 16.4   Arrangement of (a) single-plate and (b) double-plate concave lens arrays

been proposed. As illustrated in Figure 16.4a, concavities that serve as concave lenses are formed in celluloid, glass, or other such transparent materials. Another configuration in which two cylindrical lens arrays that have a concave side are used, as shown in Figure 16.4b, has been proposed for ease of fabrication.

The elemental images of the subject formed by the concave lens array are projected onto a photosensitive plate by a photographic lens. A converging lens is placed adjacent to the concave lens array. The focal distance of the converging lens is set to be equal to the distance from the concave lens array to the principal point of the photographic lens. When the lens array is configured with the concave lens in the arrangement shown in Figure 16.4, the elemental images are virtual images. The converging lens serves to efficiently direct the light rays contributed by the elemental image array of those virtual images to the photograph lens. Because the IP imaging system proposed by Lippmann formed the lens array from a convex lens (Figure 16.1), reproducing the spatial image as an orthoscopic image required the kind of reversal shown in Figure 16.2a. When the subject is placed in front of the convex lens array, individual elemental images formed by the plate are inverted, as shown in Figure 16.5a. When the subject is placed in front of the concave lens array, on the other hand, the resulting individual elemental images are not inverted, as shown in Figure 16.5b.

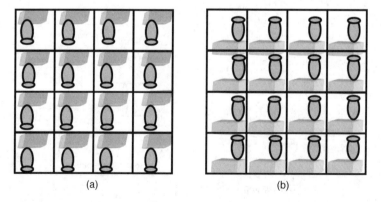

(a)   (b)

Figure 16.5   Elemental image arrays produced by (a) a convex lens array and (b) a concave lens array

In other words, the configuration of the imaging system shown in Figure 16.3 has already performed the inversion of the elemental images. The orthoscopic spatial image can be seen by viewing the elemental image array obtained through the convex lens plate as shown in Figure 16.2b. As a practical configuration, a 10-inch square lens plate that comprises two celluloid plates that have 25 600 concave lenses has been introduced (Ives, 1939).

16.2.3 Holocoder Hologram

According to Valyus, IP experiments using a convex lens array IP were conducted by S.P. Ivanov and L.V. Akimakina in 1948 (Valyus, 1966). The diameter of the elemental lens was 0.3 mm and the focal distance was 0.5 mm. The number of elemental lenses is reported to be about 2 million. Although the significance of using a lens array for the first time is important, according to Valyus, the 3D image reproduced by displaying the negative image was pseudoscopic, so the result was not a complete lens array IP.

Experiments conducted by R.V. Pole, a hologram researcher, used the convex lens array proposed by Lippmann to obtain 3D information on the subject (Pole, 1967). Those experiments used an IP method in which white light was used to create a hologram. First, the subject is imaged as an elemental image array with exactly the same configuration as shown in Figure 16.1. That elemental image array is what Pole described as a "holocoder." The method for generating a hologram from the holocoder is illustrated in Figure 16.6. The imaged holocoder is first transferred to a positive image and then illuminated with a coherent light source. At that time, a diffuser screen is placed behind the holocoder so that the light rays from the individual elemental images (the light rays from P_{1j} in Figure 16.6, for example) are spread over the entire aperture of the convex lens. In that case, the 3D real image is reproduced, but a plate for producing the hologram is placed between the reconstructed image and the holocoder and the hologram is produced by illumination with a reference beam from the left side or right side of the plate. In either case, however, it is necessary to create the hologram with coherent light. In Pole (1967) there is a photograph of a 3D image of two miniature cars. It shows,

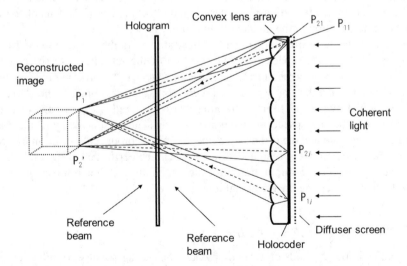

Figure 16.6 Hologram produced with a holocoder. Reprinted with permission from Pole, R.V., *Applied Physics Letters*, Vol. 10, pp. 20–22, (1967), © 2013 American Institute of Physics

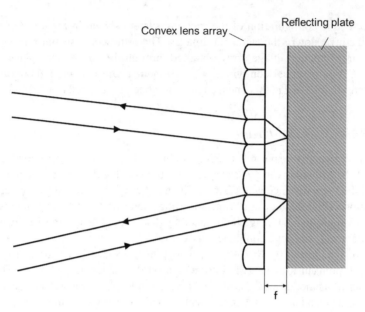

Figure 16.7 Configuration of the retrodirective screen. Reproduced with permission from Burckhardt *et al.* (1968), Formation and inversion of pseudoscopic images, Applied Optics 7 (3), 627–631, © 1968 The Optical Society

albeit indirectly, that a 3D image can be generated by IP using a convex lens array, therefore, it is an important experiment in the technical evolution of IP.

16.2.4 IP using a Retrodirective Screen

Burckhardt, Collier, and Doherty (1968) proposed a solution to the pseudoscopic image problem that uses a simple structure involving a retrodirective screen and half-mirror. The retrodirective screen is a reflecting plate placed on the focal plane of the lens plate to reflect light in the same direction as the incident light (Figure 16.7).

The effect of the retrodirective screen is illustrated in Figure 16.8. Viewing the three points A, B, and C in real space from the position of the retrodirective screen, A is the nearest. The retrodirective screen forms an image in which A, B, and C are in the same locations as in actual space, but, as we see in Figure 16.8, the locations are the mirror-image reflections formed by the half-mirror. In other words, a pseudoscopic image of the actual space is formed. When that pseudoscopic image is imaged and displayed by IP, the elemental images are not reversed and the orthoscopic image can be reproduced. Furthermore, this conversion by the retrodirective screen and half-mirror can be performed at the time the spatial image is viewed. With the configuration illustrated in Figure 16.9, the orthoscopic image can be viewed without reversing the elemental images.

16.2.5 Avoiding Pseudoscopic Images

Section 16.2 describes methods of avoiding pseudoscopic images that were proposed between Lippmann's invention and the 1970s. Those methods can be classified into two types: methods that reverse individual elemental images with point symmetry and methods that first

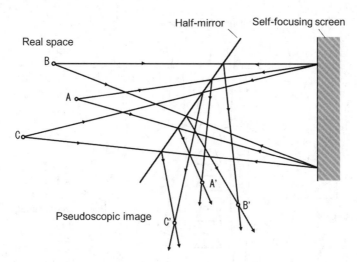

Figure 16.8 Formation of pseudoscopic image of actual space. Reproduced with permission from Burckhardt *et al.* (1968), Formation and inversion of pseudoscopic images, *Applied Optics* 7 (3), 627–631, © 1968 The Optical Society

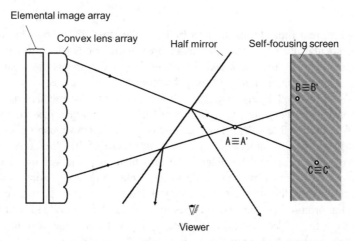

Figure 16.9 Pseudoscopic image reversal by retrodirective screen and half-mirror. Reproduced with permission from Burckhardt *et al.* (1968), Formation and inversion of pseudoscopic images, *Applied Optics* 7 (3), 627–631, © 1968 The Optical Society

generate a pseudoscopic image and then apply IP to that pseudoscopic image to again reverse the depth relationships and obtain an orthoscopic image. Those methods have continued to be important in the technological development of integral imaging since the 1970s.

16.3 Real-Time System

16.3.1 Orthoscopic Conversion Optics

Since the 1970s, improvements in the resolution of cameras and spatial optical modulators have stimulated development of systems that display spatial images from information acquired

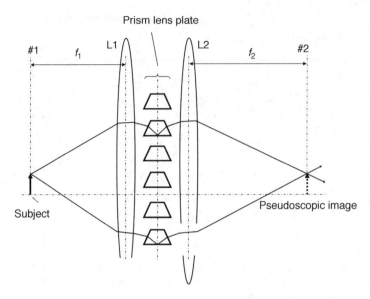

Figure 16.10   Orthoscopic conversion optics

from 3D space in real time. From the late 1970s to the late 1980s, Higuchi and Hamasaki reported experimental use of orthoscopic image conversion optics (ICO) to capture and display 3D images (Higuchi and Hamasaki, 1978). In those experiments, the conversion optics first generated a pseudoscopic image of the subject in the same way as when a retrodirective screen is used (see Section 16.2.4), and then IP was applied to the pseudoscopic image. The orthoscopic conversion optics are shown in Figure 16.10. In the figure, f_1 is the focal length of lens L1 and f_2 is the focal length of lens L2. The orthoscopic conversion optics generate the pseudoscopic image of the subject near plane #2 in Figure 16.10. Therefore, pseudoscopic elemental images are generated by placing the lens plate at plane #2, and the orthoscopic spatial image can be obtained by viewing the array of elemental images obtained through the lens plate. With the retrodirective screen, the subject and the pseudoscopic image are the same size, but with the orthoscopic conversion optics the size can be controlled by adjusting f_1 and f_2. In this method, the subject must, in principle, be in the focal plane in front of lens L1. If that condition is not satisfied, then the resolution of the generated pseudoscopic image is reduced, which affects the resolution of the 3D image that can be viewed.

If the prism lens array of the orthoscopic conversion optics is arranged in two dimensions, a pseudoscopic image that has vertical and horizontal parallax can be generated. In Higuchi and Hamasaki (1978), orthoscopic conversion optics are configured with the prism array arranged only in the horizontal direction, and a television camera and monitor that have resolution equivalent to 350 TV lines in the horizontal direction are used to display the reproduced image with parallax in the horizontal direction. In generating the reproduced image, a horizontal array of 35 cylindrical lens arranged at a pitch of 3 mm is used as the lens array.

In the period from the late 1980s to early 1990s, N. Davies and M. McCormick and coworkers proposed a method in which a newly developed autocollimating transmission screen is used for orthoscopic spatial image acquisition and display without reversal of the elemental images (Davies *et al.*, 1988; McCormick, 1995). The transparent autocollimating screen described in Davies *et al.* (1988), which appeared in 1988, comprises two lens plates and a

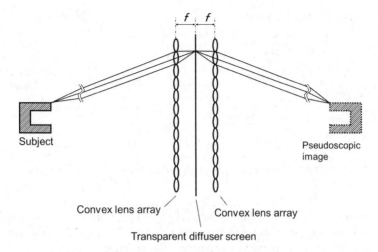

Figure 16.11 Transparent autocollimating screen

transparent diffusion screen, as shown in Figure 16.11. The pseudoscopic image is generated by the bending of light in the direction that is mirror symmetrical to the incident light. In Figure 16.11, f is the focal length of the lens that constitutes the lens array. That was followed by a proposal for an improvement in which two lens plates are placed in front of the transparent autocollimating screen to allow control of the pseudoscopic image magnification factor (McCormick, 1995). Another lens array is used in the imaging system to generate the elemental image array of the pseudoscopic image. Because the imaging results in a pseudoscopic image, in which the depth relationships of objects are reversed, a reconstructed image with correct depth relationships is reproduced when the obtained elemental image array is input to a projection or direct-viewing display system through the lens array.

McCormick (1995) presented a conceptual diagram for a 3D video capturing and display system that uses optics to generate pseudoscopic images, as shown in Figure 16.12. Furthermore, an attempt to compress integral 3D TV video data using the correlation between

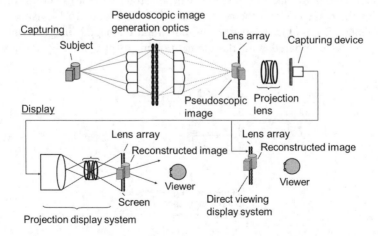

Figure 16.12 Imaging and display system that uses optics for generating pseudoscopic images

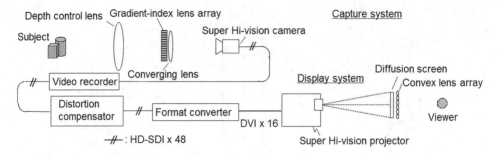

Figure 16.13 Configuration of integral 3D TV system using Super Hi-vision video system

elemental images is reported in McCormick (1995). There are reports of subjective and objective evaluation experiments performed using integral 3D TV video that has parallax in only one dimension showing that 25 : 1 compression is possible.

16.3.2   Applications of the Ultra-High-Resolution Video System

A large amount of data is required in generating high-quality spatial images by integral imaging. A full-resolution Super Hi-vision (full resolution SHV) video system (Nagoya *et al.*, 2008; Yamashita *et al.*, 2009) is currently capable of displaying motion pictures with the most pixels. The configuration of an integral 3D TV system using a full-resolution SHV camera and projector is shown in Figure 16.13 (Arai *et al.*, 2010a).

The imaging system comprises a full-resolution SHV camera, gradient-index lens array, converging lens, and a depth control lens. In imaging, first the depth control lens is used to create a real image of the subject. By doing so, the depth position of the displayed spatial image can be adjusted. For example, if the real image of the subject is generated on the camera side of the lens array, the displayed reconstructed image will be generated in front of the lens array. The camera acquires the image from the elemental image array for that real image. To avoid the problem of the depth-reversed pseudoscopic image in the imaging system, gradient-index lenses are used in the lens array. A gradient-index lens is formed of a material whose refractive index decreases radially by the square of the distance from the center. As shown in Figure 16.14, using a gradient-index lens makes it possible to reverse the individual elemental images symmetry with respect to a point compared with the elemental image produced by the convex lens

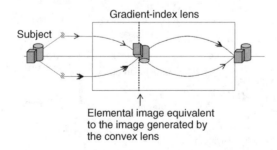

Figure 16.14 Elemental image reversal using gradient-index lens

Table 16.1 Specifications for image capturing equipment

Television camera	Pixel count	7680(H) × 4320(V) × RGB
	Frame frequency	Focal length: 61.59 mm
Lens array	Lens type	Gradient-index lens
	Number of lenses	400(H) × 250(V)
	Lens pitch	1.14 mm (horizontal direction)
	Focal length	−2.65 mm
	Arrangement	Delta array

Table 16.2 Specifications for image display equipment

Projection device	Pixel count	7680(H) × 4320(V) × RGB
	Time-division multiplexing	Complementary field-offset method
	Projection lens	Focal length: approx. 60 m
	Projection size	Diagonal angle: approx. 26 inches
Lens array	Lens type	Convex lens
	Number of lenses	400(H) × 250(V)
	Lens pitch	1.44 mm (horizontal direction)
	Focal length	2.745 mm
	Arrangement	Delta array

(Okano *et al.*, 1999). A converging lens is placed between the lens plate and the camera for efficient delivery of the light from the lens plate to the camera. The specifications of the imaging system are listed in Table 16.1. There are 7680 (H) × 4320 (V) effective pixels for each color and the number of lenses in the lens array is 400 (H) × 250 (V).

The display system creates the reconstructed image by using a projector to project the elemental image array acquired by the camera onto the diffuser screen and placing a lens array of convex lens in front of the screen. If there is distortion in the video images projected onto the diffusion screen, there will be degradation in the reproduced image, so this system corrects the distortion electrically to avert degradation of the reconstructed image (Kawakita *et al.*, 2010). The lens array is positioned so that its distance from the diffusion screen is roughly the focal length of the convex lens. The specifications of the display system are listed in Table 16.2. In the same way as for the imaging system, there are 7680 (H) × 4320 (V) effective pixels for each color and lens plate has 400 (H) × 250 (V) lens.

We conducted capture and display experiments using an experimental system on ordinary subjects. The image reproduced by the display system is shown in Figure 16.15 and an

Figure 16.15 Example of a reconstructed image. From Arai *et al.* (2010b), reproduced with permission from IET (see Plate 19 for the colored figure)

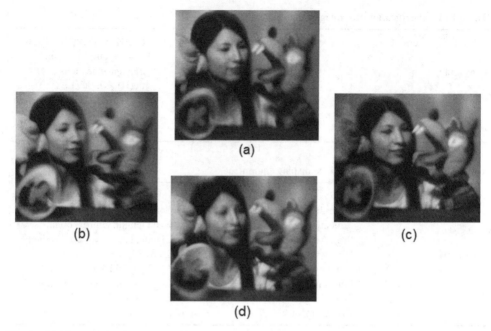

Figure 16.16 Partial enlargement of the reconstructed image: (a) upper viewpoint; (b) left viewpoint; (c) right viewpoint; (d) lower viewpoint. From Arai *et al.* (2010b), reproduced with permission from IET (see Plate 20 for the colored figure)

enlarged part of the image as seen from various viewpoints is shown in Figure 16.16, confirming that the view of the spatial image changes smoothly as the viewing position changes. The effect of placement of the diffuser plate on the position of the displayed spatial image is shown in Figure 16.17, where (a) shows the placement on the lens plate ("human face") and (b) shows the placement in front of the lens plate ("NHK"). Only the spatial image that is generated at the same depth position as the diffuser plate is clearly projected onto the diffuser plate, so we can see that the spatial image is created in space. A measured viewing angle of 24° was achieved.

Figure 16.17 Reconstructed image projected onto the diffuser plate. (a) Diffuser plate placed on the lens array. (b) Diffuser plate placed in front of the lens array. From Arai *et al.* (2010b), reproduced with permission from IET (see Plate 21 for the colored figure)

16.4 Properties of the Reconstructed Image

16.4.1 Geometrical Relationship of Subject and Spatial Image

It is useful to know the geometrical relationship of the subject in 3D space and the generated spatial image when constructing an integral imaging system. The geometrical relationships for integral imaging are illustrated in Figure 16.18 (Arai *et al.*, 2004).

In Figure 16.18, to represent the positions of the subject and the reproduced image, the center of the lens plate is the origin and the $z_p - x_p$ coordinates (for imaging) and the $z_d - x_d$ coordinates (for display) are set. The points z_p and z_d are orthogonal to the lens plate, and the positive direction is to the right in the figure. The points x_p and x_d are in the planar direction relative to the lens plate and upward is the positive direction. Taking a point light source as the subject, let the position of the subject be (z_{p1}, x_{p1}). In this case, of the light emitted from the point light source, the light that passes through the focal point of each elemental lens is given by

$$x_p = \frac{mp_{Lp} - x_{p1}}{-z_{p1}} z_p + mp_{Lp} \tag{16.1}$$

where m is a number of the elemental lens counting from the center of the lens plate and p_{Lp} is the pitch of the elemental lens in the imaging lens plate. Accordingly, the elemental image k_{mp1} of the mth elemental lens relative to the point light source (z_{p1}, x_{p1}) can be expressed by

$$k_{mp1} = \frac{mp_{Lp} - x_{p1}}{-z_{p1}} g_p \tag{16.2}$$

Here, the distance from the lens plate to the imaging plane is denoted by g_p. Next, the display plane for the elemental image is placed at the position shown in Figure 16.18 so that the reproduced image can be viewed from the viewing direction. In this case, the light from the elemental image of the mth elemental lens that passes through the focal point of the mth elemental lens is given by

$$x_d = \frac{-k_{md1}}{-g_d} z_d + mp_{Ld} \tag{16.3}$$

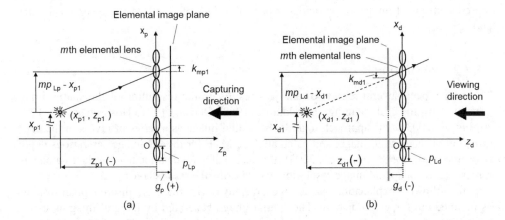

Figure 16.18 Geometrical relationship for integral imaging: (a) image capture; (b) display

where k_{md1} is the elemental image of the mth elemental lens and the distance from the lens plate to the imaging plane is $-g_d$. If we solve Equation 16.3 for x_{d1} and z_{d1}, taking elemental lens m and $(m+1)$ into account, we can obtain the position of the reproduced image. We then know that the conditions for the elemental image k_{md1} of the mth elemental lens for which the position of the reproduced image is $x_{d1} = x_{p1}$ and $z_{d1} = z_{p1}$ are given by

$$k_{md1} = -k_{mp1} = \frac{mp_{Ld} - x_{p1}}{-z_{p1}} g_d \qquad (16.4)$$

In the equation, $p_{Lp} = p_{Ld}$ and $g_p = -g_d$. Equation 16.4 shows that the spatial image (orthoscopic spatial image) will be generated at the same position as the subject if the elemental images obtained from the convex lens are converted to point symmetry relative to the centers of the elemental images. That fact can also be applied to the case in which the position of the subject is either to the left or right of the lens plate.

Next, the geometrical relationship of the subject and the reproduced spatial image is shown for the following general cases:

$$p_{Ld} = ap_{Lp} \qquad (16.5)$$

$$g_d = -bg_p \qquad (16.6)$$

$$k_{md1} = -dk_{mp1} \qquad (16.7)$$

In cases such as these, from Equations 16.1 and 16.2, the geometrical relationship of the subject and the reproduced spatial image can be expressed by the following equations:

$$x_d = \frac{adg_p}{z_p(d-a) + dg_p} x_p \qquad (16.8)$$

$$z_d = \frac{abg_p}{z_p(d-a) + dg_p} z_p \qquad (16.9)$$

Accordingly, a, b, and d in Equations 16.5, 16.6 and 16.7 determine the position of the reproduced spatial image. In particular, when the values of a and d are the same, $x_d = ax_p$ and $z_d = bz_p$.

16.4.2 Resolution

There have been several reports of analysis results on the resolution of the spatial image generated by integral imaging (Burckhardt, 1968; Okoshi, 1971; Hoshino et al., 1998; Arai et al., 2003). One approach regards the spatial image as a series of layers in the depth direction of the spatial image, and calculates the MTF for the 2D images at various depths (Hoshino et al., 1998; Arai et al., 2003). Because that analytical method produces a concise representation of spatial image resolution, we describe the results here.

In the following explanation, we assume a display device that has a discrete pixel structure as the medium for video display. The spatial image generated by integral imaging can be regarded as a series of planar images stacked in the depth direction. Therefore, we first consider elemental images projected to a distance z from the lens array by the elemental lens as

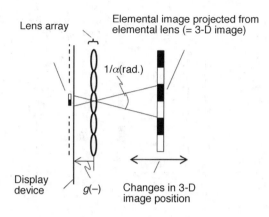

Lens array

Elemental image projected from elemental lens (= 3-D image)

$1/\alpha$(rad.)

Display device

$g(-)$

Changes in 3-D image position

Figure 16.19 Projection spatial frequency

shown in Figure 16.19. For the projected image, we denote the maximum number of lines per radian as α, which we refer to as the projection spatial frequency. The spatial frequency α is affected by aberrations from displacement from the focal point of the elemental lens as well as by the pixel pitch and the diffraction limit of the elemental lens. If the pixel pitch of the display device is p, the spatial frequency of the elemental image projected by the elemental lens α_p is given by

$$\alpha_p = \frac{|g|}{2p} \tag{16.10}$$

where g is the distance from the lens plate to the display device. We denote the diffraction-limited spatial frequency as α_d and the spatial frequency taking the aberration of the elemental lens into account as α_e. The limiting spatial frequency of the elemental image projected by the elemental lens, which is to say the spatial frequency α, is then expressed by

$$\alpha = \min[\alpha_p, \alpha_d, \alpha_e] \tag{16.11}$$

The spatial frequency α determined in this way is used to integrate the projected elemental images to generate the spatial image.

Next, when the generated spatial image is viewed at distance L from the lens plate as shown in Figure 16.20, we denote the maximum number of lines per radian as β, which is referred to as the viewing spatial frequency. That value is also affected by the depth position in the spatial image in the same way as the projection spatial frequency. The relationship is expressed by the following equation:

$$\beta = \alpha \frac{L - z}{|z|} \tag{16.12}$$

where L is the viewing distance (the distance from the lens array to the viewer) and z is the spatial image distance (the distance from the lens array to the spatial image). The positive direction is to the right of the lens array and negative values lie to the left of the lens array.

Furthermore, when displaying a spatial image by integral imaging, the pitch of the elemental lens in the lens array must be considered. As shown in Figure 16.21, the display device for the elemental image is placed at a distance from the lens array that is about the

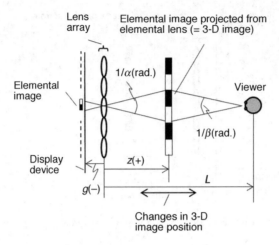

Figure 16.20 Viewing spatial frequency

same as the focal length of the elemental lens. In that case, parallel light rays are emitted from each elemental lens, as shown in Figure 16.22. Those groups of light rays are then integrated to form the spatial image. The result is that the spatial image is sampled at the pitch of the elemental lens when seen from the viewer's position. Because of that, considering the geometrical optics, the maximum spatial frequency of the spatial image is limited by the Nyquist frequency, which is determined by the pitch of the elemental lens P_L, as

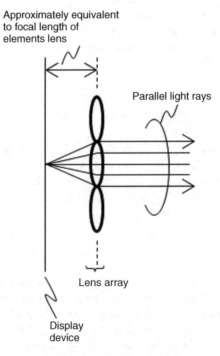

Figure 16.21 Output of parallel light rays

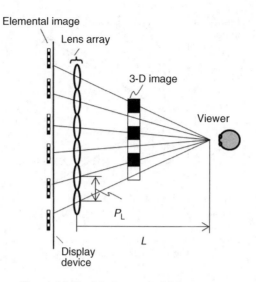

Figure 16.22 Maximum spatial frequency

expressed by the following equation:

$$\beta_n = \frac{L}{2P_L} \qquad (16.13)$$

where the spatial frequency β_n is referred to as the maximum spatial frequency.

From the discussion above, the upper limiting spatial frequency of a spatial image generated at an arbitrary depth is the lower of the maximum spatial frequency β_n, which is determined only by the elemental lens pitch and the viewing spatial frequency β, which is calculated with Equation 16.12. That relationship is expressed by

$$\gamma = \min[\beta_n, \beta] \qquad (16.14)$$

The spatial frequency γ in Equation 16.14 is referred to as the upper limit of the spatial frequency.

16.4.3 Viewing Area

For integral imaging, spatial images can be viewed according to the position of the viewer, which may vary in either the up–down or left–right directions. However, the range in which the viewer can move (the viewing area) is limited to the area in which the light from any one elemental image is emitted by the single elemental lens that corresponds to that elemental image. A cross-section of the viewing area formed by the elemental image and the elemental lens is shown in Figure 16.23. In that figure, angle Ω indicates the width of the viewing area and is referred to here as the viewing angle. The value of Ω is given by

$$\Omega \cong 2\tan^{-1}\left(\frac{P_L}{2|g|}\right) \qquad (16.15)$$

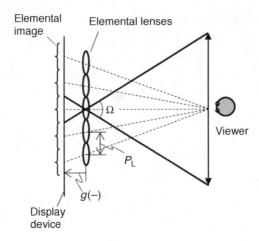

Figure 16.23   Viewing area.

Accordingly, if the distance from the lens array to the display device $(-g)$ is shorter, the viewing angle Ω can be larger. On the other hand, the relationship shown by Equation 16.10 means that the spatial frequency α_p becomes smaller. From Equations 16.11 and 16.14, we know that the resolution of the spatial image decreases when the upper limiting spatial frequency γ is limited by the spatial frequency α_p. To increase the viewing angle Ω while maintaining the spatial frequency γ, it is necessary to reduce the pixel pitch P_L as the distance $(-g)$ is reduced. Also, with the same elemental image area, the number of pixels that compose a single elemental image increases.

16.5   Research and Development Trends

16.5.1   Acquiring and Displaying Spatial Information

Usually, integral imaging uses an imaging device and lens array to acquire spatial information. When obtaining information on a large space, the size of the lens array becomes a constraint. One method proposed to address that problem involves imaging an elemental image array while moving a capturing device and lens array set in a certain direction, thus acquiring the same spatial information as can be acquired with a large lens array (Jang and Javidi, 2002) (Figure 16.24). However, it is noted that this method is not suitable for capturing moving subjects.

   As a means of increasing the pixel count for the elemental image array to be displayed, a configuration of nine projectors (three horizontal by three vertical) and a lens plate has been proposed (Liao *et al.*, 2005) (Figure 16.25). The total pixel count for the projected video is 2868 horizontal by 2150 vertical; the number of elemental lenses is 240 horizontal by 197 vertical. When the video is projected by multiple projectors that are offset from the desired position, the quality of the spatial image generated is reduced. The proposed method, therefore, measures the offset of the video projected onto the lens array with a camera. The measured displacement values are then used to calculate correction factors that are applied to the projected video to achieve projection at the desired location.

   One proposed method of expanding the viewing area is to use a curved screen and lens array (Kim *et al.*, 2005) (Figure 16.26). In that method, an elemental image array that has

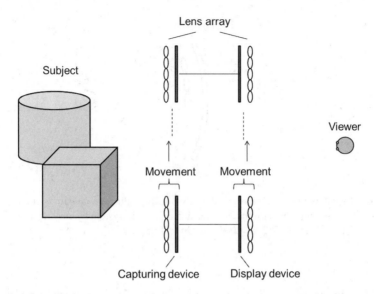

Figure 16.24 Concept of synthetic aperture integral imaging. Reproduced with permission from Jang, J.-S. and Javidi, B. (2002), "Three-dimensional synthetic aperture integral imaging," *Optics Letters*, 27 (13), 1144–1146, © 2002 The Optical Society

been corrected for distortion is projected onto a curved screen so that there is no positional displacement of the elemental images and elemental lens. A horizontal viewing area of 66° has been reported.

16.5.2   Elemental Image Generation from 3D Object Information

For integral imaging, a method of generating an elemental image array from 3D object information has been proposed (Igarashi *et al.*, 1978) (Figure 16.27). Given the shape information of a 3D object, the process of using a lens array and capture device to generate an elemental

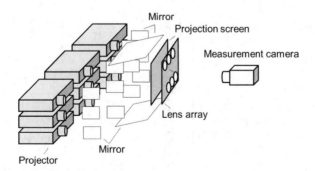

Figure 16.25 Display with multiple projectors. Reproduced with permission from Liao *et al.* (2005), "Scalable high-resolution integral videography autostereoscopic display with a seamless multi-projection system," *Applied Optics*, 44, 305–315, © 2005 The Optical Society

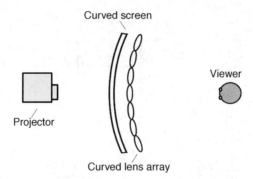

Figure 16.26   Display with curved screen and lens array. Reproduced with permission from Kim *et al.* (2005), "Wide-viewing-angle integral three-dimensional imaging system by curving a screen and a lens array," *Applied Optics*, 44, 546–552, © 2005 The Optical Society

image array is simulated with a computer. In relation to that approach, a means of generating elemental image arrays at high speed has been proposed. (Iwadate and Katayama, 2011). In that proposed method (Figure 16.28) a 3D object is imaged from many directions by a virtual camera and the resulting image is used to compose the elemental image array. Because 3D objects can be generated from information obtained from computer-generated graphics or multiple cameras, this method is expected to diversify the subjects of capturing.

16.5.3   Three-Dimensional Measurement

A method to generate cross-sectional images of a subject by using a computer has been proposed. It reverses the process of starting with the subject and generating an elemental image via the elemental lens. One reported application of that technique is to extract subject movement in 3D space by calculating the correlations between multiple cross-section images.

The elemental image array contains multiple parallax information, so a depth map of the subject can be generated by applying the multi-baseline stereo matching method. (Park *et al.*,

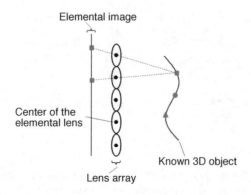

Figure 16.27   Elemental image generation from 3D object information

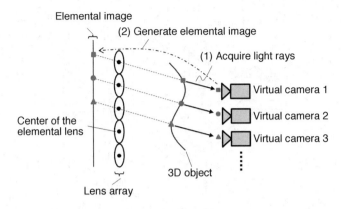

Figure 16.28 Oblique projection of elemental images

2004) The elemental image array or parallax images generated from the elemental image array can serve as parallax information for this purpose (Figure 16.29).

16.5.4 Hologram Conversion

Ordinarily, hologram generation requires a coherent light source, such as a laser beam. Pole has proposed a method to holograms from an IP elemental image array acquired using natural light as shown in Figure 16.6 (Pole, 1967). That method does not require coherent light for the imaging, but a coherent light is required when the hologram is produced. To eliminate the need for coherent light in producing the hologram, a method in which the light propagation shown in Figure 16.30 is simulated on a computer has been proposed (Mishina *et al.*, 2006).

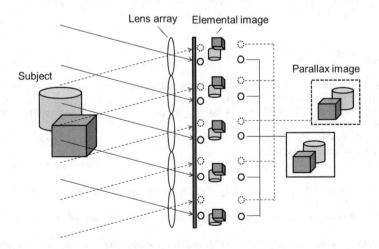

Figure 16.29 Elemental image array and parallax images

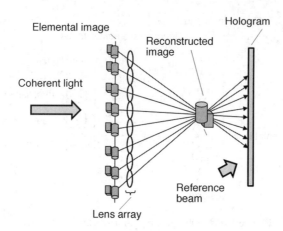

Figure 16.30   Hologram generation using elemental images

16.6   Conclusion

This chapter explains integral imaging from the viewpoint of technology for representing 3D space as a spatial image. Technical development of IP for improvement of the quality of the spatial image has continued from Lippmann's invention up to the present. Because the subject of integral imaging is 3D space, the amount of data to be handled by the system is much greater than 2D video. At this time, data limitations prevent the display of spatial images of sufficient quality.

After the year 2000, on the other hand, advances in camera and spatial optical modulation device technology and increasing computer processing capabilities have not only improved the display of spatial images, but also expanded the range of applications for integral imaging to include conversion to a hologram, 3D measurement, the tracking of a specified subject in 3D space, and so on.

In the future, it is expected that the synergy between the development of technology to improve the quality of the spatial image and the expansion of new application fields would drive Integral Imaging a more sophisticated technology.

References

Arai, J., Hoshino, H., Okui, M., and Okano, F. (2003) Effects of focusing on the resolution characteristics of integral photography. *J. Opt. Soc. Am. A*, **20**, 996–1004.

Arai, J., Okui, M., Kobayashi, M., and Okano, F. (2004) Geometrical effects of positional errors in integral photography. *J. Opt. Soc. Am. A*, **21**, 951–958.

Arai, J., Okano, F., Kawakita, M. *et al.* (2010a) Integral three-dimensional television using a 33-megapixel imaging system. *J. Disp. Technol.*, **6** (10), 422–430.

Arai, J., Kawakita, M., Sasaki, H. *et al.* (2010b) Integral 3D television using a full resolution Super Hi-vision. *IET J.*, **2**, 12–18.

Burckhardt, C.B., Collier, R.J., and Doherty, E.T. (1968) Formation and inversion of pseudoscopic images. *Appl. Opt.*, **7** (3), 627–631.

Burckhardt, C.B. (1968) Optimum parameters and resolution limitation of integral photography. *J. Opt. Soc. Am.*, **7**, 627–631.

Davies, N., McCormick, M., and Yang, L. (1988) Three-dimensional imaging systems; a new development. *Appl. Opt.*, **27** (21), 4520–4528.

Higuchi, H. and Hamasaki, J. (1978) Real-time tranmission of 3D images formed by parallax panoramagrams. *Appl. Opt.*, **17** (24), 3895–3902.

Hoshino, H., Okano, F., Isono, H., and Yuyama, I. (1998) Analysis of resolution limitation of integral photography. *J. Opt. Soc. Am. A.*, **15** (8), 2059–2065.

Igarashi, Y., Murata, H., and Ueda, M. (1978) 3-D display system using a computer generated integral photograph. *Jpn. J. Appl. Phys.*, **17** (9), 1683–1684.

Ives, H.E. (1939) Optical device, US Patent No. 2.714.003.

Iwadate, Y. and Katayama, M. (2011) Generating integral image from 3D object by using oblique projection. Proceedings of International Display Workshops 2011, pp. 269–272.

Jang, J.-S. and Javidi, B. (2002) Three-dimensional synthetic aperture integral imaging. *Opt. Lett.*, **27** (13), 1144–1146.

Kawakita, M., Sasaki, H., Arai, J. *et al.* (2010) Projection-type integral 3-D display with distortion compensation. *J. Soc. Inform. Disp.*, **18** (9), 668–677.

Kim, Y., Park, J.-H., Min, S.-W. *et al.* (2005) Wide-viewing-angle integral three-dimensional imaging system by curving a screen and a lens array. *Appl. Opt.*, **44**, 546–552.

Liao, H., Iwahara, M., Koike, T. *et al.* (2005) Scalable high-resolution integral videography autostereoscopic display with a seamless multiprojection system. *Appl. Optics*, **44**, 305–315.

Lippmann, M.G. (1908) Épreuves réversibles donnant la sensation du relief. *J. Phys.*, **4**, 821–825.

McCormick, M. (1995) Integral 3D imaging for broadcast. Proceedings of the Second International Display Workshops 1995, pp. 77–80.

Mishina, T., Okui, M., and Okano, F. (2006) Calculation of holograms from elemental images captured by integral photography. *Appl. Opt.*, **45**, 4026–4036.

Nagoya, T., Kozakai, T., Suzuki, T. *et al.* (2008) The D-ILA device for the world's highest definition (8K4K) projection systems, in *Proceedings of the 15th International Display Workshops*, ITE, pp. 203–206.

Norling, J.A. (1953) The stereoscopic art – a reprint. *J. SMPTE*, **60** (3), 286–308.

Okano, F., Arai, J., Hoshino, H., and Yuyama, I. (1999) Three-dimensional video system based on integral photography. *Opt. Eng.*, **38** (6), 1072–1077.

Okoshi, T. (1971). Optimum parameters and depth resolution of lens-sheet and projection-type three-dimensional displays. *Appl. Opt.*, **10**, 2284–2291.

Park, J.-H., Jung, S., Choi, H. *et al.* (2004) Depth extraction by use of a rectangular lens array and one-dimensional elemental image modification. *Appl. Opt.*, **43**, 4882–4895.

Pole, R.V. (1967). 3-D imagery and holograms of objects illuminated in white light. *Appl. Phys. Lett.*, **10** (1), 20–22.

Valyus, N.A. (1966). *Stereoscopy*, The Focal Press, London.

Yamashita, T., Huang, S., Funatsu, R. *et al.* (2009) Experimental color video capturing equipment with three 33-megapixel CMOS image sensors, in *Sensors, Cameras, and Systems for Industrial/Scientific Applications X* (eds E. Bodegom and V. Nguyen), Proceedings of the SPIE, Vol. **7249**, SPIE, Bellingham, WA, pp. 72490H1–72490H10.

17

3D Light-Field Display Technologies

Péter Tamás Kovács and Tibor Balogh

Holografika, Hungary

17.1   Introduction

Three-dimensional (3D) images contain much more information than a two-dimensional (2D) image showing the same scene. To generate perfect 3D images one will need systems actually capable of processing and presenting this amount of information to the viewer. Such systems contain many more pixels or apply higher speed modulation since the generation and control of independent light beams building up the 3D view requires some kind of tangible means. Using more pixels is a parallel approach, often referred to as spatial multiplexing, while time multiplexing covers higher operational speeds. True 3D display systems are generally based on these or the combination thereof.

To display proper 3D images, 3D systems should produce a sufficient number of light rays/pixels in a given time frame; that is, they should provide an appropriate rate of pixels per second. In case the system is unable to supply the sufficient number of independent light rays or pixels per second, the resulting 3D image cannot be compromise free.

The light field (Levoy and Hanrahan, 1996) is a commonly used representation of a natural 3D view. When reconstructing the view, it is necessary to produce the light beams with the same visible properties. In order to reduce the complexity of the light field, it is worthwhile considering what humans can perceive, what the significant elements are, and what is invisible to the human eye. First of all, we need to reconstruct the intensity, color, position, and direction of the light beams, up to reasonable sampling limits. But we can omit those properties to which the eye is not sensitive, like the polarization or phase of the light. These represent a huge amount of redundant information, and can even result in disturbing effects, like speckle.

Sampling is widely employed in existing displays when it comes to representing intensities in the image: typically in 8–10-bit resolution per primary color, or more for high dynamic range images. Sampling is also employed in terms of spatial (*XY*) resolution;

Emerging Technologies for 3D Video: Creation, Coding, Transmission and Rendering, First Edition.
Frédéric Dufaux, Béatrice Pesquet-Popescu, and Marco Cagnazzo.
© 2013 John Wiley & Sons, Ltd. Published 2013 by John Wiley & Sons, Ltd.

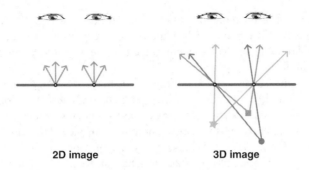

Figure 17.1   Direction-selective light emission

however, it is not straightforward that sampling can also be applied in terms of light ray directions, both horizontally and vertically.

Direction-selective light emission is the commonly required feature for any 3D system having a screen. If we could possess a light-emitting surface on which each point is capable of emitting multiple light beams of various intensity and color to multiple directions in a controlled way, the challenge of 3D displaying would be solved. As shown in Figure 17.1, the main difference between 2D and 3D displays is the direction-selective light emission property.

The rest of this chapter is structured as follows. First, we explain the fundamental parameters of 3D displays, providing means to quantify different technologies. Then we introduce one specific implementation technique of light-field displays, the HoloVizio, and continue with the discussion of its design considerations. Finally, we highlight existing implementations of this technique and close the chapter with an approximation of the capabilities of a hypothetical perfect 3D display.

17.2   Fundamentals of 3D Displaying

When comparing 3D displays with their 2D counterparts, there is an additional independent variable to take into account on top of horizontal x and vertical y resolution; namely, angular resolution ϕ. It is possible to characterize 3D displays with light-emission properties, particularly focusing on the additional key parameters important for spatial reconstruction, related to the light beams emitted from each point or pixel of the screen (Figure 17.2): the emission

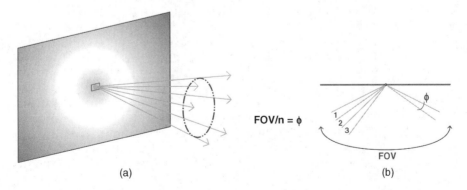

$$\text{FOV}/n = \phi$$

(a)   (b)

Figure 17.2   (a) Emission range of a pixel. (b) FOV and angular resolution

range, where the angle of this cone determines the field of view (FOV); and the number of independent beams in this range, which is the angular resolution, that also determines the field of depth (FOD). FOD shows the depth range that the 3D display can present with reasonable image quality.

FOV is a crucial measure when examining 3D displays. The bigger the FOV, the more views or directions should be processed, which poses increasing demand on the complexity of the display system, its supporting systems (such as renderers, video decoding modules, etc.), as well as content creation (the equivalent amount of views/directions also need to be captured). When designing 3D displays, screen size is just one of the factors: the viewing freedom given to users is equally important, and has an even bigger impact on complexity and cost in projection-based systems.

A reasonable simplification for reducing the number of independent light rays emitted by the 3D display is the omission or simplification of vertical parallax (Figure 17.3). Horizontal parallax is much more appreciated by viewers than vertical parallax is, since our eyes are horizontally displaced and our movements are most of the time horizontal. Our experience shows that most viewers do not event recognize the lack of vertical parallax in 3D displays, unless the content invites them to do vertical movements. The importance of horizontal motion parallax in providing depth cues is underlined when viewers do head movements, which are usually larger than the 6.5 cm average eye distance. However, it is possible to treat horizontal and vertical parallax separately and to build 3D displays with different angular resolution in the horizontal and vertical directions.

Checking common display types against these criteria reveals what we can reasonably expect from each of them, compared with true light-field reconstruction.

Stereoscopic displays (Holmes, 1906; Ezra et al., 1995) emit only two distinguishable light beams from each pixel (possibly time multiplexed or using a slightly different origin for the two light rays). Because of two views/directions, 3D perception is realistic only from a single viewing position. The creation of motion parallax requires tracking the viewer and updating image contents accordingly.

Multiview systems (van Berkel et al., 1996) reproduced the equivalent of five to nine views providing horizontal motion parallax to some extent, but the small number of transitions is clearly visible, and the limited content FOV is repeated all over the display FOV. The

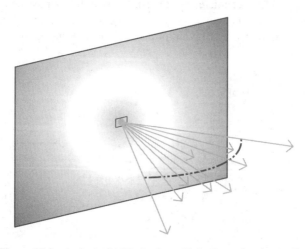

Figure 17.3 A pixel of a 3D display with horizontal-only parallax

latter is the reason for the invalid zones in the display's viewing zone. As most multiview systems (except the time multiplexed ones) use 2D panels for emitting light rays and these panels have limited resolution, there is a trade-off between the number of directions and the resolution of each direction.

Integral imaging systems (Davies and McCormick, 1992) introduce vertical parallax on top of horizontal parallax, so they can be considered the extension of multiview systems for two directions. As a consequence, resolution loss is more dramatic when providing even a small number of horizontal and vertical views.

Volumetric systems (Favalora *et al.*, 2002) are different from traditional displays in that they provide the possibility of looking into a volume instead of looking out of a virtual window (the screen). The same analysis of the number of light rays emitted also applies; however, volumetric systems possess addressable spatial spots, while in other systems the same spatial positions are represented by the combinations of light beams originated elsewhere.

The conclusion from this analysis is that for all 3D displays that have a screen the number of pixels or light rays needed is in the range of

$$\text{number of pixels} = \frac{\text{image resolution} \times \text{FOV}}{\text{angular resolution}} \tag{17.1}$$

This determines the overall quality of the 3D image. For practical 3D displays, we expect to see a 100-fold increase in terms of pixel count compared with present 2D screens.

17.3 The HoloVizio Light-Field Display System

In the HoloVizio system (Balogh, 1997), light beams are generated in optical modules that hit the points of a screen with multiple beams under various angles of incidence, which at the same time determine the exit angle (Figure 17.4). This angle depends only on the display's

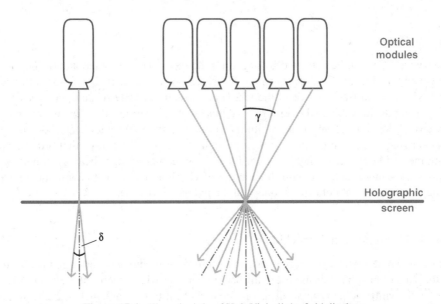

Figure 17.4 The principle of HoloVizio light-field displays

geometry, on the relative position of the given screen point, and the relevant modules. The holographic screen performs the necessary optical transformation to compose these beams into a continuous view, but makes no principal change in the direction of the light rays. The screen has a direction-selective property with angular-dependent diffusion characteristics. The screen diffusion angle δ is substantially equal to the angle γ between the neighboring optical modules; that is, the angle between the adjacent emitting directions.

In this approach, there are no optical roadblocks (like at Fresnel, lenticular, or integral screens, being single-element optical systems) that could lead to resolution and stray light issues, which in turn limit the maximum FOV of those displays to 20–30°. As 2D displays today practically show 180°, user expectation is rather 150–160°. In the HoloVizio system, as there is no deflection present at the screen by refractive elements, it is possible to reach high angles simply by the proper arrangement of the optical modules.

17.3.1   Design Principles and System Parameters

17.3.1.1   Image and Angular Resolution, Field of View, Field of Depth

Image resolution, angular resolution, FOV, and FOD are all key parameters for characterizing 3D displays. During the design process of light-field displays, the target performance for each of these parameters can be fulfilled by configuring the number of optical modules, the module–screen distance, the period of modules, the physical size and arrangement, the resolution and output angle of modules, and so on. If the optical modules are closer to the screen, then image resolution becomes higher; but if we increase the module distance, then angular resolution increases. If we increase the module angle, then FOV increases; however, image resolution and angular resolution decrease. We can compensate for this by increasing the number of modules or decreasing the module period. We can keep the module period low by making them have a smaller physical size. Setting all these parameters is an interesting optimization task in the design phase of light-field displays.

17.3.1.2   Field of View

A very wide FOV can be achieved by adjusting the module optics angle and/or the module arrangement. Inside this FOV, every spot is a "sweet spot". That is, there are no invalid zones inside; viewers can freely move in the whole area, seeing the 3D scene continuously with fixed object positions. Since the light rays address spatial locations, the view is not bound to the viewers' positions. There is a possibility to increase the FOV without introducing more optical engines, as placing side mirrors allows us to recover light rays that are otherwise hitting the inside of the display's casing. By thus collecting all light beams emitted by the modules, we can virtually increase the number of visible modules (with some modules projecting two partial light field slices, one straight and one mirrored).

17.3.1.3   Horizontal and Vertical Parallax

In the case where vertical parallax is omitted, the optical modules are arranged horizontally and the screen must have asymmetric diffusion characteristics, where the vertical FOV has large angle, while the horizontal diffusion angle is equal to the angle between the optically neighboring modules and thus it corresponds to the angular resolution.

By omitting vertical parallax we can simplify the full parallax display by a factor of 100; that is, emitting 100 different light beams horizontally which are vertically dispersed, instead of emitting 100×100 horizontal and vertical directions inside the FOV's cone.

As vertical parallax is missing in current HoloVizio implementations, the 3D image "follows" the viewers vertically. A prototype light-field display with vertical motion parallax is under development (3D VIVANT Consortium, 2010).

17.3.1.4 Screen Characteristic

The holographic screen used in these light-field displays can be a refractive or diffractive diffuser. Owing to horizontal-only-parallax configurations, the precision of the horizontal diffusion angle is critical. If it is too wide, FOD decreases. If it is too narrow, in-homogeneities may appear in the perceived image.

The proper angular light distribution characteristics are produced by the custom holographic screen, with a wide plateau and steep Gaussian slopes, precisely overlapped in a narrow region, resulting in highly selective, low-scatter hat-shaped diffuse profile eliminating crosstalk and providing proper intensity uniformity.

17.3.1.5 Angular Resolution and Field of Depth

As seen in holograms, it is possible to emit a large number of different light beams from a very small screen area; hundreds of light beams may cross a pixel's area. This number is in direct relation with the angular resolution, which determines the depth range of content the display can show with acceptable resolution. The smallest feature a display can show depends on the distance from the screen and on its angular resolution:

$$p = p_0 + s \tan \Phi \qquad (17.2)$$

where p is feature size, p_0 is pixel size, s is the distance from the screen, and Φ is the angular resolution of the display. That is, the resolution of the screen is decreasing with the distance from it. Thus, we can define the acceptable lower limit of resolution, and determine the FOD of a display using the above formula. This loss of resolution away from the screen is typically perceived as blurriness when content is too far out of the screen plane.

This formula holds for any kind of 3D display based on any technology, as it is a consequence of the fact that the same number of light beams that leave the screen is distributed over a widening FOV area. Holograms have the same behavior, especially if poorly lit.

17.3.2 Image Organization

HoloVizio light-field displays use a distributed image organization method, which is quite different from the flat-panel-based systems typically used for implementing 3D displays. Owing to the way the light rays emitted by the optical engines are distributed in the final 3D image, a single module is not associated with a specific direction, nor a single 2D view of the scene. The image projected by a single optical module is rather a "slice" of the light field, composed of a set of light rays which finally cross the screen at different positions and different directions, according to the display's geometry. Even a single view of a scene is made up by a large number of different modules. The consequence of this is that transitions from one

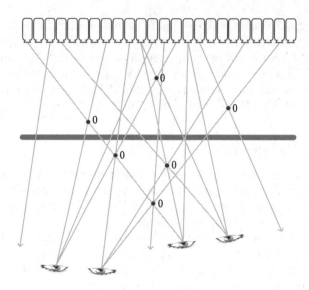

Figure 17.5 Individual light rays emitted by different optical modules addressing spatial positions

view to another do not occur abruptly, but the change is smooth, resulting in continuous motion parallax (Figure 17.5).

17.4 HoloVizio Displays and Applications

Being a scalable technique for implementing light-field displays, different models of HoloVizios have been built and used in various settings.

17.4.1 Desktop Displays

The first desktop light-field displays had a 32-inch screen size, 16 : 9 aspect ratio with 10 megapixels (that is, 10 million light rays in total); there was also a smaller 26-inch screen size, 4 : 3 aspect screen with 7.4 megapixels. These displays have the appearance and dimensions of a CRT TV set. The light rays are generated by 128 and 96 LCD microdisplay optical modules respectively. These optical modules are densely arranged in a horizontal-only-parallax configuration behind the holographic screen and all of them project their specific image onto the holographic screen over a total of 50° horizontal FOV. The 10 megapixel display roughly corresponds to 60 views, each of 512 × 320 in conventional terms. 3D content is generated by means of a high-end GPU-equipped PC (or multiple PCs), connected though multiple DVI connections. Multiple sub-images corresponding to the individual optical modules are transmitted over these connections in time-multiplex.

Later desktop monitors (Figure 17.6) have been vastly improved in terms of resolution, brightness, and FOV, with the latest implementation controlling 73 megapixels in total, shown on a 30-inch screen with approximately HD-Ready 2D-equivalent resolution. FOV has also been increased to almost 180° using a novel arrangement of optical modules, the screen and mirrors, which is exceptionally wide especially considering the capabilities of other state-of-the-art 3D displays.

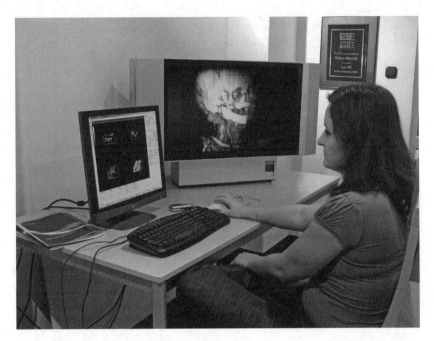

Figure 17.6 Desktop light-field display showing volumetric data (see Plate 22 for the colored figure)

17.4.2 Large-Scale Displays

To facilitate truly multi-user collaboration and interaction with 3D models, we also created large-scale light-field displays (Figure 17.7). The light-ray count of these ranges from 34.5 million to 73 million on a 72-inch screen diagonal.

The optical modules in these displays are implemented by means of compact projection modules, arranged in a horizontal-only-parallax arrangement. These systems have sub-degree angular resolution, and thus high depth range (±50 cm). Approximately 50 independent light beams originate from each pixel of the screen. The 2D equivalent image resolution is approximately 1344×768 pixels.

These large-scale displays have been implemented both using lamp-based and LED-based illumination.

As a single computer is incapable of generating the necessary pixel count in real time, large-scale displays are driven by PC-based render clusters, also containing network components, and power-related components. These rendering clusters consist of 10 to 18 rendering nodes, filled with high-end GPUs.

17.4.3 Cinema Display

All the display implementations described before are back-projected; that is, the optical modules and the viewers are on the opposite side of the holographic screen. It is also possible to implement front-projected light-field displays using a different screen structure and placing the optical modules above the viewers. This has been implemented in the 140-inch screen diagonal C80 display, which has the optical module array and the screen as separate units, so that it can be installed in flexible arrangements. This display, similar to other large-scale displays, is controlled by a rendering cluster.

Figure 17.7 Large-scale light-field display showing pre-rendered 3D data (see Plate 23 for the colored figure)

17.4.4 Software and Content Creation

Most users expect that, when introducing 3D displays in their workflow, they can continue using their existing software, instead of learning to use new software which is "3D compatible." This is possible using an OpenGL wrapper layer, which intercepts OpenGL calls coming from the application and transfers them to the 3D display. The application is not aware of the presence of the wrapper, but the 3D model that is created by the application using OpenGL is also displayed on the 3D screen.

Showing image-based content is the other major use case. This can be either pre-rendered content, or can also come from live multi-camera shots. In this case, owing to the unusual relation between the arrangements of pixels in the optical modules, a conversion process that combines images taken from different viewpoints is necessary. The result is a format that is specific for the display in question, and can be played on the display in real time. This conversion process can also run in a real-time setting, having a sufficiently high number of cameras (Balogh and Kovács, 2010).

17.4.5 Applications

Light-field displays have been employed in a range of professional use cases that require collaborative work in a 3D space without glasses, in extended use periods of 3D visualization without side effects, or when the number of views to be shown is too large for other 3D display technologies. It has been used in innovative design review and industrial design processes, oil and gas exploration tasks, medical/dental education, entertainment shows, and virtual reality. We also expect to see 3D telepresence applications in the near future, while research on the future 3DTV is also taking place.

17.5 The Perfect 3D Display

How could we just define the perfect 3D display? Assuming that human eye resolution is 1 arcmin and that the viewing distance is 25 cm, the smallest practical pixel size is 0.1 mm. For 180°, approximately 628 distinguishable light rays should be emitted. Generalizing for screen surface area, for 1 mm^2, 6×10^3 pixels are required and for a full parallax system the square of this, in the region of 10^6, which is again in good coincidence with hologram and holographic recording material resolutions of 1000 line pairs/mm.

17.6 Conclusions

We described why 3D display systems must control many more pixels (or higher speed components) than 2D displays with the same resolution. We also identified direction-selective light emission as a required feature for any 3D display having a screen. We introduced the light-field display principle along with implementation considerations that aim to fulfill these criteria.

3D displays are already commercial products, but the technologies employed in these are quite limited. Higher-end 3D displays are currently in the reach of professional users, but should eventually become mainstream products with new 3D display generations.

The good news is that 3D data are more widely used than we can imagine, and there is an obvious need for 3D displays. 3D is considered the next big step in visualization, but it comes at a cost. Transition from black-and-white images to color images means an increase of three times in the amount of information to be captured and transmitted, while true full-parallax 3D displays would require an orders-of-magnitudes increase (10^2 to 10^4) in the information content and display capabilities, which requires several intermediate steps.

References

3D VIVANT Consortium (2010) Live immerse video–audio interactive multimedia, EU IST-FP7 project no. 248420, http://www.3dvivant.eu/.

Balogh, T. (1997) Method & apparatus for displaying 3D images. US Patent 6,201,565, European patent EP0900501.

Balogh, T. and Kovács, P.T. (2010) Real-time 3D light field transmission, in *Real-Time Image and Video Processing 2010* (eds N. Kehtarnavaz and M.F. Carlsohn), Proceedings of the SPIE, Vol. **7724**, SPIE, Bellingham, WA, p. 772406.

Davies, N. and McCormick, M. (1992) Holoscopic imaging with true 3D-content in full natural colour. *J. Photon. Sci.*, **40**, 46–49.

Ezra, D., Woodgate, G.J., Omar, B.A. *et al.* (1995) New autostereoscopic display system, in *Stereoscopic Displays and Virtual Reality Systems II* (eds S.S. Fisher, J.O. Merritt, and M.T. Bolas), Proceedings of the SPIE, Vol. **2409**, SPIE, Bellingham, WA, pp. 31–40.

Favalora, G.E., Napoli, J., Hall, D.M. *et al.* (2002) 100 million-voxel volumetric display, in *Cockpit Displays IX: Displays for Defense Applications* (ed. D.G. Hopper), Proceedings of the SPIE, Vol. **4712**, SPIE, Bellingham, WA, pp. 300–312.

Holmes, O.W. (1906) *The Stereoscope and Stereoscopic Photographs*, Underwood & Underwood, New York, NY.

Levoy, M. and Hanrahan, P. (1996) Light field rendering, in *SIGGRAPH '96 Proceedings of the 23rd Annual Conference on Computer Graphics and Interactive Techniques*, ACM, New York, NY, pp. 31–42.

Van Berkel, C., Parker, D.W., and Franklin, A.R. (1996) Multiview 3D-LCD, in *Stereoscopic Displays and Virtual Reality Systems III* (eds M.T. Bolas, S.S. Fisher, and J.O. Merritt), Proceedings of the SPIE, Vol. **2653**, SPIE, Bellingham, WA, pp. 32–39.

Part Five

Human Visual System and Quality Assessment

18

3D Media and the Human Visual System

Simon J. Watt and Kevin J. MacKenzie

Wolfson Centre for Cognitive Neuroscience, School of Psychology, Bangor University, UK

18.1 Overview

Ideally, viewing stereoscopic three-dimensional (S3D) media should be much like viewing the real world, in that it creates a compelling and immersive sense of the three-dimensional structure of the portrayed scene, without producing unwanted perceptual artefacts or aversive symptoms in the viewer (such as fatigue, discomfort or headaches). With the recent growth in S3D cinema, television and computer gaming, 'three-dimensional (3D) visualization' has gone from being used primarily in specialist applications (e.g. virtual reality, computer-aided design, scientific visualization, medical imaging and surgery) to a mainstream activity. This has increased the need for a better understanding of how the human visual system interacts with S3D media, if the effectiveness and comfort of S3D viewing are to be optimized.

18.2 Natural Viewing and S3D Viewing

At first glance, the advantages of using S3D media over conventional 'two-dimensional (2D)' images might seem obvious: we receive depth information from binocular vision in real-world viewing, and so surely our viewing experience will be more 'natural' if this information is also presented in visual media. In fact, however, aspects of the creation, presentation and viewing of S3D media can all introduce unnatural stimuli to the visual system, and these sometimes result in undesirable perceptual outcomes, and/or fatigue and discomfort in the viewer.

In considering the unnatural stimulus features that can cause problems in S3D media, we draw a distinction between *technical* issues and *fundamental* issues. Technical issues are ones that are generally well understood, and for which reasonable solutions exist (though these may not be straightforward). Fundamental issues are ones that are inherent in current S3D media

Emerging Technologies for 3D Video: Creation, Coding, Transmission and Rendering, First Edition.
Frédéric Dufaux, Béatrice Pesquet-Popescu, and Marco Cagnazzo.
© 2013 John Wiley & Sons, Ltd. Published 2013 by John Wiley & Sons, Ltd.

technologies and cannot reasonably be eliminated or corrected for in typical viewing situations. In this chapter, we briefly describe several technical issues (and the possible perceptual consequences) that arise from viewing typical S3D media, but we concentrate on the fundamental issues. Since these latter issues cannot be solved, it is necessary instead to determine tolerances for the visual system in each case (i.e. by how much the stimulus can deviate from real-world viewing before problems arise). A complete analysis of each topic is beyond the scope of a single chapter, and it should be recognized that in many cases the current understanding of the relevant constraints is insufficient to provide quantitative guidelines. Nonetheless, it is hoped that improving understanding of these basic constraints will help S3D media to be developed that are better suited to the properties of the human visual system.

18.3   Perceiving 3D Structure

Use of the term '3D' (as in 3D movies, etc.) is potentially misleading because it obscures the fact that almost all conventional '2D' visual media present the visual system with information about the depth structure of the portrayed scene – that is, the third dimension. Consider the photograph of a street in Figure 18.1. We perceive depth in this picture because the 3D structure of the scene is specified by a range of so-called monocular depth cues (available with only one eye) that includes perspective/texture, occlusion, the familiar size of objects (such as people and cars), the relative sizes of these same objects at different depths, the

Figure 18.1   A photograph of California Street, San Francisco, illustrating the large number of monocular depth cues available in a conventional '2D' image

height of objects in the scene (far-away objects are generally higher up), differential blurring of objects at different depths, shading/shadows and even the reduced contrast and increasingly blue appearance of objects in the far distance (the bridge in the background, for example), known as aerial perspective (Sedgwick, 1986; Howard and Rogers, 2002). Thus, what is *different* about 3D media is the addition of information about the 3D structure of the portrayed scene from binocular stereopsis (derived from binocular disparity and the convergence angle of the two eyes; see Section 18.3.1), and all the normally available monocular depth cues remain. For this reason we use the term S3D throughout this chapter. (Note also that when conventional '2D' images are viewed with both eyes the binocular depth cues are not 'switched off', but instead specify the 3D structure of the screen surface.)

This point is significant because, while the terms 'binocular vision', '3D vision' and 'depth perception' are frequently used synonymously, depth perception is the result of many different cues, most of which are available monocularly. Indeed, experiments suggest that depth perception results from the combination of depth information from *all* of the available signals. This process, referred to in the vision-science literature as *cue combination*, or *cue integration*, is beneficial because by exploiting the statistical redundancy in multiple sources of information the brain can estimate the 3D structure of the world more precisely, and robustly, than would otherwise be possible (Clark and Yuille, 1990; Landy *et al.*, 1995; Yuille and Bülthoff, 1996; Oruç *et al.*, 2003; Knill and Pouget, 2004). Humans have been shown to be close to statistically optimal at integrating different visual cues to depth (perspective/texture and disparity cues to surface orientation, for example; Knill and Saunders, 2003; Hillis *et al.*, 2004) and at combining information from vision with other senses (Ernst and Banks, 2002; Alais and Burr, 2004). To do this correctly, the brain must weight different cues according to how informative, or *reliable*, they are in a given instance. Critically, these 'cue weights' cannot be fixed values, or learned for certain classes of situations, because the relative reliabilities of depth information from different cues (and so the weights they should receive in the overall estimate) depend on the particular viewing situation, and may vary substantially even within a single scene (Hillis *et al.*, 2004; Keefe *et al.*, 2011). Consider the reliabilities of perspective/texture and binocular disparity cues to surface orientation, for example. They depend differently on geometrical factors such as the orientation of the surface relative to the observer and on viewing distance (Knill, 1998; Knill and Saunders, 2003; Hillis *et al.*, 2004). They also depend differently on the distance of the image points from where the eyes are fixated (Greenwald and Knill, 2009). Thus, perceived depth in S3D media cannot be thought of as the result only of depth from binocular stereopsis. Moreover, the continuous variations in cue reliability make it difficult to predict how different cues will be weighted in the complex scenes typically used in S3D media.

The statistical properties of multiple signals can also be used to decide how to combine different cues when they produce significantly discrepant estimates of scene structure. In general, the brain should integrate signals that refer to the same object in the world and avoid integrating signals that do not (Ernst, 2006). Assuming that estimates from each cue are correctly calibrated, the similarity of their estimates is directly related to the probability that they were caused by the same object, and experiments have shown that the brain uses this information in deciding if different cues should be integrated (Gepshtein *et al.*, 2005; Shams *et al.*, 2005; Bresciani *et al.*, 2006; Roach *et al.*, 2006; Ernst, 2007; Knill, 2007; Körding *et al.*, 2007; Girschick and Banks, 2009; Takahashi *et al.*, 2009). Significant discrepancies between estimates can also indicate errors in the estimate from one or both cues (due to inaccurate calibration, for example). Sensitivity to this information allows the visual system to be robust in estimating 3D scene properties across a diverse range of natural viewing

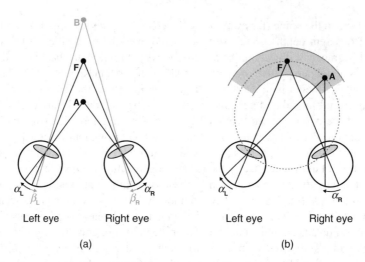

Figure 18.2 Basic geometry of horizontal binocular disparities. (a) The pattern of relative horizontal binocular disparities projected on the retinas by points at depths other than the fixation distance F (see Equations 18.1 and 18.2). (b) The horopter (dashed line) shows the locus of points in space that project zero binocular disparity at the retinas (for the case of point A, $\alpha_L = \alpha_R$, thus, $\alpha_L - \alpha_R = 0°$. Points within the horopter result in so-called crossed disparities, while points outside the horopter result in uncrossed disparities. The grey zone around the horopter represents Panum's fusion area

situations. As we shall see, however, there are several factors that cause S3D media to routinely introduce conflicting information from different depth cues, further complicating attempts to predict the perceptual outcome of viewing S3D media.

18.3.1 Perceiving Depth from Binocular Disparity

To understand some of the ways in which viewing S3D media might result in an unnatural stimulus to the visual system it is necessary to briefly consider some fundamental aspects of depth perception from binocular vision.

The basic geometry of binocular vision is schematized in Figure 18.2, for the simplified case of viewing objects positioned at different depths on a horizontal plane intersecting the nodal points of the two eyes (i.e. considering only horizontal binocular disparities). Consider viewing the scene depicted in Figure 18.2a. Convergence eye movements operate to place the same *fixation point* (F) on the centre of each eye's fovea. The lateral separation of the two eyes in the head (the *inter-ocular distance*, or IOD; typically about 6.3 cm; French, 1921; Dodgson, 2004) means that point A projects to different positions in the left- and right-eye's images; in this case, point A is 'shifted', relative to F, by an equal and opposite amount in the two eyes. This difference – between the angular positions of image points in the two eyes, relative to the fixation point – is relative horizontal binocular disparity. Clearly, the magnitude of this binocular disparity varies with the separation in depth of points F and A, providing the basic signal for binocular stereopsis. More formally, the angular disparity $\eta = \alpha_L - \alpha_R$ can be expressed (in radians) as

$$\eta = \frac{I\Delta d}{D^2 + D\Delta d} \tag{18.1}$$

where I is the IOD, Δd is the distance between the fixation point F and the point A, and D is the fixation distance (the distance to F). Since $D\Delta d$ is usually small compared with D^2 (Howard and Rogers, 2002), we can write

$$\eta \approx \frac{I\Delta d}{D^2} \tag{18.2}$$

For the visual system to compute the magnitude of binocular disparities correctly it must first solve the so-called *correspondence problem*: it must correctly determine which points in the left- and right-eye's images correspond to the same points in the world. The fact that depth from binocular disparity is visible in random-dot stereograms indicates that this process does not require specific image features to be identified and matched across the two eyes' images (Julesz, 1960). Instead, the correspondence problem is thought to be solved 'statistically', by cross-correlating the two eyes' images to extract their similarity in each region (Tyler and Julesz, 1978; Cormack *et al.*, 1991; Banks *et al.*, 2004). Consistent with this, physiological studies suggest that the visual system is 'tuned' to be best at computing disparities at or around zero, where image regions in the two eyes are maximally correlated (Nienborg *et al.*, 2004), and human sensitivity to binocular disparity is best around zero disparity and falls off thereafter (Blakemore, 1970). Indeed, when binocular disparities are sufficiently large (i.e. when the two eyes' images are sufficiently de-correlated) the visual system cannot compute disparity, and so depth perception from binocular disparity does not occur (Banks *et al.*, 2004). This is experienced as a failure of binocular fusion, resulting in double vision (diplopia). Figure 18.2b shows the locus of points that project zero binocular disparity, for a given fixation. This is referred to as the *horopter*, and is defined by a circle drawn through the fixation point and the nodal points of the two eyes (see Howard and Rogers (2002)). The range of binocular disparities around the horopter within which binocular fusion occurs is called *Panum's fusion area*, and has been estimated empirically to be approximately 0.25–0.5° (Ogle, 1932; Schor *et al.*, 1984; see also the related concept of the *disparity-gradient limit*; Tyler, 1973; Burt and Julesz, 1980; Banks *et al.*, 2004). However, the size of this area depends on several factors, such as how much blur is in the retinal image (Kulikowski, 1978).

Panum's fusion area is an example of a neural limitation on the computation of binocular disparity that has significant implications for the design of S3D content. Other key points emerge from considering basic geometrical aspects of binocular vision. Critically, there is no unique relationship between a depth in the world and the resulting binocular disparity at the eyes. Thus, the same depth in the world results in different disparities at the eyes, depending on the distance it is viewed at (for a fixed depth interval, the disparities created at the eyes fall off approximately as a function of the reciprocal of the *square* of distance; Equation 18.2). This can be seen in Figure 18.2a, where, despite the fact that F–A and F–B are separated by the same amount in depth, the resulting binocular disparities are very different. Conversely, the same disparities at the eyes can result from viewing very different depths at different distances, as shown in Figure 18.3.

To disambiguate these situations, and estimate the metric structure of the scene, binocular disparities must be 'scaled' by an estimate of fixation distance (Gårding *et al.*, 1995). A primary source of this information is thought to come from the position sense of the eyes indicating the angle between the two lines of sight (the vergence angle), and therefore the distance at which the two lines of sight are converged (Longuet-Higgins, 1981; Mon-Williams *et al.*, 2000). Thus, manipulating the eyes' vergence angle while keeping binocular

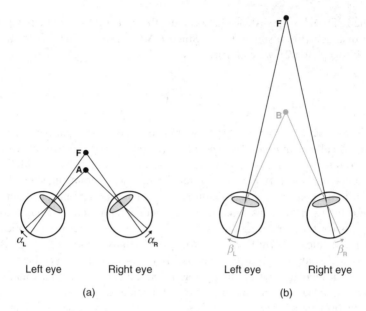

Figure 18.3 The fundamental ambiguity of depth from binocular disparity. Different depths in the world, at different distances from the viewer, can result in the same binocular disparity at the eyes ($\alpha_L - \alpha_R = \beta_L - \beta_R$). This illustrates how, in order to recover metric 3D properties of the scene, the visual system must 'scale' disparities by an estimate of viewing distance (Gårding *et al.*, 1995)

disparities the same (to move S3D content close to the surface of the display screen, for example) will have significant effects on the amount of perceived depth from disparity, shrinking or stretching the range of depths perceived in the scene, and similarly changing the shape-in-depth (e.g. 'roundness') of individual objects. Vertical differences between the two eyes' images (vertical disparities), which arise in natural viewing, have also been shown to be a signal to viewing distance that is used to scale horizontal disparities (Rogers and Bradshaw, 1993, 1995; Backus *et al.*, 1999; see Gårding *et al.* (1995) for a comprehensive geometrical explanation of this and other points). Perceived depth from disparity can also be influenced by other depth cues, including familiar size and perspective (O'Leary and Wallach, 1980) and the eye's focusing response (Watt *et al.*, 2005a).

As we explain in later sections, in S3D media all of these signals commonly specify fixation distances that are inconsistent with the portrayed scene, and so may result in depth percepts from binocular disparity that depart, sometimes substantially, from those intended.

Readers interested in a comprehensive overview of the vision-science literature on binocular vision are recommended to consult Howard and Rogers (2002).

18.4 'Technical' Issues in S3D Viewing

In principle, to capture natural S3D content one simply needs to place two cameras at the points from which the two eyes would view the scene. Then, to display that content, one needs a system for separating the two images so that the left eye sees only its image and vice versa. In practice, of course (and as discussed in other parts of this book), there are numerous reasons why this apparently straightforward process is in fact hugely challenging. Here, we briefly review some of the common perception-related problems that can arise, and for which technical solutions may be required.

18.4.1   Cross-Talk

Cross-talk is when one eye's image is (at least partially) visible to the other eye. It is difficult to avoid completely and can arise for a variety of hardware-related reasons, such as imperfect filtering of the two eyes' images (in systems using anaglyph, polarization, or 'comb' wavelength filters) or display persistence and imperfect shuttering in shutter-glasses systems. Small amounts of cross-talk do not have catastrophic effects but can reduce the amount of perceived depth. For example, Tsirilin *et al.* (2011) found that 8% cross-talk, resulting from using shutter glasses and a liquid-crystal display, reduced perceived depth by as much as 20%. At high levels (25%), cross-talk has been shown to lead to 'extreme' reductions in reported comfort (Kooi and Toet, 2004).

18.4.2   Low Image Luminance and Contrast

Compared with conventional media, S3D images often have low overall luminance and low contrast. There are several reasons for this. First, S3D content is frequently filmed with beam-splitter camera rigs (which allow the camera separation, or *inter-axial distance*, of relatively bulky cameras to be kept small), leading to significant light losses. Further losses are then incurred by the process of separating the two eyes' images. In filter-based systems, light is lost at filters on the projector(s) and the user's glasses. For shutter-glasses systems, some light is lost even when the lens is in the 'open' state. And effective (time-averaged) luminance is also reduced with frame-sequential stereoscopic presentation because the duty cycle of each frame is a maximum of 50% (and often less, in practice). Reductions in image contrast result in reduced stereoacuity – the ability to resolve fine stereoscopic detail (Legge and Gu, 1989).

For this and other reasons, scene lighting can also have substantial effects on perceived depth in the final 'product' (Benzeroual *et al.*, 2011).

18.4.3   Photometric Differences Between Left- and Right-Eye Images

Different filters used to separate left- and right-eye images can also introduce differences in the luminance, contrast and spectral properties of the two eyes' images (Kooi and Toet, 2004; Devernay and Beardsley, 2010). These can also be introduced at the capture stage, when using beam-splitter camera rigs, due to the different properties of beam-splitters in reflectance and transmission (they are polarization sensitive, for example). Large differences in luminance (25%) between the two eyes' images can lead to discomfort (Kooi and Toet, 2004), and significant spectral differences will be noticeable to the viewer. Contrast differences between the two eyes' images can reduce stereoacuity (Schor *et al.*, 1989). In principle, these differences can be 'calibrated out' (e.g. in post-production) if they are adequately understood. Note, however, that perfect photometric calibration of the two eyes' images may not be possible if the same S3D images will be viewed using several different display technologies.

Of course, similar differences can also potentially be introduced when separate projectors are used to present each eyes' image.

18.4.4   Camera Misalignments and Differences in Camera Optics

It is very difficult to maintain precise alignment of two cameras in a stereoscopic camera rig. Misalignments can result in a variety of differences between the two eyes' images, including rotations, vertical and horizontal offsets, keystoning and so on. Even if the cameras are perfectly aligned, differences in the optics and operation of individual cameras can be significant (e.g. resulting in magnification differences due to differential zooming; Mendiburu, 2009). All of these factors can affect the pattern of disparities at the retina, and so may affect the viewer's perception of 3D structure. Vertical image magnification in one eye, for example, causes *the induced effect*, where perceived surface orientation is changed (Ogle, 1938), and horizontal magnification can also create spatial distortions (Burian, 1943; see also Section 18.7). These effects can cause unpleasant symptoms. For example, magnification of one eye's image by just 2.5% has been shown to result in significantly increased reports of viewer discomfort (Kooi and Toet, 2004). And vertical offsets in the images require the viewer to make vertical vergence eye movements, which are also known to cause discomfort (Kane *et al.*, 2012). These effects can sometimes be corrected (e.g. in post-production) using image-processing algorithms that estimate the differences between camera parameters.

Another common source of differential distortion in the two eyes' images is the practice of filming with converged or 'toed-in' cameras. We discuss this in Section 18.7. Also, as with photometric properties, when using different projectors for each eye's image, care must be taken to make sure they are spatially calibrated.

18.4.5   Window Violations

Occlusion, or interposition, is a powerful cue to depth order because in natural viewing the occluded object is always farther from the viewer than the occluding object. This 'rule' can readily be broken in S3D media. The edge of the frame (or screen) occludes all parts of the image, irrespective of their position in depth, and so the viewing 'window' can occlude objects that are nearer than it. This is immediately noticeable to the viewer and creates an unpleasant conflict (Mendiburu, 2009). These so-called *window violations* can be avoided by not positioning nearer-than-screen objects at the edges of the frame. In the case of the ground plane this may be impossible, however. It is also common practice to use a so-called *floating window* (Gardner, 2012) – an artificial frame that is positioned nearer in stereoscopic depth than the content (and may move and tilt) to ensure that the occlusion relationships are correct.

18.4.6   Incorrect Specular Highlights

Because the position of specular highlights (reflections of light sources from glossy surfaces) depends on the relative position of the observer's viewpoint and the light source, they naturally have a disparity relative to the surface, appearing behind convex surfaces and in front of concave ones (Blake and Bülthoff, 1990). Various capture parameters (changing camera parameters, differential polarization sensitivity of beam-splitter rigs) can introduce errors into the position of specular highlights in S3D images. Although specular highlights may seem a small aspect of S3D content, the visual system is surprisingly sensitive to whether they are correct, and there is evidence that errors can significantly alter the perceived spatial properties of surfaces, their apparent reflectance and the perceived realism of glossy surfaces

(Blake and Bülthoff, 1990; Wendt *et al.*, 2008). Correcting these errors is likely to be challenging.

18.5 Fundamental Issues in S3D Viewing

We now turn to several fundamental perceptual and human factors issues that can arise from viewing S3D media, and for which no straightforward solution exists. As suggested earlier, for these issues the challenge is to understand the tolerances of the visual system to these problems, and to try to work within the resulting limits when producing S3D media. We also hope that a better understanding of these issues will inform the development of future S3D technologies.

18.6 Motion Artefacts from Field-Sequential Stereoscopic Presentation

In cinema, the principal technologies for presenting separate images to the left and right eyes can be divided according to whether they use active or passive glasses. Active-glasses systems typically use liquid-crystal shutter-glasses to control each eye's view. The left- and right-eye's images are presented on alternate frames, and the glasses open and close synchronously with the projector, to allow each eye to see only its image. Passive-glasses systems include (*i*) 'comb' wavelength interference filters that divide the visible spectrum into separate RGB components for each eye, which are then 'decoded' by matching filters in viewers' glasses (Dolby/Infitec), and (*ii*) polarization filters (circular or linear) that differently polarize the left- and right-eye images, again 'decoded' by suitably matched glasses (e.g. IMAX, RealD). S3D televisions frequently use active liquid-crystal shutter-glasses (e.g. XPAND). Other passive systems display the left- and right-eye images on alternate rows of the television screen and separate them by covering the rows with different polarization filters. There are also a number of 'glasses-free' autostereoscopic computer displays and televisions that use parallax barriers or lenticular-lens arrays to render different pixels visible to each eye.

Presenting the left- and right-eye images in a time-multiplexed or field-sequential manner is an inherent aspect of liquid-crystal shutter-glasses systems (Figure 18.4). In principle, systems that use 'comb' wavelength filters or polarization filters could present the two eyes' images simultaneously, using separate projectors for each. In practice, however, the most commonly used cinema projection systems use an active 'switchable' filter placed in front of a single projector, to also present left- and right-eye images field sequentially (Figure 18.4; IMAX is a notable exception: it uses two projectors with polarization filters and simultaneous left/right eye presentation). Using a single projector – and, therefore, presenting left- and right-eye images on alternate frames – has some clear advantages. It offers a cost saving over separate projectors for each eye and eliminates the need to cross-calibrate luminance, colour and spatial properties of a pair of projectors. Field-sequential stereoscopic presentation does, however, present an unnatural stimulus to the visual system, and we explore the effects of this here (Hoffman *et al.*, 2011).

There are several important differences between the visual stimulus that results from viewing the real world and viewing the same scene via field-sequential stereoscopic presentation, particularly for moving objects. First, blank frames are inserted into the stream of images that each eye receives (Figure 18.4). Second, the capture rate of images can be quite low – typically 24 frames per second (fps) in cinema – and so motion is captured, and displayed, rather discontinuously (Figure 18.5). Third, left- and right-eye images are usually captured

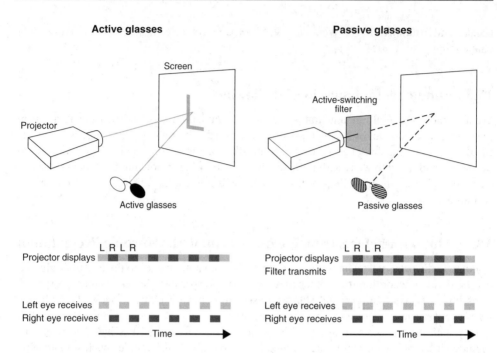

Figure 18.4 Field-sequential presentation in typical 'active-glasses' and 'passive-glasses' cinema S3D projection systems. In 'active' systems, each eye necessarily receives a stream of images that is interrupted by blank frames, every other frame. Although not necessary, passive systems also typically present left- and right-eye images on alternate frames, achieving stereo separation by using an active, switchable filter in front of the projector. Thus, in practice, the two approaches typically present a similar stimulus to the visual system

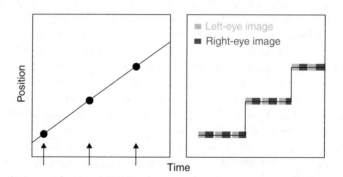

Figure 18.5 Capture and field-sequential stereoscopic presentation of a moving stimulus, using a triple-flash protocol. The line in the left panel shows continuous motion of an object, at a constant speed. The arrows/circles, represent the points at which each eye's image is captured. There are single arrows/circles for each frame because the two eyes' images are captured simultaneously. The right panel shows when each eye's image is presented and the corresponding position of the stimulus in that eye's image. The solid black line in the right panel denotes the actual motion trajectory presented to the viewer. Here, the image capture process is treated as instantaneous when in fact the camera shutter is open for a finite amount of time, leading to motion blurring in the image. This aspect is considered in the Discussion in Hoffman *et al.* (2011). The figure was adapted (with the author's permission) from one presented by David Hoffman (UC Berkeley and MediaTek, USA), at 3DStereo Media 2010, Liege, Belgium

simultaneously, but displayed in counter phase, with the right-eye presentation delayed (see Figures 18.4 and 18.5).

The risks of field-sequential presentation, therefore, are that it could result in visible flicker, perception of unsmooth (juddering) motion and distortions in perceived depth from binocular disparity. A recent analysis by Hoffman *et al.* (2011) examines these issues in detail, in light of the spatiotemporal contrast sensitivity of the visual system, and provides insights into how each of these problems might be minimized. There is insufficient space in this chapter to provide more than a summary of the main points of this very comprehensive study, and readers interested in this topic are strongly recommended to consult the original source material.

18.6.1 Perception of Flicker

The human visual system is sensitive to flicker – variations in the perceived brightness of an image – below the so-called *critical flicker fusion rate* of ~60 Hz (although this is lower in central vision and under dim lighting; Landis, 1954; Rovamo and Raminen, 1988). Thus, presenting conventional (non-stereoscopic) images at the capture rate of 24 fps will result in visible flicker. A common and effective solution is to use multi-flash protocols such as *triple-flash*, in which each frame is presented three times in a row (i.e. flashed three times), at three times the capture rate, effectively making the presentation rate 72 Hz. In triple-flash S3D presentation, each eye receives the same frame three times, but with blank frames inserted in between (see Figure 18.5). One might imagine that inserting dark, blank frames into the image stream makes the perception of flicker more likely to occur. Hoffman *et al.* (2011) analysed the spatial and temporal frequencies introduced by a moving stimulus presented in this manner and made predictions based on the visual system's spatiotemporal contrast sensitivity, which they confirmed using psychophysical experiments. Overall, they found that flicker visibility is not very different for S3D and non-stereoscopic presentation. Moreover, it is primarily determined by presentation, not capture rate, and so the triple-flash protocol effectively eliminates perceived flicker in S3D.

18.6.2 Perception of Unsmooth or Juddering Motion

Discontinuous capture and display of motion is schematized in Figure 18.5, for field-sequential stereoscopic presentation using a triple-flash protocol. As the right panel indicates, continuous motion in the world is presented as discontinuous motion on the cinema screen, and so will likely be perceived as such under some conditions. In addition, viewers frequently move their eyes to track moving objects (making so-called smooth pursuit eye movements), attempting to maintain no movement of the object on the retina. When making smooth, continuous eye movements, with a multi-flash protocol, each repeat of the flashed image of the object will fall on a different part of the retina because the eye has moved but the image is not yet updated. There is, therefore, a risk with multi-flash protocols of visible 'banding' of the image (Hoffman *et al.*, 2011). Hoffman *et al.* (2011) again examined the spatial and temporal frequencies introduced as a function of different presentation protocols, capture rates and the speed of the moving object and considered these in light of the known spatiotemporal sensitivity of the visual system. Key findings of their analysis (again, supported by experiments) were that motion artefacts are (*i*) increasingly visible with increasing movement speed and (*ii*) increasingly visible with *decreasing* capture rate. The authors estimate that for images captured at 60 Hz, visible artefacts will occur above speeds of ~12–14 degrees of visual angle per second. With typical 24 Hz capture, and triple-flash presentation, artefacts will be visible at just 3–5 deg/s.

18.6.3 Distortions in Perceived Depth from Binocular Disparity

Field-sequential stereoscopic presentation also typically introduces a relative temporal delay in presentation of the left- and right-eye images, because they are captured simultaneously, but displayed in counter phase. Hoffman *et al.* (2011) examined the possible effects of this for perceived depth from disparity in S3D because a delay in the input to one eye is known to result in biases in perceived depth from disparity (Burr and Ross, 1979; Morgan, 1979).

As outlined earlier, the visual system must perform binocular matching between the two eyes' images in order to compute disparities. Consider the triple-flash protocol for field-sequential S3D shown in Figure 18.5. Because the left- and right-eye images are not available simultaneously, there is uncertainty about whether a given left eye's image should be matched with the previous or next right eye's image. It can be seen that if the system always did the same thing (e.g. match each image with the next one from the eye) it would occasionally match images captured at different times, rather than on the same frame. For an object moving in depth, this would result in an incorrect disparity, because one eye's image would correspond to when the object was at a different position. Disparity processing operates quite slowly, and integrates over time (Tyler, 1974), and so this could lead to biases in the average disparity that is computed over time (Hoffman *et al.*, 2011).

Hoffman *et al.* (2011) examined the magnitude of these distortions in a psychophysical experiment, as a function of the number of flashes used in the display protocol, the capture rate and the speed of lateral motion of the stimulus. Smaller numbers of flashes would be expected to result in increased distortions because, with fewer flashes, the probability of matching a given eye's image to one captured in a subsequent or previous frame increases. Slow capture rates should cause larger depth distortions because, for a given speed of motion, the change in an object's position across erroneously matched frames is greater. Higher stimulus speeds should result in larger distortions for similar reasons – in subsequent or previous frames the object will have moved farther, introducing larger errors to the time-average disparity estimate (Hoffman *et al.*, 2011). The results of the experiment confirmed these predictions, demonstrating that, under many circumstances, field-sequential presentation can make the position-in-depth of objects moving laterally appear different than was intended (whether the distortion is towards or away from the viewer depends on the direction of motion).

In some instances, these distortions in depth may be relatively inconsequential. However, one can imagine that it is particularly problematic for S3D content in which perceiving accurate trajectories of objects is important (e.g. ball sports; Miles *et al.*, 2012). Unpleasant perceptual cue conflicts can also be created, similar to the window violations described earlier, where a moving object (such as a football) can appear to be below, or inside, the surface it is moving on.

18.6.4 Conclusions

Because all moving pictures must sample the scene at some capture rate, they necessarily present a stimulus that differs fundamentally from the real world, and this is exacerbated by frame-sequential stereoscopic presentation. The analysis and experiments carried out by

Hoffman *et al.* (2011) suggest that motion artefacts and stereoscopic depth distortions are worse with higher speeds of motion. They also suggest these problems can be significantly reduced by increasing capture rates from the typical 24 fps.

18.7 Viewing Stereoscopic Images from the 'Wrong' Place

When portraying any scene pictorially, it is possible to create a retinal image at the eye that is geometrically identical to the one that would result from viewing the same scene in the real world, by positioning the optical centre (nodal point) of the eye at the centre of projection (CoP) of the image. In the case of S3D media, creating a pair of such retinal images, and therefore the same pattern of binocular disparities that would be generated by viewing the real scene, requires of course that both eyes are positioned at the CoP of their respective images. There are several fundamental reasons, at each stage of the process of S3D production, why this is rarely achieved in practice, and we explore these in this section.

Of course, creating geometrically identical binocular disparities to those that would result from viewing a real scene (sometimes called orthostereoscopic presentation) may not be the goal of S3D producers. There are specialist uses of stereoscopic viewing, such as virtual reality, where this may be the case, but S3D content makers frequently intentionally deviate from natural viewing geometry for artistic reasons. Nonetheless, the content maker is still presumably attempting to convey a particular sense of 3D scene structure, and the factors discussed here can cause the pattern of binocular disparities at the viewers' eyes to be significantly different than is intended.

18.7.1 Capture Parameters

One of the key parameters that can be varied in stereoscopic filming is the lateral separation of the cameras, or *inter-axial distance*. As Figure 18.6 shows, the magnitude of disparities projected by a given scene increases with increasing separation between the two viewpoints (Equation 18.1). Thus, variations in camera inter-axial distance are used to vary the overall

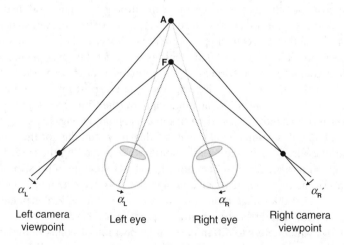

Figure 18.6 Changes in horizontal binocular disparity introduced by changing the camera inter-axial distance

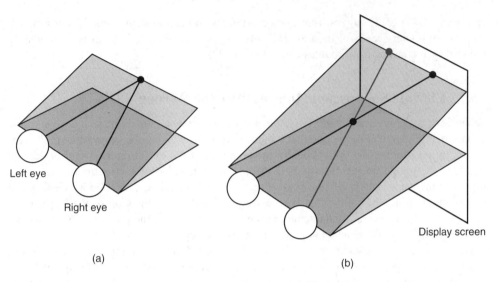

Left eye

Right eye

Display screen

(a)   (b)

Figure 18.7 Epipolar geometry in natural binocular vision and S3D. (a) In natural viewing, the eyes and an image point lie on an epipolar plane. (b) Creating the same epipolar plane in S3D viewing. The black and dark-grey dots show the screen positions of the same point in the left- and right-eye images. There is horizontal parallax, consistent with the point's position nearer than the screen, but zero vertical parallax. This analysis assumes that the viewer's inter-ocular axis is parallel to the ground (i.e. the head is upright) and parallel to the screen (i.e. the screen is viewed 'straight on'). See also Banks *et al.* (2009, 2012)

range of perceived depth from disparity in a scene (e.g. increasing it for distant objects, that would normally project very small disparities, or reducing it for very close-up viewing) and to vary the perceived 3D shape (roundness) of objects (Mendiburu, 2009). Of course, we view the resulting S3D content with a fixed distance between our eyes (our IOD), and so nonnatural disparities are necessarily introduced, with risks such as the portrayed scene appearing like a scale model (the *puppet-theatre effect*) or grossly oversized (*gigantism*).

Another stereoscopic filming parameter that can introduce an unnatural disparity stimulus to the viewer is the degree of convergence, or 'toe-in', of the cameras. It is sometimes assumed that because the eyes converge to a greater or lesser extent to fixate objects, stereoscopic cameras should also converge. This logic is incorrect, however, owing to differences in the projection geometry in the two cases. The retina is a roughly spherical surface and so the eye rotates 'behind' its image, leaving the relative positions of points essentially unchanged. In contrast, the image surface in a camera (the film or sensor) is planar, and so rotation introduces trapezoidal transformations in the captured images.

To understand how this results in an unnatural disparity stimulus for the S3D viewer it is helpful first to consider how image points should be displayed on the screen to recreate a natural stimulus. Figure 18.7a illustrates how, in natural viewing, a point in the world and the nodal points of the two eyes lie on an epipolar plane. To recreate this situation in an S3D display (Figure 18.7b) the image points must lie on the same epipolar plane. Thus, there should be no vertical parallax between image points on the screen (note the distinction between *parallax*, which refers to differences in the position of points in the left- and right-eye images on the screen, and *disparity*, which refers to differences in the position of points in the retinal images).

Shooting with converged cameras   **Shooting with parallel cameras**

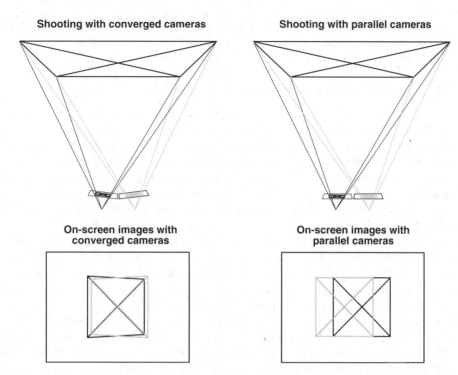

On-screen images with   **On-screen images with**
converged cameras   **parallel cameras**

Figure 18.8 Inappropriate vertical parallax introduced by filming with converged cameras, as opposed to parallel cameras. The figure was adapted (with the author's permission) from figures presented by Jenny Read (Newcastle University, UK), at 3DStereo Media 2010, Liege, Belgium

Camera convergence leads to violations of this epipolar geometry by introducing patterns of vertical parallax. Consider the simple situation, schematized in Figure 18.8, of filming a 'fronto-parallel' square. With converged cameras there is an equal and opposite trapezoidal distortion in the resulting images, leading to a systematic pattern of vertical parallax on the screen (the shape should be a square on the screen in both the left- and right-eye's image). See Allison (2007) for a detailed analysis of image changes resulting from variations in camera parameters.

As noted earlier, vertical parallax per se can cause viewer discomfort (Kooi and Toet, 2004; Kane *et al.*, 2012). Less well recognized is the effect that the introduction of systematic patterns of vertical parallax may have on perceived depth from binocular disparity. Natural viewing results in a pattern of vertical *retinal disparities*. Consider again viewing the square in Figure 18.8. The left edge is nearer to the left eye than to the right eye, and so will be taller in the left-eye's image than in the right-eye's image, and vice versa. Thus, there is a horizontal gradient of vertical disparity across the retinal images, which depends on the distance to the stimulus (Rogers and Bradshaw, 1993, 1995; Backus *et al.*, 1999). As noted earlier, the visual system uses this signal as part of the distance estimate used to scale binocular disparities (see Section 18.3.1). The addition of a pattern of vertical parallax introduced by camera convergence changes the pattern of vertical disparities at the retina, and so would be expected to change perceived depth from disparity (Allison, 2007; Banks *et al.*, 2012).

These problems can in principle be avoided simply by shooting S3D content with parallel camera axes. This is reasonable when the camera inter-axial distance is close to, or smaller than, the IOD. Where large inter-axial distances are used, however, the cameras may have to be converged to preserve sufficient field-of-view either side of the point of interest.

18.7.2 Display Parameters and Viewer Parameters

In S3D content, differences between the positions of points in the left- and right-eye images are usually described in terms of horizontal *parallax*, expressed as a percentage of the screen (or image) size. The pattern of binocular disparities at the eyes is a product not only of this parallax, but also of how the images are displayed and viewed. These latter two factors are frequently not under the control of the content maker and so the pattern of binocular disparities (which are fundamentally angular in nature), and therefore the resulting 3D percept, can differ substantially from what was intended (Woods *et al.*, 1993; Banks *et al.*, 2009; Benzeroual *et al.*, 2012).

Most obviously, content is typically produced for multiple platforms, including cinema, television, computer screens and handheld devices, which all have very different-sized screens and viewing distances. A well-known problem is that when images are scaled up to larger than intended screen sizes it can be necessary to *diverge* the eyes (i.e. beyond parallel visual axes) to achieve stereoscopic fusion. This should be avoided because humans can achieve only about 1° of divergence. Other systematic effects occur too, however. Changing the angular size of the screen (viewing different-sized screens at the same distance, for example) changes the depth specified by binocular disparity (Figure 18.9a). And even if the *angular* size of the screen is held reasonably constant as viewing distances changes (a not wholly unreasonable assumption because we tend to view smaller screens at nearer distances; Ardito, 1994) the range of physical depths specified by binocular disparity nonetheless still changes (Figure 18.9b). Benzeroual *et al.* (2012) provide a detailed analysis of the implications of this and explore how to scale content between cinema and television, for example. Moreover, our viewing position is frequently even less well controlled than these situations. Multiple viewers in cinema and television, for example, will view the same screen at different distances (Figure 18.9c) and eccentricities (Figure 18.9d), also causing changes in the 3D structure specified by binocular disparity (in the latter case the viewer's inter-ocular axis may also deviate from parallel to the screen, which introduces violations of epipolar geometry; Banks *et al.*, 2009).

18.7.3 Are Problems of Incorrect Geometry Unique to S3D?

Figure 18.9 shows several reasons why, in S3D viewing, each eye typically receives a geometrically 'incorrect' retinal image (or one other than was intended). Of course, the same is true when we view conventional images, because the viewer is rarely positioned exactly at the CoP. Here, however, the visual system seems to compensate for being 'in the wrong place' because we experience very little apparent distortion in 3D structure. One might reasonably ask, therefore, whether such compensation will also occur for S3D media, in which case we need not worry about this issue. Unfortunately, however, there are at least two reasons why compensation for being in the wrong place is unlikely to be so effective in S3D.

First, most monocular depth cues (see the discussion around Figure 18.1) provide non-metric information about depth. Occlusion, for example, provides ordinal depth information: it specifies that one object is in front of another, but provides no information at all about the relative distances between those objects. Texture/perspective provides relative depth

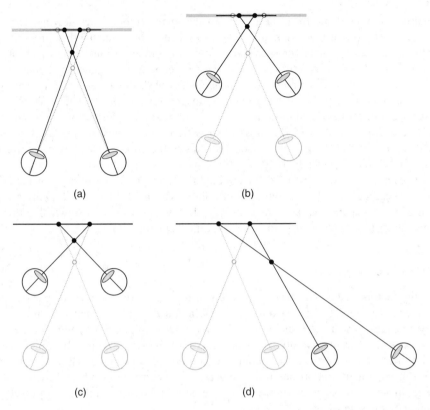

(a)   (b)

(c)   (d)

Figure 18.9 Common display and viewer parameters that lead to incorrect (or different than intended) patterns of binocular disparities at viewers' eyes. In each case the semi-transparent image shows the same intended binocular geometry, and the solid image shows the consequences of the changes. (a) Changing screen size. (b) Changing screen size and viewing distance, to preserve a constant angular screen size. (c) Moving the viewer to a different distance. (d) Moving the viewer to one side

information that can be used to extract the orientation of a surface relative to the viewer, but not its distance (Knill, 1998). In contrast, information from binocular vision (horizontal disparity, convergence, vertical disparity) can provide the complete metric structure of the 3D scene (Gårding *et al.*, 1995), including the absolute distance to image points. When this information is incorrect it can, therefore, result in large perceptual distortions such as the entire scene appearing miniaturized or gigantic. Similar effects are caused by artificial manipulations of the blur gradient in conventional images. The blur gradient contains information about the absolute distances in a scene, and manipulating it (in tilt–shift photography, for example) can result in powerful effects of such as large scenes appearing as scale models (Held *et al.*, 2010; Vishwanath and Blaser, 2010).

Second, experiments suggest that the binocular disparity signal from non-S3D images, which here specifies the orientation of the picture or display surface, may be the critical signal used by the brain to compensate for incorrect viewing position. Vishwanath *et al.* (2005) measured the perception of 3D shape from monocular depth cues at different viewing angles from the CoP when (*i*) no information was available about the picture-surface orientation (monocular viewing through a pinhole aperture), (*ii*) monocular cues to surface orientation were available from the outline of the image (monocular viewing without an aperture) and (*iii*) the surface orientation was also specified by binocular disparity (binocular viewing, no aperture). Observers showed little or no compensation for being at incorrect positions in the first two conditions, but showed perceptual invariance (compensation for incorrect position) for a large range of viewing angles when binocular disparity was available. This finding suggests that, for conventional images, binocular information about the picture surface allows the visual system to 'undo' the geometric distortions induced by not being at the CoP. In S3D viewing, this signal is 'removed' (because it is used instead to specify 3D structure of the portrayed scene) and so such compensation may not occur, or at least may be much reduced (Banks *et al.*, 2009).

More recently, Banks *et al.* (2009) examined this question more directly, using stereoscopic images, and found little evidence of compensation for incorrect viewing position. They used scenes containing few monocular depth cues, however, and so whether this finding generalizes to more typical, complex S3D scenes remains to be determined.

18.7.4   Conclusions

All of the factors discussed above can result in a pattern of binocular disparities at the S3D viewers' eyes that differs from what was intended. The fact that S3D media often work well suggests that the visual system has reasonable tolerances to many of the resulting issues. And certainly, variations in the percept of 3D structure in S3D cannot be predicted entirely by 'errors' in the geometric content of the images (Benzeroual *et al.*, 2011). But significant distortions in perceived 3D structure do occur, and more research is needed if we are to accurately predict the percept of 3D structure that will result from viewing a given S3D image.

This work is likely to be extremely challenging for vision science because it necessarily involves the consideration of S3D scenes that contain a large number of cues to depth, which can be differently altered by the factors discussed above. This can introduce cue conflicts unlike those experienced in the real world. Indeed, S3D content producers sometimes successfully use composite shots, where different stereo filming parameters are used for different parts of the same scene. Perception of 3D structure in these circumstances poses a particularly interesting set of vision-science questions.

18.8   Fixating and Focusing on Stereoscopic Images

In the real world, when we look at objects at different depths our eyes make two principal *oculomotor responses*. If the retinal image is blurred, the ciliary muscles in the eye act to change the shape of the crystalline lens to bring the image into focus (or more precisely, to optimize retinal-image contrast; Alpern, 1958; Fender, 1964; Toates, 1972; Heath, 1956; Owens, 1980; Manny and Banks, 1984; Kotulak and Schor, 1986; MacKenzie *et al.*, 2010). This process is called *accommodation*. If the desired fixation point is at noncorresponding points on the two retinas (i.e. there is an *absolute disparity*) then the eyes make *vergence eye movements* (they converge or diverge relative to their current position) to place the point on the centre of both eyes' retinas. In S3D media, the relationship between the stimulus to accommodation (retinal blur) and the stimulus to vergence (absolute disparity) is unnatural, sometimes resulting in unwanted consequences for the viewer. We explore this below.

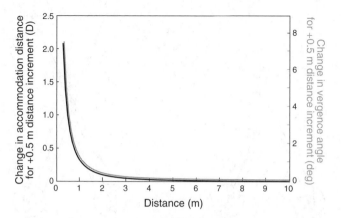

Figure 18.10 The relationship between viewing distance and (i) accommodation and (ii) vergence. The black curve plots the change in accommodation state, at each distance, required to refocus the eye following a 0.5 m increment in viewing distance (left y-axis). The grey curve plots the change in vergence, at each distance, required to fixate an object with the two eyes following a 0.5 m increment in viewing distance (right y-axis)

18.8.1 Accommodation, Vergence and Viewing Distance

We first consider how accommodation and vergence are related to viewing distance. This will be relevant later for understanding how screen distance affects the likelihood of unwanted effects in the viewer.

The amount of blur in the retinal image, and so the magnitude of the required focusing response, is proportional to how defocused the eye is in units of dioptres (D), where dioptres are the reciprocal of distance in metres. The relationship between distance and dioptres is summarized in Figure 18.10, which plots the change in dioptric distance (i.e. the required accommodation response) that corresponds to an increment in depth of 0.5 m at all physical distances. At far distances, even large changes in stimulus depth require only small focusing responses. At near distances, even quite small changes in depth require large focusing responses.

Vergence is described in terms of the angle between the two visual axes. When fixating a near object the vergence angle is large, and for fixating at infinity the vergence angle is zero (the two visual axes are parallel). Figure 18.10 also plots the change in vergence angle (i.e. the required vergence response) that corresponds to an increment in depth of 0.5 m. It can be seen that vergence follows the same relationship with distance as accommodation does. This is because the geometry of light passing through an entrance pupil of a certain size – the cause of defocus – and of two eyes viewing the world from two different positions is equivalent (Held et al., 2012).

18.8.2 Accommodation and Vergence in the Real World and in S3D

In real-world viewing, as we look around scenes the stimuli to accommodation and vergence are at the same distance, and so the two responses co-vary. In conventional stereoscopic displays, the stimulus to vergence (binocular disparity) is manipulated to simulate different depths in the world, but the stimulus to accommodation remains at the display surface (the screen) because the image is always displayed there, independent of the

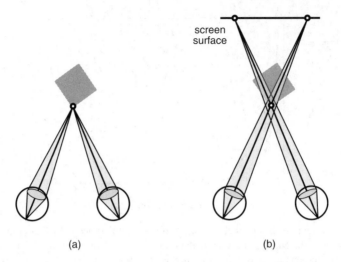

Figure 18.11 The vergence–accommodation conflict in S3D. (a) Real-world viewing: consistent stimuli to accommodation and vergence. (b) S3D viewing: conflicting stimuli to accommodation and vergence

portrayed depth structure in the scene (Figure 18.11). This is the well-known *vergence–accommodation conflict*. In order to converge and accommodate accurately to image points at distances other than the screen, the viewer must converge at one distance while accommodating at another. Vergence and accommodation responses are neurally coupled, acting synergistically to make responses more efficient in the real world (Fincham and Walton, 1957; Martens and Ogle, 1959). In S3D viewing, however, these two responses must be decoupled. This can be effortful, and is not always possible, leading to the risk of a blurry retinal image, failure of binocular fusion, or both (Miles *et al.*, 1987; Wöpking, 1995; Wann and Mon-Williams, 1997; Peli, 1998; Rushton and Riddell, 1999; Hoffman *et al.*, 2008).

There is now considerable evidence that vergence–accommodation conflicts per se can cause a range of unwanted perceptual effects such as increases in the time taken to achieve stereoscopic fusion and reduced stereoacuity – the degree to which fine modulations in depth from binocular disparity are visible (Akeley *et al.*, 2004; Watt *et al.*, 2005b; Hoffman *et al.*, 2008). Moreover, they have also been shown to cause fatigue and discomfort (Wann and Mon-Williams, 1997; Yano *et al.*, 2004; Emoto *et al.*, 2005; Häkkinen *et al.*, 2006; Ukai, 2007; Hoffman *et al.*, 2008; Ukai and Howarth, 2008; Shibata *et al.*, 2011).

18.8.3 Correcting Focus Cues in S3D

One way to address these problems is to build a display in which the focal distance to points in the image matches the convergence distance, thereby eliminating the vergence–accommodation conflict. A successful approach has been to build so-called multiple-focal-planes displays, in which each eye sees the sum of images presented on several image planes (Akeley *et al.*, 2004; Liu and Hua, 2009; Love *et al.*, 2009; MacKenzie *et al.*, 2010, 2012). Because the sensitivity of the human visual system to focal depth is relatively poor

(Campbell, 1957), it is sufficient to sample the range of focal distances relatively coarsely, compared with the spatial resolution (i.e. screen pixel size) required (Rolland *et al.*, 1999). This can be further optimized using a technique called *depth filtering*, in which continuous variations in focal distance are simulated by distributing image intensity across focal planes (Akeley *et al.*, 2004; MacKenzie *et al.*, 2010; Ravikumar *et al.*, 2011). This is critical to the practicality of this approach because there are limits on the number of image planes that can be used. We have recently shown that depth-filtered images, created using image planes spaced 0.6 D apart, result in performance that is indistinguishable from real-world viewing in terms of accommodation and vergence responses, and stereoscopic depth perception (fusion time, stereoacuity) (MacKenzie *et al.*, 2010, 2012; Watt *et al.*, 2012). This suggests that a display with just five image planes could cover the distance range 40 cm to infinity.

Such displays are promising, but they are only useful in specialist applications, where there is a single viewer, at a fixed viewpoint relative to the display (virtual reality head-mounted displays, or devices with a viewing aperture such as some surgical devices, for example). For general use, in multi-viewer applications such as television and cinema, there are currently no solutions to the vergence–accommodation conflict. In these cases it is necessary instead to understand the tolerances of the visual system to this situation. We turn to this next.

18.8.4 The Stereoscopic Zone of Comfort

Accommodation and vergence need to be reasonably accurate, although not perfectly so, in order for the viewer to perceive a clear and binocularly fused stereoscopic percept. The required accuracy of accommodation is specified by the eye's effective depth of focus, which has been estimated to be approximately 0.25–0.3 D (Campbell, 1957; Charman and Whitefoot, 1977). The required accuracy of vergence is given by Panum's fusion area, which, as mentioned earlier, is approximately 0.25–0.5°. As well as these tolerances, it is possible to decouple accommodation and vergence responses to some degree, although the ability to do so seems to be quite idiosyncratic (Saladin and Sheedy, 1978). Together, this gives rise to a concept called the *zone of clear, single binocular vision* (ZCSBV; Fry, 1939). As the name suggests, this describes the range of conflicting accommodation and vergence stimuli for which viewers generally perceive a clear, single (i.e. non-diplopic) stimulus. This represents an upper limit on the ability to decouple accommodation and vergence responses, however, and a much smaller range of conflicts is required if the viewer is to be comfortable and avoid fatigue – what might be termed the stereoscopic zone of comfort (ZoC; see also Percival (1920)).

Shibata *et al.* (2011) estimated the extent of ZoC by asking participants to make subjective ratings of fatigue and discomfort after periods of viewing stimuli with various degrees of vergence–accommodation conflict. They examined the effect of the screen distance and the sign of the conflict (disparity-defined objects nearer versus farther than the screen surface). Critically, they used a multiple-image-planes display (see Section 18.8.3) to isolate the vergence–accommodation conflict and to allow comparisons to equivalent 'real-world' stimuli (i.e. the same variations in stereoscopic depth, but with a correctly varying stimulus to accommodation). The main result of their study is replotted in Figure 18.12. The figure shows the typical estimate of the size of the ZCSBV (light grey), as well as the Shibata *et al.* (2011) estimate of the ZoC (dark grey). Both accommodation and vergence distances are plotted in units of dioptres so that they can be directly compared. In these units, which are presumably near linear in terms of accommodative and vergence effort, the ZoC is approximately the same size at different distances (though slightly smaller at far screen distances).

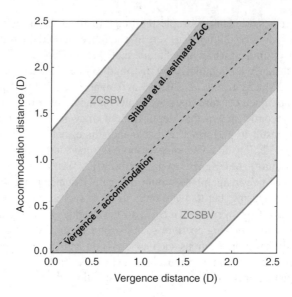

Figure 18.12 The ZCSBV (in light grey), and the Shibata *et al.* (2011) estimate of the stereoscopic ZoC (in dark grey) plotted in units of dioptres. The dashed diagonal line shows the relationship between the stimuli to accommodation and vergence in real-world viewing, with zero conflict. Our thanks to Shibata and colleagues for kindly supplying the data used to produce this figure

There are subtle asymmetries, however, with the ZoC generally extending farther beyond the screen surface for near screen distances, and vice versa.

18.8.5 Specifying the Zone of Comfort for Cinematography

To understand the practical implications of this estimated ZoC for S3D content production, it is helpful to plot the data in units of physical distance (Figure 18.13). Because of how dioptres are related to physical distance (Figure 18.10), the ZoC is large at far distances and becomes very small at near distances. This suggests that exceeding it in cinema will be quite unlikely (though not impossible, for objects coming out of the screen towards the audience), whereas at nearer screen distances, including typical television- and computer-viewing distances, the ZoC is actually quite small.

It should also be noted that this way of specifying the ZoC is not compatible with the typical industry practice of specifying the ZoC in terms of percentage of screen parallax (2–3% of the screen width for objects nearer than the screen and 1–2% of screen width for objects farther away; Mendiburu, 2009; ATSC, 2011). As outlined in previous sections, units of screen parallax do not uniquely relate to disparities (and, therefore, vergence–accommodation conflicts) at the viewer's eyes because of uncontrolled variations in screen size, viewing position and so on. Indeed, for these reasons, any rule based on screen parallax must be technically incorrect. However, given that content makers must typically work in image-based units, and that we tend to view large screens at farther distances than small screens (Ardito, 1994), this working rule undoubtedly has practical value. In principle, it could be refined, however, to reflect the asymmetries observed at different distances by Shibata *et al.* (2011). See Shibata *et al.* (2011) for an exploration of this and related issues.

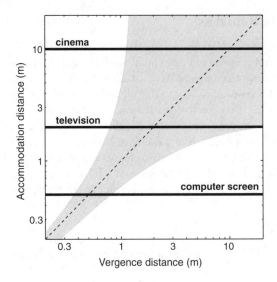

Figure 18.13 Shibata *et al.*'s (2011) estimate of the stereoscopic ZoC in units of physical distance. The dashed diagonal line again shows the relationship between the stimuli to accommodation and vergence in real-world viewing. The solid horizontal lines show the ZoC at representative viewing distances of 10, 2 and 0.5 m. Our thanks to Shibata and colleagues for kindly supplying the data used to produce this figure

18.8.6 Conclusions

The above research is a valuable start, but much remains to be determined in this area if comprehensive guidelines are to be produced. Unlike in most experiments, in actual S3D content such as movies the range of disparity-specified depths, and therefore the degree of vergence–accommodation conflict, is often deliberately controlled throughout according to a 'depth script' (Mendiburu, 2009). The effects of these temporal variations (Do viewers recover during rest periods, etc.?) remain largely unknown (although it does seem that higher frequency changes cause more discomfort than lower frequency ones; Kim *et al.*, 2012). There appear to be large individual differences in people's susceptibility to unpleasant symptoms caused by vergence–accommodation conflicts (and indeed other causes; Lambooij *et al.*, 2009), and these are hardly at all understood. The age of the viewer may well be important too. Older viewers, with less ability to accommodate, effectively live with vergence–accommodation conflicts all the time, so might be expected to be less susceptible to problems (Yang *et al.*, 2011; Banks *et al.*, 2012). Relatively large-scale 'population' studies are needed to address these questions.

Other oculomotor factors may also play a role in discomfort. For example, tilting the head to one side (so the inter-ocular axis is no longer horizontal to the ground) introduces vertical parallax, with respect to the viewer, requiring a vertical vergence eye movement in order to fuse the stimulus. Kane *et al.* (2012) recently showed that this produces significant levels of discomfort and fatigue.

We also know little about whether there are any significant longer term effects of being exposed to unnatural oculomotor stimuli. Currently, there is no evidence to support specific problems, but neither have long-term studies been carried out. As society shifts from S3D viewing being a niche activity to a mainstream one, the precautionary principle suggests that the industry, clinicians and researchers should be alert to potential problems, particularly while the visual system is still developing (Rushton and Riddell, 1999).

18.9 Concluding Remarks

Undoubtedly, S3D media can be compelling. But they can also be unpleasant and tiring to watch, as well as creating unwanted perceptual artefacts and departures from what the content producer intended. Understanding these issues from the perspective of human vision is challenging, not least because it relates to stimuli and situations far more complex than typical experimental studies. Perhaps for this reason, and the fact that mainstream success of S3D media is relatively recent, we currently lack sufficient understanding in several key areas to suggest more than qualitative guidelines for content producers. Addressing these issues is worthwhile, however, as it is likely to be of benefit not just to the S3D industry, but also to the general population, who are the users of S3D, and to the scientific community, by adding to our understanding of the fundamental principles of human vision.

Acknowledgments

This material is supported by a grant from the Engineering and Physical Sciences Research Council to SJW.

References

Akeley, K., Watt, S. J., Girshick, A. R. and Banks, M. S. (2004) A stereo display prototype with multiple focal distances. *ACM Trans. Graphic.*, **23**, 1084–1813.

Alais, D. and Burr, D. (2004) The ventriloquist effect results from near-optimal cross-modal integration, *Curr. Biol.*, **14**, 257–262.

Allison, R. S. (2007) An analysis of the influence of vertical disparities arising in toed-in stereoscopic cameras. *Imaging Sci. Technol.*, **51**, 317–327.

Alpern, M. (1958) Variability of accommodation during steady state fixation at various levels of illuminance. *J. Opt. Soc. Am.*, **48**, 193–197.

Ardito, M. (1994) Studies of the influence of display size and picture brightness on the preferred viewing distance for HDTV programs. *SMPTE J.*, **103**, 517–522.

ATSC (2011) Advanced Television Systems Committee (ATSC) report on 3D digital television, ATSC Planning Team 1 Interim Report, http://www.atsc.org/PT1/PT-1-Interim-Report.pdf.

Backus, B. T., Banks, M. S., van Ee, R. and Crowell, J. A. (1999) Horizontal and vertical disparity, eye position, and stereoscopic slant perception. *Vision Res.*, **39**, 1143–1170.

Banks, M. S., Gepshtein, S. and Landy, M. S. (2004) Why is spatial stereoresolution so low? *J. Neurosci.*, **24**, 2077–2089.

Banks, M. S., Held, R. T. and Girshick, A. R. (2009) Perception of 3-D layout in stereo displays. *Inform. Disp.*, **25**, 12–16.

Banks, M. S., Read, J. R., Allison, R. S. and Watt, S. J. (2012) Stereoscopy and the human visual system. *SMPTE Motion Imag. J.*, **121** (4), 24–43.

Benzeroual, K., Wilcox, L. M., Kazimi, A. and Allison, R. S. (2011) On the distinction between perceived & predicted depth in S3D films. Proceedings of the International Conference on 3D Imaging.

Benzeroual, K., Allison, R., and Wilcox, L. M. (2012) 3D display matters: compensating for the perceptual effects of S3D display scaling, in *2012 IEEE Computer Society Conference on Computer Vision and Pattern Recognition Workshops (CVPRW)*, IEEE, pp. 45–52.

Blake, A. and Bülthoff, H. H. (1990) Does the brain know the physics of specular reflections? *Nature*, **343**, 165–168.

Blakemore, C. (1970) Range and scope of binocular depth discrimination in man. *J. Physiol.*, **211**, 599–622.

Bresciani, J. P., Dammeier, F. and Ernst, M. O. (2006) Vision and touch are automatically integrated for the perception of sequences of events. *J. Vision*, **6** (5), 554–564.

Burian, H. M. (1943) Influence of prolonged wearing of meridional size lenses on spatial localization. *Arch. Ophthamol.*, **30**, 645–668.

Burr, D. C. and Ross, J. (1979) How does binocular delay give information about depth. *Vision Res.*, **19**, 523–532.

Burt, P. and Julesz, B. (1980) A disparity gradient limit for binocular fusion. *Science*, **208**, 615–617.

Campbell, F. W. (1957) The depth of field of the human eye. *Opt. Acta*, **4**, 157–164.

Charman, W. N. and Whitefoot, H. (1977) Pupil diameter and the depth-of-field of the human eye as measured by laser speckle. *Opt. Acta*, **24**, 1211–1216.

Clark, J. J. and Yuille, A. L. (1990) *Data Fusion for Sensory Information Processing Systems*, Kluwer, Boston, MA.

Cormack, L. K., Stevenson, S. B. and Schor, C. M. (1991) Interocular correlation, luminance contrast and cyclopean processing. *Vision Res.*, **31**, 2195–2207.

Devernay, F. and Beardsley, P. (2010) Stereoscopic cinema, in *Image and Geometry Processing for 3-D Cinematography* (eds R. Ronfard and G. Taubin), Geometry and Computing, Vol. **5**, Springer, Berlin, pp. 11–51.

Dodgson, N. (2004) Variation and extrema of human interpupillary distance, in *Stereoscopic Displays and Virtual Reality Systems XI* (eds A. J. Woods, J. O. Merritt, S. A. Benton and M. T. Bolas), Proceedings of SPIE, Vol. **5291**, SPIE, Bellingham, WA, pp. 36–46.

Emoto, M., Niida, R., and Okano, F. (2005) Repeated vergence adaptation causes the decline of visual functions in watching stereoscopic television. *J. Disp. Technol.*, **1**, 328–340.

Ernst, M. O. (2006) A Bayesian view on multimodal cue integration, in *Human Body Perception From The Inside Out* (eds G. Knoblich, I.M. Thornton, M. Grosjean and M. Shiffrar), Oxford University Press, New York, NY.

Ernst, M. O. (2007) Learning to integrate arbitrary signals from vision and touch. *J. Vision*, **7** (7), 1–14.

Ernst, M. O. and Banks, M. S. (2002) Humans integrate visual and haptic information in a statistically optimal fashion. *Nature*, **415**, 429–433.

Fender, D. H. (1964) Control mechanisms of the eye. *Sci. Am.*, **211**, 24–33.

Fincham, E. F. and Walton, J. (1957) The reciprocal actions of accommodation and convergence. *J. Physiol.*, **137**, 488–508.

French, J. W. (1921) The interocular distance. *Trans. Opt. Soc.*, **23**, 44–55.

Fry, G. A. (1939) Further experiments on the accommodation–convergence relationship. *Am. J. Optom. Arch. Am. Acad. Optom.*, **16**, 325–336.

Gårding, J., Porrill, J., Mayhew, J. E. W. and Frisby, J. P. (1995) Stereopsis, vertical disparity and relief transformations. *Vision Res.*, **35**, 703–722.

Gardner, B. R. (2012) Dynamic floating window: new creative tool for three-dimensional movies. *J. Electron. Imag.*, **21** (1), 011009.

Gepshtein, S., Burge, J., Ernst, M. and Banks, M. S. (2005) The combination of vision and touch depends on spatial proximity. *J. Vision*, **5**, 1013–1023.

Girschick, A. R. and Banks, M. S. (2009) Probabilistic combination of slant information: weighted averaging and robustness as optimal percepts. *J. Vision*, **9** (9), 1–20.

Greenwald, H. S. and Knill, D. C. (2009) Cue integration outside central fixation: a study of grasping in depth. *J. Vision*, **9** (2), 1–16.

Häkkinen, J., Pölönen, M., Takatalo, J. and Nyman, G. (2006) Simulator sickness in virtual gaming: a comparison of stereoscopic and non-stereoscopic situations, in *MobileHCI '06 Proceedings of the 8th Conference on Human–Computer Interaction with Mobile Devices and Services*, ACM, New York, NY, pp. 227–230.

Heath, G. G. (1956) Components of accommodation. *Am. J. Optom.*, **33**, 569–579.

Held, R.T., Cooper, E., O'Brien, J. and Banks, M.S. (2010) Using blur to affect perceived distance and size. *ACM Trans. Graphic.*, **29** (2), article 19.

Held, R. T., Cooper, E. A. and Banks, M. S. (2012) Blur and disparity are complementary cues to depth. *Curr. Biol.*, **22**, 1–6.

Hillis, J. M., Watt, S. J., Landy, M. S. and Banks, M. S. (2004) Slant from texture and disparity cue: optimal cue combination. *J. Vision*, **4**, 967–992.

Hoffman, D. M., Girshick, A. R., Akeley, K. and Banks, M. S. (2008) Vergence–accommodation conflicts hinder visual performance and cause visual fatigue. *J. Vision*, **8** (3), 1–30.

Hoffman, D. M., Karasev, V. I. and Banks, M. S. (2011) Temporal presentation protocols in stereoscopic displays: flicker visibility, perceived motion, and perceived depth. *J. Soc. Inf. Disp.*, **19** (3), 255–281.

Howard, I. P. and Rogers, B. J. (2002) *Seeing in Depth*, University of Toronto Press, Toronto, ON.

Julesz, B. (1960) Binocular depth perception of computer-generated patterns. *Bell Labs Tech. J.*, **39**, 1125–1162.

Kane, D., Held, R. T. and Banks, M. S. (2012) Visual discomfort with stereo 3D displays when the head is not upright, in *Stereoscopic Displays and Applications XXIII* (eds A. J. Woods, N. S. Holliman and G. E. Favalora), Proceedings of the SPIE, Vol. **8288**, SPIE, Bellingham, WA, p. 828814.

Keefe, B. D., Hibbard, P. B. and Watt, S. J. (2011) Depth-cue integration in grasp programming: no evidence for a binocular specialism. *Neuropsychogia*, **49**, 1249–1257.

Kim, J., Kane, D. and Banks, M. S. (2012) Visual discomfort and the temporal properties of the vergence-accommodation conflict, in *Stereoscopic Displays and Applications XXIII* (eds A. J. Woods, N. S. Holliman and G. E. Favalora), Proceedings of the SPIE, Vol. **8288**, SPIE, Bellingham, WA, p. 828811.

Knill, D. C. (1998) Surface orientation from texture: ideal observers, generic observers and the information content of texture cues. *Vision Res.*, **38**, 1655–1682.

Knill, D. C. (2007) Robust cue integration: a Bayesian model and evidence from cue-conflict studies with stereoscopic and figure cues to slant. *J. Vision*, **7** (7), 1–24.

Knill, D. C. and Pouget, A. (2004) The Bayesian brain: the role of uncertainty in neural coding and computation. *Trends Neurosci.*, **27**, 712–719.

Knill, D. C. and Saunders, J. (2003) Do humans optimally integrate stereo and texture information for judgments of surface slant? *Vision Res.*, **43**, 2539–2558.

Kooi, F. L. and Toet, A. (2004) Visual comfort of binocular 3D displays. *Displays*, **25**, 99–108.

Körding, K. P., Beierhold, U., Ma, W. J. *et al.* (2007) Causal inference in multisensory perception. *PLoS ONE*, **2**, e943.

Kotulak, J. C. and Schor, C.M. (1986) A computational model of the error detector of human visual accommodation. *Biol. Cybern.*, **54**, 189–194.

Kulikowski, J. J. (1978) Limit of single vision depends on contour sharpness. *Nature*, **276**, 126–127.

Lambooij, M., Ijsselsteijn, W., Fortuin, M. and Heynderickx, I. (2009) Visual discomfort and visual fatigue of stereoscopic displays: a review. *J. Imaging Sci. Technol.*, **53** (3), 1–14.

Landis, C. (1954) Determinants of the critical flicker-fusion threshold. *Physiol. Rev.*, **34**, 259–286.

Landy, M. S., Maloney, L. T., Johnston, E. B. and Young, M. (1995) Measurement and modeling of depth cue combination: in defense of weak fusion. *Vision Res.*, **35**, 389–412.

Legge, G. E. and Gu, Y. (1989) Stereopsis and contrast. *Vision Res.*, **29**, 989–1004.

Liu, S. and Hua, H. (2009) Time-multiplexed dual-focal plane head-mounted display with a liquid lens. *Opt. Lett.*, **34** (11), 1642–1644.

Longuet-Higgins, H. C. (1981) A computer algorithm for reconstructing a scene from two projections. *Nature*, **293**, 133–135.

Love, G. D., Hoffman, D. M., Hands, P. J. W. *et al.* (2009) High-speed switchable lens enables the development of a volumetric stereoscopic display. *Opt. Express*, **17**, 15716–15725.

MacKenzie, K. J., Hoffman, D. M. and Watt, S. J. (2010) Accommodation to multiple-focal-plane displays: implications for improving stereoscopic displays and for accommodation control. *J. Vision*, **10** (8), 1–20.

MacKenzie, K. J., Dickson, R. A. and Watt, S. J. (2012) Vergence and accommodation to multiple-image-plane stereoscopic displays: "real world" responses with practical image-plane separations? *J. Electron. Imaging*, **21** (1), 011002.

Manny, R. E. and Banks, M. S. (1984) A model of steady-state accommodation: II. Effects of luminance. *Invest. Ophthalmol. Vis. Sci.*, **25**, 182.

Martens, T. G. and Ogle, K. N. (1959) Observations on accommodative convergence; especially its nonlinear relationships. *Am. J. Otolaryngol.*, **47**, 455–462.

Mendiburu, B. (2009) *3D Movie Making: Stereoscopic Digital Cinema from Script to Screen*, Focal Press, Burlington, MA.

Miles, F. A., Judge, S. J. and Optican, L. M. (1987) Optically induced changes in the couplings between vergence and accommodation. *J. Neurosci.*, **7**, 2576–2589.

Miles, H. C., Pop, S. R., Watt, S. J. *et al.* (2012) A review of virtual environments for training in ball sports. *Comput. Graph.*, **36** (6), 714–726.

Mon-Williams, M., Treisilan, J. R. and Roberts, A. (2000) Vergence provides veridical depth perception from horizontal retinal image disparities. *Exp. Brain Res.*, **133**, 407–413.

Morgan, M. J. (1979) Perception of continuity in stroboscopic motion: a temporal frequency analysis. *Vision Res.*, **19**, 523–532.

Nienborg, H., Bridge, H., Parker, A. J. and Cumming, B. G. (2004) Receptive field size in V1 neurons limits acuity for perceiving disparity modulation. *J. Neurosci.*, **24**, 2065–2076.

Ogle, K. N. (1932) An analytical treatment of the longtitudnal horopter, its measurement and application and application to related phenomena, especially to the relative size and shape of the ocular images. *J. Opt. Soc. Am.*, **22**, 665–728.

Ogle, K. N. (1938) Induced size effect. I. A new phenomenon in binocular space perception associated with the relative sizes of the images of the two eyes. *Arch. Ophthalmol.*, **20**, 604–623.

O'Leary, A. and Wallach, H. (1980) Familiar size and linear perspective as distance cues in stereoscopic depth constancy. *Percept. Psychophys.*, **27**, 131–135.

Oruç, I., Maloney, L. T. and Landy, M. S. (2003) Weighted linear cue combination with possibly error. *Vision Res.*, **43**, 2451–2468.

Owens, D. A. (1980) A comparison of accommodative responses and contrast sensitivity for sinusoidal gratings. *Vision Res.*, **20**, 159–167.

Peli, E. (1998) The visual effects of head-mounted display (HMD) are not distinguishable from those of desk-top computer display. *Vision Res.*, **38**, 2053–2066.

Percival, A. S. (1920) *The Prescribing of Spectacles*, J. Wright, Bristol.

Ravikumar, S., Akeley, K. and Banks, M. S. (2011) Creating effective focus cues in multi-plane 3D displays. *Opt. Express*, **19** (21), 20940–20952.

Roach, N. W., Heron, J. and McGraw, P. V. (2006) Resolving multisensory conflict: a strategy for balancing the costs and benefits of audio–visual integration. *Proc. R. Soc. Lond. B Biol. Sci.*, **273**, 2159–2168.

Rogers, B. J. and Bradshaw, M. F. (1993) Vertical disparities, differential perspective and binocular stereopsis. *Nature*, **361**, 253–255.

Rogers, B. J. and Bradshaw, M. F. (1995) Disparity scaling and the perception of frontoparallel surfaces. *Perception*, **24**, 155–179.

Rolland, J. P., Goon, A. A. and Krueger, M. W. (1999) Dynamic focusing in head-mounted displays, in *Stereoscopic Displays and Virtual Reality Systems VI*, (eds J. O. Merritt, M. T. Bolas and S. S. Fisher), Proceedings of the SPIE, Vol. **3639**, SPIE, Bellingham, WA, pp. 463–470.

Rovamo, J. and Raminen, A. (1988) Critical flicker frequency as a function of stimulus area and luminance at various eccentricities in human cone vision: a revision of Granit–Harper and Ferry–Porter laws. *Vision Res.*, **28**, 785–790.

Rushton, S. K. and Riddell, P. M. (1999) Developing visual systems and exposure to virtual reality and stereo displays: some concerns and speculations about the demands on accommodation and vergence. *Appl. Ergon.*, **30**, 69–78.

Saladin, J. J. and Sheedy, J. E. (1978) A population study of relationships between fixation disparity, hetereophoria and vergences. *Am. J. Optom. Phys. Opt.*, **55**, 744–750.

Schor, C.M., Wood, I. and Ogawa, J. (1984) Binocular sensory fusion is limited by spatial resolution. *Vision Res.*, **24**, 661–665.

Schor, C.M., Heckman, T. and Tyler, C.W. (1989) Binocular fusion limits are independent of target contrast, luminance gradient and component phase. *Vision Res.*, **29**, 821–835.

Sedgwick, H.A. (1986) Space perception, in *Handbook of Perception and Human Performance, Vol. 1, Sensory Processes and Perception* (eds K. R. Boff, L. Kaufman and J. P. Thomas), John Wiley & Sons, Inc., New York, NY.

Shams, L., Ma, W. J. and Beierholm, U. (2005) Sound-induced flash illusion as an optimal percept. *Neuroreport*, **16**, 1923–1927.

Shibata, T., Kim, J., Hoffman, D. M. and Banks, M. S. (2011) The zone of comfort: predicting visual discomfort with stereo displays. *J. Vision*, **11** (8), 1–29.

Takahashi, C., Diedrichsen, J. and Watt, S. J. (2009) Integration of vision and haptics during tool use. *J. Vision*, **9**, 1–15.

Toates, F. M. (1972) Accommodation function of the human eye. *Physiol. Rev.*, **52**, 828–863.

Tsirilin, I., Wilcox, L. M. and Allison, R. S. (2011) The effect of crosstalk on the perceived depth from disparity and monocular occlusions. *IEEE Trans. Broadcast.*, **57**, 445–453.

Tyler C. W. (1973) Stereoscopic vision: cortical limitations and a disparity scaling effect. *Science*, **181**, 276–278.

Tyler, C. W. (1974) Depth perception in disparity gratings. *Nature*, **251**, 140–142.

Tyler, C. W. and Julesz, B. (1978) Binocular cross-correlation in time and space. *Vision Res.*, **18**, 101–105.

Ukai, K. (2007) Visual fatigue caused by viewing stereoscopic images and mechanism of accommodation. Proceedings of the First International Symposium on Universal Communication, Kyoto, Vol. 1, 176–179.

Ukai, K. and Howarth, P. A. (2008) Visual fatigue caused by viewing stereoscopic motion images: background, theories and observations. *Displays*, **29**, 106–116.

Vishwanath, D. and Blaser, E. (2010) Retinal blur and the perception of egocentric distance. *J. Vision*, **10** (10), 1–16.

Vishwanath, D., Girshick, A. and Banks, M. S. (2005) Why pictures look right when viewed from the wrong place. *Nat. Neurosci.*, **8**, 1401–1410.

Wann, J. P. and Mon-Williams, M. (1997) Health issues with virtual reality displays: what we do know and what we don't. *ACM SIGGRAPH Comput. Graph.*, **31**, 53–57.

Watt, S. J., Akeley, K., Ernst, M. O. and Banks, M. S. (2005a) Focus cues affect perceived depth. *J. Vision*, **5**, 834–886.

Watt, S. J., Akeley, K., Girshick, A. R. and Banks, M. S. (2005b) Achieving near-correct focus cues in a 3-D display using multiple image planes, in *Human Vision and Electronic Imaging X* (eds B. E. Rogowitz, T. N. Pappas and S. J. Daly), Proceedings of SPIE, Vol. **5666**, SPIE, Bellingham, WA, pp. 393–401.

Watt, S. J., MacKenzie, K. J. and Ryan, L. (2012) Real-world stereoscopic performance in multiple-focal-plane displays: how far apart should the image planes be? in *Stereoscopic Displays and Applications XXIII* (eds A. J.

Woods, N. S. Holliman and G. E. Favalora), Proceedings of the SPIE, Vol. **8288**, SPIE, Bellingham, WA, p. 82881E.

Wendt, G., Faul, F. and Mausfeld, R. (2008) Highlight disparity contributes to the authenticity and strength of perceived glossiness. *J. Vision*, **8** (1): 14, 1–10.

Woods, A. J., Docherty, T. and Koch, R. (1993) Image distortions in stereoscopic video systems, in *Stereoscopic Displays and Applications IV*, (eds J. O. Merritt and S. S. Fisher), Proceedings of the SPIE, Vol. **1915**, SPIE, Bellingham, WA, p. 36–48.

Wöpking, M. (1995) Viewing comfort with stereoscopic pictures: an experimental study on the subjective effects of disparity magnitude and depth of focus. *J. Soc. Inf. Disp.*, **3**, 101–103.

Yang, S., Schlieski, T., Selmins, B. *et al.* (2011) Individual differences and seating positions affect immersion and symptoms in stereoscopic 3D viewing. Technical Report, Vision Performance Institute, Pacific University.

Yano, S., Emoto, M. and Mitsuhashi, T. (2004) Two factors in visual fatigue caused by stereoscopic HDTV imges. *Displays*, **25**, 141–150.

Yuille, A. and Bülthoff, H. (1996) Bayesian decision theory and psychophysics, in *Perception as Bayesian Inference* (eds D. C. Knill and W. Richards), Cambridge University Press, Cambridge, UK.

19

3D Video Quality Assessment

Philippe Hanhart, Francesca De Simone, Martin Rerabek, and
Touradj Ebrahimi
Multimedia Signal Processing Group (MMSPG), Ecole Polytechnique Fédérale de Lausanne (EPFL), Switzerland

19.1 Introduction

Quality is a fundamental concept closely linked to measuring performance in systems, products, and services. Any compression and representation technique in audiovisual signals needs metrics of quality, not only in their design stage but also in demonstrating their superior performance when compared with state of the art. In recent years, we have been witnessing a renewed interest in 3D video applications, especially for consumers. Hollywood movies have already made the move to 3D, and 3DTV devices have started appearing in homes. 3D gaming consoles and mobile phones with 3D capabilities announce even more dazzling progress on this front, along with next-generation television sets without the need for cumbersome 3D glasses. At present, it is not clear if this renewed wave of interest in 3D will fade as its predecessors did, or if it is the prelude to a new era that will revolutionize the multimedia applications of tomorrow. The answer to this question is neither trivial nor easy, as it does not only depend on technological and scientific considerations. In this chapter we examine some parts of this complex puzzle by mainly focusing on the important issue of how to measure 3D visual quality, as an essential factor in the success of 3D technologies, products, and services. Some believe that the main reason for the failure of previous waves of 3D was the lack of a minimum quality requirement in 3D. It is important, therefore, to consider mechanisms of 3D vision in humans, and their underlying perceptual models, in conjunction with the types of distortions that today and tomorrow 3D video processing systems produce. In particular, there is a need to understand how known two-dimensional (2D) concepts can be extended to 3D, and where they need to be entirely reinvented from scratch, in order to lead to more efficient methods for assessing subjective and objective quality of 3D visual content. We will start by providing an overview of different types of distortions that occur in a typical 3D video processing chain, from capture to consumption by end users. Then we will discuss how 3D visual quality subjective evaluations and objective metrics are achieved in the state

Emerging Technologies for 3D Video: Creation, Coding, Transmission and Rendering, First Edition.
Frédéric Dufaux, Béatrice Pesquet-Popescu, and Marco Cagnazzo.
© 2013 John Wiley & Sons, Ltd. Published 2013 by John Wiley & Sons, Ltd.

of the art, by analyzing their various strengths and weaknesses. In each case, we conclude with a discussion of open problems and trends in 3D visual quality assessment, and some of the challenges still ahead.

19.2   Stereoscopic Artifacts

Quality assessment in the conventional video processing chain takes into account many characteristic 2D artifacts (Yuen, 2005). When extended to 3D video, the human visual system (HVS) further processes additional monocular and binocular stimuli. Thus, the resulting video quality at the end of the 3D video processing chain depends also on the level of stereoscopic artifacts or binocular impairments affecting the depth perception. In fact, stereo artifacts can cause unnatural changes in structure, motion, and color vision of the scene and distort the binocular depth cues, which result in visual discomfort and eyestrain.

At each particular stage of a 3D video processing chain (see Figure 19.1), different stereoscopic artifacts can be identified and described (Boev *et al.*, 2009). According to the HVS layer interpretation of the visual pathway, all artifacts can be divided into four groups: structural, color, motion, and binocular. While structural artifacts affect contours and textures and distort spatial vision, the motion and color artifacts negatively influence motion and color perception. The most important binocular artifacts that have an impact on depth perception are discussed below. Nevertheless, it is very important to consider the interaction between individual groups and artifacts.

During 3D capture, several distortions occur (keystone distortion, depth field curvature, cardboard effect, etc.), which are mostly optical and due to camera setup and parameters. Assuming a toed-in convergent camera setup where each camera has a different perspective of the scene, the quality of binocular depth perception is affected by the vertical and unnatural horizontal parallax caused by *keystone distortion* and *depth plane curvature*, respectively. Keystone distortion, as a result of the position of the two cameras in slightly different planes, causes a vertical difference between homologous points, called vertical parallax. Vertical parallax is bigger in the corners of the image, proportional to camera separation, and inversely proportional to convergence distance and focal length (Woods *et al.*, 1993). The same principle leads, in the horizontal plane, to depth field curvature. This causes an unnatural horizontal parallax, which results in wrong representation of relative object distances on the screen. Keystone distortion and depth plane curvature can be suppressed by using a parallel camera configuration in stereoscopic video acquisition.

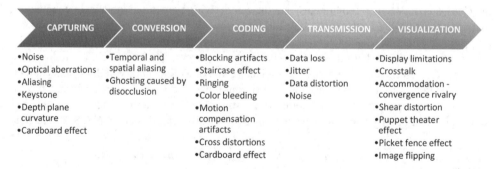

Figure 19.1   3D video processing chain artifacts

The *cardboard effect* is related to the availability of the proper disparity information, without which the viewer sees wrong objects–screen distance and, therefore, the perceived size and depth of objects do not correspond one to another. This results in objects appearing flat, as if the scene was divided into discrete depth planes. While keystone distortion and depth plane curvature are introduced only at the capture stage, the cardboard effect occurs also at the conversion and coding phase due to the sparse depth quantization.

Depending on the capture and rendering formats used and their mutual adaptation, various 3D artifacts occur in the conversion stage. The most common artifact here is *ghosting*, which is caused by disocclusion. It occurs when video plus depth representation and rendering are used, mainly due to the interpolation of occluded areas needed for view synthesis.

For individual coding of stereo image pairs, binocular suppression (Julesz, 1971) as a property of the HVS can be exploited. This ability to compensate the loss of information in one of the stereo views is particularly suitable for asymmetric coding (Fehn *et al.*, 2007). In asymmetric coding, the left and right views are encoded with different quality. However, if the qualities of two encoded stereo channels differ too much, the resulting spatial (resolution) and temporal (frame rate) mismatch between them, commonly referred to as a *cross-distortion*, produce wrong depth perception.

Crosstalk, also known as image ghosting, one of the stereo artifacts with the largest influence on the image quality, is caused by visualization of 3D content. It comes from imperfect left and right image separation when the view for the left eye is partially visible by the right eye, and vice versa. Crosstalk usually results in ghosting, shadowing, and double contours perception. *Picket fence*, *image flipping*, and *shear distortion*, as other artifacts arising during visualization, are related to the display technology used and when an observer changes his position. Shear distortion occurs with a change of position of the viewer resulting in wrong head parallax and distorted perspective vision. While shear distortion is typical for stereoscopic displays allowing only one correct viewing position, picket fence and image flipping are experienced with autostereoscopic displays exclusively. More specifically, picket fence is introduced by the spatial multiplexing of parallax barrier-based autostereoscopic displays only. It is noticeable as vertical banding when the observer moves laterally in front of the screen. Image flipping is basically the consequence of parallax discretization. It is observed as a leap transition between viewing zones.

3D video quality of experience (QoE) is influenced by many parameters, including stereoscopic distortions and impairments, properties of the HVS, and their reciprocal interactions. All these aspects need to be studied and understood thoroughly in order to quantify the overall quality of 3D visual content, either subjectively or objectively.

19.3 Subjective Quality Assessment

In a subjective video quality assessment experiment, a group of viewers is presented with a set of video sequences and asked to judge the perceived quality. The viewers are commonly referred to as *subjects*, while the data presented to them is referred to as *stimuli*. The responses provided by the subjects can be considered as measures of perceptual impressions only if the experimenter controls a set of *initial conditions* (Bech and Zacharov, 2006). These consist of dependent and independent experimental variables. The *dependent variable* is the answer provided by the subject. The experimenter can control it to a certain extent since they design the evaluation methodology to collect the subjective responses. The evaluation methodology includes the question that is presented to the subject and the rating scale that is used to answer. The *independent variables* are those that can be controlled directly by the experimenter, such as: the stimuli presented to viewers; the subjects themselves – the experimenter decides how many and with which profile; the experimental setup, which includes the hardware and software components used to reproduce the stimuli, the characteristics of the environment where the experiment takes place, and so on; and the test plan, which specifies the duration of the test session, the number of stimuli repetitions, and so on. The statistical analysis of the subjective results determines if the influence of the controlled independent variables is above chance and thus contributes to the variation of the dependent variable.

The design of the subjective experiment depends on the *research question* that the study is aiming at answering. In general, multimedia quality subjective experiments aim at studying a specific attribute of the stimulus (via so-called *perceptual measurements*) and its overall impression on the subject (via so-called *affective measurements*).

The experimental methodologies documented in the literature, targeting both perceptual and affective measurements, can be clustered in the following groups (Meesters *et al.*, 2004; Jumisko-Pyykko and Strohmeier, 2009):

- **Psycho-perceptual (or psychophysical) approaches:** These consist of experiments performed in highly controlled environments and according to strictly defined protocols, usually following internationally standardized guidelines. These tests are performed in order to derive a preference order in the responses provided by a limited set of subjects (typically not more than 20). Such a preference order allows making conclusions about the impact of test parameters on subjective quality. The conclusions achieved for the limited set of subjects under consideration are assumed to be statistically generalizable to other subjects having similar characteristics.
- **Descriptive (or explorative) approaches:** These aim at identifying and understanding what perceptual attributes are relevant to the users in order to determine their quality preferences. Descriptive approaches include methods that assume a close relationship between a given sensation and the verbal descriptors used by the subject to describe it (so-called *direct elicitation methods*), such as consensus vocabulary techniques and individual vocabulary techniques, and nonverbal elicitation techniques (so-called *indirect elicitation methods*).
- **Hybrid approaches:** These combine conventional quantitative psycho-perceptual evaluation and qualitative descriptive quality evaluation.

In the following we provide a review of the work available in the literature concerning 3D visual stimuli and employing the different testing approaches described above. Open issues and future directions are discussed at the end of the section.

19.3.1 *Psycho-perceptual (or Psychophysical) Experiments*

Guidelines for psycho-perceptual subjective assessment of stereoscopic television pictures are provided in the ITU-R BT.2021 Recommendation. These are mainly based on the guidelines detailed in the ITU-R BT.500 Recommendation for the assessment of 2D television pictures. Three main categories of stimulus presentation and rating procedure are defined: double stimulus methods (sequential presentation of each pair of a reference and a test stimulus), single stimulus methods (presentation and rating of the test stimuli only), and stimulus comparison methods (presentation of pairs of stimuli and rating of relative quality). The rating scale can be either continuous or discrete, and either numerical or categorical. Finally, the rating can be performed after each stimulus presentation for assessing the overall quality, or continuously during presentation for assessing temporal quality variations. The guidelines to process the collected subjective scores resulting from some of the test methodologies include: first, screening the scores to detect and exclude outliers whose ratings significantly deviate from the panel behavior; then, summarizing the scores of the subjects for each stimulus by computing the mean opinion score (MOS) or differential mean opinion score (DMOS) for single or double stimulus methods, respectively, and corresponding confidence interval.

Most of the studies available in literature concerning 3D subjective quality assessment rely on these methodologies or slightly modified versions of them. The most frequently used test methods are:

- The single stimulus (SS) or absolute category rating (ACR) method with five-point numerical or categorical quality or impairment scale; that is, {"bad," "good," "fair," "poor," "excellent"} or {"very annoying," "annoying," "slightly annoying," "perceptible but not annoying," "imperceptible"}, respectively. For example, Seuntiens *et al.* (2007), applied the SS method to ask the subjects to rate the *naturalness*, *sense of presence*, and *viewing experience* in relation to image quality, depth, and environmental lighting conditions. The SS method was used by Wang *et al.* (2012) for the evaluation of the *overall quality* of stereoscopic video transmissions with different scalability options, in an error-free scenario, and of the overall quality of video sequences transmitted in an error-prone scenario, when error concealment techniques were applied. Xing *et al.* (2012) used the SS method with an impairment scale asking the subjects to rate the *level of crosstalk* in stereoscopic video sequences.
- The single stimulus continuous quality scale (SSCQS) or double stimulus continuous quality scale (DSCQS) methods, where a test video sequence or a pair of reference and test video sequences are played and rated on a continuous scale spanning {"bad," "good," "fair," "poor," "excellent"}. For example, in the study by Saygili *et al.* (2011), the DSCQS method was applied to compare the *overall quality* of stereoscopic video sequences compressed by asymmetric and symmetric coding. Stelmach *et al.* (2000a,b) studied the impact of mixed-resolution stereoscopic video sequences on *overall quality*, *depth*, and *sharpness*, as well as the impact of asymmetric quantization and low-pass filtering, via the DSCQS method. Aksay *et al.* (2005) studied the impact of downsampling the chrominance components of stereoscopic images on overall quality via the DSCQS method. In the study by IJsselsteijn *et al.* (2000) the subjects were asked to rate the *quality of depth* and the *naturalness of depth* of stereoscopic images varying in camera separation, focal length, and convergence distance using a modified SSCQS method with numerical continuous scale. Different display duration conditions were analyzed.
- The single stimulus continuous quality evaluation (SSCQE) method, where a test video sequence is played and rated while being watched, using a continuous scale. In the studies by IJsselsteijn *et al.* (1998) and Yano *et al.* (2004) the subjects were asked to rate the *sense of presence* and the *visual fatigue*, respectively, using the SSCQE method.
- The stimulus comparison (SC) method, where two stimuli are shown and the relative quality of one against the other is assessed; for example, {"worse," "same," "better"}. An example of a study employing the SC method is that by Lee *et al.* (2012), where the effect of the camera distance used during the acquisition of stereoscopic images on the *overall quality* was studied. In this study, the results of the SC method were also compared with those of the SSCQS method, demonstrating that the SC methodology improves the quality discriminability between stimuli, at the cost of increased overall test duration.

While many studies focus on quantifying the overall quality of 3D visual stimuli, the following specific perceptual attributes have been studied in the literature by means of psycho-perceptual experiments:

- naturalness and quality of depth
- visual sharpness
- sense of presence
- visual discomfort (or fatigue)
- annoyance of crosstalk, as well as of other artifacts described in Section 19.2.

Mainly, the following sources of quality degradation have been considered:

- display technology
- viewing conditions
- view synthesis
- symmetric and asymmetric coding
- transmission with packet loss
- spatio-temporal video scalability.

19.3.2 Descriptive (or Explorative) Approaches

Since 3D digital images and video sequences are multidimensional stimuli, several perceptual attributes are excited by a single visual stimulus. In the descriptive subjective evaluation approaches, no direct subjective ratings are acquired. Instead, the feelings and reactions caused by the stimuli on the user are collected and analyzed.

In *direct elicitation methods*, the goal of the experiment is to identify and define the set of independent attributes that are relevant to the user and affect the quality judgment. This kind of evaluation is useful to identify the added value of new technologies. An example of such an approach applied to 3D stimuli is the study by Freeman and Avons (2000), who used focus groups to study viewers' reactions to conventional 2D TV versus 3D stereoscopic TV and their content dependency. As a result of the experiment, attributes such as sense of "being there," realism, naturalness, and involvement were identified by the subjects with respect to 3DTV. More recently, Jumisko-Pyykko et al. (2008) and Strohmeier et al. (2008) combined an online survey, focus groups, and a probe study to identify user requirements, and thus quality expectations, when considering mobile 3D video. Differently from the results for home or cinema viewing, instead of presence, the increased realism and the closer emotional relation were identified as the most important added values to the content.

An example of *indirect elicitation methods* is provided by Boev et al. (2010), who designed a task-based subjective test where the subjects were asked to judge the relative depth of an object in the scene. From the answers collected and the time needed for making the decision, the authors investigated the impact of different depth cues on the perception of depth.

19.3.3 Hybrid Approaches

The study by Häkkinen et al. (2008) represents a first simple example of the hybrid approach, where elements of psycho-perceptual evaluation and descriptive evaluation are both applied. In that study, viewers were presented with pairs of stereoscopic and non-stereoscopic versions of the same video and, after each pair presentation, were asked to indicate which version they preferred (psycho-perceptual approach) and describe why (descriptive approach).

Recently, more sophisticated methods have been proposed. Strohmeier *et al.* (2011) introduced the open profiling of quality (OPQ) method, which combines evaluation of quality preferences and elicitation of individual quality factors. The method has been used to build a perceptual model of the experienced quality factors when considering mobile 3D audiovisual content. The results show that deeper understanding on QoE is achieved by applying such an approach with respect to conventional ones.

Additionally, in order to include a human behavioral level in the evaluation, real systems, services, and environmental conditions should be used. Within this scope, Jumisko-Pyykko and Utriainem (2010) proposed a hybrid user-centered QoE evaluation method for measuring quality in the context of use. Quantitative preference ratings, qualitative descriptions of quality and context, characterization of context in the macro and micro levels, and measures of effort were included in the methodology. This new method was applied to evaluate mobile 3DTV on a portable device with parallax barrier display technology, considering different field settings and conventional laboratory settings. The results showed significant differences between the different field conditions and between field and laboratory measures, highlighting the importance of context when performing quality evaluation.

19.3.4   Open Issues

In most studies available in the literature, psycho-perceptual methodologies designed for 2D stimuli have been slightly adapted and used for 3D quality evaluation. The appropriateness of such an approach has not been accurately analyzed in the literature, but doubts may exist. For example, in the current recommendations, there is a clear lack of up-to-date guidelines concerning the viewing distance and viewer's position to be used when considering different 3D displaying technologies. Also, Chen *et al.* (2010) discussed the fact that the minimum number of observers, as recommended in ITU-BT.500, may not be sufficient for 3D evaluation owing to a higher variability in viewers' perception and assessment. Furthermore, they highlighted the importance of test material and display technology used for the evaluation, as these are a priori very critical factors impacting the quality perception. This problem is linked to the lack of publicly available 3D content, representative of typical content in real-life 3D applications, to be used by the research community. The usage of unrealistic test material may limit the generalizability of the findings of most of the state-of-the-art studies.

In addition to the test material, it is important to mention that very few results of 3D subjective quality assessment studies are publicly available. Currently, the only public databases of 3D test material and related subjective quality annotations available are those proposed by Goldmann *et al.* (2010) and Urvoy *et al.* (2012). Public databases of subjective data are important to allow comparison of different test methodologies, cross-validate findings of different studies, and validate the performance of objective quality metrics.

Finally, guidelines for 3D audiovisual quality evaluation and for user-centered subjective test methodologies are also very much needed. Until now, few studies have been conducted in the real context of use of 3D technologies, even if the impact of context on subjective quality perception has been proven to be significant (Jumisko-Pyykko *et al.*, 2011).

In order to address some of these open issues, standardization activities have been initiated in recent years and are currently ongoing. In particular, the ITU-R Working Party 6C has started identifying requirements for broadcasting and subjective testing of 3DTV (ITU-R Study Group 6, 2008), the ITU-T Study Group 9 has recently added 3D video quality into its scope (ITU-T Study Group 9, 2009), and the Video Quality Experts Group 3DTV ad-hoc group is working towards the definition of new methodologies for reliable subjective assessment of different quality attributes in 3DTV. The outcome of such activities is expected to be the definition of internationally recognized guidelines for 3D subjective quality assessment, mainly considering as applied, but not limited, to 3DTV.

19.3.5   Future Directions

In terms of future research trends, considering the rapid spread of ultrahigh definition (UHD) 2D content and display technologies, it can be expected that subjective quality perception of 3D UHD content will be a relevant topic of investigation in the coming years. Recently, 4K 3D consumer displays have started to appear on the market.

Owing to the fact that at such high resolutions the entire field of view of the user will be covered by the content, an important aspect of human visual perception to be studied is visual attention. Huynh-Thu *et al.* (2011) discussed the importance of visual attention to improve the QoE for 3DTV. They highlighted the fact that 3D viewing conditions may vary a lot, for example, when comparing mobile viewing, home environment, and cinema theaters. Thus, depending on the viewing conditions, the 3D content needs to be adapted differently. This content repurposing can be more efficiently achieved by taking into account the 3D visual attention.

Finally, it is important to mention that, depending on the evolution of commercial 3D applications for markets other than the entertainment, new subjective quality assessment methods targeting, for example, 3D viewing medical application or 3D video-conferencing services may be needed in the future.

19.4   Objective Quality Assessment

19.4.1   Objective Quality Metrics

Depending on the amount of information required about the original video, objective metrics can be classified into three categories:

- Full-reference metrics, which compare the test video with an original video, usually on a frame-by-frame basis. The majority of objective metrics fall into this category. However, it is impossible to implement such metrics in practical situations where the original video is not available.
- No-reference metrics, which do not use any information about the original video. Therefore, these metrics can be used anywhere. Nevertheless, they are more complicated owing to the difficulty of distinguishing between distortions and actual content.
- Reduced-reference metrics, which have access to a number of features from the original video, extract the same features from the test video, and compare them. Thus, this category lies in between the two extremes.

From a different perspective, techniques for objective quality assessment can be classified into two approaches: bottom up and top down. The bottom-up approach, also referred to as the psychophysical approach, attempts to model the HVS to predict image quality. Based on psychophysical experiments, aspects of the HVS, such as color perception, contrast sensitivity, and pattern masking, are incorporated in the metric. The top-down approach, also referred to as the engineering approach, only makes some high-level assumption on the mechanisms of the HVS. These metrics primarily extract and analyze features, such as structural elements, contours, and so on.

19.4.2 From 2D to 3D

Quality metrics have been widely investigated for 2D quality assessment. Some of the proposed metrics have shown high correlation with perceived quality, outperforming a simple peak signal-to-noise ratio (PSNR), which is still commonly used by video compression experts. Among these metrics, one can mention the structural similarity index (SSIM; Wang *et al.*, 2004) and the video quality metric (VQM; ITU-T, 2004).

It is common to solve a complex problem using techniques that were designed for a simpler problem, such as using image quality metrics on a frame-by-frame basis and averaging the results among the frames to perform video quality assessment. This approach can give fair results but does not capture the full nature of the problem, such as the temporal aspect in this case. Based on this idea, a simple solution to assess the quality of a stereoscopic video is to use a 2D objective metric on the left and right views separately and to combine the two scores.

There are several possible ways of combining the scores from the left and right views of a stereo pair. Averaging is the most commonly used. However, humans generally do not have the same visual acuity between the left and the right eye and it is known that about two-thirds of the population is right-eye dominant. Therefore, we do not perceive the same quality with each eye, which is important for 3D content since the two views are fused in our brain. Campisi *et al.* (2007) compared the "average" approach against the "main eye" approach and the "visual acuity" approach. For the "main eye" approach, only the objective score of the view corresponding to the dominant eye was taken into account for each subject. For the "visual acuity" approach, the left and right scores were weighted with the visual acuity of the left and right eyes, respectively. Different distortions, such as blurring and JPEG compression, were applied to the left and right views of the stereo pair. The results have shown no performance improvement over the "average" approach. However, in this study, the same distortion strength was applied to the left and right views.

When the stereo pair is asymmetric, namely when the two views have different types and amounts of distortions, specific properties of the HVS, such as binocular suppression (the masking of low-frequency content in one view by the sharp visual content in the other view), should be taken into account in order to predict the overall quality of the rendered asymmetric stereo pair. Indeed, previous studies (e.g., Stelmach *et al.*, 2000b; Seuntiens *et al.*, 2006) have shown that the perceived quality of an asymmetric stereo pair is, in general, close to the average quality of the two views, but in some cases it is closer to the highest quality, depending on the type of artifact and the difference in quality between the two views.

Hewage *et al.* (2009) investigated objective quality assessment of 3D videos represented in the video plus depth (2D + Z) format. Two experiments were conducted: the first on symmetric/asymmetric coding of color and associated depth map video and the second on transmission errors over an IP network. In that study, PSNR, SSIM, and VQM were used as objective quality metrics. They were computed on the color video and on the rendered left and right views. It was established that VQM had the highest correlation with perceived quality. In the second experiment, a slightly better performance was obtained when using the average quality of the left and right views as opposed to when using the quality of the color video. However, in the first experiment, the metrics showed lower correlation with perceived quality when using the average quality of the left and right views when compared with using the quality of the color video. This effect is particularly strong for PSNR, where the correlation coefficient drops from 0.81 to 0.74. Unfortunately, these results were not discussed, nor investigated further by the authors. However, recent studies (see Section 19.4.5) have shown that traditional 2D metrics fail at predicting the quality of synthesized views.

Only a few 2D metrics have been developed for 3D quality assessment, such as the perceptual quality metric proposed by Joveluro *et al.* (2010) that quantifies the distortion in the luminance component and weights the distortion based on the mean of each pixel block. In that study, a set of 3D videos represented in the $2D + Z$ format was assessed after being encoded with different compression factors. The results show that the proposed metric significantly outperforms VQM.

19.4.3 Including Depth Information

The simple approach presented in Section 19.4.2 can be easily extended by taking into account the quality of the depth information. The disparity map can be estimated using a stereo correspondence algorithm. The original and degraded disparity maps are estimated from the original and degraded stereo pairs, respectively. The two can be compared using a standard 2D quality metric to estimate the quality of the depth information. However, disparity maps have different properties when compared with natural images. They are mostly composed of large piecewise-smooth regions, corresponding to the different objects, with sharp boundaries between the objects. Therefore, it is not trivial to find efficient quality metrics for disparity maps. Nevertheless, standard image-quality metrics have been used in many studies. The disparity estimation process is also nontrivial for asymmetric stereo pairs. Indeed, most stereo correspondence algorithms try to match a block of N-by-N pixels in the left view with a block of N-by-N pixels in the right view using a correlation metric such as the sum of absolute differences, the sum of squared differences, and so on.

The image quality and the depth quality can be combined in many different ways. Benoit *et al.* (2008) used the SSIM and C4 metrics to evaluate image quality of the left and right views of stereo pairs degraded with blurring and JPEG/JPEG2000 compression. Then, the two scores were averaged to produce the image quality score. The depth quality score was computed using the correlation coefficient between the original and the degraded disparity maps. Results showed that C4 had a better correlation with perceived quality than SSIM did when considering only the image quality and when considering both the image and depth quality. Two different disparity estimation algorithms were used in that study: one based on graph cuts and one based on belief propagation. The results showed that better performance is obtained with the algorithm based on belief propagation and that the depth quality is significantly less correlated with perceived quality than the image quality. Similar or slightly lower performance is obtained when considering both image and depth quality as opposed to considering only the image quality.

You *et al.* (2010), conducted a similar study, including also the quality of the disparity map. In that study, only the right view of the stereo pair was degraded with blurring, JPEG/JPEG2000 compression, or white noise, while the left eye image was kept undistorted. Eleven 2D metrics were used for both the image and depth quality. To measure the depth quality, the global correlation coefficient, mean square error, and mean absolute difference (MAD) were also included. Authors have reported a strong correlation between the perceived quality and the measured quality of the degraded view for sequences with the same distortion types. When different sequences with different distortion types were considered, the performance dropped drastically. However, in this scenario, the performance of the metrics on the depth maps was significantly better than on the degraded view. It was found that an appropriate combination of the image and depth quality performs better than using only the image quality or depth quality alone. The best results were obtained using SSIM to measure the image quality and MAD for the depth quality.

Instead of producing only one single quality score for the whole image, some metrics, such as SSIM, are capable of producing one score per pixel. Thus, the image and depth quality scores can be combined at the pixel level, which can further improve the correlation with perceived quality. This approach is used by Benoit *et al.* (2008), You *et al.* (2010), and Wang *et al.* (2011).

The HVS fuses the two images received from the two eyes into a single mental image, the so-called cyclopean image. Researchers have found that the HVS solves the correspondence problem between the left and the right images, which results in the cyclopean image and an associated perceptual depth map. The cyclopean image is modeled for a neighborhood of each pixel by taking a local window from the left view and averaging it with the disparity-matched window from the right view. The disparity map models the perceptual depth map. The cyclopean images extracted from the original and distorted stereo pairs can be compared using a 2D metric. A similar technique can be used for the perceptual depth map. This approach was used by Boev *et al.* (2006), combined with a pyramidal decomposition. In that study, monoscopic and stereoscopic quality scores were produced based on the cyclopean image and perceptual depth map, respectively. Maalouf and Larabi (2011) used a metric based on a wavelet transform and contrast sensitivity function to measure the quality of the cyclopean image and a coherence measure for the perceptual depth map.

Many different approaches using depth information have been proposed for objective quality assessment of 3D video. For example, Jin *et al.* (2011) divided the left image into blocks of M-by-N pixels. For each block, the algorithm found the most similar block in the left image and the two most similar blocks in the right image (one of them being the disparity-compensated block). A 3D discrete cosine transform (DCT) transform was then applied to the 3D structure formed from the original block and the three corresponding blocks. A contrast sensitivity function and a luminance mask were used to keep only the relevant DCT coefficients. The coefficients extracted from the original stereo pair were compared with those of the distorted pair using a PSNR-like function. The authors reported a significant gain of the proposed metric over state-of-the-art 2D metrics for simulcast MPEG-4-encoded 3D video sequences.

19.4.4 Beyond Image Quality

3D is an immersive technology. The goal is to reproduce the action as close as possible to reality, such that the viewer can experience the event as if he/she was physically there. Thus, in subjective tests, researchers often measure the QoE, through visual discomfort and depth perception, and not only the image quality. Many factors can degrade the QoE, such as large disparity values outside the so-called comfort zone or crosstalk artifacts. A few metrics have recently been designed to estimate QoE. For example, Mittal *et al.* (2011) proposed a no-reference metric based on spatial and temporal statistical properties of the estimated disparity map and statistical properties of the spatial activity. These properties were used to produce a large set of features. Only the principal features were kept to predict QoE.

Xing *et al.* (2011) proposed two full-reference objective metrics to estimate QoE, based on significant factors such as depth, screen size, and crosstalk. To simulate crosstalk, the authors added a fraction (corresponding to the crosstalk level) of the right image to the left image and vice versa. SSIM was used between the original left image and the crosstalk-distorted image. Perceived crosstalk was obtained by weighting the SSIM map using the depth map. This index was combined with screen size. The results showed that the proposed metrics have a significantly higher correlation with subjective scores than PSNR and SSIM do to predict the QoE.

19.4.5   Open Issues

3D video can be represented in many different formats: stereoscopic video (side-by-side, top-and-bottom, frame compatible, etc.), video plus depth (2D + Z), multiview video, multiview video plus depth (MVD), and so on. The 3D display technologies also have different characteristics:

- active stereoscopic displays, which use temporal multiplexing by alternating between left and right images;
- passive stereoscopic displays, which use either spatial multiplexing (for example, by displaying the left and the right images on the even and odd rows, respectively) or color multiplexing (usually chromatically opposite colors);
- autostereoscopic displays, which use spatial multiplexing to project different views to different positions.

To be displayed on a stereoscopic monitor, a video represented in the 2D + Z format has to be first rendered into left and right views. This process is typically performed using depth-image-based rendering, which introduces artifacts in the stereo pair. Therefore, it is essential to evaluate the quality on the final 3D video. However, for a passive stereoscopic monitor using spatial multiplexing, only half of the pixels of each frame are actually displayed on the screen. It was shown in 2D quality assessment that factors such as viewing distance and screen size impact perceived quality and some metrics take into account such parameters. Therefore, in 3D quality assessment, both the video format and the display technology should be taken into account by the objective metrics. Nevertheless, most stereoscopic quality metrics require the left and right views of the stereo pair in full resolution, regardless of the video format and the display technology.

Bosc *et al.* (2011) showed that state-of-the-art 2D metrics fail at predicting perceived quality of synthesized views. It is known that most metrics better handle specific types of artifacts, such as compression artifacts, rather than other types of artifacts. The performance usually drops when different types of artifacts are combined. In that study, a central view was synthesized from the left and right views and compared against the corresponding original view. No coding artifacts were considered. Hanhart *et al.* (2012) investigated the PSNR-based quality assessment of stereo pairs formed from a decoded view and a synthesized view. The 3D videos considered in that study were represented in the MVD format and compressed. At the decoder side, the displayed stereo pair was formed from one of the decoded views and a virtual view, which was synthesized using the decoded MVD data. It was reported that the PSNR of the decoded view had the highest correlation with perceived quality, while the PSNR of the synthesized view had a significantly lower correlation. Understanding the effect of view synthesis on perceived quality, in conjunction with compression, is particularly important for multiview autostereoscopic displays, which usually synthesize N views from a limited number of input views, and stereoscopic displays that modify the baseline to adjust the depth perception based on viewing distance and viewing preferences.

19.4.6   Future Directions

A few studies have been conducted on the perception of mismatched stereo pairs to try to understand the masking effect, which depends on the type and strength of the artifacts between the left and the right views. However, most objective metrics do not take into account the masking properties of the HVS. A good metric should consider this effect and not simply average the scores of the left and the right views. However, this problem is quite challenging and will require conducting additional subjective tests to better understand the HVS and designing efficient metrics to not only qualify but also quantify precisely the different artifacts.

As seen in Section 19.4.3, a few metrics take into account depth quality to improve the performance of 2D metrics used to measure image quality. However, the gain is not always significant. More advanced techniques have been proposed that take into account the perceptual disparity map. Nevertheless, these techniques rely on the accuracy of the disparity estimation algorithm. The mechanisms that occur in the HVS to extract the depth information are probably more complex than such simple approaches. A good 3D metric should take into account all depth cues, including monoscopic depth cues. Recently, some algorithms have been proposed to extract a depth index from a 3D video.

In Hollywood movies, many film directors exaggerate various 3D effects, which can cause visual discomfort. Even if the so-called comfort zone is respected, the video might look unnatural or unpleasant. Thus, image quality is not the only important factor. If our brain cannot (partially) fuse the two images, the viewer experience can become unpleasant. Therefore, it is important to take into account these factors and care about the overall QoE.

There is also a need for quality metrics designed specifically for new technologies such as multiview autostereoscopic displays, where a large number of views are synthesized from a limited number of input views (typically one or two) and associated depth maps. In this scenario, no reference is available to assess the quality of the synthesized views and the consistency among the views. In the case of decoded data, the original data are often used to synthesize the same views and the quality is measured between the synthesized views using decoded data and the synthesized views using original data. However, this approach assumes that the view synthesis does not add artifacts or that the view synthesis artifacts are not as significant as the compression artifacts, which has not been proven yet. Also, a metric designed for multiview should take into account all views at the same time rather than averaging the scores among the views.

References

Aksay, A., Bilen, C., and Akar, G.B. (2005) Subjective evaluation of effects of spectral and spatial redundancy reduction on stereo images. Proceedings of the 13th European Signal Processing Conference (EUSIPCO), Antalya, Turkey.

Bech, S. and Zacharov, N. (2006) *Perceptual Audio Evaluation: Theory, Method and Application*, John Wiley & Sons, Ltd, Chichester.

Benoit, A., Le Callet, P., Campisi, P., and Cousseau, R. (2008) Quality assessment of stereoscopic images. *EURASIP J. Image Vid. Process.*, **2008**, 659024.

Boev, A., Gotchev, A., Egiazarian, K. *et al.* (2006) Towards compound stereo-video quality metric: a specific encoder-based framework, in *IEEE Southwest Symposium on Image Analysis and Interpretation*, IEEE, pp. 218–222.

Boev, A., Hollosi, D., Gotchev, A., and Egiazarian, K. (2009) Classification and simulation of stereoscopic artifacts in mobile 3DTV content, in *Stereoscopic Displays and Applications XX* (eds A.J. Woods, N.S. Holliman, and J.O. Merritt), Proceedings of the SPIE, Vol. **7237**, SPIE, Bellingham, WA, p. 72371F.

Boev, A., Poikela, M., Gotchev, A., and Aksay, A. (2010) Modelling of the stereoscopic HVS. Mobile 3DTV Network of Excellence, Rep. D5.3.

Bosc, E., Pepion, R., Le Callet, P. *et al.* (2011) Towards a new quality metric for 3-D synthesized view assessment. *IEEE J. Sel. Top. Signal Process.*, **5** (7), 1332–1343.

Campisi, P., Le Callet, P., and Marini, E. (2007) Stereoscopic images quality assessment, in *15th European Signal Processing Conference, EUSIPCO 2007. Proceedings* (eds M. Domański, R. Stasiński, and M. Bartkowiak), PTETiS Poznań, Poznan.

Chen, W., Fournier, J., Barkowsky, M., and Le Callet, P. (2010) New requirements of subjective video quality assessment methodologies for 3DTV. Proceedings of the 5th International Workshop on Video Processing and Quality Metrics for Consumer Electronics (VPQM), Scottsdale, AZ, USA.

Fehn, C., Kauff, P., Cho, S. *et al.* (2007) Asymmetric coding of stereoscopic video for transmission over T-DMB, in *3DTV Conference, 2007*, IEEE, pp. 1–4.

Freeman, J. and Avons, S. (2000) Focus group exploration of presence through advanced broadcast services, in *Human Vision and Electronic Imaging V* (eds B.E. Rogowitz and T.N. Pappas), Proceedings of the SPIE, Vol. **3959**, SPIE, Bellingham, WA, pp. 530–539.

Goldmann, L., De Simone, F., and Ebrahimi, T. (2010) A comprehensive database and subjective evaluation methodology for quality of experience in stereoscopic video, in *Three-Dimensional Image Processing (3DIP) and Applications* (ed. A.M. Baskurt), Proceedings of the SPIE, Vol. **7526**, SPIE, Bellingham, WA, p. 75260T.

Häkkinen, J., Kawai, T., Takatalo, J. *et al.* (2008) Measuring stereoscopic image quality experience with interpretation based quality methodology, in *Image Quality and System Performance V* (eds S.P. Farnand and F. Gaykema), Proceedings of the SPIE, Vol. **6808**, SPIE, Bellingham, WA, p. 68081B.

Hanhart, P., De Simone, F., and Ebrahimi, T. (2012) Quality assessment of asymmetric stereo pair formed from decoded and synthesized views. 4th International Workshop on Quality of Multimedia Experience (QoMEX), Yarra Valley, Australia.

Hewage, C.T.E.R., Worrall, S.T., Dogan, S. *et al.* (2009) Quality evaluation of color plus depth map-based stereoscopic video. *IEEE J. Sel. Top. Signal Process.*, **3** (2), 304–318.

Huynh-Thu, Q., Barkowsky, M., and Le Callet, P. (2011) The importance of visual attention in improving the 3D-TV viewing experience: overview and new perspectives. *IEEE Trans. Broadcast.*, **57** (2), 421–431.

IJsselsteijn, W.A., de Ridder, H., Hamberg, R. *et al.* (1998) Perceived depth and the feeling of presence in 3DTV. *Displays*, **18**, 207–214.

IJsselsteijn, W.A., de Ridder, H., and Vliegen, H. (2000) Subjective evaluation of stereoscopic images: effects of camera parameters and display duration. *IEEE Trans. Circ. Syst. Vid.* **10** (2), 225–233.

ITU-R Study Group 6 (2008) Digital three-dimensional (3D) TV broadcasting. Question ITU-R 128/6.

ITU-T (2004) Objective perceptual video quality measurement techniques for digital cable television in the presence of a full reference, Recommendation J.144, ITU-T Telecommunication Standardization Bureau.

ITU-T Study Group 9 (2009) Objective and subjective methods for evaluating perceptual audiovisual quality in multimedia services within the terms of Study Group 9. Question 12/9.

Jin, L., Boev, A., Gotchev, A., and Egiazarian, K. (2011) 3D-DCT based perceptual quality assessment of stereo video, in *2011 18th IEEE International Conference on Image Processing (ICIP)*, IEEE, pp. 2521–2524.

Joveluro, P., Malekmohamadi, H., Fernando, W.A.C., and Kondoz, A.M. (2010) Perceptual video quality metric for 3D video quality assessment, in *3DTV-Conference: The True Vision – Capture, Transmission and Display of 3D Video (3DTV-CON), 2010*, IEEE, pp. 1–4.

Julesz, B. (1971) *Foundations of Cyclopean Perception*, University of Chicago Press.

Jumisko-Pyykko, S. and Strohmeier, D. (2009) Report on research methodology for the experiments. Mobile 3DTV Network of Excellence, Rep. D.4.2.

Jumisko-Pyykko, S. and Utriainen, T. (2011) A hybrid method for quality evaluation in the context of use for mobile (3D) television. *Multimed. Tools Appl.*, **55** (2), 185–225.

Jumisko-Pyykko, S., Weitzel, M., and Strohmeier, D. (2008) Designing for user experience – what to expect from mobile 3D television and video. Proceedings of the 1st International Conference on Designing Interactive User Experiences for TV and video (uxTV), Mountain View, CA, USA.

Lee, J.-S., Goldmann, L., and Ebrahimi, T. (2012) Paired comparison-based subjective quality assessment of stereo-scopic images. *Multimed. Tools Appl.* doi: 10.1007/s11042-012-1011-6.

Maalouf, A. and Larabi, M.-C. (2011) CYCLOP: a stereo color image quality assessment metric, in *2011 IEEE International Conference on Acoustics, Speech and Signal Processing (ICASSP)*, IEEE, pp. 1161–1164.

Meesters, L., IJsselsteijn, W., and Pieter, J.H.S. (2004) A survey of perceptual evaluations and requirements of three-dimensional TV. *IEEE Trans. Multimed.*, **14** (3), 381–391.

Mittal, A., Moorthy, A.K., Ghosh, J., and Bovik, A.C. (2011) Algorithmic assessment of 3D quality of experience for images and videos, in *2011 IEEE Digital Signal Processing Workshop and IEEE Signal Processing Education Workshop (DSP/SPE)*, IEEE, pp. 338–343.

Saygili, G., Gurler, C.G., and Tekalp, A.M. (2011) Evaluation of asymmetric stereo video coding and rate scaling for adaptive 3D video streaming. *IEEE Trans. Broadcast.* **57** (2), 593–601.

Seuntiens, P., Meesters, L., and IJsselsteijn, W. (2006) Perceived quality of compressed stereoscopic images: effects of symmetric and asymmetric JPEG coding and camera separation. *ACM Trans. Appl. Percept.*, **3**, 95–109.

Seuntiens, P., Vogels, I., and van Keersop, A. (2007) Visual Experience of 3D-TV with pixelated Ambilight. Proceedings of the 10th Annual International Workshop on Presence, Barcelona, Spain.

Stelmach, L.B., Tam, W.J., Meegan, D.V., and Vincent, A. (2000a) Stereo image quality: effects of mixed spatio-temporal resolution. *IEEE T. Circ. Syst. Vid.* **10** (2), 188–193.

Stelmach, L.B., Tam, W.J., Meegan, D.V. *et al.* (2000b) Human perception of mismatched stereoscopic 3D inputs, in *Proceedings. 2000 International Conference on Image Processing*, Vol. **1**, IEEE, pp. 5–8.

Strohmeier, D., Jumisko-Pyykkö, S., Weitzel, M., and Schneider, S. (2008) Report on user needs and expectations for mobile stereo-video. Mobile 3DTV Network of Excellence, Rep. D.4.1.

Strohmeier, D., Jumisko-Pyykkö, S., Kunze, K., and Bici, M.O. (2011) The extended-OPQ method for user-centered quality of experience evaluation: a study for mobile 3D video broadcasting over DVB-H. *EURASIP J. Image Vid. Process.*, **2011**, 538294.

Urvoy, M., Gutiérrez Sánchez, J., Barkowsky, M. *et al.* (2012) NAMA3DS1-COSPAD1: subjective video quality assessment database on coding conditions introducing freely available high quality 3D stereoscopic sequences, in *2012 Fourth International Workshop on Quality of Multimedia Experience (QoMEx)*, IEEE, pp. 109–114.

Wang, K., Barkowsky, M., Brunnström, K. *et al.* (2012) Perceived 3D TV transmission quality assessment: multi-laboratory results using absolute category rating on quality of experience scale. *IEEE Trans. Broadcast.*, **58** (4), 544–557.

Wang, X., Kwong, S., and Zhang, Y. (2011) Considering binocular spatial sensitivity in stereoscopic image quality assessment, *2011 IEEE Visual Communications and Image Processing (VCIP)*, IEEE, pp. 1–4.

Wang, Z., Bovik, A.C., Sheikh, H.R., and Simoncelli, E.P. (2004) Image quality assessment: from error visibility to structural similarity. *IEEE Trans. Image Process.*, **13** (4), 600–612.

Woods, A.J., Docherty, T., and Koch, R. (1993) Image distortions in stereoscopic video systems, in *Stereoscopic Displays and Applications IV* (eds J.O. Merritt and S.S. Fisher), Proceedings of the SPIE, Vol. **1915**, SPIE, Bellingham, WA, pp. 36–48.

Xing, L., You, J., Ebrahimi, T., and Perkis, A. (2011) Objective metrics for quality of experience in stereoscopic images, in *2011 18th IEEE International Conference on Image Processing (ICIP)*, IEEE, pp. 3105–3108.

Xing, L., You, J., Ebrahimi, T., and Perkis, A. (2012) Assessment of stereoscopic crosstalk perception. *IEEE Trans. Multimed.*, **14** (2), 326–337.

Yano, S., Emoto, M., and Mitsuhashi, T. (2004) Two factors in visual fatigue caused by stereoscopic HDTV images. *Displays*, **25** (4), 141–150.

You, J., Xing, L., Perkis, A., and Wang, X. (2010) Perceptual quality assessment for stereoscopic images based on 2D image quality metrics and disparity analysis. Fifth International Workshop on Video Processing and Quality Metrics for Consumer Electronics.

Yuen, M. (2005) Coding artifacts and visual distortions, in *Digital Video Image Quality and Perceptual Coding* (eds H. Wu and K. Rao), CRC Press.

Part Six

Applications and Implementation

20

Interactive Omnidirectional Indoor Tour

Jean-Charles Bazin, Olivier Saurer, Friedrich Fraundorfer, and Marc Pollefeys
Computer Vision and Geometry Group, ETH Zürich, Switzerland

20.1 Introduction

The development of Google Street View marked an important milestone for online map services. It provides panoramic views acquired by an omnidirectional camera, generally embedded on the roof of a moving car. From then on it was possible to virtually and interactively walk through cities along roads and experience views as if you were there. Its simplicity of use and its convincing tele-presence/immersion feeling attracted millions of users all over the world. The system was deployed on a large scale and with high-quality photographs. Key features of the system are that the photographs are aligned to road map data and that it is possible for the user to navigate and turn at road intersections. For this, the photographs are geo-referenced and aligned with satellite map data using a global positioning system (GPS). This allows a user to click on a point in the map and the corresponding street-level view shows up. The map alignment and the detection of intersections are the main challenges of such a system, and in Google Street View these are resolved using GPS-annotated photographs. In this chapter we propose a system for Street View-like virtual tours mainly dedicated to indoor environments, such as museums, historical buildings, and shopping malls. Besides the traditional entertaining applications, a potential scenario could be the training of security, rescue or defense staff to, for example, virtually explore a building before starting a rescue mission. The applications are not limited to virtual navigation: our system can also be used to guide visitors through museums or exhibition halls and provide them with additional information about the sculptures or exhibited objects they are currently viewing.

Within buildings, geo-referencing with GPS is not possible because the satellite signals cannot be received, and thus map alignment and junction detection cannot be accomplished as for the Street View application. We overcome this limitation by using structure from motion (SfM) and visual place recognition instead of GPS-annotated photographs. We

Emerging Technologies for 3D Video: Creation, Coding, Transmission and Rendering, First Edition.
Frédéric Dufaux, Béatrice Pesquet-Popescu, and Marco Cagnazzo.
© 2013 John Wiley & Sons, Ltd. Published 2013 by John Wiley & Sons, Ltd.

present a semi-automatic work flow that processes the visual data as much as possible automatically and allows manual intervention for a final polishing.

Our system design includes the following features, which are, based on literature and our own investigation, essential key components for an interactive indoor tour system: (a) omnidirectional cameras, instead of traditional cameras (i.e., with a limited field of view), to look around, (b) image stabilization to reduce shakes and misalignments if a hand-held/wearable camera is used, (c) automatic camera pose computation to estimate the location of the acquired images, (d) automatic junction detection to create consistent paths and overcome the drift, (e) image interpolation to obtain a smooth visualization, and (f) an appropriate navigation program to explore the images (e.g., rotation and zoom-in/out) and navigate in the scene (forward/backward motions and turns at bifurcations).

The system works with omnidirectional images collected from a wearable image acquisition setup. As a first step, we perform SfM to compute an initial camera path. Next, a visual place recognition system is applied to detect loops and junctions. This information permits one to improve the initial camera path by adding them as constraints into an optimization step. We also apply a rectification tool to align the unstabilized images with the vertical direction so that the images displayed to the user appear as if the camera was continuously vertically aligned; that is, straight. The next step is the alignment of the camera path with the floor plan of the building, for which an interactive authoring tool was designed. A user can specify ground control points which align the camera path to the floor plan. This process is interactive, every change is immediately incorporated, and the user can see the change instantaneously. After alignment, the virtual tour can be experienced with the viewer application or used on a mobile phone as an indoor navigation system, which provides route information to a particular point of interest.

The remainder of this chapter is organized as follows. Section 20.2 reviews some existing studies and discusses their similarities and differences. In Section 20.3 we give a general overview of the proposed system, which is further elaborated in Sections 20.4–20.7. Section 20.8 presents an application of the system. Section 20.9 is dedicated to the vertical rectification of the images. We close the chapter by showing experimental results obtained by applying the proposed method to a full floor (about 800 m) of the ETH Zürich main building in Switzerland. This building is of important historical value because it was the first one constructed for ETH and it was designed by the German architect Professor Gottfried Semper in a neoclassical style, which was unique to him. The building was then constructed under Professor Gustav Zeuner between 1861 and 1864.

20.2 Related Work

One of the first virtual tours was within the Movie Map project, developed by Lippman (1980) in the 1980s. The city of Aspen in Colorado was recorded using four 16 mm cameras mounted on a car where each camera pointed in a different direction such that they captured a 360° panorama. The analog video was then digitized to provide an interactive virtual tour through the city. Thirty years later, the process of scanning entire cities or buildings has become much more practical. Image-based rendering techniques have increased the interactivity when exploring virtual scenes. Boult (1998) developed a campus tour, allowing a user to freely look around while navigating through the campus. The images were taken from a catadioptric camera, where a curved mirror provides a 360° panoramic view. While the previous projects focused on outdoor scenes, Taylor's VideoPlus provided an indoor walk through a sparse image set (Taylor, 2002).

Recently, Uyttendaele *et al.* (2003) proposed a system to create virtual tours using six cameras tightly packed together. The six camera views are then combined into a single high-resolution omnidirectional view using image-based rendering techniques. They provide virtual tours through indoor and outdoor scenes. At junctions, the viewer can change from one path to another. Unfortunately, their system does not automatically recognize junctions. Instead, a system operator is asked to manually select an intersection range in both paths, and then the system performs an exhaustive search to find the optimal transition between both paths; that is, the transition with minimal visual distance. A topological map is created by manually augmenting the location map with the acquisition path and by associating frame numbers with different key locations along the path. At exploration time, the viewer's position on the map is updated by linearly interpolating between key locations.

Furukawa *et al.* (2009) proposed a fully automated reconstruction and visualization system for architectural scenes from a sparse set of images. Their approach is based on extracting simple 3D geometry by exploiting the Manhattan-world assumption. The extracted models are then used as a geometric proxy for view-dependent texture mapping. Their system is limited by the planar assumption of the scene, making it inappropriate to represent nonplanar scene objects.

Levin and Szeliski (2004) proposed a system to automatically align a camera trajectory, obtained from an SfM algorithm, with a rough hand-drawn map describing the recorded path. Similar panoramic views are recognized by a hierarchical correspondence algorithm. In a first step, color histograms are matched, and then the rotation invariance is verified by computing the 3D rotation between frames. The final frame correspondence is accepted if the epipolar geometry provides enough consistent feature matches. The ego motion is then matched with a hand-drawn map and optimized using loopy belief propagation. Their approach is limited to the accuracy of the hand-drawn map and does not allow user interaction to refine the alignment. Similar to the previous study, Lothe *et al.* (2009) proposed a transformation model to roughly align 3D point clouds with a coarse 3D model.

In contrast to these existing studies, we propose a system which gives a good first estimate of the camera trajectory automatically by SfM and we then provide an authoring tool to manually align the camera trajectory with a floor plan. This alignment is aided by an optimization algorithm which incorporates both manual and place-recognition constraints to solve for an optimal camera trajectory aligned with a floor plan.

20.3   System Overview

An overview of the proposed processing pipeline is illustrated in Figure 20.1. The input to our algorithm is an omnidirectional video stream acquired by a rigid cluster of six synchronized and calibrated cameras, as well as a floor plan. In the preprocessing stage of the pipeline, we first convert each of the six frames to radial undistorted images. After extracting scale-invariant feature transform (SIFT) features (Lowe, 2004), we select key-frames based on the motion of the features. The camera trajectory is estimated by an SfM algorithm. Then, during the junction detection step, we search for already-visited places using visual words and filter out false frame correspondences with a hierarchical filtering scheme followed by a geometric verification. Finally, the full camera trajectory is optimized and aligned with the floor plan by providing both place recognition and user-supplied constraints to the maximum a posteriori (MAP) optimization process. User-supplied constraints are obtained at the authoring step, during which a user may adjust, if needed, individual camera poses, which are directly incorporated into the MAP optimizer. The final result is a topological map – that is, a planar

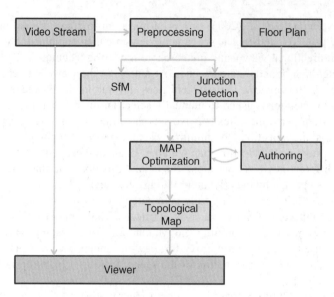

Figure 20.1 Overview of the proposed processing pipeline for semi-automatic alignment of camera trajectories with a floor plan

2D representation of the camera trajectory – which, combined with our viewer application, provides the user's current location and allows the user to freely explore the scene.

20.4   Acquisition and Preprocessing

20.4.1   Camera Model

The input to our algorithm is an omnidirectional video stream captured by a Point Grey's Ladybug 2. This omnidirectional sensor consists of six 1024×768 color cameras; five are positioned horizontally and one points upward, which permits one to acquire an omni-directional view of the scene (see Figure 20.2). We model the Ladybug unit as a single cam-era holding six image planes. The projection equation for each of the six cameras is given by

$$x_i = K_i R_i [I| - C_i]X, \quad i \in \{1, \ldots, 6\} \tag{20.1}$$

where K_i represents the intrinsic calibration matrix, R_i and C_i are the rotation and translation of the i-th camera relative to the Ladybug unit head coordinate system. This permits one to express the data of the six cameras in a common coordinate system. Similar to Banno and Ikeuchi (2009) and Tardif *et al.* (2008), we consider the Ladybug a central projection device; that is, $\|C_i\|$ gets negligible and the projection equation is rewritten as

$$x_i = K_i R_i [I|0]X, \quad i \in \{1, \ldots, 6\} \tag{20.2}$$

This assumption is especially reasonable for points relatively far from the camera, which is typically the case in our scenario. Then we adopt the spherical representation (Banno and Ikeuchi, 2009): each projected point is normalized to one (i.e., lies on the unit sphere surface).

Figure 20.2 (a-f): the six pictures acquired by the Ladybug camera (after lens distortion). (g): stitching to create an omnidirectional image (see Plate 24 for the colored figure)

20.4.2 Data Acquisition

To acquire omnidirectional images in long planar paths, such as corridors and hallways, we mount the Ladybug camera onto a wheeled platform, which provides a smooth camera motion and a natural height when virtually exploring the building. Locations not accessible with the wheeled platform, such as stairways, are recorded by a person wearing the acquisition setup in a backpack, as illustrated in Figure 20.3 (see Section 20.9 about stabilization).

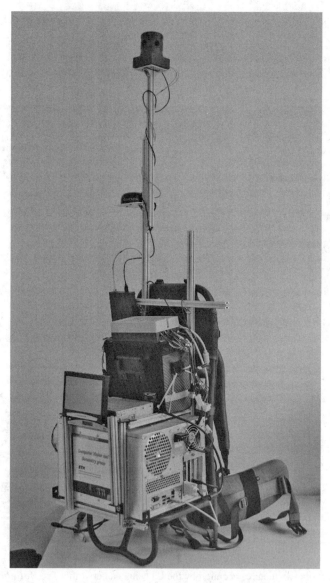

Figure 20.3 Our backpack acquisition setup, consisting of a Ladybug2 camera, a small computer to store the captured data, and a battery set

20.4.3 Feature Extraction

Once a video sequence is captured, each camera stream is converted into a sequence of images. The high lens distortion is corrected using the lens undistortion program provided by the Ladybug SDK. To extract feature points we use the SIFT implementation provided by Vedaldi and Fulkerson.[1] Owing to the high lens distortion and calibration inaccuracies towards the border of the image, only features enclosed in a bounding box of 315 pixels, centered at the principle point, are considered further.

20.4.4 Key-Frame Selection

Key-frames represent a subset of the image sequence, where two neighboring key-frames are separated such that a well-conditioned essential matrix can be estimated. A frame is selected as a key-frame if more than 10% of its features have moved over a threshold of 20 pixels. The radial undistorted and cropped image is of 512×300 pixels size. The key-frames together with the corresponding SIFT features constitute the input to the SfM and junction-detection algorithms outlined in Sections 20.5 and 20.6.

20.5 SfM Using the Ladybug Camera

The input to our SfM algorithm is the key-frames of an omnidirectional video stream. The stream holds multiple discontinuous scans, which overlap at least in one location, such that the whole camera trajectory is represented as one single connected graph.

First, we transform the feature points of all six images into the global Ladybug coordinate system and merge them into a single feature set. Then, the feature points are transformed into a spherical representation, where 2D rays are represented as unit-length 3D vectors (see Section 20.4). This increases robustness of the pose estimation algorithm. For each pair of key-frames we estimate the camera trajectory by extracting rotation and translation from the essential matrix (Hartley and Zisserman, 2004). The essential matrix is computed by the one-point algorithm of Scaramuzza *et al.* (2009). It exploits the non-holonomic constraints of our wheeled platform and the planar motion and thus requires only one point correspondence to compute the essential matrix. This makes motion estimation very fast and robust to high numbers of outliers. We omit relative scale estimation as this is usually subject to drift. Instead, we rely on the internal loop and junction constraints and the manual input to compute the relative scales of the path. The full path is then obtained by concatenating consecutive transformations.

20.6 Loop and Junction Detection

Camera pose estimation usually suffers from drift; for instance, the estimated start and end locations of a closing loop trajectory would not necessary coincide. We overcome this limitation by introducing loop closure constraints in the final optimization process of the camera trajectory estimation. Furthermore, we use place recognition to concatenate paths which were recorded at different times.

---

[1] http://www.vlfeat.org/~vedaldi/code/siftpp.html.

In this section, we show how bifurcations are detected solely based on vision without any complementary sources of information such as encoders, inertial measurement units, or GPS. In particular, satellite signals for GPS cannot be acquired in indoor environments; therefore, an alternative approach to recognize an already-visited place is needed. The technique is required to be rotation invariant, since the camera might traverse an already-visited place pointing into a different direction. Furthermore, finding spatially adjacent frames which are temporally separated requires a framework which quickly discards a large portion of the input images to reduce frame correspondence search. We make use of the bag-of-word scheme of Nistér and Stewénius (2006) providing a first guess of a loop closure, which is verified using a geometric constraint. The visual dictionary used for quantization is trained beforehand on a dataset containing random images taken from the internet. The quantized frames are inserted into a database. Then, for each frame the database is queried for similar frames. The potential frame matches obtained from the query are further pruned down using a hierarchical filtering scheme, consisting of a visual word match, SIFT feature match, and a final geometric verification. Each stage of the hierarchical filtering scheme can either accept or reject a frame pair. If a frame pair is accepted by one stage, then it is passed on to the next stage. A frame pair is finally accepted as a true match if it satisfies the epipolar constraint $x^T \mathbf{E} x' = 0$, where $\mathbf{E}$ is the essential matrix and x and x' are the point correspondences between the two frames. The total number of inliers between two images is then stored in a similarity matrix, which encodes the similarities between image pairs.

The similarity matrix is then post-processed to remove perceptual aliasing matches, characterized by sparse clutter. We identify sparse clutter by labeling the connected components of the binarized similarity matrix and removing regions whose size is below a certain value (30 in our experiments).

For each frame, we then search for the best match in the similarity matrix. The best match is defined as the image pair with the highest ratio of inliers versus feature matches. These frame correspondences are then used as constraints in the optimization process.

20.7 Interactive Alignment to Floor Plan

20.7.1 Notation

In the following, camera poses (i.e., location and orientation) and motions are written as coordinate frame transforms. Camera poses are denoted by the coordinate frame transform from the world origin into the camera coordinate system and written as E_i for the i-th camera. The motion between the i-th and $(i+1)$-th cameras is denoted as M_i. All coordinate frame transforms consist of a 3×3 rotation matrix $\mathbf{R}$ and a 3×1 translation vector $\mathbf{t}$ and are represented by a 4×4 homogeneous transformation matrix of the form

$$\begin{pmatrix} \mathbf{R} & \mathbf{t} \\ 0_{1 \times 3} & 1 \end{pmatrix} \qquad (20.3)$$

In the following derivation, the variables E and M, as well as V and N (see Section 20.7.2), describe a homogeneous transformation and are represented as in Equation 20.3. In this work the transformations and their uncertainties are written in terms of the Lie algebra of SE(3) using the exponential map. The interested readers can refer to Smith *et al.* (2003) for an extensive discussion of this parameterization.

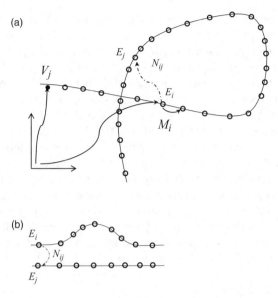

Figure 20.4 (a) A self intersecting camera trajectory and (b) two interconnected paths. V_j denotes the control points set by the user, E_i denotes the absolute camera pose (solid lines), M_i the relative transformation between neighboring frames and N_{ij} loop closure or inter-path transformations, between frames (dashed lines)

20.7.2　Fusing SfM with Ground Control Points

From the SfM algorithm we get the camera path as a sequence of transformations between subsequent images. The transformations have six degrees of freedom and are denoted as $M_0, \ldots, M_n$, where n denotes the total number of acquired images. In the interactive alignment this path needs to be fused with ground control points $V_0, \ldots, V_m$ that are specified by the user, by simply dragging the camera location to the correct position on the floor plan and adapting the viewing direction of the camera. The trajectory then needs to be updated such that it fulfills the constraints (camera location and orientation) provided by the user through the ground control points V_i. The individual camera positions of the path are denoted by $E_0, \ldots, E_n$, which are the results of the fusion. Figure 20.4 shows an illustration of a camera path and the corresponding transformations. Every transformation has an uncertainty specified by a covariance matrix C. We are seeking the MAP estimate of the transformations $E_0, \ldots, E_n$, which is obtained by minimizing the following Mahalanobis distance:

$$w = \min_{E} \left(\sum_i [M_i - (E_{i+1} - E_i)]^T C_{M_i}^{-1} [M_i - (E_{i+1} - E_i)] + \sum_i (V_i - E_i)^T C_{V_i}^{-1} (V_i - E_i) \right)$$

(20.4)

$$= \min_{E} \left((M - HE)^T \hat{C}_M^{-1} (M - HE) + (V - KE)^T \hat{C}_V^{-1} (V - KE) \right)$$

(20.5)

In the first term of Equation 20.4, E_i and E_{i+1} should be computed so that the transformation between the two camera poses matches the transformation M_i computed from SfM. At the same time, the distance between the ground control point transformation V_i to E_i needs to be minimized. Equation 20.4 can be written in matrix form without summation with H and K being incidence matrices that specify for each constraint which E, M, and V transformations are compared with each other. In general, this problem can be solved by nonlinear optimization, as shown by Agrawal (2006). A different solution to this problem was proposed by Smith *et al.* (2003). They proposed a linear time algorithm for the case of a sequential camera path with sparse ground control points. The algorithm works in three steps. First, initial estimates for the E_i are computed by concatenating the M_i transformations starting from the beginning of the sequence. The covariances are propagated accordingly. Ground control points V_i are fused by combining these two measurements if available. In the second step, the same procedure is applied again but starting with the end of the sequence. This results in two measurements for each E_i which are then combined optimally in a third step. The combination is obtained for two individual measurements only:

$$w = \min_{E_{\text{opt}}}[(E_j - E_{\text{opt}})^{\text{T}} C_{E_j}^{-1}(E_j - E_{\text{opt}}) + (E_i - E_{\text{opt}})^{\text{T}} C_{E_i}^{-1}(E_i - E_{\text{opt}})] \qquad (20.6)$$

This scheme is also used to combine a transformation E_i with a ground control point V_i:

$$w = \min_{E_{\text{opt}}}[(E_i - E_{\text{opt}})^{\text{T}} C_{E_i}^{-1}(E_i - E_{\text{opt}}) + (V_i - E_{\text{opt}})^{\text{T}} C_{V_i}^{-1}(V_i - E_{\text{opt}})] \qquad (20.7)$$

For our system we extended the scheme by adding internal loop constraints N_{ij}. These N_{ij} are transformations between frames i and j that are computed by place recognition. The illustration in Figure 20.4 depicts a loop constraint N_{ij}. In our fusion, these constraints need to be fulfilled too. For this, Equation 20.5 is extended by an additional term:

$$w = \min_E \left(\sum_i [M_i - (E_{i+1} - E_i)]^{\text{T}} C_{M_i}^{-1}[M_i - (E_{i+1} - E_i)] + \sum_i (V_i - E_i)^{\text{T}} C_{V_i}^{-1}(V_i - E_i) \right.$$
$$\left. + \sum_{i,j} [N_{ij} - (E_j - E_i)]^{\text{T}} C_{N,ij}^{-1}[N_{ij} - (E_j - E_i)] \right) \qquad (20.8)$$

$$= \min_E \left((M - HE)^{\text{T}} \hat{C}_M^{-1}(M - HE) + (V - KE)^{\text{T}} \hat{C}_V^{-1}(V - KE) + (N - LE)^{\text{T}} \hat{C}_N^{-1}(N - LE) \right) \qquad (20.9)$$

To solve Equation 20.9 we extend the original algorithm proposed by Smith *et al.* (2003) as follows. Our data consist of multiple discontinuous path sequences, which are interconnected by place-recognition constraints. The sequences are optimized independently and sequentially. Figure 20.4b illustrates the case of optimizing two connected sequences. The illustration contains two paths p_1 and p_2. In a first step, p_1 is optimized. When computing the value for E_j, the position of E_i from path p_2 is fused with path p_1. Next, path p_2 is optimized, and here the transformation E_j is used to be fused into E_i. This process iterates so that updates in poses and covariances are propagated sufficiently. Place-recognition constraints from self-intersecting paths are treated in the same way. This extension allows us to use the initial sequential algorithm of Smith *et al.* (2003) for paths with intersections and loops. It does not guarantee to find the global minimum of Equation 20.9, but experiments showed that it is an efficient and practical approach which provides a satisfying solution almost instantly and, therefore, is adapted for interaction.

20.8   Visualization and Navigation

In this section, we present an authoring tool to align the camera trajectory with the floor plan and a viewer application to explore the virtual tour afterwards.

20.8.1   Authoring

The authoring tool provides a simple and efficient way to align the camera trajectory to a floor plan. The user can adjust both camera position and rotation. A preview of the selected camera is provided to help the user localize the corresponding position on the floor plan. The core algorithm to support the alignment process is outlined in Section 20.7. We show in our experiments that a user can align a full floor plan within a couple of minutes using a reasonable number of control points.

20.8.2   Viewer

The viewer tool consists of two windows: a main window showing the environment which is currently explored by the user and a mini-map for location information. To display the environment, we first render a flat panoramic image from all six camera views which is used to texture a sphere. The mini-map is provided as a navigation help. It displays the current position and viewing direction of the user on top of a floor plan. The mini-map also provides an easy way to quickly change from one place to another by clicking the new location of interest.

When exploring the virtual tour, the user can freely move forward, backward, and change direction at places where two paths intersect. Therefore, we cannot rely on the sequential storage of the video stream to display the next frame, since neighboring frames, especially at junctions, have been recorded at different time instances. Instead, we use the camera position to select the next frame. Depending on the user's current viewing direction and position, we search the neighborhood for the closest frame located in the user's viewing frustum. The new frame's rotation is adapted such that it matches the current frame's viewing direction.

The distance between two consecutive images is around 50 cm. Therefore, when the user navigates from one location to the next, the visualization of the images might not look continuous. To achieve a smooth and realistic transition, we apply view interpolation. This is achieved by first aligning the neighboring panoramic images such that the image centers are aligned. Then, we extract the dense optical flow between pairs of

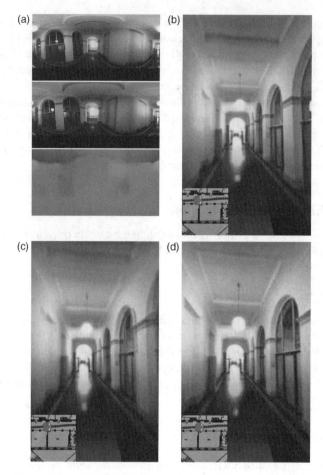

Figure 20.5 View interpolation for smooth visualization. (a) Two spatially adjacent panorama views (top) and the corresponding optical flow (bottom). (c) Interpolated view between the two existing images (b) and (d) issued from the two spatially adjacent panorama views (see Plate 25 for the colored figure)

adjacent panoramas as in Liu (2009). The optical flow is then used to create a series of synthetic views between two existing panoramas on the fly. The view morphing process linearly interpolates the color between the two existing panoramas; see Figure 20.5. This permits one to get a visually smooth transition, but also to emphasize the motion feeling/experience.

To further enhance the exploration of a virtual tour, the viewer application provides a functionality to guide the user to a certain point of interest. Given a starting location, the application computes the shortest path to the desired destination and provides navigation directions, such as turns and distance. For this purpose, the virtual tour was manually augmented by associating camera poses with locations of interest, such as lecture room, cafeteria, and student registration office. Representative screenshots of the viewer application are shown in Figure 20.6.

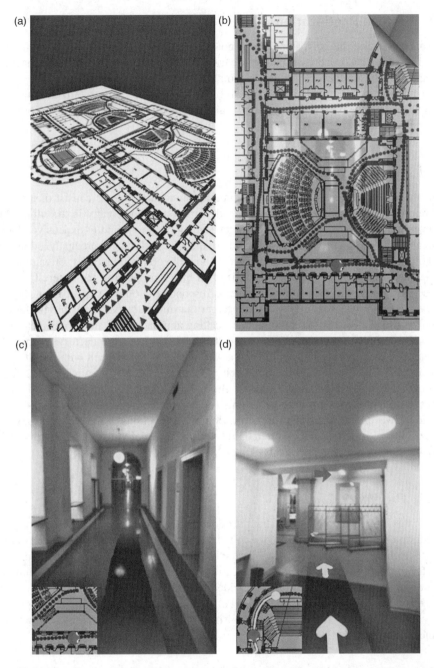

Figure 20.6 Visualization application to interactively explore the virtual tour. (a, b) A mini-map shows the user's current position and viewing direction (large dot with a small arrow) on the floor plan as well as the mapped locations (small dots). (c, d) Display of the image corresponding to the user's current position and viewing direction from the panoramic view together with exploration information. The wide lines show a possible exploration direction. In guiding scenario (d), the lower arrows indicate the shortest path to a location of interest and the top arrow indicates turns at bifurcations (see Plate 26 for the colored figure)

20.9 Vertical Rectification

As explained in Section 20.4, some places not accessible with a wheeled platform, like stairways, have been recorded by a person wearing the acquisition setup in a backpack. As a consequence, these recorded sequences contain shakes and misalignments that make the visualization inconvenient. This can be solved by applying a video stabilization technique. In the following, we review the existing studies on video stabilization and then present the approach applied.

20.9.1 Existing Studies

To obtain smooth and stabilized videos, some professional solutions consist of moving the camera on a rail or using a mechanical steadycam system, especially in the film industry. However, they are inconvenient for consumer-level applications and/or expensive. An interesting alternative is a software approach where the acquired images are analyzed and processed offline (Liu *et al.*, 2009). Stabilization of omnidirectional videos by software has been little studied. Albrecht *et al.* (2010) stabilized an image cropped in the sphere by a virtual perspective camera, but the returned image corresponded to only a small part of the sphere and the algorithm was required to track a target that must always be visible in the image. Moreover, the authors had to combine this tracking approach with an inertial measurement unit to obtain satisfying results. Torii *et al.* (2010) applied camera trajectory estimation and subsequently rectified the images in such a way that they got aligned with the first frame. Kamali *et al.* (2011) extended the approach of Liu *et al.* (2009), originally developed for traditional perspective cameras, towards the sphere space to stabilize omnidirectional images. These methods provide satisfying results, in the sense that the shakes can be reduced. However, they are not aware of the dominant directions of the scene; thus, even the stabilized image outputs can look tilted or not straight, especially in our challenging scenario where the camera is embedded in a backpack.

20.9.2 Procedure Applied

To solve this issue, we propose to extract the dominant vertical direction of the scene and use this information to align the images with the vertical direction. Concretely, the stabilized images will look as if the camera was perfectly vertical; that is to say, aligned with gravity. We refer to this process as vertical rectification. We applied the stabilization technique of Bazin *et al.* (2012), which maintains the video in a certain orientation in a fully automatic manner. It is based on vanishing points (VPs), which are the intersections of the projection of world parallel lines in the image. It consists of two main steps: (i) line extraction and (ii) line clustering and VP estimation.

20.9.3 Line Extraction

The lines are extracted by a generalization of the polygonal approximation towards sphere space. It is based on the geometric property that a line segment in the world is projected onto a great circle in the equivalent sphere space (see Figure 20.7). We refer to this property as the great circle constraint. A great circle is the intersection between the sphere and a plane passing through the sphere center, and can be represented by a unit normal vector. The algorithm

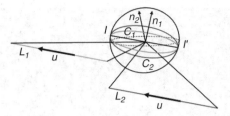

Figure 20.7 Line projection: a world line (L_1) is projected onto the sphere as a great circle (C_1) defined by a normal vector n_1 and the projection of parallel lines (L_1 and L_2) intersect at two antipodal points (I and I')

starts by detecting edges in the image (see Figure 20.8a) and building chains of connected edge pixels (see Figure 20.8b). Then these chains are projected on the equivalent sphere and we check whether they verify the great circle constraint; that is to say, whether they correspond to the projection of world lines. For this, we perform a split-and-merge algorithm based on the distance between the spherical chain points and the plane defining a great circle. Let P_s be a spherical point and n the normal vector of the plane corresponding to a great circle. We compute the geodesic distance from each spherical point to the great circle: $|\arccos(P_s \cdot n) - \pi/2|$. If the distance is greater than a certain threshold, then the chain is split at the farthest point from the great circle and we apply the procedure on the two sub-chains. If the distance is small, then the chain is considered the projection of a world line. Finally, since a world line might correspond to several extracted chains (because of occlusions or edge detection), we merge the great circles having a similar orientation. A representative result of this split-and-merge step is shown in Figure 20.8c. This procedure returns a list of normal vectors defining great circles and corresponding to the detected lines.

20.9.4   Line Clustering and VP Estimation

The second step aims to cluster the extracted lines into sets of world parallel lines and compute the associated VPs. Readers are invited to refer to Bazin *et al.* (2012) for a review of existing methods. An interesting geometric property states that the great circles associated with a pencil of 3D parallel lines intersect in two antipodal points in the sphere (i.e., the so-called VPs), as illustrated in Figure 20.7. To cluster the lines with respect to the unknown-but-sought VPs, we apply the top-down approach of Bazin *et al.* (2012). This is based on the observation that (i) the tasks of line clustering and VP estimation are tightly joined, in the sense that knowing one provides the other one, and (ii) searching for orthogonal VPs is equivalent to searching for a rotation. The approach directly searches for a rotation: a multi-scale sampling hypothesizes a set of rotations and then the consistency of the associated hypothesized VPs is evaluated. The consistency is defined as the number of clustered lines with respect to the VPs. Finally, the algorithm returns the solution maximizing the consistency (consensus set maximization): the rotation, the associated orthogonal VPs, and the line clustering (i.e., which line belongs to which VP). By applying the inverse of this rotation to the image, we can, for example, align the computed vertical VP with the gravity direction so that the resulting image is now vertically aligned and looks as if the camera was straight. Experimental results are shown in Section 20.10.

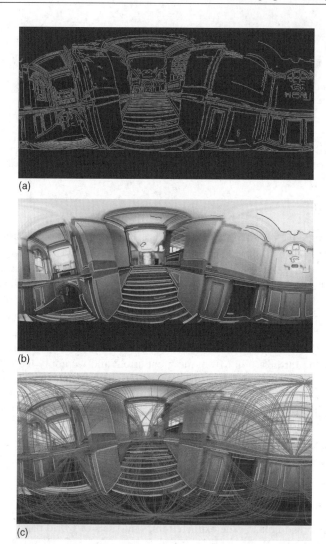

Figure 20.8 The three main steps of line extraction: (a) edge detection (edge pixels have been enlarged for a better display); (b) edge chaining; (c) line detection by split-and-merge (see Plate 27 for the colored figure)

20.10   Experiments

To demonstrate our algorithm we recorded a full floor (about 800 m) of the ETH Zürich main building in Switzerland, excluding offices and rooms not accessible to the public. The stream was recorded at 10 Hz and holds over 14 000 omnidirectional frames, resulting in a total of over 84 000 megapixel images. After preprocessing, about 4000 key-frames are extracted. They form the input to the junction detection and the structure-from-motion algorithm.

20.10.1   Vertical Rectification

Figure 20.9 shows a representative result of image vertical rectification when the acquisition setup was carried in a backpack. This experiment contains two interesting

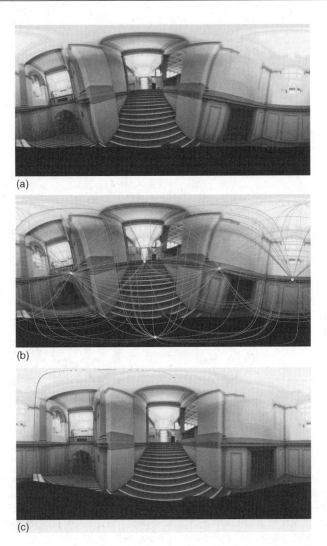

Figure 20.9 Vertical rectification: (a) original image with a tilted camera in a stairway; (b) automatic VP extraction by Bazin *et al.* (2012); (c) rectified image where the camera becomes aligned with gravity and the vertical lines of the world become vertical straight lines (see Plate 28 for the colored figure)

challenges: the Ladybug camera is not exactly central and the camera orientation has a relatively strong pitch. Despite these difficulties, the rectified image looks as if the camera was aligned with gravity and the vertical lines of the world now appear as vertical straight lines in the image, which provides a pleasant visualization for the interactive tour.

20.10.2   *Trajectory Estimation and Mapping*

The similarity matrix obtained from the junction detection represents correspondences between frames which lay temporally apart; see Figure 20.10. Clusters parallel to the

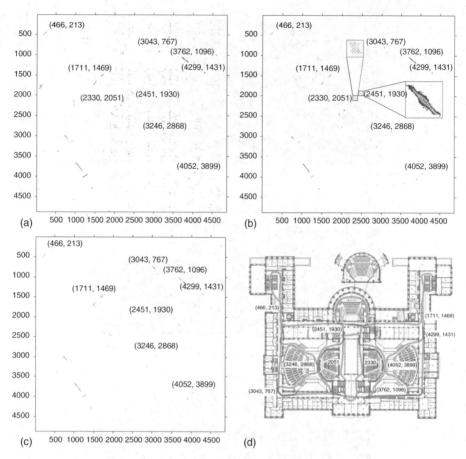

Figure 20.10 The similarity matrix of the processed data. (a) Similarity matrix after visual word matching and (b) after geometric verification. Note that the sparse clutter around frame $(2040, 2330)$ in (b) represents false frame correspondences due to perceptual aliasing. (c) The final similarity matrix after post-processing; that is, after removing sparse clutter. (d) The floor plan showing the full aligned camera trajectory with the according frame correspondences

diagonal represent paths that were re-traversed in the same direction, while clusters orthogonal to the diagonal were traversed in the opposite direction. We only show off-diagonal frame matches, since frames temporally close to each other always look similar. Figure 20.10a shows the similarity matrix after visual word matching and Figure 20.10b after geometric verification. The sparse clutter around frame $(2040, 2330)$ in Figure 20.10b represents false frame matches which are due to perceptual aliasing. Figure 20.10c illustrates the similarity matrix after identifying and removing sparse clutter. The loop-closure algorithm identified 10 loop closures, which are annotated in Figure 20.10c and d.

The SfM algorithm is run on the whole stream, which holds multiple discontinuous paths. The beginning of each sub-path can then be detected by analyzing the number of feature matches and their corresponding inlier ratio. Neighboring frames, which are spatially apart

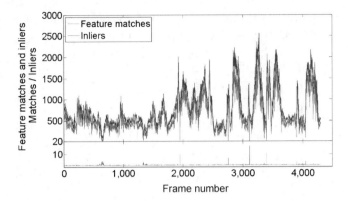

Figure 20.11 The stream holds multiple noncontinuous path sequences. To find the beginning of a new path, we compute the ratio between the number of feature matches (main plot: top trace) and the number of inliers (main plot: bottom trace). The ratio is shown in the subplot. Peak values represent the beginning of a new path sequence, five in the processed dataset

and, therefore, represent the start or end of a new path, will have a few feature matches but almost no inliers satisfying the epipolar constraint. Figure 20.11 shows the total number of feature matches between two consecutive frames and their number of inliers. The sub-graph represents the ratio of feature matches and inliers. Single peaks in the ratio graph indicate the start of a new path.

We then automatically extract frame correspondences from the similarity matrix to append noncontinuous paths and introduce loop-closure constraints into our optimization. In our experiments, 695 loop-closing constraints were introduced automatically. Figure 20.12 shows the SfM result after optimization together with the final alignment on top of the floor plan.

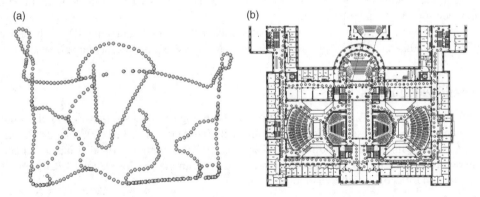

Figure 20.12 Camera path and alignment (in both images only every 10th camera pose is illustrated with dots for a better visualization). (a) The output of the SfM algorithm after applying recognition constraints, obtained from the similarity matrix, without any user interaction. (b) The final result after aligning the camera trajectory with the floor plan. The point correspondences used to align the camera trajectory with the underlying floor plan are shown by the darker dots

When combining the SfM trajectory with control points provided by the user, the error uncertainty of the SfM can be guided through the covariance matrix C

$$
C = \begin{pmatrix}
10^{-6} & 0 & 0 & 0 & 0 & 0 \\
0 & 10^{-6} & 0 & 0 & 0 & 0 \\
0 & 0 & \alpha \cdot 10^{-3} & 0 & 0 & 0 \\
0 & 0 & 0 & x \cdot 10^{-2} & 0 & 0 \\
0 & 0 & 0 & 0 & y \cdot 10^{-2} & 0 \\
0 & 0 & 0 & 0 & 0 & 10^{-6}
\end{pmatrix}
\tag{20.10}
$$

where a strong motion in one direction provides more uncertainty than a small motion; and likewise for the rotation, where a large rotation holds more uncertainty than a small one. We therefore linearly adapt the variance in the x and y directions depending on the motion. Similarly for the rotation around the z-axis, the variance is increased with increasing rotation angle α.

20.11   Conclusions

In this chapter, we presented a system to create topological maps from an omnidirectional video stream, which can be aligned with a floor plan with little user interaction. We have built our own scanning device and particularly focused on the mapping of indoor environments. Since our system does not rely on GPS information, it is suitable not only for indoor scenes but also for general GPS-challenging environments such as urban canyoning. The camera trajectory is obtained from a state-of-the-art one-point algorithm. Already-scanned places are automatically recognized and used to append noncontinuous path sequences and to introduce loop-closure constraints when optimizing the camera trajectory. The trajectory optimization can be guided by the user to align the camera trajectory with a floor plan. Furthermore, we provide two tools: an authoring tool to align the camera trajectory with the floor plan and a visualization tool to virtually explore the scene. Thanks to its generality, our system offers potential applications in various contexts, such as virtual tourism, tele-immersion, tele-presence, and e-heritage.

In future work, we would like to investigate further automation to align the SfM with a floor plan, using additional sources of information such as a sparse 3D reconstruction and VPs.

Acknowledgments

This research, which is partially carried out at the BeingThere Centre, is supported by the Singapore National Research Foundation under its International Research Centre @ Singapore Funding Initiative and administered by the IDM Programme Office. It has also been funded in part by the EC's Seventh Framework Programme (FP7/2007–2013)/ERC grant #210806 4D Video, as well as a Google Award.

References

Agrawal, M. (2006) A Lie algebraic approach for consistent pose registration for general Euclidean motion, in *IEEE/RSJ International Conference on Intelligent Robots and Systems*, pp. 1891–1897.

Albrecht, T., Tan, T., West, G., and Ly, T. (2010) Omnidirectional video stabilisation on a virtual camera using sensor fusion, in *International Conference on Control, Automation, Robotics & Vision (ICARCV)*, pp. 2067–2072.

Banno, A. and Ikeuchi, K. (2009) Omnidirectional texturing based on robust 3D registration through Euclidean reconstruction from two spherical images. *Comput. Vis. Image Und.*, **114** (4), 491–499.

Bazin, J.C., Demonceaux, C., Vasseur, P., and Kweon, I. (2012) Rotation estimation and vanishing point extraction by omnidirectional vision in urban environment. *Int. J. Robot. Res.*, **31** (1), 63–81.

Boult, T.E. (1998) Remote reality via omnidirectional imaging DARPA Image Understanding Workshop, pp. 1049–1052.

Furukawa, Y., Curless, C., Seitz, S., and Szeliski, R. (2009) Reconstructing building interiors from images, in *IEEE Conference on Computer Vision and Pattern Recognition*, pp. 80–87.

Hartley, R.I. and Zisserman, A. (2004) *Multiple View Geometry in Computer Vision*, 2nd edn, Cambridge University Press.

Kamali, M., Banno, A., Bazin, J.C. *et al.* (2011) Stabilizing omnidirectional videos using 3D structure and spherical image warping. *IAPR Conference on Machine Vision Applications*, pp. 177–180.

Levin, A. and Szeliski, R. (2004) Visual odometry and map correlation, in *IEEE Conference on Computer Vision and Pattern Recognition*, Vol. **1**, pp. 611–618.

Lippman, A. (1980) Movie-maps: an application of the optical videodisc to computer graphics. *SIGGRAPH*, **14** (3), 32–42.

Liu, C. (2009) Beyond pixels: exploring new representations and applications for motion analysis, PhD Thesis.

Liu, F., Gleicher, M. and Agarwala, A. (2009) Content-preserving warps for 3D video stabilization, in *ACM SIGGRAPH*, pp. 44:1–44:9.

Lothe, P., Bourgeois, S., Dekeyser, F. *et al.* (2009) Towards geographical referencing of monocular SLAM reconstruction using 3D city models: application to real-time accurate vision-based localization, in *IEEE Conference on Computer Vision and Pattern Recognition*, pp. 2882–2889.

Lowe, D. (2004) Distinctive image features from scale-invariant keypoints. *Int. J. Comput. Vision*, **60** (2), 91–110.

Nistér, D. and Stewénius, H. (2006) Scalable recognition with a vocabulary tree, in *IEEE Conference on Computer Vision and Pattern Recognition*, pp. 2161–2168.

Scaramuzza, D., Fraundorfer, F., and Siegwart, R. (2009) Real-time monocular visual odometry for on-road vehicles with 1-point RANSAC, in *IEEE International Conference on Robotics and Automation*, pp. 4293–4299.

Smith, P., Drummond, T., and Roussopoulos, K. (2003) Computing MAP trajectories by representing, propagating and combining PDFs over groups, in *IEEE International Conference on Computer Vision*, Vol. **2**, pp. 1275–1282.

Tardif, J.-P., Pavlidis, Y., and Daniilidis, K. (2008) Monocular visual odometry in urban environments using an omnidirectional camera, in *IEEE/RSJ International Conference on Intelligent Robots and Systems*, pp. 2531–2538.

Taylor, C. (2002) Videoplus: a method for capturing the structure and appearance of immersive environments. *IEEE Trans. Vis. Comput. Graph.*, **8** (2), 171–182.

Torii, A., Havlena, M., and Pajdla, T. (2010) Omnidirectional image stabilization for visual object recognition. *Int. J. Comput. Vision*, **91** (2), 157–174.

Uyttendaele, A., Criminisi, A., Kang, S.B. *et al.* (2003) High-quality image-based interactive exploration of real-world environments. Technical Report MSR-TR-2003-61, Microsoft Research.

21

View Selection

Fahad Daniyal and Andrea Cavallaro

Queen Mary University of London, UK

21.1 Introduction

When multiple vantage points for viewing a scene exist, some camera views might be more informative than others. These views can be displayed simultaneously for playback (e.g. on a video wall) or the best view might be selected by an operator (e.g. a director). However, given the growing availability of cameras, monitoring multiple videos is an expensive, monotonous and error-prone task.

Automated view selection is a fundamental task in various applications such as object localization (Ercan *et al.*, 2007), target tracking (Han *et al.*, 2011), video summarization (Yamasaki *et al.*, 2008) and autonomous video generation (Li and Bhanu, 2011). For instance, automated surveillance systems necessitate remote identification of targets of interest to automatically recognize a person by acquiring a high-resolution image. The goal in this case is to identify the targets/regions that can be of interest in the scene across all the cameras. For autonomous video generation, it is important to maintain continuity in the resulting video by avoiding frequent inter-camera switches to generate a pleasant and intelligible stream that shows the appropriate camera view over time. To this end, it is desirable to automate the task with tools that

1. automatically analyse the content of each video stream;
2. rank views based on their importance for a specific task; and
3. identify the best camera view at a particular time instant.

In this chapter, we address the following questions for solving the view-selection problem:

- Can we define and extract a set of features that are suitable to represent the *information* in a view?

Emerging Technologies for 3D Video: Creation, Coding, Transmission and Rendering, First Edition.
Frédéric Dufaux, Béatrice Pesquet-Popescu, and Marco Cagnazzo.

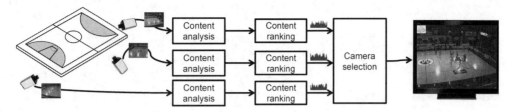

Figure 21.1 Block diagram of a content-based camera selection system

- Can we devise methods to quantify the quality of view (QoV) of each camera based on its content?
- Can we select a view over time based on its QoV to generate a pleasant video that is comparable to the one generated by an amateur director?

We investigate the use of automatic solutions for content analysis, content ranking and camera selection (Figure 21.1). The first stage is the analysis of the content associated with each video stream. This analysis involves the automatic extraction of information from each camera view that is represented in terms of features associated with each view. The choice of these features is dependent on the task at hand. This is followed by the ranking of the analysed content that allows us to assign a QoV score to the frames of each camera. Finally, view selection is achieved given the ranking and task parameters while limiting the number of inter-camera switches.

21.2   Content Analysis

Selecting a view requires a representation framework that maps the information conveyed by the videos/images into scores. These scores measure the information content in the scene. The significance of a scene can be measured in terms of the observable features of interest, which are dictated by the task at hand (Tarabanis *et al.*, 1995a). These features define representative instances of the targets or the scene. For example, object pose can be a relevant feature in face recognition because a pose constraint on the recognition input can enhance the system reliability and performance. Similarly, in surveillance tasks, if we want to extract representative images of people entering or leaving a scene, then the location of the object with reference to the exit and entry points is a relevant feature.

A review of the state-of-the-art methods yields a range of features that are repeatedly used in varying scenarios. These features include the visibility of the face (Shen *et al.*, 2007), size of the targets (Del-Bimbo and Pernici, 2006; Jiang *et al.*, 2008; Qureshi and Terzopoulos, 2007), their location ((Javed *et al.*, 2000; Krishnamurthy and Djonin, 2007), amount of motion they undergo (Snidaro *et al.*, 2003) and their audio-visual activity (Kubicek *et al.*, 2009; Fu *et al.*, 2010). The various representations of these features are discussed below.

21.2.1   Pose

To capture representative instances of a face, the *view angle* or *gaze direction* can be used to determine the QoV (Trivedi *et al.*, 2005; Wallhoff *et al.*, 2006; Shen *et al.*, 2007; Kim and Kim, 2008). The frontal view of a person is in general more informative than the side views; the view angle can be represented as the angle θ_1 the head of a person subtends with the

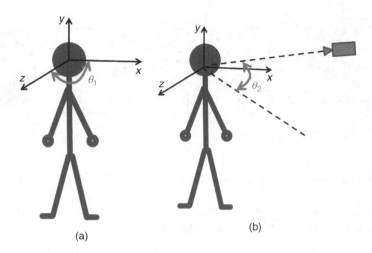

Figure 21.2 Definition of the parameters describing the viewing angle and gaze direction. (a) Angle between the head and the image plane. (b) Angle between the head and the ground plane

image plane and identifies whether a person is looking to their left or to their right (Figure 21.2a). Another measure of the view angle is the angle θ_2 that the person's head subtends with the ground plane. This angle identifies whether the person is looking up or down (Figure 21.2b). At $\theta_1 = 0°$ the camera has the frontal view of the person, while at $\theta_2 = 45°$ the camera is assumed to be at the head level.

We can estimate $\theta_2 = 45°$ by assuming that all the cameras are fixed at head height level and then use θ_1 to represent the quality of the captured face (Trivedi *et al.*, 2005; Wallhoff *et al.*, 2006). The value of θ_1 can be determined by fitting an ellipse on the segmented face pixels, where θ_1 is the angle between the ellipse major-axis and the image vertical axis (Trivedi *et al.*, 2005). The accuracy of the estimation of this angle tends to drop in cluttered backgrounds owing to errors in the face segmentation step. Also, θ_1 can extracted using a neural network (NN) for face detection (Rowley, 1999) by expanding the NN's hidden layer and the output vector with probabilities for the view angles. Thus, the output includes the face location as well as the probability of the view angle (Wallhoff *et al.*, 2006). However, such formulation assumes that $\theta_1 \in \{\pm90°, \pm45°, \pm22.5°, 0°\}$, thus limiting the accuracy in the angle estimates.

Both angles θ_1 and θ_2 can be used together to quantify the quality associated with a view (Shen *et al.*, 2007). θ_1 can be extracted by fitting an elliptical pillar to the area of the image occupied by the person and then setting the origin of the image axes to the centre of gravity of the person's head. Then, θ_1 is defined as the rotation between the image axis and the ellipse axis. The second angle θ_2 is extracted between the ground plane and the line passing through the camera centre and the centre of gravity of the person's head. This approach requires accurate head detection and reliable camera calibration, which is not always possible in realistic and crowded scenarios. Extending the view angle, information about the face orientation can be determined from the pose of the person (Li and Bhanu, 2011). The pose with respect to a camera can be estimated using silhouettes for frontal and side pose discrimination (Prince *et al.*, 2005), appearance (Agarwal and Triggs, 2004) or motion cues (Fablet and Black, 2002). In Li and Bhanu (2011), pose is derived from the ratio between face pixels and body pixels. In this case, an error in the calculation of the size of any of the two regions (face or object) can lead to pose errors.

21.2.2 Occlusions

We can also measure visibility (or the lack of it) in terms of the occlusions of the features of interest (Ercan et al., 2007; Gupta et al., 2007; Chen and Davis, 2008; Chen et al., 2010). Occlusions can be divided into self-occlusions and inter-object occlusions (Chen and Davis, 2008). In self-occlusions, the object itself occludes the features of interest. For instance, if we are interested in the faces in the scene, occlusions can be caused by a hand covering a face or by a person turning their face away from the camera view. In case of inter-object occlusions we are more concerned with capturing the whole object; hence, the layout of the scene (Ercan et al., 2007) as well as the location of other objects (Gupta et al., 2007) need to be taken into account. Owing to these occlusions, we need to select a view of the target that contains sufficient features for it to be recognized unambiguously. This view selection can be over time (selection of a frame from a single camera) or across cameras. This occlusion can be determined as the visibility probability of each part of a person in each camera based on probabilistic estimates of the poses of all other people in the scene (Gupta et al., 2007). The pose is recognized using a part-based approach where each part of the body is represented as a group of voxels (volumetric picture element representing a value on a regular grid in 3D space), where no two voxels can occupy the same space at the same time. Object detection is achieved by segmenting and tracking multiple people on a ground plane (Mittal and Davis, 2002). The algorithm alternates between using segmentation to estimate the ground-plane positions of people and using ground-plane position estimates to obtain segmentations. This approach relies heavily on accurate object detection and tracking. Moreover, it is assumed that only people are present in the scene; therefore, a person with a bag could lead to erroneous detections.

We can also assume that a camera can distinguish between the object and the occluders based on the layout of the scene (Ercan et al., 2007; Chen and Davis, 2008). In such cases only static occlusions are considered. For dynamic occlusions it is in general assumed that, at any given time, at least one camera sees the feature of interest and the cameras that do not see it are subject to the occlusions (Chen and Davis, 2008). However, this is a very strong assumption that fails in realistic scenarios.

In the case of partially overlapping cameras, objects can be detected on the ground plane using a multi-view multi-planar approach (Khan and Shah, 2006; Delannay et al., 2009). In such scenarios the occlusion of an object is defined to be the number of pixels of the object that are hidden by other objects when back-projected onto the camera plane (Chen et al., 2010). However, in this implementation, the height of the person is assumed to be estimated using calibration information. Moreover, no compensation for homography errors that can lead to errors in the occlusion estimation is included.

21.2.3 Position

The position of objects in the scene is also a feature of interest to estimate the QoV (Figure 21.3). The position of an object carries information regarding its contextual location in the scene (Trivedi et al., 2005; Ba and Odobez, 2011) or in terms of proximity to other objects or regions of interest (Shen et al., 2007), such as a pedestrian on the road, when the road is defined a priori (Daniyal et al., 2008). Alternatively, a set of predefined discrete locations, each with an importance score (Ba and Odobez, 2011), can be used to define a set of rules to capture representative object instances (Trivedi et al., 2005). For example, it can be assumed that when a person enters the camera view they are facing the camera. Otherwise the location of the target view angle calculated for each camera can be used to estimate face visibility.

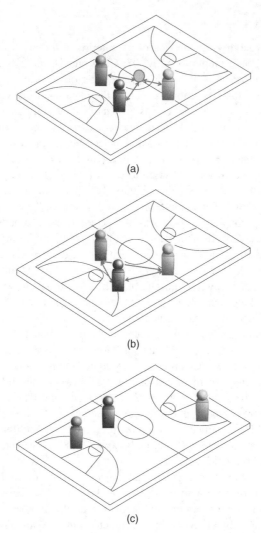

Figure 21.3 Definition of the positions of objects with respect to (a) a point of interest, (b) each other and (c) their contextual location in the scene

When considering *proximity*, one can measure the distance between the frame centre and the body centroid on the image plane (Li and Bhanu, 2011), the distance between the object and the camera centre (Shen *et al.*, 2007; Tessens *et al.*, 2008) or the distance between the object and a point of interest in the scene (Chen and De Vleeschouwer, 2010b; Daniyal and Cavallaro, 2011).

Some interesting information that can be extracted from the distance is the *deadline* for an object, which is defined as the minimum time a target will take to exit the scene. The calculation of the deadline for an object is based on its distance from the (known) exit points and its motion model. Deadlines are used in camera networks that have active (pan, tilt and zoom) cameras, which focus on a single target at each time in order to capture representative target instances before they leave the scene (Del-Bimbo and Pernici, 2006; Qureshi and Terzopoulos, 2007). These studies usually model the motion through linear propagation of trajectories. Similarly, a frame that contains an object about to exit or an object that has just entered the scene can be identified via deadlines and are deemed more important than other frames.

21.2.4 Size

The most commonly used feature for determining the QoV is the *size* of targets (Snidaro *et al.*, 2003; Goshorn *et al.*, 2007; Chen *et al.*, 2010; Li and Bhanu, 2011). Various definitions for size exist based on the task at hand and on the selected object representation strategy.

Size can be measured as the ratio between the number of pixels inside the bounding box representing a person and the size of the image (Li and Bhanu, 2011), the area (width times height) of the bounding box around a face (Goshorn *et al.*, 2007) or an object (Jiang *et al.*, 2008), or simply the number of segmented object pixels (Snidaro *et al.*, 2003; Chen and De Vleeschouwer, 2010b).

In scenarios with partially overlapping camera views and when objects are detected on the ground plane, people are considered to have a predefined height. When their corresponding regions are back-projected onto the image plane using calibration information, different heights are generated due to the perspective (Chen and De Vleeschouwer, 2010a). The distance from the camera, which affects the size, can also be used (Shen *et al.*, 2007). However, this distance calculation requires accurate knowledge of the 3D geometry of the scene, which is difficult to obtain in complex scenarios.

21.2.5 Events

Events have also been used for quantifying the information content in a scene (Park *et al.*, 2008; Fu *et al.*, 2010). Events can be detected by employing a Bayesian network (BN) within each camera (Park *et al.*, 2008). Basic features such as object location, pose and motion information are manually annotated and then employed to train the BN for event recognition. The predefined probability and the priorities of the recognized events are then used as a feature to rank views. The detected events can then also be merged with the object and face detection results (Fu *et al.*, 2010). These high-level semantics are fused with low-level features to represent the information in each video segment and view. In particular, low-level features include appearance-based features; for example, colour histograms (Benmokhtar *et al.*, 2007), edge histograms (Gong and Liu, 2003), and wavelets (Ciocca and Schettini, 2006). However, when dealing with such low-level features, the system accuracy can be reduced due to noisy observations.

Audio activity and audio-visual events can also be used to find the visual focus of attention (VFOA) in meetings and classrooms (Trivedi *et al.*, 2005; Kubicek *et al.*, 2009; Ba and Odobez, 2011). In this case it is assumed that the source of sound is an important cue and that viewers tend to focus on the audio source when visible.

21.3 Content Ranking

Once the feature extraction process is complete, the next step is to assign a QoV to each frame based on the observed features to enable content ranking. Content ranking can be object centric or view centric. The object-centric content ranking refers to the quality of an object instance at any given time and is constrained by the relationship between the camera and the object under the contextual information drawn from the task at hand. View-centric content ranking refers to the quality of the entire view and depends on the visibility of the features in the frame.

21.3.1   Object-Centric Quality of View

The QoV of an object is related to its features. Assigning a QoV to an object ensures that these features satisfy particular constraints and criteria, which tend to vary from various vantage points or as the object moves within a camera view (Tarabanis *et al.*, 1995b; Cai and Aggarwal, 1999; Motai, 2005; Ercan *et al.*, 2007). Different constraints and criteria, mainly defined manually (Trivedi *et al.*, 2005; Kubicek *et al.*, 2009), need to be used at various locations to define the best view; for example, if we want to capture a face of the target as it enters the scene, leaves and is in certain locations (Trivedi *et al.*, 2005). Camera selection and camera-network planning techniques exist that aim to automatically map constraints to algorithms. In this case a search is performed in time over all object instances, where each instance is hypothesized as the highest ranking view. These hypotheses are assessed according to a matching criterion, which is usually based on the cross-correlation between the object features and the required constraints. The matching confidence can be used as a measure of the QoV (Hutchinson and Kak, 1989; Cameron and Durrant-Whyte, 1990; Lee, 1997). This is a recursive process that stops when a local maximum is reached or another stopping condition is met (Gavrila, 2007). In order to limit the search in the instance space within this *hypothesize-and-verify paradigm*, a discrete approximation of this space is commonly employed (Wheeler and Ikeuchi, 1995), which takes into account the contextual layout of the scene. These techniques draw on a considerable amount of *prior* knowledge of objects, cameras and task requirements. Since the identities and poses of the viewed objects are known, the imaging constraints are usually planned off-line and then used on-line when objects are actually observed, thus rendering these methods ineffective in cluttered scenarios.

For methods using a single static criterion or feature for identifying the view of an object, the derivation of QoV is straightforward. For instance, the size of an object of interest can be used as the only feature (Jiang *et al.*, 2008). In terms of object pose, views are ranked as the inner product between the direction that the object is facing and the principal axis of each camera that observes the object, as shown in Figure 21.4 (Trivedi *et al.*, 2005).

When the goal is to acquire images of faces in the scene, usually a master–slave configuration is used: the master camera performs feature extraction, while the slave camera is adjusted based on the parameters extracted from the master camera. The master camera can be a ceiling-mounted omnidirectional camera (Greiffenhagen *et al.*, 2000) or a wide field-of-view camera (Zhou *et al.*, 2003; Costello *et al.*, 2004). In the former case the QoV is defined as the percentage of the head visible to the master camera and time-based ranking is performed for multiple targets. In the latter case the master camera provides the position information to the slave camera that captures targets at high resolution. The speed of the person as they move along a predefined path determines the release times (i.e. when a person enters the field of view) and deadlines (i.e. when a person leaves the field of view) for each person. There is a strong assumption that each person enters the field of view only once and they should be imaged for a certain duration before they leave the field of view of the master camera.

A common approach for combining multiple features is the weighted average (Goshorn *et al.*, 2007; Shen *et al.*, 2007). The two view angles θ_1 and θ_2 can be combined with the distance from

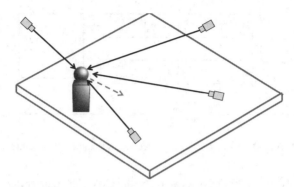

Figure 21.4 Object-centric QoV estimation based on the object direction (dotted arrow) with respect to the camera principal axis (solid arrows)

the camera and the weights are empirically set based on the task at hand. Similarly, each object can be assigned a unit score that is adjusted using the weighted sum of the features based on the object size and its distance from the centre of the image (Chen *et al.*, 2010).

More sophisticated frameworks for feature fusion have also been studied and use, for instance, size and appearance in the form of an appearance ratio (Snidaro *et al.*, 2003). The appearance ratio is defined as the sum of all the pixel values in the segmented blob of an object normalized by the number of pixels. Alternatively, each object can be assigned a score that is viewed as a marginal contribution to the global utility. The global utility is the summation over all camera QoVs that, in turn, is the summation of the QoVs of all the objects (Li and Bhanu, 2011).

Content ranking has also been studied for target localization using position estimation (Ercan, 2006; He and Chong, 2006; Krishnamurthy and Djonin, 2007; Rezaeian 2007). Some studies use a partially observable Markov decision process (POMDP) to estimate the state (location and velocity) of the target. Content ranking is such that the highest ranked camera minimizes the estimation error in detecting the state of the target (He and Chong, 2006; Krishnamurthy and Djonin, 2007). Two approaches are available for modelling the POMDP, namely Monte Carlo estimation using a particle filter to estimate the belief state of the target (He and Chong, 2006) and structured threshold policies that consider the camera network as a simplex (Krishnamurthy and Djonin, 2007).

Objects can also be localized using the object and dynamic occluder priors, the positions and shapes of the static occluders, the camera fields of view and the camera noise parameters. In this case the minimum mean square error of the best linear estimate of the object position is used as a measure for the localization error (Ercan, 2006). Active tracking with a simple behaviour (policy) with a finite state machine can also be used (Batista *et al.*, 1998). The finite state machine is defined in order to give continuity when the currently tracked target is changed using information about (single) target locations and camera placements, and does not take into account target interactions with the environment, thus limiting the adaptability to more complex scenarios.

21.3.2 View-Centric Quality of View

The sum of objects scores in a view can be used as a QoV for that camera (Li and Bhanu, 2011), where the object score may be accumulated over a time window to obtain a smoother

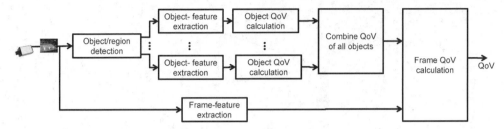

Figure 21.5   QoV estimation using object-centric and camera-centric features

QoV evolution (Kelly *et al.*, 2009). Alternatively, the QoV can be extracted based only on the observed frame-level information such as event probability and priority (Park *et al.*, 2008). In this case the view with the highest event score is selected as the optimal one if only one event is captured within a time segment. If two or more events are captured during the same segment, views of events with the highest priorities are selected as candidates and then the view with the highest probability is selected amongst the candidates. The visibility of objects is not considered in this formulation.

The method of Tarabanis and Tsai (1991) measures view-centric QoV in terms of observable features of interest dictated by the task at hand. Features of interest in the environment are required to simultaneously be visible inside the field of view, in focus and magnified. These features extracted from the track information are fused to obtain a global score to mark regions as interesting on the basis of the quantization of the QoV.

The rank for each camera can also be estimated as a weighted sum of individual object features such as face visibility, blob area and the direction of motion of the target (Goshorn *et al.*, 2007). However, without incorporating high-level scene analysis, the highest rank may be given to the view with the largest number of targets while ignoring other views with fewer targets but which contain interesting or abnormal behaviours.

The view-centric QoV can be calculated by fusing features using predefined weights using contextual knowledge (Kubicek *et al.*, 2009) or a distribution-based fusion model (Fu *et al.*, 2010; Daniyal and Cavallaro, 2011). For instance, low-level features and high-level semantics can be fused using a Gaussian entropy model (Fu *et al.*, 2010). Low-level features include appearance-based features (colour histogram, edge histogram and wavelets; Gong and Liu, 2003; Ciocca and Schettini, 2006; Benmokhtar *et al.*, 2007), while high-level features include events. Similarly, multivariate Gaussians can be used for feature fusion (Daniyal and Cavallaro, 2011). In this case, each feature is normalized in its own feature space, thus allowing the features to be weighted either equally or based on contextual knowldge. Finally, based on the frame-level information and the object-centric QoV, the view-centric QoV is calculated (Figure 21.5).

21.4   View Selection

The goal of camera selection is to choose a view such that QoV is maximized over time while the number of inter-camera switches is controlled. The first constraint implies that the best view is selected at a certain time t or over a temporal window $[t_1, t_2]$, where $t_1 \leq t \leq t_2$. The most common approach is to select the camera with the maximum value of QoV at each time t (Snidaro *et al.*, 2003; Goshorn *et al.*, 2007; Park *et al.*, 2008; Tessens *et al.*, 2008; Kubicek *et al.*, 2009). However, this approach introduces frequent camera switches.

Limitating the number of switches whilst selecting the view has been tackled as a scheduling problem or as an optimization problem.

21.4.1   View Selection as a Scheduling Problem

Given a camera network and the objects in its field of view, scheduling strategies aim at assigning cameras to objects. The assignment is restricted under the object-centric QoV and camera switching is triggered by certain activities detected in the scene. Activities can be related to audiovisual cues, such as an object entering the field of view (Qureshi and Terzopoulos, 2007) or a specific sound being detected (Kubicek *et al.*, 2009). These systems usually target single-person and low-activity scenarios, and perform selection based on naive explicit rules.

To restrict frequent view switching, a predefined scheduling interval can be used that maintains for fixed intervals camera–object pairs once they are created (Del-Bimbo and Pernici, 2006; Goshorn *et al.*, 2007; Krishnamurthy and Djonin, 2007; Tessens *et al.*, 2008; Rudoy and Zelnik-Manor, 2012). The duration of the scheduling interval is specified a priori: too short scheduling intervals can cause frequent switches, while too long scheduling intervals can cause the loss of information in dynamic scenarios.

Under the scheduling paradigm, the camera-selection problem also draws from the resource allocation framework (Zhou *et al.*, 2003; Costello *et al.*, 2004; Qureshi and Terzopoulos, 2007). In this framework the camera network aims to serve a maximum number of targets, given constraints on the scene dynamics and the task at hand. With these strategies, views or targets are assigned priorities based on their features. For example, the distance from the camera (Zhou *et al.*, 2003), the time in which a target entered the scene (Qureshi and Terzopoulos, 2007) or a deadline (Costello *et al.*, 2004) can be used to assign priorities to targets. Greedy scheduling policies, such as round robin, first-come first-served, earliest deadline first and current minloss throughput optimal, can be implemented when multiple targets are present. However, these approaches do not include a cost that allows one to limit frequent inter-camera switches.

21.4.2   View Selection as an Optimization Problem

Rather than defining explicit rules, several methods adaptively select the view by optimizing a cost function over a temporal window (Ercan *et al.*, 2007; Rezaeian, 2008). For example, the cost function can depend on a multi-feature view-quality measure (Jiang *et al.*, 2008). The cost function can also include a penalizing factor to avoid frequent switches.

The optimization process can also be included in a game theoretic framework (Li and Bhanu, 2011). To be selected in this framework (i) each view is considered a *player* that bargains with other views or (ii) views compete with each other to be selected for tracking a specific target. Various bargaining and competing mechanisms for collaboration and for resolving conflicts amongst views can be used with view-centric and object-centric QoVs.

The optimal camera can also be found by evaluating completeness, closeness and occlusions in the scene, under specified user preferences (Chen and De Vleeschouwer, 2010b). Each view is ranked according to the (quality of its) completeness/closeness trade-off, and to its degree of occlusions (Figure 21.6). The highest rank corresponds to a view that makes most objects of interest visible (occlusion and completeness) and presents important objects with a high resolution (closeness). The smoothing process is implemented based on the definition of two Markov random fields (Szeliski *et al.*, 2008).

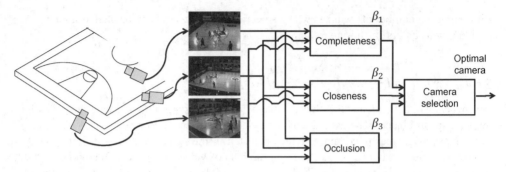

Figure 21.6 View selection using a user-defined weighted combination of completeness, closeness and occlusion in the scene

Within the optimization paradigm, view selection can be modelled as a POMDP, where the decision to select a camera is based on all the past measurements (Krishnamurthy and Djonin, 2007; Rezaeian, 2008; Spaan and Lima, 2009; Daniyal and Cavallaro, 2011). This selection is regarded as the control action within the POMDP. This action is not for the control of the state process, which is autonomous, but for influencing the measurement of the process via the selection of a camera (Rezaeian *et al.*, 2010). The cost/reward for a policy can be the estimation entropy (Rezaeian, 2008), the sensor usage and the performance costs (Krishnamurthy and Djonin, 2007) or a cost function that can be modified based on the task at hand (Spaan and Lima, 2009). This modelling is used for radars and not for cameras. In camera networks, the cost function can be modelled to maximize the feature observability and to minimize the camera switches (Daniyal and Cavallaro, 2011). This framework learns the camera connections via ground truth and shall be adapted for changes in the position or number of the cameras.

21.5 Comparative Summary and Outlook

A comparative summary of representative state-of-the-art methods based on the set of features used for content analysis and ranking is presented in Table 21.1, which lists methods for camera selection that minimize the number of switches. The most common features are size and location of targets and related appearance features. Other features include pose, face visibility, deadlines, motion and object number. Based on the task requirements, the choice of features is adapted for the specific camera selection approach. Most methods either work with a single feature or merge multiple features with predefined weights, whereas recent approaches combine features adaptively.

A class of the studies presented does not perform camera scheduling (Senior *et al.*, 2005; He and Chong, 2006; Gupta *et al.*, 2007; Shen *et al.*, 2007) and focuses on content ranking only. These methods are used for acquiring representative instances of targets in a multi-camera environment or when using an active camera.

Based on the ranking, view-selection methods aim at reducing the number of switches either by employing a fixed scheme or by making the decision based on the past, current and/or future QoV of each frame. Most methods assume that the dynamics are predictable and do not ensure that minimal loss of information occurs in the views that are not selected while minimizing the switches. A decision can be taken by selecting the camera

Table 21.1 Comparative summary of content-ranking and view selection methods

Reference	Features								Ranking strategy	MO	Selection strategy	Smoothing strategy
	Object					Deadlines	Camera					
	Size	Appearance	Location	Pose	Velocity		Event	Change				
Senior et al. (2005)	✓			✓					First come first served			
He and Chong (2006)		✓			✓				Q-value estimation			
Shen et al. (2007)	✓		✓	✓					Weighted feature sum			
Gupta et al. (2007)	✓		✓	✓					Heuristic-based greedy prog.			
Snidaro et al. (2003)	✓		✓	✓				✓	Colour channel average	✓	Argmax based	
Kubicek et al. (2009)	✓		✓	✓		✓		✓	Rule-based weighted fusion	✓	Argmax based	
Park et al. (2008)							✓		Event probability and priority		Argmax based	
Qureshi and Terzopoulos (2007)	✓					✓			Rule-based weighted fusion		Weighted round-robin techniques	
Li and Bhanu (2011)	✓		✓	✓					Weighted feature fusion	✓	Game theoretic framework	
Fu et al. (2010)		✓						✓	Gaussian entropy fusion		Multi-objective optimization	

(*continued*)

Table 21.1 (*Continued*)

Reference	Features — Object						Features — Camera		Ranking strategy	MO	Selection strategy	Smoothing strategy
	Size	Appearance	Location	Pose	Velocity	Deadlines	Event	Change				
Rezaeian (2007)			✓						POMDP policies		Estimation entropy minimization	
Del-Bimbo and Pernici (2006)			✓		✓				Time-dependent orienting		Argmax based	Scheduling interval
Tessens *et al.* (2008)	✓		✓	✓	✓	✓			Weighted fusion of features	✓	Maximization of QoV	Scheduling interval
Goshorn *et al.* (2007)	✓		✓	✓			✓	✓	Weighted feature sum	✓	Argmax based	Scheduling interval
Krishnamurthy and Djonin (2007)			✓						POMDP policies		Minimization of sensor-usage and -activation costs	Scheduling interval
Jiang, Fels, and Little (2008)	✓								Sum of object area	✓	Dynamic programming	Temporal QoV filtering
Chen and De Vleeschouwer (2010b)	✓		✓	✓		✓			Weighted probabilities	✓	Closeness/completeness trade-off	Markov random fields
Rudoy and Zelnik-Manor (2012)				✓				✓	Dot product of spatio-temporal shape measures		Argmax based	Scheduling interval

Key: MO: formulation for multiple objects; POMDP: partially observable Markov decision process.

with the highest score (Snidaro *et al.*, 2003; Park *et al.*, 2008; Kubicek *et al.*, 2009) or by applying a ranking criterion (Qureshi and Terzopoulos, 2007; Rezaeian, 2007; Fu *et al.*, 2010; Li and Bhanu, 2011). These studies focus on view selection and then feed other vision tasks such as object tracking, video summarization, object recognition and identification.

The last category of methods mostly generates videos for human observers, thus restricting frequent inter-camera switches. Frequent switches can be avoided by using scheduling intervals (Del-Bimbo and Pernici, 2006; Goshorn *et al.*, 2007; Krishnamurthy and Djonin, 2007; Tessens *et al.*, 2008) or dynamic approaches (Jiang *et al.*, 2008; Chen and De Vleeschouwer, 2010b). A Turing test can be performed to evaluate view selection (Daniyal and Cavallaro, 2011): users are shown videos generated automatically and manually and are asked to differentiate between them.

Open issues in view selection include enabling systems to adapt to various scenarios, to moving cameras and to a change in the number of sensors. In fact, features are in general extracted relying on the knowledge of the camera positioning and of areas of interest in the observed scene. Moreover, feature and score fusion strategies for content ranking generally require manual setting of the weights that should instead be determined dynamically to favour adaptability and scalability.

References

Agarwal, A. and Triggs, B. (2004) 3D human pose from silhouettes by relevance vector regression, in *Proceedings of the IEEE International Conference on Computer Vision and Pattern Recognition, 2004. CVPR 2004*, vol. 2, IEEE, pp. II-882–II-888.

Ba, S. and Odobez, J. (2011) Multiperson visual focus of attention from head pose and meeting contextual cues. *IEEE T. Pattern Anal.*, **33**(1), 101–116.

Batista, J., Peixoto, P. and Araujo, H. (1998) Real-time active visual surveillance by integrating peripheral motion detection with foveated tracking, in *1998 IEEE Workshop on Visual Surveillance, 1998. Proceedings*, IEEE, pp. 18–25.

Benmokhtar, R., Huet, B., Berrani, S.A. and Lechat, P. (2007) Video shots key-frames indexing and retrieval through pattern analysis and fusion techniques, in *2007 10th International Conference on Information Fusion*, IEEE, pp. 1–6.

Cai, Q. and Aggarwal, J. (1999) Tracking human motion in structured environments using a distributed-camera system. *IEEE Trans. Pattern Anal.*, **21**(11), 1241–1247.

Cameron, A. and Durrant-Whyte, H.F. (1990) A Bayesian approach to optimal sensor placement. *Int. J. Robot. Res.*, **9**(5), 70–88.

Chen, F. and De Vleeschouwer, C. (2010a) Personalized production of basketball videos from multi-sensored data under limited display resolution. *Int. J. Comput. Vision Image Und.*, **114**, 667–680.

Chen, F. and De Vleeschouwer, C. (2010b) Automatic production of personalized basketball video summaries from multi-sensored data, in *2010 17th IEEE International Conference on Image Processing (ICIP)*, IEEE, pp. 565–568.

Chen, F., Delannay, D., De Vleeschouwer, C. and Parisot, P. (2010) Multi-sensored vision for autonomous production of personalized video summary, in *Computer Vision for Multimedia Applications: Methods and Solutions*, IGI Global, pp. 102–120.

Chen, X. and Davis, J. (2008) An occlusion metric for selecting robust camera configurations. *Mach. Vision Appl.*, **19**(4), 217–222.

Ciocca, G. and Schettini, R. (2006) An innovative algorithm for key frame extraction in video summarization. *J. Real-Time Image Process.*, **1**, 69–88.

Costello, C.J., Diehl, C.P., Banerjee, A. and Fisher, H. (2004) Scheduling an active camera to observe people, in *VSSN '04 Proceedings of the ACM International Workshop on Video Surveillance & Sensor Networks*, ACM, New York, NY, pp. 39–45.

Daniyal, F. and Cavallaro, A. (2011) Multi-camera scheduling for video production, in *2011 Conference for Visual Media Production (CVMP)*, IEEE, pp. 11–20.

Daniyal, F., Taj, M. and Cavallaro, A. (2008) Content-aware ranking of video segments, in *Second ACM/IEEE International Conference on Distributed Smart Cameras, 2008. ICDSC 2008*, IEEE, Piscataway, NJ, pp. 1–9.

Delannay, D., Danhier, N. and Vleeschouwer, C.D. (2009) Detection and recognition of sports(wo)men from multiple views, in *Third ACM/IEEE International Conference on Distributed Smart Cameras, 2009. ICDSC 2009*, IEEE, Piscataway, NJ, pp. 1–7.

Del-Bimbo, A. and Pernici, F. (2006) Towards on-line saccade planning for high-resolution image sensing. *Pattern Recogn. Lett.*, **27**(15), 1826–1834.

Ercan, A.O. (2006) Camera network node selection for target localization in the presence of occlusions Proceedings of ACM Workshop on Distributed Cameras, Boulder, CO, USA.

Ercan, A.O., Gamal, A.E. and Guibas, L.J. (2007) Object tracking in the presence of occlusions via a camera network, in *IPSN '07 Proceedings of the 6th International Conference on Information Processing in Sensor Networks*, ACM, New York, NY, pp. 509–518.

Fablet, R. and Black, M.J. (2002) Automatic detection and tracking of human motion with a view-based representation, in *ECCV '02 Proceedings of the 7th European Conference on Computer Vision – Part I*, Springer-Verlag, London, pp. 476–491.

Fu, Y., Guo, Y., Zhu, Y. *et al.* (2010) Multi-view video summarization. *IEEE Trans. Multimed.*, **12**(7), 717–729.

Gavrila, D. (2007) A Bayesian, exemplar-based approach to hierarchical shape matching. *IEEE Trans. Pattern Anal.*, **29**(8), 1408–1421.

Gong, Y. and Liu, X. (2003) Video summarization and retrieval using singular value decomposition. *Multimed. Syst.*, **9**, 157–168.

Goshorn, R., Goshorn, J., Goshorn, D. and Aghajan, H. (2007) Architecture for cluster-based automated surveillance network for detecting and tracking multiple persons, in *First ACM/IEEE International Conference on Distributed Smart Cameras, 2007. ICDSC '07*, IEEE, pp. 219–226.

Greiffenhagen, M., Ramesh, V., Comaniciu, D. and Niemann, H. (2000) Statistical modeling and performance characterization of a real-time dual camera surveillance system, in *Proceedings. IEEE Conference on Computer Vision and Pattern Recognition, 2000*, IEEE, pp. 335–342.

Gupta, A., Mittal, A. and Davis, L.S. (2007) COST: an approach for camera selection and multi-object inference ordering in dynamic scenes, in *IEEE 11th International Conference on Computer Vision, 2007. ICCV 2007*, pp. 1–8.

Han, B., Joo, S.W. and Davis, L.S. (2011) Multi-camera tracking with adaptive resource allocation. *Int. J. Comput. Vision*, **91**(1), 45–58.

He, Y. and Chong, E.K. (2006) Sensor scheduling for target tracking: a Monte-Carlo sampling approach. *Digit. Signal Process.*, **16**(5), 533–545.

Hutchinson, S. and Kak, A. (1989) Planning sensing strategies in robot work cell with multi-sensor capabilities. *IEEE Trans. Robot. Automat.*, **5**(6), 765–783.

Javed, O., Khan, S., Rasheed, Z. and Shah, M. (2000) Camera handoff: tracking in multiple uncalibrated stationary cameras, in *Proceedings. Workshop on Human Motion*, IEEE, pp. 113–118.

Jiang, H., Fels, S. and Little, J.J. (2008) Optimizing multiple object tracking and best view video synthesis. *IEEE Trans. Multimed.*, **10**(6), 997–1012.

Kelly, P., Conaire, C., Kim, C. and O'Connor, N. (2009) Automatic camera selection for activity monitoring in a multi-camera system for tennis, in *Third ACM/IEEE International Conference on Distributed Smart Cameras, 2009. ICDSC 2009*, IEEE, Piscataway, NJ, pp. 1–8.

Khan, S.M. and Shah, M. (2006) A multiview approach to tracking people in crowded scenes using a planar homography constraint, in *ECCV'06 Proceedings of the 9th European Conference on Computer Vision – Volume Part IV*, Springer-Verlag, Berlin, pp. 133–146.

Kim, J. and Kim, D. (2008) Probabilistic camera hand-off for visual surveillance, in *Second ACM/IEEE International Conference on Distributed Smart Cameras, 2008. ICDSC 2008*, IEEE, pp. 1–8.

Krishnamurthy, V. and Djonin, D. (2007) Structured threshold policies for dynamic sensor scheduling – a partially observed markov decision process approach. *IEEE Trans. Signal Process.*, **55**(10), 4938–4957.

Kubicek, R., Zak, P., Zemcik, P. and Herout, A. (2009) Automatic video editing for multimodal meetings, in *ICCVG 2008 Proceedings of the International Conference on Computer Vision and Graphics: Revised Papers*, Springer-Verlag, Berlin, pp. 260–269.

Lee, S. (1997) Sensor fusion and planning with perception action network. *J. Intell. Robot. Syst.*, **19**(3), 271–298.

Li, Y. and Bhanu, B. (2011) Utility-based camera assignment in a video network: a game theoretic framework. *IEEE Sensors J.*, **11**(3), 676–687.

Mittal, A. and Davis, L.S. (2002) M2tracker: a multi-view approach to segmenting and tracking people in a cluttered scene using region-based stereo. *Int. J. Comput. Vision*, **51**(3), 189–203.

Motai, Y. (2005) Salient feature extraction of industrial objects for an automated assembly system. *Comput. Ind.*, **56** (8–9), 943–957.

Park, H.S., Lim, S., Min, J.K. and Cho, S.B. (2008) Optimal view selection and event retrieval in multi-camera office environment, in *IEEE International Conference on Multisensor Fusion and Integration for Intelligent Systems, 2008. MFI 2008*, IEEE, pp. 106–110.

Prince, S.J.D., Elder, J.H., Hou, Y. and Sizinstev, M. (2005) Pre-attentive face detection for foveated wide-field surveillance, in *Seventh IEEE Workshops on Application of Computer Vision, 2005. WACV/MOTIONS '05*, vol. 1, IEEE, pp. 439–446.

Qureshi, F.Z. and Terzopoulos, D. (2007) Surveillance in virtual reality: system design and multi-camera control, in *IEEE Conference on Computer Vision and Pattern Recognition, 2007. CVPR '07*, IEEE, pp. 1–8.

Rezaeian, M. (2007) Sensor scheduling for optimal observability using estimation entropy, in *Fifth Annual IEEE International Conference on Pervasive Computing and Communications Workshops, 2007. PerCom Workshops '07*, IEEE, pp. 307–312.

Rezaeian, M. (2008) Estimation entropy and its operational characteristics in information acquisition systems, in *11th International Conference on Information Fusion, 2008*, IEEE, pp. 1–5.

Rezaeian, M., Vo, B.N. and Evans, J. (2010) The optimal observability of partially observable Markov decision processes: discrete state space. *IEEE Trans. Autom. Control*, **55**(12), 2793–2798.

Rowley, H.A. (1999) Neural network-based face detection, PhD thesis, School of Computer Science, Carnegie Mellon University.

Rudoy, D. and Zelnik-Manor, L. (2012) Viewpoint selection for human actions. *Int. J. Comput. Vision*, **97**(3), 243–254.

Senior, A., Hampapur, A. and Lu, M. (2005) Acquiring multi-scale images by pan–tilt–zoom control and automatic multi-camera calibration, in *Seventh IEEE Workshops on Application of Computer Vision, 2005. WACV/ MOTIONS '05*, vol. **1**, IEEE, pp. 433–438.

Shen, C., Zhang, C. and Fels, S. (2007) A multi-camera surveillance system that estimates quality-of-view measurement, in *IEEE International Conference on Image Processing, 2007. ICIP 2007*, vol. **3**, IEEE, pp. III-193–III-196.

Snidaro, L., Niu, R., Varshney, P. and Foresti, G. (2003) Automatic camera selection and fusion for outdoor surveillance underchanging weather conditions, in *IEEE Conference on Advanced Video and Signal Based Surveillance, 2003. Proceedings*, IEEE, pp. 364–369.

Spaan, M.T.J. and Lima, P.U. (2009) A decision-theoretic approach to dynamic sensor selection in camera networks. Proceedigs of International Conference on Automated Planning and Scheduling, Thessaloniki, Greece, pp. 1–8.

Szeliski, R., Zabih, R., Scharstein, D. *et al.* (2008) A comparative study of energy minimization methods for Markov random fields with smoothness-based priors. *IEEE Trans. Pattern Anal.*, **30**(6), 1068–1080.

Tarabanis, K. and Tsai, R.Y. (1991) Computing viewpoints that satisfy optical constraints, in *IEEE Computer Society Conference on Computer Vision and Pattern Recognition '91*, IEEE Computer Society Press, Washington, DC, pp. 152–158.

Tarabanis, K.A., Allen, P.K. and Tsai, R.Y. (1995a) A survey of sensor planning in computer vision. *IEEE Trans. Robot. Automat.*, **11**(1), 86–104.

Tarabanis, K.A., Tsai, R.Y. and Allen, P.K. (1995b) The MVP sensor planning system for robotic vision tasks. *IEEE Trans. Robot. Automat.*, **11**(1), 72–85.

Tessens, L., Morbee, M., Lee, H. *et al.* (2008) Principal view determination for camera selection in distributed smart camera networks, in *Second ACM/IEEE International Conference on Distributed Smart Cameras, 2008. ICDSC 2008*, IEEE, pp. 1–10.

Trivedi, M., Huang, K. and Mikic, I. (2005) Dynamic context capture and distributed video arrays for intelligent spaces. *IEEE Trans. Syst. Man Cybernet. Part A: Syst. Hum.*, **35**(1), 145–163.

Wallhoff, F., Ablaßmeier, M. and Rigoll, G. (2006) Multimodal face detection, head orientation and eye gaze tracking, in *IEEE International Conference on Multisensor Fusion and Integration for Intelligent Systems, 2006*, IEEE, pp. 13–18.

Wheeler, M. and Ikeuchi, K. (1995) Sensor modeling, probabilistic hypothesis generation, and robust localization for object recognition. *IEEE Trans. Pattern Anal.*, **17**(3), 252–265.

Yamasaki, T., Nishioka, Y. and Aizawa, K. (2008) Interactive retrieval for multi-camera surveillance systems featuring spatio-temporal summarization, in *MM '08 Proceedings of the 16th ACM International Conference on Multimedia*, ACM, New York, NY, pp. 797–800.

Zhou, X., Collins, R.T., Kanade, T. and Metes, P. (2003) A master–slave system to acquire biometric imagery of humans at distance, in *IWVS '03 First ACM SIGMM International Workshop on Video Surveillance*, ACM, New York, NY, pp. 113–120.

22

3D Video on Mobile Devices

Arnaud Bourge and Alain Bellon

STMicroelectronics, France

22.1 Mobile Ecosystem, Architecture, and Requirements

Smartphones are mobile connected multimedia devices. Beyond audio and video calls, they are now also (or even mainly) used for web browsing, social networking, shooting and playing photos and videos, playing games, storing multimedia content, and so on. Tablets have the same mobile usage while targeting a home appliance level of performance. Both devices support an ever extending range of applications that benefits from cross-breeding between new technologies: positional sensors (magnetometer, accelerometers, gyroscope, GPS), communication (3G/4G, Wi-Fi, NFC, etc.), connectivity (USB, Bluetooth, HDMI, etc.). 3D stereoscopy is one of these bricks that could be used for multiple applications: as consumer and producer of 3D content, and also as provider of depth information on the surrounding environment. Like other new technologies, its introduction can be eased by the fast replacement rate of the mobile market, and by a panel of early adopters that are fond of new applications.

Consequently, stereoscopic 3D (S3D) impacts all mobile multimedia requirements: new hardware at both ends (stereoscopic camera and display) and new processing stages (S3D imaging, video/still codec and rendering). For this purpose, mobile application processors provide (Figure 22.1):

- powerful CPU and GPU;
- hardware accelerators for imaging, encoding, decoding, display, and rendering.

Besides functional aspects, S3D-specific hardware must comply with the demands of the mobile consumer market:

- low part prices (a few dollars for camera and display);
- small silicon area for the application processors, for price and packaging reasons;

Emerging Technologies for 3D Video: Creation, Coding, Transmission and Rendering, First Edition.
Frédéric Dufaux, Béatrice Pesquet-Popescu, and Marco Cagnazzo.
© 2013 John Wiley & Sons, Ltd. Published 2013 by John Wiley & Sons, Ltd.

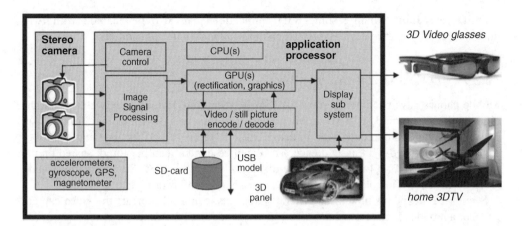

Figure 22.1 3D mobile device hardware architecture. Pictures of 3D panel and home 3DTV from Shutterstock (http://www.shutterstock.com), © STMicrolectronics. Picture of cinemizer® video glasses is courtesy of Carl Zeiss

- low power consumption, to save battery lifetime;
- high performance (e.g., high-definition video).

On the software side, S3D functions must fit memory architecture, size, and bandwidth that may be shared between multiple applications running in parallel on real-time operating systems (iOS, Android, Windows phone, etc.).

Moreover, mobile usages bring additional constraints compared with home appliances. Indeed, the device is:

- pocket sized, which challenges both S3D rendering and stereo-camera design;
- hand-held, which brings mechanical stress over its lifetime;
- touch-screen capable, which must remain true for a 3D panel;
- generally used by a single user, who can move it to get the best view point on an auto-stereoscopic display but who also does not want to be stuck in a fixed position.

In the following sections of this chapter, we first list and describe the various applications enabled by stereoscopy on a mobile device (Section 22.2). Then, we enter into technical considerations for the image and video capture system (Section 22.3) and the display system (Section 22.4), both in terms of hardware architecture and visual processing. We subsequently extend the pure stereo considerations to the usage of depth and disparity maps in the mobile context (Section 22.5). Finally, Section 22.6 concludes the chapter.

22.2 Stereoscopic Applications on Mobile Devices

Stereoscopy is primarily introduced as an extension of *still/video recording and playback* applications to S3D. Gaming is also a major mobile application domain that drives the S3D rendering of *3D graphics*, for which the game play benefits from positional information reported by embedded sensors (gyroscopes, accelerometers). These features are also used in increasingly popular *interactive applications* on mobile phones like augmented reality (AR) and 3D navigation. Finally, some S3D applications can also be enabled with two-dimensional (2D) mobile devices: we will call these *monoscopic applications*.

22.2.1 3D Video Camcorder

Mobile phones have become the most-used image recording devices, even well beyond the digital still camera market. A 3D mobile phone must thus be a 3D camcorder. From a user perspective, this should come with no compromise in terms of performance and quality of experience compared with a 2D device. The captured content can be stored, shared on the internet, and played on local and remote displays. It must be 1080p30 minimum, contain sharp details and vivid colors, and generate no visual discomfort or fatigue.

Section 22.3 explains how these general statements technically impact the video capture chain of a mobile phone, from camera to codecs. In addition, the captured stereo images can be processed by computer-vision algorithms, in particular to retrieve the depth of the surrounding environment (see Sections 22.2.5 and 22.5).

22.2.2 3D Video Player

A 3D mobile phone is expected to support the usual applications, like pictures gallery, movie playback, video streaming, and videoconferencing. Some 3D content is already available on the internet, and a major source will be the publication of user-generated 3D clips on popular video sharing web sites. All these applications require the usage of standard compression formats, for interoperability and also for performance reasons (hardware acceleration). The trends in this domain are detailed in Section 22.3.5. Another important source is the existing huge database of 2D still pictures and videos. The resulting 3D quality can vary a lot depending on the scenes since the depth can only be guessed and occlusions are not available, but the current lack of 3D content makes 2D-to-3D conversion a "must have" for 3D terminals.

22.2.3 3D Viewing Modalities

Regarding 3D viewing modalities, the local display of a 3D phone is necessarily auto-stereoscopic. Indeed, mobile usages are very different from quietly watching a movie in the living room or at the theater, and one cannot expect people to constantly carry polarized glasses with their phone. Moreover, auto-stereoscopy fits well with mobile devices because the viewer will instinctively place the display at the optimal distance and viewing angle to get the best 3D viewing comfort and image quality. Finally, a two-view panel is sufficient for a single user, which makes the usual resolution loss brought by lenticular or parallax-barrier technologies less of a problem.

Head-up displays could also become a popular peripheral for 3D mobile devices, for occasional circumstances like long-distance travel in planes or trains. When the technology is mature enough, it will provide high-quality experience, as video glasses naturally host one view for each eye, in full resolution without any crosstalk between views. Obstructive or see-through video glasses favor respectively movie playback or AR usage.

Besides these single-user applications, people want to share the visual content stored on their mobile device with friends and family in "get together" situations. As stand-alone 3D displays become widespread on the market (high-definition 3DTVs, but also auto-stereoscopic 3D photo frames), 3D playback to an external display must thus be enabled. Proper connectivity is available on smartphones for this purpose, with support of 3D formats signaling (HDMI1.4a notably).

The technical aspects of 3D rendering on mobile devices in all these 3D playback cases are explored in the Section 22.4. Moreover, the user typically gets some partial control through the interface: typically, the selection of 2D/3D mode, the parallax setting, and the depth volume if the content allows it. These aspects are also detailed in the Section 22.4.

22.2.4   3D Graphics Applications

Stereoscopic rendering is also requested for 2D and 3D graphics.

A typical 2D-based graphic is the GUI, which is the main user interface on a mobile device: desktops are much smarter with 3D icons, and a more complex user interface like a carousel brings a "wow" effect as well. Components are often defined as 2D graphic objects with a depth.

3D graphics are used in applications like navigation and 3D tools. But the largest application domain of 3D graphics for S3D is gaming, either on a games console (Nintendo 3DS) or on S3D smartphones. Typically, left and right views are generated from virtual left and right cameras shooting the same scene, with almost the same tools as for a PC, thanks to support of DirectX or OpenGL-ES by the embedded GPU.

For both 2D-based and 3D-based graphics, the user should have a control on the depth-volume to accommodate the "strength" of the 3D effect.

22.2.5   Interactive Video Applications

Besides the traditional video playback and record use cases, a new family of applications has emerged in recent years, where the image content is not only captured but also analyzed to bring a new level of semantic feedback to the user. They are usually referred to as AR. Initially restricted to research laboratories and workstation-based environments, AR has migrated to the mobile area thanks to four complementary factors:

- the jump in quality of mobile phone cameras;
- the processing power available on modern smartphones for both signal processing and graphics (as described in Section 22.1);
- the generalization of motion and position sensors on these devices (accelerometers, gyroscopes, compass, GPS);
- the fast exchanges with remote servers enabled by the deployment of 3G networks and terminals.

Many mobile AR applications are already operational and successful without S3D. The main point of stereoscopy is thus to further enrich AR by adding extra information, typically the depth map.

22.2.6   Monoscopic 3D

In this section we describe some of these mobile applications which provide a 3D viewing experience while either the capturing or the viewing process is not a stereo one; hence the term "monoscopic 3D." Their compatibility with regular hardware and their nondependency

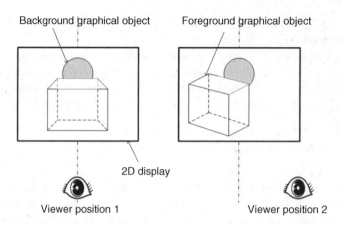

Background graphical object   Foreground graphical object

2D display

Viewer position 1   Viewer position 2

Figure 22.2   3D motion parallax on a 2D display with eye tracking

on stereo cameras or auto-stereoscopic displays makes them attractive and cheap solutions, not as anecdotic as they could sound.

On the capture side, 3D with one camera consists of mimicking a spatial stereo set-up by temporally moving the device. The most straightforward modality is to take two shots of the same scene, where for the second shot the user shifts the camera by a few centimeters. Another modality consists in taking a panoramic capture in video mode and generating one panorama for the left eye and a second one for the right eye. The former solution allows the capture of full-resolution pictures, while the latter one has the advantage of being more user friendly because the horizontal camera shift is continuous and thus does not require one to "stop and shoot" after a given number of centimeters. In both cases the main difficulty is to obtain a proper vertical alignment relying solely on the shaky hand of the user. Some visual aid is thus required, either with a simple overlay of the past capture over the camera preview, or with on-screen indications such as up/down/left/right. Although the resulting image pair cannot match the typical requirements of mobile stereo capture (see Section 22.3), these simple methods are an effective and low-cost way for making 3D shots.

On the viewing side, the wide usage of stereoscopy in 3D movie theaters should not make us forget the other depth cues. Among them, motion parallax is so important that it can bring the illusion of 3D even in monoscopic conditions. Face detection and eye tracking allow giving this illusion in graphics-use cases (typically games). The view point of the synthetic scene is dynamically changed by adjusting the position of the virtual camera according to the viewer's eyes coordinates (Figure 22.2). The same principle can apply for stereoscopic displays, enabling a mix of binocular disparity and motion parallax cues (Surman *et al.*, 2008).

22.3   Stereoscopic Capture on Mobile Devices

22.3.1   Stereo-Camera Design

Stereoscopic camera modules for mobile devices are built upon a pair of same low-cost CMOS imagers, with plastic lens and autofocus actuators, mounted in a light mechanical chassis with small lens-spacing, in order to fit into the restricted phone form-factor. Sensors being identical to those of 2D products (high definition and fast readout), each view could have same resolution as in the 2D case. But the subsequent on-the-fly image processing and memory bandwidth requirements generally lead to halving the picture resolution: typically 2×5 megapixels still picture and $2 \times 720p30$ video for an application processor capable of handling 10 megapixels and 1080p30 in 2D.

The dual camera can be viewed by the application processor either as separate streams connected to two ports (like in the LG Optimus 3D), or directly as a single sequence of dual views sent to one camera port (like in the HTC Evo 3D). In the latter case, a dedicated processor composes the left and right camera outputs and could rescale the views to half-resolution frame-compatible format (typically side by side) to reduce the amount of data sent to the application processor. To still keep full-resolution pictures, the trend is only to merge multiple camera data-streams in virtual channels as defined in the MIPI CSI2 standard (MIPI Alliance, 2005), and still let the application processor perform all the imaging tasks.

In terms of control, timing synchronization is required between sensors in order to get the left and right views at the exact same time. They are currently synchronized through external camera commands, but the trend of manufacturers is to add master–slave communications directly between the sensors, embedding clocks and time bases. Also, for multi-camera control purposes, multiple or reprogrammable I2C addresses are introduced so that camera configurations can be either broadcasted to all cameras or issued separately. The latter is mandatory in order to use specific calibration data for each camera and tune camera parameters like focus, taking into account manufacturing part-to-part variations.

22.3.2   Stereo Imaging

The term "imaging" refers to the processing steps that turn raw camera sensor data (usually in BAYER format) into a viewable and encodable image (usually in YUV $4:2:0$ format). These steps include defect pixel correction, demosaicking, color matrices, noise reduction, tone mapping, scalers, and so on, but also image statistics (histograms, color maps) used by the "3A" feedback loops: auto-focus (AF), auto-exposure (AE), and auto-white-balance (AWB). They are accelerated by the image signal processor (ISP), a hardware integrated circuit managing throughputs of several hundreds of megapixels per second in recent systems. For S3D on mobile devices, we do not duplicate the ISP and keep the same one as for 2D products. But the image dataflow is adapted to transform the simultaneous image capture from the two cameras into a sequential processing by the ISP. This is simply enabled by a memory dump of the right-view raw data while the left-view one is being processed (Figure 22.3). The right view is then read from memory (instead of the camera serial interface channel directly), before a next pair of images is captured. In such a system, the performance for one view is naturally halved.

This is not the only imaging difference between the 2D and the 3D modes. Quality-wise, inter-view consistency must be ensured. Consequently, the camera 3A should convey identical lighting (exposure time, analog and digital gains from the AE), focus distance (from the AF), and color rendition (from the AWB). Running them separately induces a risk of divergence, as they do not observe exactly the same scene. Two main strategies are possible and implemented: a master–slave system, where the camera 3A commands derived from the master-view statistics are replicated for the slave view; a joint system, where the camera 3A uses statistics from both cameras. The issue then resides in part-to-part variations and mechanical uncertainties: sending the same command does not guarantee the same result. For instance, the position of a focus actuator is not very precise; hence, the usage of hyper-focal mode or extended depth-of-field systems represents easy-to-manage alternatives and

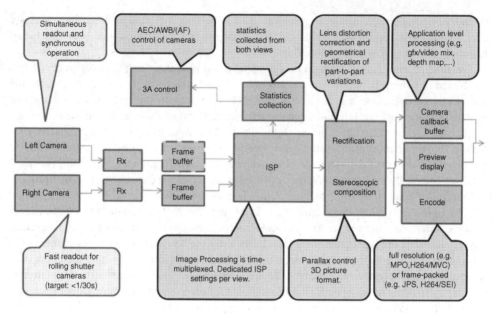

Figure 22.3   Stereoscopic capture dataflow in mobile architecture

can sometimes be preferred to a genuine stereo AF. For the pixel processing pipe itself, part-to-part sensor variations must also be managed, in particular for colorimetric and defect-pixel maps. The latter are typically static in 2D mode, while in 3D it requests fast context switching of the ISP: its parameters must be reloaded and the processing must be rescheduled at each frame.

22.3.3   Stereo Rectification, Lens Distortion, and Camera Calibration

On a mobile phone, the stereo spatial configuration is necessarily fixed, with parallel camera axes and a constrained baseline of a few centimeters maximum. Nevertheless, the mechanical placement of low-cost consumer camera modules on the device suffers from variations up to several pixels in the vertical direction and several degrees in the three axes (Figure 22.4). Picture quality perception and comfort of vision is subjectively much better with good correction of these mechanical misalignments. We know that $10'$ in rotation in the front plane is

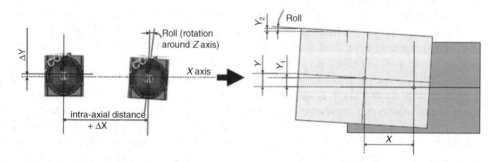

Figure 22.4   Effect of mechanical camera misalignments

noticeable and that higher values can lead to fatigue, and even eye strain (Lipton, 1982). Stereo computer vision raises these requirements to less than one pixel in the vertical direction and 0.1° in roll, particularly because it allows limiting the search area to the horizontal axis, hence increasing pattern-matching success rate and simplifying the processing complexity at the same time. The digital post-processing that aligns the epipolar lines of each view is called "stereo rectification" and can be modeled by a global perspective transform.

The camera lenses introduce geometric distortions that must also be compensated. This step is generally omitted in 2D, but is preferably done in 3D; otherwise, it adds another cause of stereo misalignment, more pronounced at the image borders than in the center. In practice, lens distortion is approximated by a mesh of triangles (typically for GPU programming) or tiles (typically for dedicated hardware acceleration). Experiments show that a few hundred vertexes is sufficient to get a good lens model for an image of several megapixels.

As stereo rectification and lens distortion correction are purely geometrical operations, they can be factorized. This brings the advantage of a one-pass processing on the GPU or on a dedicated hardware accelerator, limiting latencies and memory transfers on the mobile platform.

For applying these corrections, their parameters must be known; that is the goal of stereo camera calibration methods (Figure 22.5). Calibration can be performed on the production line, using dedicated patterns (Zhang, 2000). But mobile devices are subject to shocks and temperature variations that cause noticeable changes in mechanical alignment, leading to the need of regular calibration updates.

For this purpose, the trend is to use self- or auto-calibration techniques that could be run by the user or the system during normal usage. Auto-calibration techniques use natural scene capture, identifying and matching so-called "feature points" in left and right views, in order to determine the extrinsic calibration parameters (Figure 22.6). Much research is performed to increase success rate (with low number of feature points) and robustness (to light conditions, etc.); one way is to perform bundle adjustment of previously known calibration data, coming for instance from a pattern-based technique deployed on the production line (Lourakis and Argyros, 2009).

Note that stereo rectification should preferably be the first applied spatial processing, since other algorithms in the chain (e.g., depth estimation or jitter estimation) generally assume a perfectly aligned and parallel stereo camera. But, by construction, it necessarily comes after optical zoom. Since each zoom position changes the lens distortion and the intrinsic camera

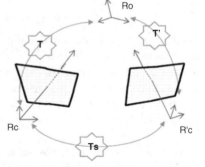

Stereoscopic calibration computes parameters of the camera model
- The 5 intrinsic parameters of both cameras
  - focal in pixels (fx,fy) optical center in pixels (cx,cy)
  - Theta angle between axis
- The 6 intrinsic parameters
  - Rotation that puts the right camera plane in the left one
  - Translation between camera references origins

Lens distortion is also measured on each camera

Figure 22.5   The objectives of stereoscopic calibration

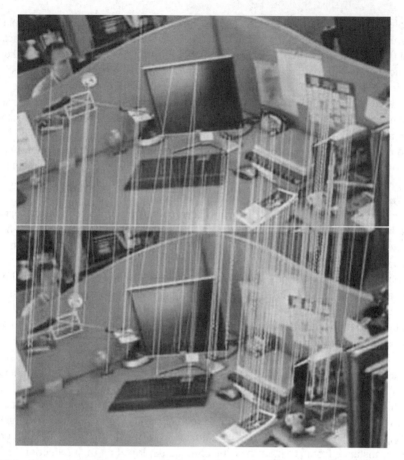

Figure 22.6 Point matching on stereoscopic images for auto-calibration of the ST-Ericsson platform (see Plate 29 for the colored figure)

parameters, the calibration must be performed at various optical zoom factors, and the correction parameters are interpolated from two known sets.

22.3.4 Digital Zoom and Video Stabilization

Digital zoom and video stabilization are appreciated features on mobile consumer devices, and thus must remain available for 3D products. But special care has to be taken: since they act on the pixel coordinates of each view, they could impact stereo vision and ruin the rectification effort described in the previous paragraph.

Indeed, conventional two-axis video stabilization independently applied on both views can convey two types of issues causing eyestrain:

1. vertical stereo misalignment because of different vertical jitter components;
2. temporal variation of the horizontal disparity because of a varying difference of horizontal jitter components.

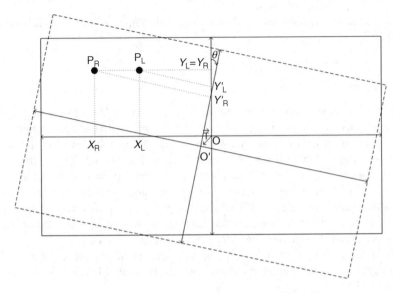

Figure 22.7 Video stabilization in stereo 3D mode

Consequently, the exact same global motion compensation must be applied to both views. Additionally, three-axis video stabilization (Auberger and Miro, 2005) became a new requirement on mobile devices. In Figure 22.7, P_L and P_R represent the projection of an observed point P respectively on the left and on the right camera. If the stereo pair is correctly rectified prior to stabilization, P_L and P_R have the same vertical coordinate. Ensuring this property after stabilization by an angle θ and a vector **T** implies

$$T_{R,y} - T_{L,y} = (x_L - x_R)\sin\theta \tag{22.1}$$

where $(x_L - x_R)$ is the disparity of the point P, which varies inside the image depending on the content. The other terms in the equation are constant, so it can be respected only if

$$T_{R,y} = T_{L,y}; \theta = 0 \tag{22.2}$$

In summary, video stabilization is restricted to two axes (i.e., translation only) in 3D mode. Like for the AF case, two approaches might be adopted: a master–slave model, where the jitter vector from the first view is directly reused for the second view; or a central algorithm optimization of the stereo jitter compensation. The former has a complexity advantage, while the latter is preferred quality-wise.

The case of digital zoom is different. As long as it is fixed, it does not introduce vertical misalignment and remains horizontally stable. However, a popular application of digital zoom is to lock the vision on the speaker's face in video telephony. In this case, both the position of the cropping window and the scaling factor may vary in time, according to the speaker's motion. Whereas in 2D it gives a sensation of stability, in stereo it can lead to uncontrolled and uncomfortable parallax variations. Like in the video stabilization case, dynamic digital zooming (e.g., centered on a region of interest) must be constrained: same cropping window size and scaling factor for each view of course, same vertical position, and a horizontal position ensuring that the tracked object remains visible in both views.

Then, fine horizontal adjustment is similar to global disparity control in full-view mode (Section 22.4.3).

22.3.5 Stereo Codecs

In the mobile consumer market, a big emphasis is put on standardization of codecs since visual content must be encoded/decoded on the fly, in various sizes (high-definition recording, low-resolution for MMS or internet sharing), and various bit rates (constant bit rate for videoconferencing, variable bit rate for optimal image quality). All these settings must be handled by standardized codecs to guarantee interoperability of mobile devices between themselves and between mobile devices and the internet and computer world.

Still picture codecs are based on JPEG, either stereo-jpeg for historical reasons (i.e., jpeg encoding of a frame containing side-by-side left/right views), or MPO file format that packages a jpeg picture of the left view and a jpeg picture of the right view.

De facto frame-packing format is also used with various video codecs; for example, YouTube3D accepts side-by-side and top–bottom formats. But only MPEG4/AVC extensions are deployed in the consumer market:

- H264/SEI signaling for frame-packing format, adopted by DVB-step1 (ITU-T, 2010: Annex D)
- H264/MVC for multiview full-resolution encoding, adopted by 3D Blu-ray (ITU-T, 2010: Annex H).

Theses codecs are under scrutiny by mobile standardization bodies (3GPP, etc.) with discussions around supported levels and profiles to handle with mobile applications of movie record/playback, video streaming, and videoconferencing.

More details on 3D video representation and formats can be found in the Chapter 6.

22.4 Display Rendering on Mobile Devices

22.4.1 Local Auto-Stereoscopic Display

As stated previously, dual-view auto-stereoscopic technologies have their best usage as a mobile 3D panel, since the viewer can instinctively place the display at the best distance and viewing angle for a correct 3D rendering. For this use case, 3D displays are designed so that almost no left–right crosstalk is perceptible when the display is in front of the viewer at about half-arm distance. Several manufacturers are developing 3D panels using various technologies: parallax barrier, lenticular screen, directional or time-sharing backlight. The challenge here is to provide a display module combining good 3D picture quality, both in portrait and landscape mode, integrated with touch-screen capabilities, with robustness and brightness required by outdoor usage.

The MIPI consortium is defining specific display commands (2D/3D, landscape/portrait modes, etc.) and picture formats (pixel interleaved, frame packed, time interleaved) for the mobile internal display serial interface DSI (MIPI Alliance, 2011).

At display module level, integrated-circuit display driver manufacturers are working on integration of 2D and 3D picture management, as well as 2D-to-3D conversion. The challenge is still to keep power consumption as low as possible, possibly avoiding the need of a frame-buffer for picture format conversion, and possibly using memory compression techniques.

22.4.2 Remote HD Display

To share 3D content, the mobile phone can be connected to an external large 3D display (typically a 3D HDTV) and used as a multimedia player and remote controller. For this purpose, 3D signaling is sent along with 3D content, so the external display can switch automatically in proper 2D or 3D picture format. The HDMI de-facto standard paved the way with S3D signaling (HDMI, 2010) for frame-compatible formats (half-resolution side-by-side and top–bottom formats) and full-resolution formats (frame packing); other wired and wireless technologies (DisplayPort, WifiDisplay, etc.) endorsed the same approach. A specific challenge in the mobile market is to lower the power consumption of the digital interface, giving the huge amount of data of dual-view video in high definition, at high frame rate. This use case pushes the whole mobile multimedia architecture design in terms of performances: the display subsystem, but also the graphics subsystem (for gaming and user interface), the camera and the codec sub-subsystems to generate/playback contents for external HD displays.

22.4.3 Stereoscopic Rendering

Some adaptation of the 3D scene has to be performed for rendering it on a stereoscopic display. An obvious one is the conversion, if necessary, from the original content format to the appropriate display format (side by side, top–bottom, portrait/landscape orientation, etc.), as described in the previous paragraphs. Although conceptually trivial, this operation can require an extra memory-to-memory pass, hence stressing the system a little more. The other adaptations are related to the 3D viewing comfort. This is due to several factors, an important and generic one being the change of stereo paradigm between capture and display.

Figure 22.8 depicts a simplified 3D display viewing situation. For a given object point present in the left and right views respectively at positions P_L and P_R, the observer sees a point at position P with depth $z(P)$. This is the basic stereopsis mechanism. The disparity disp is the signed distance between P_L and P_R. D is the observer's distance to the screen and b_{eyes} is the observer's interocular distance (distance between the left and right eye, typically 65 mm).

If the observer's eyes are perpendicular to the display panel, then the perceived depth and the disparity are linked by the following equation:

$$z(P) = \frac{D}{1 - \dfrac{\text{disp}(P)}{b_{eye}}} \tag{22.3}$$

which can also be expressed in depth relative to the screen plane:

$$z_{screen}(P) = z(P) - D = \frac{D}{\dfrac{b_{eye}}{\text{disp}(P)} - 1} \tag{22.4}$$

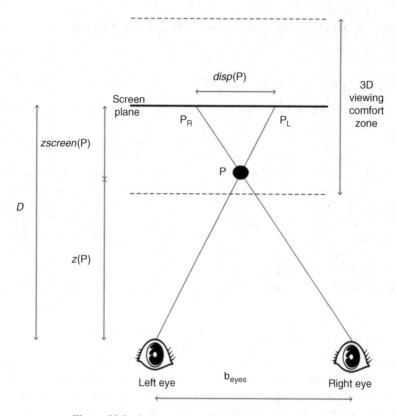

Figure 22.8 Stereo triangulation in a display situation

For small disparities (i.e., for a depth next to the screen plane), a linear approximation is sometimes used:

$$\text{zscreen}\,(P) = -\text{disp}\,(P)\,\frac{D}{b_{\text{eye}}} \tag{22.5}$$

Knowing that the stereo camera set-up is parallel on a mobile device, objects located at infinite distance during the capture have a null disparity. Consequently, they appear by default on the display screen plane, and the rest of the scene appears in front of it. To compensate for this effect, a constant disparity shift is typically applied. As long as the conditions for the linear approximation above are valid, it is perceived as a translation of the displayed scene on the depth axis. This disparity shift implies a cropping of the original images at the borders, and hence a slight reduction of the field of view. In professional 3D films production, this parameter is manually controlled by the stereographer, who can adjust the distance and angle between the cameras according to the recorded scene, and refined during the post-production stage. For consumer devices, the stereo configuration is mechanically fixed and users expect an easy 3D experience.

Several strategies can be implemented. The simplest one is to let the user choose the disparity shift manually, at application level. But even then an automatic algorithm needs to be applied first; otherwise the content would always appear completely in front of the screen the first time. A typical way is to place the main object of interest on the display plane. Indeed, this is where the viewing comfort is optimal, because the accommodation–convergence mismatch does not exist on this plane. The camera focus distance is relevant for this purpose, as the focus is made on the main character. Because other camera parameters like the focal length and inter-axial distance also have to be known to translate a focus distance into a disparity value, the best place for performing this automatic display disparity adjustment is thus the capture subsystem.

More elaborate strategies would take into account the characteristics of the display to make use of its whole 3D comfort zone. This especially makes sense for mobile displays: they are small (from 3 inches for mid-end smartphones to 9 inches for tablets) and viewed from a short distance (arm's length), so that the 3D budget is limited (Pockett and Salmimaa, 2008). The best disparity shift would then be the one minimizing the number of "out of comfort range" points. The solution is not unique, and must be coupled with other empiric considerations (e.g., favoring foreground or background rendering). The literature on the subject remains limited, but one thing is certain: it requires knowledge of the disparity histogram.

Going one step further in this optimization of 3D rendering, the depth range of the content should be reduced or stretched to fully match the screen capabilities. This is equivalent to virtually changing the camera inter-axial distance, which is desired for at least two other reasons. First, to respect depth proportions: these are distorted when applying a mere rescaling of a stereo pair in the horizontal and vertical dimensions. It is often quoted as a drawback of repurposing stereo content to smaller 3D displays. Second, users might want to turn the "3D volume" button up or down depending on the situation and on their own preference. These three objectives (3D range optimization, respect of depth proportions, user's 3D volume) can be contradictory, but they have one technical point in common: they require digital view synthesis, which is detailed next.

22.5 Depth and Disparity

22.5.1 View Synthesis

A direct usage of depth (or disparity) maps in the stereoscopy domain resides in view synthesis. Ideally, it should be based on a complete 3D space model derived not only from the observed views and their depth maps (here, we assume that they are available), but also the intrinsic and extrinsic parameters of the real cameras they are attached to. New views are then generated by arbitrarily placing virtual cameras in this space. A good description of this principle can be found in Shin *et al.* (2008), where an efficient parallel implementation on a general-purpose GPU (GPGPU) is also presented. However, this approach remains too complex for mobile devices. Even after optimizations, the authors report a processing time of 148 ms for a 1024×768 resolution, with a PC architecture including a CPU clocked at 1.86 GHz and a GPGPU clocked at 1.64 GHz.

A simpler approach has been implemented in the scope of the projects TRISCOPE and CALDER (Bourge *et al.*, 2009). The algorithm is reduced to a horizontal "pixel position shift" from a single view and depth map. Intrinsic and extrinsic camera parameters are ignored and depth values are directly interpreted as disparities; thus, the virtual camera position is implicitly constrained to a translation perpendicular to the optical axis (Figure 22.9).

The process is adapted to backward mapping on the RENOIR, a hardware rendering IP deployed on mobile platforms. As described in Pasquier and Gobert (2009), the first step consists of projecting disparities from the original-view to the output-view domain. Instead

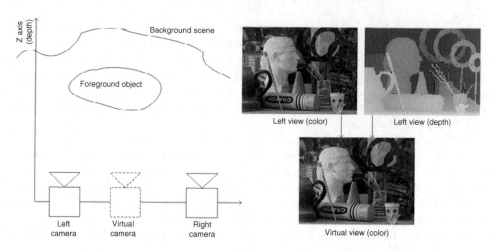

Figure 22.9   Virtual camera and view synthesis. Example input image from Scharstein and Pal (2007)

of explicitly controlling the virtual camera distance, a *gain* and an *offset* are applied:

$$\begin{cases} D(x) = (\text{depth}_{\text{input\_view}}(x) - \text{offset}) \times \text{gain} \times v \\ \quad \text{disp}_{\text{virtual\_view}}(x - D(x)) = D(x) \end{cases}, \qquad x = 0 \dots \text{width} - 1 \qquad (22.6)$$

v is negative for a view located on the left and positive otherwise. This is followed by an occlusion/de-occlusion management pass. Finally, the destination view is scanned pixel by pixel, using the projected disparities:

$$\text{color}_{\text{vitual\_view}}(x) = \text{color}_{\text{input\_view}}(x + \text{disp}_{\text{virtual\_view}}(x)), \qquad x = 0 \dots \text{width} - 1 \quad (22.7)$$

In the CALDER project, such a view synthesis was added to the RENOIR IP and ported on a field-programmable gate array running at 300 MHz. The measured performance exceeds 1080p60. A software implementation of the same algorithm adapted to forward mapping runs four times more slowly on an ARM Cortex A9 clocked at 1 GHz, a typical processor for a mobile platform.

22.5.2   *Depth Map Representation and Compression Standards*

The interest in depth maps for view synthesis has been well studied in the literature, not only for the 3D rendering controls it allows (see previous section), but also for addressing multi-view auto-stereoscopic displays at a reasonable transmission cost. Indeed, only some of the views need to be transmitted with their depth maps; the others can be synthesized. Such a system was described by Redert *et al.* (2002) and Fehn (2004) and led to the MPEG-C part 3 specification (ISO/IEC 23002-3), finalized in 2007 with primarily a single view plus depth representation in mind (but not only). As stated in Bourge *et al.* (2006), "this specification gives a representation format for depth maps which allows encoding them as conventional 2D sequences." Its main advantage resides in its flexibility regarding transport and compression techniques: any conventional codec can be used (including ones that did not exist at that time, like HEVC), complemented by some lightweight signaling and metadata.

The objective and subjective quality measures conducted by Strohmeier and Tech (2010) and also reported in Gotchev *et al.* (2011) show the competitiveness of the MPEG-C part 3 approach for stereo video on mobile devices, including when channel errors are taken into account. Nevertheless, they also stress the importance of having accurate depth maps in the first place: with erroneous or imprecise depth, the visual artifacts can become unacceptable.

The current work by the 3DV group of MPEG follows the same mindset. It can be viewed as the fusion of the MVC and the MPEG-C part 3 principles, but it also targets a compression breakthrough through two main axes (Müller *et al.*, 2011):

1. specific coding techniques for the depth;
2. joint texture/depth coding approaches.

These evolutions, if confirmed, would raise two challenges for a mobile implementation: respectively, the development of specific hardware acceleration (where MPEG-C part 3 seamlessly relies on conventional video compression) and the increase of memory accesses – hence power consumption – to perform cross-domain predictions (where MPEG-C part 3 treats the texture and depth signals separately and a mere multiplexing is performed at transport level). In comparison, an evolution of MVC incorporating the emerging HEVC standard and the independent coding of depth channels would be more straightforward for industrial deployment. This is also one of the tracks envisioned by the 3DV group, and only a major coding efficiency gain should justify a completely different system.

More technical insights on these matters can be found in Chapter 7 and Chapter 8.

22.5.3 Other Usages

Besides view synthesis and multiview compression, depth maps can improve or enable several other applications on a mobile device. On the capture side, stereo AF (Section 22.3.2) can be made more consistent by shifting the focus window of the slave camera according to the depth of the master focus region. On the rendering side, some evolved disparity offset strategies may be deployed according to the principles mentioned in Section 22.4.3.

In the scope of AR, one challenge is to reach a seamless integration of virtual objects in a captured natural scene. At least two technical difficulties arise: occlusion handling and lighting. Both aspects deal with the 3D geometry of the scene, which is the reason why stereoscopy and depth are helpful in this domain (Lansing and Broll, 2011).

Finally, human–machine interfaces are evolving towards gesture-based paradigms. The first step has been the generalization of touch interfaces on all smartphones and tablets. The second phase appeared in the home market and more particularly the gaming console with the Kinect, above which body segmentation and gesture-recognition algorithms are built. On a 3D mobile device, the depth estimated from the stereo camera could be an alternative such infrared projection-based ranging sensor. Besides cost, an advantage would be the functioning in the outdoor environment, a must for any mobile application.

These applications do not necessarily require precise depth estimation at high resolution and frame rate. However, having a real-time, robus, and low-power algorithm remains a challenge. Even with the increasing CPU power on smartphones, it probably requires hard-wired acceleration. Among the numerous techniques described in the literature, those relying on block-based motion estimation are attractive (De Haan *et al.*, 1993), since mobile platforms already embed such modules in their hardware codecs.

22.6   Conclusions

S3D makes its way in the mobile devices market with almost the same level of expectation in terms of performance and picture quality as for home appliances. However, major differencing constraints are the size and the power consumption of the device. The consequences in terms of design, from camera to display, have been covered throughout this chapter.

The first products have reused existing mobile application processors for time-to-market reasons, enabling basic 3D capture and rendering. The upcoming generation of mobile platforms takes 3D into account from the architecture definition. Efficient camera control and on-the-fly stereo rectification increase the 3D picture quality, and the optimization of the stereo processing chain increases the performance of the platform, allowing some headroom for complex use cases. Indeed, minimizing pixel format conversions and memory buffers between the various hardware subsystems (camera, encoder, rendering) and software layers (firmware, middleware, OS, application) is critical in terms of performance (picture resolution and frame rate, memory requirements, power consumption).

In terms of applications, the availability of stereo-cameras for still-picture and video recording should promote 3D smartphones as a significant source of 3D visual content, while mobile 3D rendering should become the main market for auto-stereoscopic display technologies. Moreover, it will bring a new dimension to booming domains: videoconferencing, gaming, augmented and virtual reality, navigation, user interfaces. In this respect, real-time depth estimation represents a promising opportunity.

Acknowledgments

We would like to thank the following for their work with us on stereoscopic 3D:

- Our past and present colleagues at STMicroelectronics, namely Antoine Chouly, Antoine Drouot, Jean Gobert, Laurent Pasquier, Selim Ben Yedder, Victor Macela, Olivier Gay-Bellile, Jean-Marc Auger, Eric Auger, Maurizio Colombo, Rebecca Richard, Laurent Cuisenier and Jacques Dumarest.
- Our partners from the Moov3D and CALDER projects, namely Telecom ParisTech, Telecom SudParis, NXP Software, CEA-LETI, Grenoble_INP, MicroOLED, Visioglobe, and Pointcube.More particular acknowledgment is due to Yves Mathieu from Telecom ParisTech, for the joint work on the Renoir IP and hardware view synthesis over the past years.

References

Auberger, S. and Miro, C. (2005) Digital video stabilization architecture for low cost devices, in *ISPA 2005. Proceedings of the 4th International Symposium on Image and Signal Processing and Analysis, 2005*, IEEE, pp. 474–479.

Bourge, A., Gobert, J., and Bruls, F. (2006) MPEG-C Part 3: enabling the introduction of video plus depth contents. Proceedings of the 3D Workshop on Content Generation and Coding for 3DTV, Eindhoven, The Netherlands, June.

Bourge, A., Pasquier, L., Gobert, J. *et al.* (2009) Applications and challenges for 3D on mobile. 3D Media Conference, Liège, Belgium, December.

De Haan, G., Biezen, P.W.A.C., Huijgen, H., and Ojo, O.A. (1993) True-motion estimation with 3-D recursive search block matching. *IEEE Trans. Circ. Syst. Vid. Technol.*, **3** (5), 368–379, 388.

Fehn, C. (2004) A 3D-TV system based on video plus depth information, in *Conference Record of the Thirty-Seventh Asilomar Conference on Signals, Systems and Computers*, Vol. **2**, IEEE, pp. 1529–1533.

Gotchev, A., Akar, G.B., Capin, T. *et al.* (2011) Three-dimensional media for mobile devices. *Proc. IEEE*, **99** (4), 708–741.

ITU-T (2010) ITU-T H.264 – Advanced video coding for generic audiovisual services, 2010-03-09, http://www.itu.int.

HDMI (2010) High-definition multimedia interface – specification version 1.4a – extraction of 3D signaling portion, March 4, http://www/HDMI.org.

Lansing, P. and Broll, W. (2011) Fusing the real and the virtual: a depth-camera based approach to mixed reality. Proceedings of IEEE International Symposium on Mixed and Augmented Reality (ISMAR), Basel, Switzerland, October.

Lipton, L. (1982) *Foundations of the Stereoscopic Cinema – A Study in Depth*, Van Nostrand Reinhold.

Lourakis, M.I.A. and Argyros, A.A. (2009) SBA: a software package for generic sparse bundle adjustment. *ACM Trans. Math. Software*, **36** (1), article no. 2.

MIPI Alliance (2005) MIPI Alliance Standard for Camera Serial Interface 2 (CSI-2), version 1.00, 29 November, http://www.mipi.org/specifications/camera-interface.

MIPI Alliance (2011) MIPI Alliance Standard for Display Serial Interface (DSI), version 1.1, 22 November, http://www.mipi.org/specifications/display-interface.

Müller, K., Merkle, P., and Wiegand, T. (2011) 3-D video representation using depth maps. *Proc. IEEE*, **99** (4), 643–656.

Pasquier, L. and Gobert, J. (2009) Multi-view renderer for auto-stereoscopic mobile devices, in *ICCE '09. Digest of Technical Papers International Conference on Consumer Electronics, 2009*, IEEE.

Pockett, L.D. and Salmimaa, M.P. (2008) Methods for improving the quality of user created stereoscopic content, in *Stereoscopic Displays and Applications XIX* (eds A.J. Woods, N.S. Holliman, and J.O. Merritt), Proceedings of the SPIE, Vol. **6803**, SPIE, Bellingham, WA, p. 680306.

Redert, A., Op de Beeck, M., Fehn, C. *et al.* (2002) Advanced threedimensional television system technologies, in *Proceedings. First International Symposium on 3D Data Processing Visualization and Transmission, 2002*, IEEE, pp. 313–319.

Scharstein, D. and Pal, C. (2007) Learning conditional random fields for stereo, in *CVPR' 07. IEEE Conference on Computer Vision and Pattern Recognition, 2007*, IEEE, pp. 1–8.

Shin, H.-C., Kim, Y.-J., Park, H., and Park, J.-I. (2008) Fast view synthesis using GPU for 3D display. *IEEE Trans. Consum. Electron.*, **54** (4), 2068–2076.

Strohmeier, D. and Tech, G. (2010) On comparing different codec profiles of coding methods for mobile 3D television and video. Proceedings of International Conference on 3D Systems and Applications (3DSA).

Surman, P., Sexton, I., Hopf, K. *et al.* (2008) European research into head tracked autostereoscopic displays, in *3DTV Conference: The True Vision – Capture, Transmission and Display of 3D Video, 2008*, IEEE, pp. 161–164.

Zhang (2000) A flexible new technique for camera calibration. *IEEE Trans. Pattern Anal. Mach. Intell.*, **22** (11), 1330–1334.

23

Graphics Composition for Multiview Displays

Jean Le Feuvre and Yves Mathieu

Telecom ParisTech, France

23.1 An Interactive Composition System for 3D Displays

In this chapter, an integrated system for displaying interactive applications on multiview screens is presented, hereafter called a "multiview compositor." The design goal of this system is to provide a platform capable of rendering various 3D multimedia applications, such as 3D video, 3D pictures, interactive maps, or games, while still allowing enhancing classic 2D content (slide shows, user interfaces, 2D games) with various depth effects. One major requirement of this system is to support most existing 3D display technologies, from simple stereoscopic in side-by-side mode to complex screens with a large number of views. A second important design choice is to focus on embedded systems, where resource constraints are problematic for real-time generation of a large number of views. Finally, the focus of the work is not on disparity estimation from a stereo couple; therefore, 3D video and images used in the multiview compositor always have disparity/depth map information associated.

The multiview compositor is available in two versions:

- A version relying on generic GPU (graphics processing unit) hardware and standard OpenGL/OpenGL ES (http://www.khronos.org/) pipeline programming for test purposes; this version runs on both desktop computers and mobile phones, including 3D stereo phones available in stores.
- A version relying on ad-hoc hardware implemented in FPGA (field-programmable gate array), performing on-the-fly generation of the different views required by the display in a single pass; this platform is designed with constrained devices in mind, and has much less computing power than a simple smartphone.

The interactivity and presentation logic are handled by a software layer; this software, used in both versions, is in charge of the various tasks of a graphics composition engine:

Emerging Technologies for 3D Video: Creation, Coding, Transmission and Rendering, First Edition.
Frédéric Dufaux, Béatrice Pesquet-Popescu, and Marco Cagnazzo.
© 2013 John Wiley & Sons, Ltd. Published 2013 by John Wiley & Sons, Ltd.

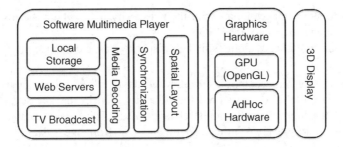

Figure 23.1   High-level view of the multiview compositor architecture

fetching the resources (from local storage or from web servers), decoding the media resources (audio, video, images), synchronizing the various media and animations, and then driving the underlying hardware to generate the video output. A high-level view of the architecture is given in Figure 23.1. The multimedia player used is GPAC (Le Feuvre *et al.*, 2007) and can handle various multimedia languages as well as media coding formats (video, audio, images, subtitles, etc.). Multimedia data can be read from various sources such as RTP (Real-Time Transport Protocol), MPEG-2 Transport Stream and ISO Base Media files (MP4, 3GP, MJ2K), commonly used in digital TV and the mobile world.

23.2   Multimedia for Multiview Displays

23.2.1   Media Formats

As seen in Chapters 6–8, many 3D video coding standards are compressing the disparity map or depth map along with color information, to simplify depth reconstruction at the client side: the process of computing virtual views from texture and depth is called view synthesis, or more generally DIBR (depth-image based rendering), as explained in Fehn (2003).

Although less well known than their video counterparts, there are also specific media formats for images describing stereoscopic or multiview data. The underlying image coders do not change, however; JPEG, PNG, and JPEG2000 are the most common ones. The most common formats are:

- JPS: a stereo pair in side-by-side mode encoded as a single JPEG file.
- PNS: a stereo pair in side-by-side mode encoded as a single PNG file.
- MPO: the Multi Picture Object (CIPA DC-007, http://www.cipa.jp/english/hyoujunka/ kikaku/pdf/DC-007_E.pdf) image enables storing in a single file a collection of images representing different views/versions of the same scene. While most devices using this format only store stereo pairs in the file, the format allows for more than two images to be stored (camera arrays), and it also supports disparity images. The detail of the optical capture system can be provided, such as the baseline distance between the cameras, camera convergence angle, camera orientation, rotation angles around the X, Y, and Z axes, and so on.

There is currently no specification for multiview images based on the PNG format, since the MPO format requires the exchangeable image file format (EXIF). During the development of the multiview compositor, internal pixel packing conventions have been defined for PNG-based 3D images:

- *PND*, used to store images along with their disparity map, where the disparity map is stored as the alpha channel of the PNG;
- *PNDS*, used to store images along with their disparity map and a shape mask.

Using side-by-side representation of color and depth has also been envisaged, but for compactness reasons the format stored the disparity map in the low seven bits and the shape mask in the high bit of the alpha channel of the PNG. This compactness is most useful to transfer on-the-fly synthesized image data from the system memory to the FPGA memory.

In a similar approach to MPEG-C (Bourge *et al.*, 2006), depth or disparity meta-data – that is, near and far parameters of the depth map, reference viewing distance and display size – can be stored in the PNG format using text chunks.

23.2.2 Multimedia Languages

There are many existing languages designed for the presentation of multimedia content. From the successful proprietary language Flash to mainstream web standards as HTML (HyperText Markup Language) or SVG (Scalable Vector Graphics), most of these languages fit the basic requirements of a multimedia presentation: describing the location and formats of media resources and the spatiotemporal relationship between these resources. Some formats also allow usage of mathematical descriptions of 2D or 3D drawings, through geometry (points, lines, triangles, Bezier curves and splines, etc.) and colors (gradient, texture/image mapping, shading with or without lighting), thereby enabling precise reconstruction of pixel data at any resolution or view scaling factor.

The multiview compositor has been designed as a test bed for interactive applications on 3D displays, and it is therefore a logical choice to use existing languages for describing applications rather than defining a new one. With the increasing importance of web browsers as an application platform on embedded devices, the choice is to use existing web languages and evaluate how they fulfill 3D use cases. Without going deeper into 3D displays specificities, the languages can be classified as follows:

- SVG: well suited for 2D applications but cannot be used for 3D ones; moreover, the language deals with a 2D coordinate system only, and has no notion of "depth", which means objects are behind/in front of each other without any notion of distance (*how "far" behind . . .*).
- HTML family: well suited for 2D applications, but has the same drawbacks as SVG. Recent additions to the CSS specification have seen the adoptions of 3D perspective and matrices (CSS Transforms, http://www.w3.org/TR/css3-transforms/), which allow applying a perspective transformation matrix to some part of the HTML content. These features are well suited for 3D effects on menus and other user interfaces, and thereby make HTML a good candidate for 3D displays. One problem is the lack of support for 3D data (models, virtual worlds). With recent work on WebGL (https://www.khronos.org/registry/webgl/specs/1.0/), some browsers now have support for hardware-accelerated 3D through a JavaScript version of the OpenGL ES API; once embedded web browsers integrate support for these two technologies, together with SVG support for describing 2D content, the web platform HTML + CSS + JavaScript + SVG + WebGL offers all the tools needed for building compelling applications on 3D displays.

Figure 23.2 2D/3D mixing of SVG, X3D, and MPEG-4

- VRML (Virtual Reality Modeling Language) and X3D: well suited for 3D models and virtual worlds, their 2D features are however less developed than what can be found in other languages.
- MPEG-4: derived from VRML, the MPEG-4 BIFS (binary format for scene) language also features advanced tools for 2D graphics composition, 3D model animation and compression, as well as interactivity, without scripting or through ECMAScript. This makes the MPEG-4 platform quite suited for deploying applications on 3D displays.

It should be noted that most of these languages can more or less be combined together in modern multimedia browsers, whether through native drawing support or through JavaScript-based drawing using Canvas2D or WebGL. Figure 23.2 shows an example of mixing a 2D SVG interface in an MPEG-4 world with an X3D model.

While classifying these languages is fairly simple, identifying the missing pieces of technology needed for 3D display compatibility is slightly more complex.

23.2.3 Multiview Displays

3D displays enable perception of objects in front or behind the screen plane; however, the range of the perceived depth position of objects is limited by the display physical characteristics. This volume is usually referred to as the "stereoscopic box," which can be compared to a matchbox theater on small screens (mobile devices). As seen in Chapter 18, attempting to place objects too close to the viewer (e.g., in front of the box) will cause eyestrain and viewing discomfort. It is therefore extremely important for the display system to know the maximum and minimum disparities that will be used during the lifetime of the content; not knowing it could result in bad calibration and lead to bad user experience.

With natural media, additional information is usually sent along with the depth/disparity map, as for example in MPEG-C (Bourge *et al.*, 2006), in order to explain how the encoded depth/disparity is mapped to a physical disparity on the display; these indications ensure the content creator/provider that the 3D media will be viewable on all devices, regardless of their physical characteristics. Unfortunately, such indications are not present in existing

Box 23.1 Simple SVG augmented with depth information

```
<?xml version="1.0" ?>
<svg width="160" height="160" depth-viewbox="-100 100">
  <title>Depth UI item</title>
  <desc>Shows usage of depth extension tools.</desc>
  <g id="button1" depth-offset="30" depth-scale="0.5">
    <rect width="100" height="30" fill="blue"/>
    <g depth-offset="30">
      <text x="20" y="20">Click!</text>
    </g>
  </g>
</svg>
```

multimedia languages. Moreover, in the case of 2D-only multimedia content, languages (BIFS-2D, SVG, HTML) must be extended to add depth information. This information will, for example, define the depth of each object in a menu; however, like rendering viewports define how the global coordinate system is mapped to the rendering area for X and Y dimensions, an author may also need to indicate how the depth values match the stereoscopic box of the screen by specifying the minimum and maximum disparities used in the content. An example of SVG extended for 3D displays is given in Box 23.1; in this example, depth is chosen as the 3D unit, but disparity could be used as well.

When designing 3D applications for multiview displays, the existing languages (BIFS, VRML, X3D) may be used as is but, however, lack describing camera setup, as will be seen in the following section.

23.3 GPU Graphics Synthesis for Multiview Displays

23.3.1 3D Synthesis

The process of generating the different views to be presented to the user, regardless of the stereo device type, does not differ much from single-view synthesis. Instead of rendering a single frame from a centered camera, the virtual world is rendered once per view with a camera shifted horizontally. As shown by Bourke and Morse (2007), there are two ways to compute the camera displacement in virtual worlds. The first approach consists of shifting the camera position by the inter-ocular distance while still looking at the same point in the virtual space, and rendering the scene with a symmetric perspective projection, as shown in Figure 23.3. While simple, this method has the drawback of having a nonparallel Z-axis, thereby introducing vertical disparity between views. This means that an object of the scene would appear higher in the view where it is closer to the camera than in the other. As vertical disparity increases eyestrain, this method should be used carefully. The second method consists of using parallels cameras (and Z-axis) corrected by an asymmetric perspective projection, as shown in Figure 23.3.

The camera displacement should reflect the inter-ocular distance (distance between the pupils) of the viewer. This value is usually mapped to an average inter-ocular distance of 65 mm and translated into virtual world coordinates using the viewing distance of the user; in other words, the focal length of the virtual camera. Most of this information is not present

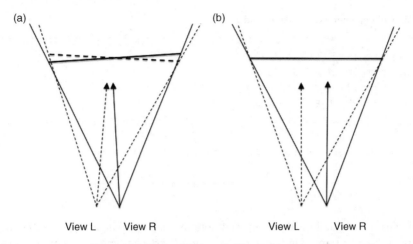

Figure 23.3   Toe-in (a) and off-axis (b) cameras

in the 3D virtual content (VRML, BIFS, etc.) and has to be computed by the rendering engine, with regard to user viewing position and screen size (as seen in previous chapters, larger screen size can benefit a larger zone of viewing comfort).

When using automatic mapping of the stereoscopic box of the display in a virtual world space, the near and far clipping planes become the point of minimal and maximal disparity. During the development of the multiview compositor, it appeared clearly that, especially for mobile devices, the disparity range is quite limited and trying to map the entire virtual world into this range is not always useful; some content only has a few points of interest for 3D, while the rest of the world can be seen as "background information" and could, therefore, be placed at background depth (minimal disparity). Additionally, an author should be able to define where the physical screen is located in the virtual world (point of zero disparity). Specifying the convergence of the virtual cameras could achieve this; however, the convergence angle should be re-estimated by the renderer based on the environment characteristics (screen size, viewing distance, user preferences, etc.). The multiview compositor instead introduced extensions to BIFS/VRML/X3D formats by allowing the content creator to specify:

- type of camera (toe-in or off-axis);
- distance D of the zero-depth plane from the camera;
- target depth range d_w, instructing the stereo renderer to map the disparity interval used to $[D - d_w, D + d_w]$.

These settings can be dynamically modified to provide new visual effects during animations.

23.3.2   View Interleaving

Multiview displays are based on mechanical or optical systems filtering the pixel visibility according to the viewer position. These mechanisms require that all the pixels of all the views are present on the screen; this is referred to as view interleaving or view multiplexing. As seen in previous chapters, there are as many types of interleaving as there are 3D display

Box 23.2   Column interleaving stereo shader

```
uniform sampler2D gfView1;
uniform sampler2D gfView2;
void main(void) {
  if ( int ( mod(gl_FragCoord.x, 2.0) ) == 0)
    gl_FragColor = texture2D(gfView1, gl_TexCoord[0].st);
  else
    gl_FragColor = texture2D(gfView2, gl_TexCoord[0].st);
} ]]>
```

technologies, from side-by-side or top-and-bottom image packing, to pixel sub-component (red, green, blue) mixing. Once the camera parameters can be established, the algorithm used by the multiview compositor is as follows:

- Create a GPU texture for each view to be generated
- For each different viewpoint
    - set up camera displacement
    - render the scene
    - store the result of the scene in the associated GPU texture
- Draw a rectangle with multiple textures and use a pixel shader to perform the view interleaving.

Since the view interleaving introduces a loss of resolution (most of the time horizontal), the off-screen view generation can be performed at a lower resolution than the native one (e.g., half the final width for two-view displays). Box 23.2 shows a fragment shader code writing the left (right) view pixel in the odd (even) pixel of the video output surface. Figure 23.4 shows the result of the shader.

Box 23.3 shows a fragment shader interleaving five views at sub-pixel level.

Note how the mapping view/sub-components is shifted at each line, because of the oblique parallax barrier of the screen used. Figure 23.5 shows the resulting sub-pixel interleaving of the shader.

The advantage of using a shader-based approach for the generic GPU version is that only a few lines of code have to be changed (shader code) to render any view configuration, since the main renderer can produce as many views as desired with the same code base. In the TriScope (TRISCOPE, http://triscope.enst.fr) project, a simple interactive carousel menu is designed in SVG and MPEG-4 BIFS to show the different possibilities of the platform

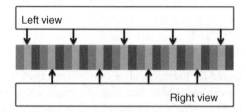

Figure 23.4   Pixel-column interleaving of stereo views

Box 23.3 Five-view sub-component interleaving shader

```
uniform sampler2D gfView1;
uniform sampler2D gfView2;
uniform sampler2D gfView3;
uniform sampler2D gfView4;
uniform sampler2D gfView5;
void main(void) {
  vec4 color[5];
  color[0] = texture2D(gfView5, gl_TexCoord[0].st);
  color[1] = texture2D(gfView4, gl_TexCoord[0].st);
  color[2] = texture2D(gfView3, gl_TexCoord[0].st);
  color[3] = texture2D(gfView2, gl_TexCoord[0].st);
  color[4] = texture2D(gfView1, gl_TexCoord[0].st);
  float pitch = 5.0 - mod(gl_FragCoord.y, 5.0);
  int col = int( mod(pitch+3.0*(gl_FragCoord.x),5.0));
  int Vr = int(col);
  int Vg = int(col) + 1;
  int Vb = int(col) + 2;
  if (Vg >= 5) Vg -= 5;
  if (Vb >= 5) Vb -= 5;
  gl_FragColor.r = color[Vr].r;
  gl_FragColor.g = color[Vg].g;
  gl_FragColor.b = color[Vb].b;
}
```

(video, still images, 3D models, games, etc.); it is shown in Figure 23.6 drawn in five-view, sub-pixel interleaved mode.

23.3.3 3D Media Rendering

Generating views from color and depth information is the main task of the FPGA version of the multiview compositor, as explained later on in this chapter. However, in order to test the generic GPU version, simple rendering techniques for image and depth rendering are used. All these techniques are based on defining each pixel in the texture as a vertex with:

- x and y coordinates derived from pixel location (u, v) in the color texture;
- z coordinate derived from the value at pixel location (u, v) in the depth texture.

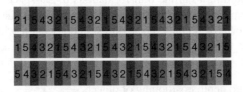

Figure 23.5 Sub-pixel interleaving of a five-view 3D display

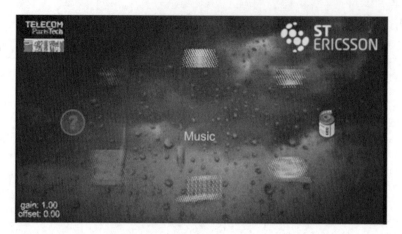

Figure 23.6   Five-view interleaving of an interactive menu (see Plate 30 for the colored figure)

Obviously, depth information has to be updated at each video frame, along with color information. The vertices are pushed to the GPU using VBO (Vertex Buffer Object), which allow storing vertex data on the GPU. The rendering still works without VBO support, but requires sending all the vertex data at each rendering pass, which becomes very costly when generating several views for each frame (typically five or more on multiview displays). The data can then be rendered:

- As a point cloud, where each point has a size of one pixel. In this case, depth discontinuities are seen as holes in the generated views.
- As a triangle strip. In this case, depth discontinuities result in wrong interpolated colors in the triangles drawn.

Other techniques could be used but have not yet been integrated in the multiview compositor. One can mention usage of new GPUs with tessellation units, which could be used to generate the vertices information on the fly. The GPU implementation of the multiview compositor is far from complete for view synthesis, especially since it does not handle de-occlusion at all. Usage of a GPU for view synthesis is quite common in the literature (Rogmans *et al.*, 2009), and a good overview of a video player using such techniques can be found in Duchêne *et al.* (2012).

23.4   DIBR Graphics Synthesis for Multiview Displays

23.4.1   Quick Overview

The multiview compositor has been ported to an FPGA-powered platform in charge of the view synthesis and pixel interleaving. The hardware is designed to compose multiple objects together on screen, according to affine transformations in 2D space and depth information. An overview of the complete architecture is given in Figure 23.7. The hardware only handles raster formats, in the form of rectangular textures, and covers a wide range of input pixel formats, RGB or YUV based with alpha blending support. The software is in charge of decoding video, images, and multimedia scenes as follows:

- compute spatial 2D and depth positioning of objects;
- if needed, generate 2D and 3D synthetic graphics and depth maps;

Composition using Depth

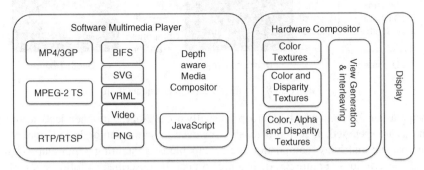

Figure 23.7   Architecture of the DIBR compositor

- send the objects to the DIBR hardware;
- execute timers and user interactions.

The main advantage of this version over the GPU-based version is its ability to perform view generation and interleaving in a single pass, when the GPU version had to render each view separately and perform the interleaving in a final stage.

23.4.2   DIBR Synthesis

Since the software module is in charge of rasterizing all synthetic content (2D and 3D vector graphics), it is also designed to retrieve the depth map associated with the rasterized content.

When playing 3D content, the entire 3D world is rendered in an off-screen surface, either through OpenGL or OpenGL ES. The OpenGL color buffers and depth buffers are read back, packed into a color plus depth texture object handled to the hardware. The OpenGL z-buffer is read back in memory and used as a basis for the depth map. However, the values of the z-buffer are not entirely suited for depth information, as they cover the depth range from the camera to infinity, which far exceeds the display capabilities (size of the stereoscopic box). With this in mind, a depth map from the OpenGL z-buffer is produced as follows:

- compute the stereoscopic box size in 3D world coordinates, either automatically or using author indications as described previously;
- normalize all z-buffer values inside the box to [0, 0xFF];
- saturate all z-buffer values further than the back side of the box to 0xFF (e.g., all objects further than the back side of the box will be pushed towards infinity);
- optionally, if the color buffer uses transparency, resample the depth map on seven bits and use the high-order bit as a shape mask.

OpenGL ES has successfully been experimented as a rendering backend, with one drawback, however: the standard does not allow for depth buffer reading, and the multiview compositor therefore needs an additional pass to render the scene, compute each pixel depth, and write the value to an off-screen color buffer. The benefit of using this approach compared with the GPU one is therefore only significant when more than two views (e.g., OpenGL rendering passes) have to be generated.

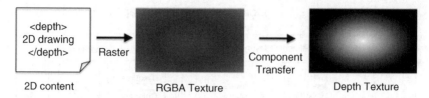

Figure 23.8 Synthetic depth maps generation

When playing 2D content, a depth map is usually not present, but rather depth affine transformations are specified, as seen previously. However, to let the author design more powerful animations, the multiview compositor is also capable of generating depth maps from synthetic 2D descriptions (Figure 23.8). The process can be described as follows: a part of the 2D content is tagged as "depth map," rendered in an off-screen RGBA canvas, and component transfer rules are given to use the generated data as a depth map of another object. The code sample in Box 23.4 shows how an SVG radial gradient can be used to define a depth map, which will be used for animating a button.

By modifying the vector graphic description of the depth map upon user interactions or through timers, the author can produce interesting, complex depth effects at a very low computational cost.

23.4.3 Hardware Compositor

DIBR composition has been successfully implemented on the Renoir hardware rendering intellectual property (IP). Originally designed to process 2D textures, the Renoir IP (Cunat

Box 23.4 Example of synthetic depth map definition using SVG filters

```
<svg .. width="160" height="160">
  <defs>
    <radialGradient xml:id="MyGradient" r="0.1">
      <stop offset="0" stop-color="blue"/>
      <stop offset="1" stop-color="red"/>
    </radialGradient>
    <rect xml:id="depthRect" fill="url(#MyGradient)"
width="100" height="30"/>
    <filter xml:id="depthFilter">
      <feImage xlink:href="url(#depthRect)" result="depth"/>
      <feDepthComponent in="SourceGraphic" in2="depth">
      <feFuncR type="linear" slope="-1" intercept="1"/>
    </feDepthComponent>
    </filter>
  </defs>
  <g filter="url(#depthFilter)" transform="translate(20,20)">
    <rect width="100" height="30" fill="blue"/>
    <text x="20" y="20">Click!</text>
  </g>
</svg>
```

et al., 2003) allows real-time composition of visual objects including still pictures, natural video, graphics, and text. It also supports affine transformation of these objects as well as color transformation between various input and output formats. The Renoir architecture, based on tiled rendering (Antochi *et al.*, 2004), requires low memory bandwidth as is usually the case for mobile or embedded devices. The rendered image is divided into small "tiles" of 16×16 pixels processed one after the other. All intermediate data needed for the rendering of a tile can be kept in local buffers of the IP. The external memory bandwidth is thereby limited to textures loading and final tile storing. Tiled rendering requires a preliminary step of determining for each object of the scene the texture area covering the tile being processed. A local texture memory is dimensioned to hold four times the area covered by a tile to deal with common transformations like rotation, and downscaling of up to a factor of 2. For other factors, a subsampling of the texture is performed before filling the local memory texture. The composition rate of this IP is one pixel per clock cycle due to a fully parallel and pipelined architecture.

The main idea leading to this solution is a unified treatment of visual effects (color effects, transparency effects, etc.) and of final composition between 2D objects and 3D objects. Note that our goal is to use only simple algorithms that fit easily on hardwired processing units.

A first implementation of DIBR processing is based on a three steps forward algorithm, as shown in Figure 23.9. Given a line of pixels in the local texture memory, pixel components of 3D objects are projected to the destination space according to their depth values modified by an offset and a gain, as explained in Chapter 22. Occlusions are handled by a priority process based on the left/right orientation of the view. In the second step, holes left in the local line buffers are filled according to simple disocclusion rules, as discussed later in this chapter.

In order to keep the one pixel per clock cycle processing rate of the basic 2D compositor, the number of hardware units needed for the first and second steps is set to the number of views N of the 3D screen. During the third step of the algorithm, color components inside the N buffers are interleaved to create the final line of pixels. Classical 2D composition with the current content of the tile is then performed. These three algorithmic steps are fully pipelined with the 2D composition. Intersection between a 3D object and a tile is computed by upscaling the bounding area of the object according to the maximum supported disparity, $M_{ax}D_{isp}$, the hardware is dimensioned for.

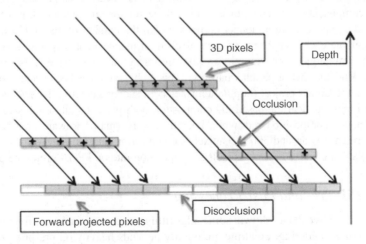

Figure 23.9 Forward processing for one view angle

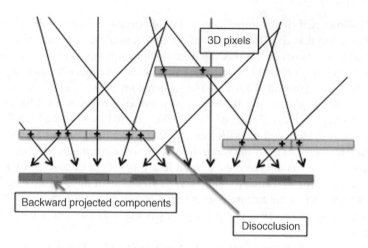

Figure 23.10   Backward processing (five-view display)

This first processing architecture, successfully demonstrated on several prototypes during the TriScope project, has a hardware complexity linear in the number of views N. This is suitable for a hardware IP intended to be used with a fixed 3D screen, where the hardware IP can be designed to fulfill the needs exactly. This is not optimal for a general-purpose IP potentially connectable to screens with different characteristics.

Our second implementation of the DIBR processing is based on a one step backward ray-tracing algorithm, as illustrated in Figure 23.10. For given screen characteristics, the view number can be computed for each component (red, green, or blue) of each pixel in the line of a tile. Given this view number, a ray can be traced from the observer to the component position. Knowing the depth of all 3D pixels of the corresponding line in the local texture memory, selection is made by computing the distance between the ray and the pixels. The nearest pixel from the observer, with a distance from the ray smaller than a fixed threshold, is selected. Disocclusion cases, detected when no distance is smaller than the threshold, are treated on the fly.

In order to keep a one pixel per cycle processing rate, all comparisons between the ray and the 3D pixels are performed in parallel. The complexity of such hardware is linear in the maximum supported disparity $M_{ax}D_{isp}$. The number of hardware units needed is equal to the number of components per pixel; for example, three for textures without transparency. Performance and complexity are no longer correlated with the number of views. Furthermore, the ability to compute each component of each pixel in an independent way allows some corrections of the view angles, for example, to take into account nonlinear effects in the optical system (e.g., lenticular net or parallax barriers). This architecture has been successfully demonstrated during the Calder project (http://artemis.telecom-sudparis.eu/2012/10/29/calder-2/).

Embedded in a classical tile-based 2D compositor, both presented algorithms can achieve a high-performance composition rate while keeping a very small hardware area. The demonstrated performance is around 300 megapixels/s composition rate on a low-cost, low-power IP for mobile platform, and a 100 megapixels/s composition rate on FPGA prototype platforms.

23.4.4   DIBR Pre- and Post-Processing

In Section 23.4.3 we have shown some algorithms and associated hardware suited to performing DIBR rendering. Final picture quality depends heavily on pre-processing of the 2D + depth input images and on post-processing of the rendered views.

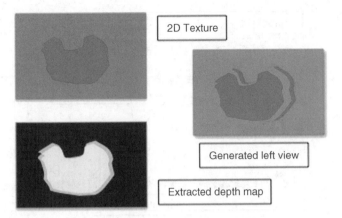

Figure 23.11   Ghost effect along object borders

Depth maps of low quality (e.g., extracted from natural stereo views), or carelessly filtered or resized 2D + depth pictures may lead to highly visible artifacts. In such pictures, borders between objects and background are not sharp enough; that is, pixels in the borders have intermediate depth values between the background depth value and the objet depth value. These intermediate values do not match any real object in the color image. This leads to a ghost effect, as shown in Figure 23.11. Furthermore, if color components and depth components are misaligned, some background pixels have the color of the foreground object and inversely.

We configured our hardware compositor to tackle these artifacts. First, the resampling unit of the compositor is configured to use point sampling for the depth component rather than bilinear interpolation as used for the color components. With such a method, no artificial depth values are generated by the compositor itself. Second, the DIBR hardware is configured to deal with isolated pixels; an isolated pixel is a pixel having a depth that is neither a foreground nor a background depth. During the computations for a line of pixels in a tile, the depth of each pixel is compared with the depth of its neighboring pixels. If the difference in depth with the previous one and the difference in depth with the following one are more than an arbitrary threshold, this pixel is defined as an isolated pixel. For such a pixel, the compositor copies the depth of the foreground object, thus suppressing the ghost effect. A more sophisticated algorithm could compute a distance between the colors of the neighboring pixels and then choose between foreground and background depth.

As far as post-processing is concerned, the main problem is disocclusion. Disocclusion occurs on the borders of objects when the view angle is such that the observer should see pixels under the objects. The standard way of dealing with disocclusion is by using padding with pixels from the neighboring background. A correct disocclusion algorithm should take into account the two following problems. First, disocclusion pixels should be consistent for the N views: when the point of view changes from view I to view $I + 1$ the observer should see a simple shift of the disocclusion zone. For this purpose, a simple mirror algorithm is implemented on the hardware platform: missing pixels are taken from the background neighbor of the object, as if the edge of the object were a mirror for the background. Figure 23.12 shows such a situation with the mirrored pixels for a left oriented view.

Second, disocclusion pixels should be carefully selected in order to avoid artifacts when the background is a thin strip of pixels between two neighboring objects. In such a case, if the difference of disparity between background and foreground is greater than the width of the strip, the number of background pixels is not enough to fill the hole, as illustrated in

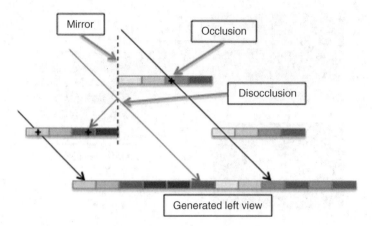

Figure 23.12   Padding with pixel mirroring

Figure 23.13. More efficient padding techniques can be used, but their algorithmic complexity does not fit with the goal of a fully hardwired compositor.

On our hardware platform, we used a simple strategy based on objet detection. Each pixel of the texture line is tagged with a flag 0/1 based on disparity comparison: if the difference in disparities between the current pixel and the previous one is less than an arbitrary threshold, the current pixel flag is set to the same value as the previous pixel flag, else the flag is inverted. With such an algorithm, areas of homogeneous flags define objects and backgrounds. When padding holes, only pixels of homogeneous flags are used. If the mirroring process fails, arbitrary pixels from the defined background area are chosen.

23.4.5 Hardware Platform

Several platforms have been designed to demonstrate several parts of our DIBR synthesis system. In order to set up the full chain of a multimedia demonstrator using hardwired multimedia composition, we designed a portable platform based on an SH4 processor coupled with a StratixII FPGA (Bourge *et al.*, 2006). All software layers are running in a Linux environment, and the 2D + depth compositor is running on the FPGA. Although the performance

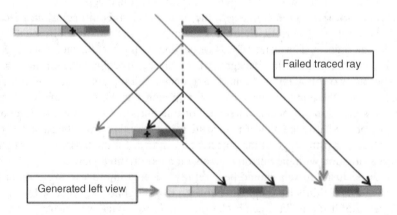

Figure 23.13   Padding failure example

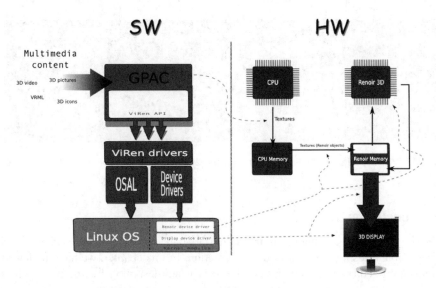

Figure 23.14   Hardware platform overview

of such a platform is relatively low (maximum 10 frames per second of 2D + depth video decoding for a five-view 800 × 480 display), this platform fruitfully helped us to define the optimal hardware–software partition. As shown in Figure 23.14, the FPGA is in charge of the video output as pixel component interleaving is closely correlated to the physical parameters of the screen.

Several use cases have been tested on this platform, including human–machine interfaces (3D carousel), 2D + depth video streaming, or 3D synthetic object visualization. Some stereoscopic views of the generated pictures are shown in Figures 23.15–23.17.

Figure 23.15   3D carousel menu (2D flat objects composition) (see Plate 31 for the colored figure)

Figure 23.16   Natural 2D + depth picture (see Plate 32 for the colored figure)

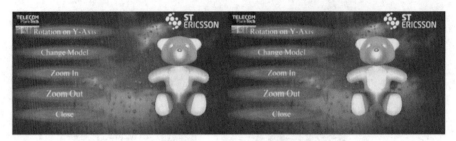

Figure 23.17 Composition of 2D menu and synthetic 2D + depth model (see Plate 33 for the colored figure)

23.5 Conclusion

Depth-based image processing is gaining more importance in many fields, such as 3D video compression and display or natural user interaction. In this chapter, we have reviewed the different multimedia languages that can be used when designing interactive services for auto-stereoscopic displays. In order to handle displays with many views, we have investigated usage of DIBR techniques in these languages. We have identified that these languages lack the ability for an author to describe depth-related information (animations, virtual camera positioning) and proposed simple extensions to solve these issues and allow for more creativity from the author by using the depth effect.

We have built open-source GPU-based and dedicated hardware prototypes for services combining video + depth, 2D graphics with depth animations, and 3D graphics. We have shown that dedicated hardware composition can be an effective alternative to GPU processing to handle multiview synthesis. This low-cost rendering IP could be used in a video back-end controller on mobile or embedded platforms to relieve the main GPU from the view synthesis task, thereby giving it more time for other processes. Such a design is becoming common in the mobile industry, where two CPUs or GPUs with very different capabilities cooperate in order to lower the overall power consumption.

Acknowledgments

We would like to thank the following for their collaboration and advice on this work around stereoscopic 3D graphics composition:

- Our past and present colleagues from Telecom ParisTech: Daniel Comalrena and Florent Lozach, and more specifically Alexis Polti for his precious help on the hardware platform design.
- Our partners from the CALDER and TriScope projects: ST Ericsson, NXP Software, and Telecom SudParis, with a special thanks to Jean Gobert and Laurent Pasquier from ST Ericsson for the initial architecture of the IP.

Part of the work presented in this chapter has been sponsored by the French Government through funded research projects TriScope and CALDER.

References

Antochi, I., Juurlink, B.H.H., Vassiliadis, S., and Liuha, P. (2004) Memory bandwidth requirements of tile-based rendering. Proceedings of SAMOS, pp. 323–332.

Bourge, A., Gobert, J., and Bruls, F. (2006) MPEG-C Part 3: enabling the introduction of video plus depth contents. Proceedings of the 3D Workshop on Content Generation and Coding for 3DTV, Eindhoven, The Netherlands, June.

Bourke, P. and Morse, P. (2007) Stereoscopy: theory and practice. Workshop at 13th International Conference on Virtual Systems and Multimedia.

Cunat, C., Gobert, J., Mathieu, Y. (2003) A coprocessor for real-time MPEG4 facial animation on mobiles. Proceedings of ESTImedia, pp. 102–108.

Duchêne, S., Lambers, M., and Devernay, F. (2012) A stereoscopic movie player with real-time content adaptation to the display geometry, in *Stereoscopic Displays and Applications XXIII* (eds A.J. Woods, N.S. Holliman, and G.E. Favalora), Proceedings of the SPIE, Vol. **8288**, SPIE, Bellingham, WA, p. 82882Q.

Fehn, C. (2003) A 3D-TV approach using depth-image-based rendering (DIBR). Proceedings of VIIP 03, Benalmadena, Spain, September.

Le Feuvre, J., Concolato, C., and Moissinac, J. (2007) GPAC: open source multimedia framework, in *MULTIMEDIA '07 Proceedings of the 15th International Conference on Multimedia*, ACM, New York, NY, pp. 1009–1012.

Rogmans, S., Lu, J., Bekaert, P., and Lafruit, G. (2009) Real-time stereo-based view synthesis algorithms: a unified framework and evaluation on commodity GPUs. *Signal Process. Image Commun.*, **24** (1–2), 49–64.

24

Real-Time Disparity Estimation Engine for High-Definition 3DTV Applications

Yu-Cheng Tseng and Tian-Sheuan Chang

Department of Electronics Engineering, National Chiao Tung University, Taiwan

24.1 Introduction

In 3D video processing, disparity estimation is one of the most important techniques to generate disparity or depth maps for virtual view synthesis. Current disparity estimation algorithms can be categorized into local or global approaches (Scharstien and Szeliski, 2002). The local approach can be easily implemented to achieve real-time processing speed for high-definition (HD) videos but suffers from degraded disparity quality (Greisen *et al.*, 2011). On the other hand, the global approach can deliver better disparity with an additional disparity optimization but it needs more computation time. The widely used disparity optimization algorithms are dynamic programming (DP), belief propagation (BP), and graph-cut (GC).

Based on GC, the MPEG 3D Video Coding (3DVC) has developed a state-of-the-art 3DTV system, called the depth estimation reference software (DERS) algorithm (Tanimoto *et al.*, 2009). However, the sequential computation of GC makes GPU and VLSI parallel acceleration hard to satisfy the required throughput of HD videos. Different from GC, the BP-based algorithms are highly parallel to compute high-quality disparity maps for the Middlebury test bed (Scharstien and Szeliski, n.d.). Their previous GPU and VLSI implementations (Yang *et al.*, 2010; Liang *et al.*, 2011) attained throughputs of $800 \times 600@0.67$ frames per second (fps) and $640 \times 480@58$ fps, respectively. However, they are far from the HDTV requirement, $1920 \times 1080@60$ fps. In addition, the BP-based algorithms demand large memory space for their message passing mechanism, even if the tile-based computation (Liang *et al.*, 2011) is applied. To sum up, previous studies did not satisfy the required performance of HDTV owing to their complicated optimization process and large memory cost due to HD videos.

Besides the main issue of disparity optimization, the occlusion problem often results in bad disparity maps and severely affects synthesized videos. Various occlusion handling methods are surveyed by Egnal and Wildes (2002). Among those methods, the left–right check (LRC) method is simple and widely used in the post-processing of disparity estimation. In addition, another approach puts the ordering and occlusion constraints into the disparity optimization to iteratively recover occluded regions.

Moreover, temporal consistency for video applications is also an important problem in disparity estimation. Previous studies (Gong, 2006; Min *et al.*, 2010a; Khoshabeh, Chan, and Nyuyen, 2011) applied different smoothing approaches in the disparity flow, which consists of many successive disparity frames. Another approach based on a temporal BP algorithm (Larsen *et al.*, 2007) included the previous and next frames in the optimization process. However, the above methods need large buffers to store the temporal data. Different from them, the DERS algorithm only propagates the previous disparity into current matching costs for still pixels (Bang *et al.*, 2009), but it results in foreground copy artifacts due to its overpropagation. Similarly, the method (Min *et al.*, 2010b) referred to motion vectors to propagate disparities for both still and motion pixels, but it needs extra computation cost for motion estimation.

Motivated by the above problems, this chapter proposes a hardware-efficient disparity estimation algorithm and its VLSI design to generate high-quality disparity maps and meet the real-time throughput of HD 3DTV. The main contributions of this chapter are as follows. At the algorithm level, for the high computation due to the HD frame size, we first downsample the computation of matching cost to generate a low-resolution disparity map and upsample it to a high-resolution one. Second, for the high memory cost due to the frame-level optimization, we propose a hardware-efficient diffusion method that only requires the memory of two frame rows. Third, we propose a no-motion registration (NMR) and still-edge preservation (SEP) method for temporal consistency, and intra/inter occlusion handling for disparity refinement. At the architecture level, we propose a three-stage pipelining architecture and efficient processing elements (PEs) design for a high-throughput disparity estimation engine to achieve a throughput of HD1080p@95fps. The disparity estimation engine can simultaneously generate three-view disparity maps to support the configuration of a nine-view display defined by MPEG 3DVC (MPEG, 2010).

The rest of this chapter is organized as follows. First, Section 24.2 reviews the state-of-the-art disparity estimation algorithms and implementations. Then, Section 24.3 presents our proposed algorithm, and Section 24.4 describes its corresponding architecture. Finally, Section 24.5 compares our performance with previous work, and Section 24.6 concludes this chapter.

24.2 Review of Disparity Estimation Algorithms and Implementations

In previous studies, state-of-the-art disparity estimation algorithms employ the GC- or BP-based optimization and well-designed post-processing to acquire high-quality disparity for different applications. In addition, the DP-based optimization is often adopted in disparity estimation because of its lower computation. In this section, we reviewed the DP-, GC-, and BP-based algorithms, and their associated fast algorithms and implementations.

24.2.1 DP-Based Algorithms and Implementations

The DP-based algorithm (Scharstien and Szeliski, 2002) finds the shortest path in a disparity space image by the two steps: forward cost accumulation and backward path tracing. The first step can also attach the occlusion cost using a greedy method, or a smoothness cost using a min-convolution method, called scanline optimization (SO). But the DP-based algorithm natively suffers from streak artifacts due to 1D optimization. To address this, previous studies (Ohta and Kanade, 1985; Kim *et al.*, 2005; Veksler, 2005) added connections between scanlines, and others (Hirschmüller, 2005; Wang *et al.*, 2006) aggregated neighboring matching costs in 2D space.

To reduce its computation, Gong and Yang (2007) proposed a local minimum searching method and a reliable match prediction method. Their GPU implementation can achieve a throughput of $352 \times 288@20$ fps. Wang *et al.* (2006) adopted a vertical cost aggregation to avoid streak artifacts and also achieve real-time performance on the GPU. In addition, Park and Jeong (2007) applied a systolic-array architecture and reached a throughput of $320 \times 240@30$ fps by VLSI design.

Different from the 1D optimization in DP-based algorithms, the following two algorithms are 2D optimizations, and deliver better disparity results as shown in the Middlebury test bed (Scharstien and Szeliski, n.d.).

24.2.2 GC-Based Algorithms and Implementations

The GC-based algorithms convert the disparity estimation to the min-cut/max-flow problem, and apply the associated optimization techniques, such as the push-relabeling (Ford and Fulkerson, 1962) and the augmenting path (Kolomogorov and Zabih, 2001). But they natively suffer from high computational cost due to large disparity range and graph size.

To address this, Cherkassky and Goldberg (1995) changed the first-in–first-out order to the highest label order for the push-relabeling, and Delong and Boykov (2008) further increased its parallelism by block-based computation. For the augmenting path, Boykov and coworkers (Boykov *et al.*, 2001; Boykov and Kolmogorov, 2004) proposed the disparity swap method that considers only one disparity in one iteration. Based on this, Chou *et al.* (2010) further predicted disparities and skipped partial optimization. However, these fast algorithms still suffer from low parallelism in data access and computation because they are sequentially computed on a tree-structural graph, and the connection of edges is often changed irregularly. In the previous implementation, a real-time scalable GC engine (Chang and Chang, 2007) only supported a small graph with 16 nodes. To sum up, the GC-based algorithms are not suitable for GPU and VLSI acceleration.

24.2.3 BP-Based Algorithms and Implementations

Sun *et al.* (2003) first applied BP to acquire accurate disparity maps. In BP optimization, the matching costs of a pixel are iteratively diffused to four neighbors by the message passing mechanism. After several iterations, the disparity maps converge to a steady state. Based on the baseline BP, advanced algorithms (Klaus *et al.*, 2006; Yang *et al.*, 2009) combined the color segmentation to deliver more accurate disparity maps. In addition, the temporal consistency can also be included in the optimization (Larsen *et al.*, 2007). However, the message passing suffers from high computational complexity, $O(\text{HW} \times \text{DR}^2 \times T)$, where T is the iteration count, HW is the frame size, and DR is the disparity range.

To reduce the complexity, Felzenswalb and Huttenlocher (2006) proposed a hierarchical BP to accelerate convergence speed, and the linear-time message passing to reduce the convolution from $O(\text{DR}^2)$ to $O(\text{DR})$. In addition, Szeliski *et al.* (2008) proposed a max-product loopy BP to reduce the iteration count by a scale. BP computation is highly parallel, and it was implemented by GPU and VLSI design (Yang *et al.*, 2006; Park *et al.*, 2007; Liang *et al.*, 2011). The state-of-the-art implementation (Liang *et al.*, 2011) can achieve a through-put of $640 \times 480@58$ fps but needs significantly high memory cost.

In summary, the BP-based and GC-based algorithms can generate more accurate disparity than the DP-based algorithm, but their main design challenges are the high computational complexity and large memory cost that are dramatically increased with the increasing video resolution and disparity range. Therefore, it is necessary to develop a new disparity estimation algorithm that can provide high-quality disparity maps with low computation and memory cost.

24.3 Proposed Hardware-Efficient Algorithm

Figure 24.1 shows the proposed disparity estimation algorithm. In which, I and D respectively refer to the image frame and the disparity map. Their superscript refers to the frame number in the time domain, and the subscripts refer to the frame resolution and the viewpoint. In this algorithm, the three view image frames $I_{H,L}{}^{t}$, $I_{H,C}{}^{t}I_{H,R}{}^{t}$ are used to compute their corresponding disparity maps $D_{H,L}{}^{t}$, $D_{H,C}{}^{t}$, $D_{H,R}{}^{t}$. In addition, the previous image frames and disparity maps are also fetched for temporal consistency enhancement. This three-view input and output configuration follows the configuration of the nine-view display defined by MPEG 3DVC (MPEG, 2010).

The proposed algorithm adopts four ideas (i.e., low-resolution matching cost, cost diffusion, temporal consistency enhancement, and intra/inter-frame-based occlusion handling) to solve the complexity and storage problems while obtaining a high-quality disparity map. First, one of the complexity problems is directly related to the processing at the HD video size, which is particularly a burden for the matching cost computation. Thus, we use the downsampled matching cost to reduce the complexity while keeping the same disparity range to minimize the impact on the disparity quality. The resulting low-resolution disparity maps are upsampled to high-resolution ones with the desired smoothing effect by the joint bilateral upsampling. Second, to further reduce the memory cost, we propose the cost diffusion method for the disparity optimization, which only needs to store a matching cost row and a disparity row. Third, for the disparity quality, we propose the NMR and SEP methods to deal with the temporal consistency problem. Furthermore, we propose the intra-frame- and inter-frame-based occlusion handling method to solve the occlusion problem.

The proposed algorithm flow is as follows. We first calculate the matching costs for the downsampled pixels, and add the temporal costs to propagate the disparities of the previous

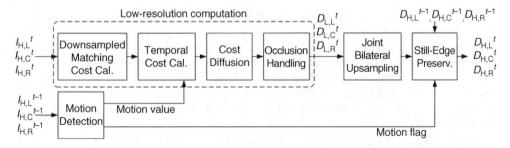

Figure 24.1 Flow of the proposed disparity estimation algorithm

frame to the current frame. Then, the cost diffusion step generates the low-resolution disparity maps, and the occlusion handling step recovers their occluded regions. Finally, the low-resolution disparity maps are upsampled by the joint bilateral upsampling. In addition, the SEP deals with the flicker artifact in the last step. In this flow, the required motion information is provided by the motion detection at the branch. The details of each step are presented in the following.

24.3.1 Downsampled Matching Cost for Full Disparity Range

To overcome the high complexity due to HD videos, we propose the downsampled matching cost with full disparity range to generate an initial low-resolution disparity map and then adopt the joint bilateral upsampling (JBU) method (Kopf *et al.*, 2007; Riemens *et al.* 2009) to generate its high-resolution disparity map. In our proposed method, only the spatial domain is downsampled, while the disparity domain is fully computed to avoid losing disparity precision. The horizontal and vertical sampling factors are set as 1/2 and 1/4 according to the quality analysis in Section 24.5. In addition, the sum of absolute difference is applied to compute the subsampled pixel's matching cost C_{SAD}. Note that the center view additionally has two side views as matching reference, and the minimal one is selected.

24.3.2 Hardware-Efficient Cost Diffusion Method

With the low-resolution C_{SAD}, we further apply the proposed disparity optimization, called cost diffusion method, to compute the low-resolution disparity map $D_L{}^t$. The proposed cost diffusion is a 1D optimization method operated row by row. The objective cost function is defined as

$$E(d) = \sum_{(x,y)\in I}\left[C_{total}(x,y,d) + \sum_{q\in N(x,y)} V(D(x,y),D(q))\right] \qquad (24.1)$$

where V is the smoothness term to enforce the disparity consistency among neighbors. In addition, C_{total} is the data term to mainly enforce the similarity of correspondence pairs using the matching cost C_{SAD}. It is defined as

$$C_{total}(x,y,d) = C_{SAD}(x,y,d) + C_{temp}(x,y,d) + C_{vert}(x,y,d) \qquad (24.2)$$

where the vertical cost C_{vert} and the temporal cost C_{temp} are additionally enforcing the disparity consistency in the spatial and temporal domains. We adopt the vertical cost

$$C_{vert}(x,y,d) = \lambda_{vert}(\Delta I_L{}')\min\{|d - D_L{}'(x,y-1)|, \tau_{vert}\} \qquad (24.3)$$

proposed by Van Meerbergen *et al.* (2002) to avoid streak artifacts in 1D optimization. The disparities $D_L{}'(x,y-1)$ in the upper row are propagated into the current processed row, and their effect is controlled by λ_{vert}, which is proportional to color consistency between the current pixel and the upper pixel. In addition, the temporal cost C_{temp} can enforce consistency between the current and previous disparities for a still object. Its details are introduced in Section 24.3.4.

To minimize the objective function, the proposed cost diffusion consists of a forward process and a backward process. The two processes diffuse the current pixel's data term C_{total} to its neighbor. This is similar to the message passing in the BP and SO algorithms, but immediately determines the best disparity for the current pixel to avoid a large memory space proportional to the disparity range. The forward process is performed from left to right through the diffused cost

$$C_{\text{fwd}}(x, y, d) = C_{\text{total}}(x, y, d) + \min_{d_s}(V(d, d_s) + C_{\text{fwd}}(x - 1, y, d_s)) - \kappa \qquad (24.4)$$

where C_{total} is at the current pixel, V is the smoothness term using the Potts model for convolution, and $C_{\text{fwd}}(x - 1, y, d)$ is the diffused cost from its neighbor. In addition, κ is the average of the convoluted $C_{\text{fwd}}(x - 1, y, d)$ for normalization to avoid overflow, and it can be a minimal cost to further avoid signed arithmetic computation (Hirschmüller, 2005).

According to the current $C_{\text{fwd}}(x, y, d)$, we immediately determine the disparity $D_{\text{fwd}}(x, y)$ and the minimal cost $C_{\text{fwd,min}}(x, y)$ by the winner-take-all manner. With the same calculation in the inverse direction, we can obtain the disparity map $D_{\text{bwd}}(x, y)$ and the minimal cost $C_{\text{bwd,min}}(x, y)$ in the backward process. Finally, the disparity with smaller cost will be selected as the final disparity map, as defined by

$$D_{\text{L}}{}^t(x, y) = \begin{cases} D_{\text{fwd}}(x, y), & C_{\text{fwd,min}}(x, y) < C_{\text{bwd,min}}(x, y) \\ D_{\text{bwd}}(x, y), & \text{otherwise} \end{cases} \qquad (24.5)$$

In other words, the final disparity is chosen from the cost

$$C(x, y, d) = C_{\text{total}}(x, y, d) + \min\left[\min_{i \in D}(V(d, i) + C_{\text{fwd}}(x - 1, y, i)), \min_{i \in D}(V(d, i) + C_{\text{bwd}}(x + 1, y, i))\right] \qquad (24.6)$$

where the smaller cost from the two diffusion directions is selected to avoid inappropriate diffusion with large cost due to occlusion.

The proposed cost diffusion method is operated row by row, and the forward and backward processes can be executed sequentially to reduce memory cost. This method only requires the memory of a disparity row and a cost row, which amounts to $2W$. On the other hand, the baseline BP algorithm (Sun *et al.*, 2003) requires a memory of $5HW \times \text{DR}$ for messages and matching costs, while the SO algorithm requires a memory of $W \times \text{DR}$ for the transition directions. Our method needs much lower memory cost and is more hardware efficient than the other two are.

24.3.3 Upsampling Disparity Maps

The low-resolution disparity maps are converted to high-resolution versions by JBU (Kopf *et al.*, 2007) in our algorithm. The main idea of JBU is to apply a high-resolution image to guide the upsampling process. For upsampling a disparity map, given the high resolution image $I_{\text{H}}{}^t$ and the low-resolution disparity map $D_{\text{H}}{}^t$, the high resolution disparity map $D_{\text{H}}{}^t$ can be computed by

$$D_{\text{H}}{}^t(i) = \frac{1}{\kappa} \sum_{j_{\text{L}} \in S} D_{\text{L}}{}^t(j_{\text{L}}) \cdot f(\|i_{\text{L}} - j_{\text{L}}\|) \cdot g(\|I_{\text{H}}{}^t(i) - I_{\text{H}}{}^t(j_{\text{H}})\|) \qquad (24.7)$$

where f is the spatial kernel with the argument of spatial distance in low resolution and g is the range kernel with the argument of color distance in two resolutions. Both kernels are Gaussian weighted functions. In our algorithm, we adopt the improved JUB (Van Meerbergen *et al.*, 2002), whose $I_H^t(j_H)$ is changed to $I_L^t(j_L)$ for better results.

After JBU, we proposed the window vote method, which considers pixel similarity in the voting process to avoid the wrong decision problem in the regional vote method (Zhang *et al.*, 2009). The original regional vote method approximates the performance of plane fitting without segment information but does not perform well in the highly textured regions. Thus, our modified method considers all the disparities in a window, instead of partial disparities in a continuously grown region, to improve the quality. Note that the window sizes of JBU and window vote are 5×5 for low hardware cost.

24.3.4   Temporal Consistency Enhancement Methods

For the temporal consistency problem, the DERS algorithm propagates previous disparities to current ones for no-motion regions by adding the temporal cost

$$C_{\text{temp}}(x, y, d) = \begin{cases} \lambda_{\text{temp}} |d - D_H^{t-1}(x, y)|, & \text{MV} < \gamma_{\text{temp}} \\ 0, & \text{otherwise} \end{cases} \qquad (24.8)$$

where MV is the motion value computed from the 16×16 block difference of current and previous frames, and γ_{temp} is a threshold to determine the no-motion block. Inconsistent disparities would result in higher temporal cost. This method solves the flicker artifact but incurs foreground copy artifact, because background objects do not have enough time to update their disparities. In addition, the JBU in our algorithm would introduce more flicker artifacts on the object boundary. To address these problems, we propose the following two methods based on Equation 24.8.

First, the NMR method can be applied when a foreground object passes, as shown in Figure 24.2. For this case, the DERS algorithm only takes a short time to update the foreground disparity to background. To provide sufficient updating time, our proposed method extends the motion interval by τ_{NMR} frames, and postpones adding the temporal cost C_{temp} into C_{total}. In addition, λ_{temp} in Equation 24.8 is changed to a function inversely proportional to MV for less sensitivity to the fixed threshold, and MV is computed using a 3×3 window for lower

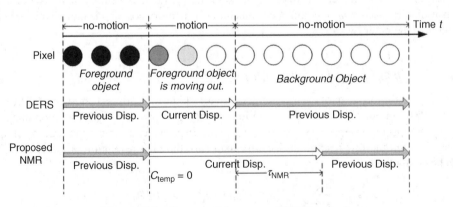

Figure 24.2   Concept of the proposed NMR method

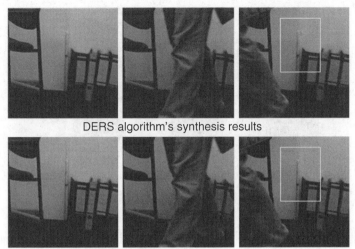

DERS algorithm's synthesis results

Proposed NMR method's synthesis results

Figure 24.3 Successive synthesis results of the DERS algorithm and the proposed NMR method (see Plate 34 for the colored figure)

hardware cost. The proposed C_{temp} is defined as

$$C_{\text{temp}}(x, y, d) = \lambda_{\text{vert}}(\text{MV})\min\{|d - D_{\text{L}}^{t-1}(x, y)|, \tau_{\text{temp}}\} \tag{24.9}$$

Figure 24.3 compares the synthesis results by the DERS algorithm and the proposed method. In our method, the door pivot is not doubled after the person in the foreground passed.

Second, we propose the SEP method to reduce flickering on object boundaries. This method keeps the previous disparities for the still-edge pixels in the current frame. The still-edge pixel is detected by a Sobel filter and the MV in the NMR method.

24.3.5 Occlusion Handling

For the occlusion problem, the proposed occlusion handling first applies the LRC to detect occluded regions and then fills them using the inter-view and intra-view references, as shown in Figure 24.4. The inter-view references are warped from the non-occluded disparities of the other two views in the disparity cross-warping step. In this step, most of occluded regions are recovered and the disparity consistency among different views is also enhanced. On the other hand, the intra-view references are from the surrounding reliable disparities, which pass the double LRC and are refined by the window vote in the good disparity detection step. Finally,

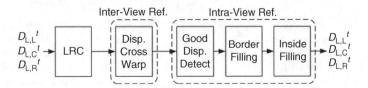

Figure 24.4 Flow of the proposed occlusion handling method

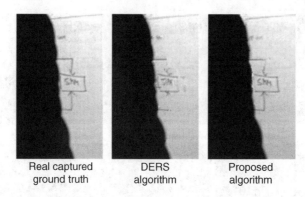

| Real captured ground truth | DERS algorithm | Proposed algorithm |

Figure 24.5 Comparison of the occluded regions

the remaining occluded regions are filled by the reliable disparities with different scan directions in the border filling and inside filling steps. Figure 24.5 shows the proposed occlusion handling that recovers the occluded regions better than the DERS algorithm does.

24.4 Proposed Architecture

In this section, we present the proposed architecture of the disparity estimation engine, which supports real-time processing of three-view HD1080p disparity maps with a disparity range of 128.

24.4.1 Overview of Architecture

In Figure 24.6, the proposed architecture consists of the main core and the I/O interface. In the I/O interface, the memory access controller serves all the requests from the main core to

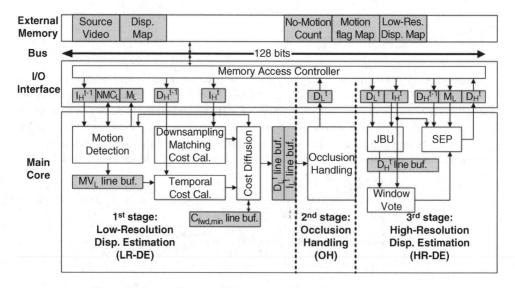

Figure 24.6 Architecture of the proposed disparity estimation engine

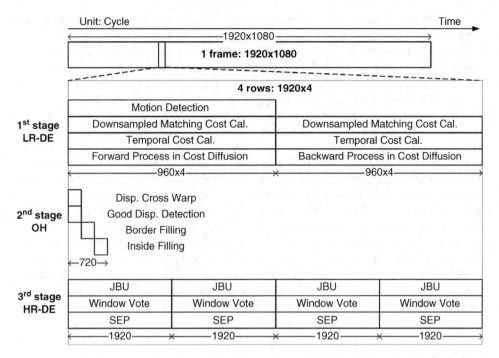

Figure 24.7 Computational schedule of the proposed disparity estimation engine

access the external memory. To decrease the idle time of the main core, we propose an efficient memory access schedule and a compact memory data configuration.

For the main core, we propose a three-stage row-level pipelining architecture according to the computational characteristics of the algorithm. The first stage computes the low-resolution disparity maps and motion information, and the second stage deals with the occlusion problem. Finally, the third stage upsamples and refines the disparity maps. Note that the sizes of the pipelining buffers are proportional to those frame widths, and those between the second and third stages are put in the external memory to decrease the internal memory cost.

Figure 24.7 shows the computational schedule of the main core for one frame that computes three-view disparity maps simultaneously. A schedule tile includes the computation of four disparity rows in high resolution, and its throughput is 1 pixel/cycle. Because of the downscaled computation in the first stage, the throughput can be 1/8 pixels/cycle. In addition, for the second stage, the latency is very short because of its low computational complexity. With the proposed computational schedule, the disparity estimation engine attains the throughput of 60 fps for three-view HD1080p disparity maps at a clock frequency of 125 MHz.

24.4.2 Computational Modules

In the first stage, the motion detection unit fetches the previous and current frames to compute the MVs stored in the internal line buffer for temporal cost calculation and the motion flags stored to external memory for the final stage. In addition, the no-motion counts for the NMR method are also updated in the external memory. The other modules in the first stage have a parallelism factor of 32 (i.e., a quarter of full disparity range) as a result of computing in the low-resolution domain. For lower hardware cost in the computation of C_{vert} and C_{temp}, the disparity differences and scaling functions in Equations 24.3 and 24.9 are implemented by look-up tables, adders, and shifters. In addition, the cost diffusion module adopts a similar parallel architecture (Liang *et al.*, 2011) for the BP algorithm, but sequentially computes Equation 24.4 in four iterations to share common operators. To sum up, the first stage generates three-view disparity maps in low resolution and takes advantages of our proposed downsampling matching cost and cost diffusion methods to achieve low hardware cost.

In the second stage, the disparity rows of three views from the previous stage are cross referred to handle their occlusion problems. The computation consists of four iterations, as shown in Figure 24.7. In the first iteration, it cross-warps disparities and detects good ones. In the second and third iterations, it fills the border and inside occluded regions. The major process of this stage is data movement among buffers, instead of heavy computation like in the other two stages. Thus, its main hardware cost is the internal memories.

In the third stage, the three modules are fully parallel to achieve a throughput of 1 pixel/ cycle. For the JBU, the spatial and range kernels in Equation 24.7 are simplified into shifters and adders, and the normalization is implemented by a highly pipelined divisor for high working frequency. In addition, the architecture of the window vote is similar to that in Zhang *et al.* (2009), and the SEP architecture is a straightforward one.

24.4.3 External Memory Access

The toughest design challenge in the disparity estimation engine is the external memory access due to too many access requests from the main cores. According to our analysis, the required average bandwidth is 507 bits/cycle without the consideration of memory row miss. Thus, we select a 128-bit bus and eight parallel DDR3 SDRAMs (Micron, n.d.) with a clock frequency of 800 MHz. This bus and external memory support a bandwidth of 1024 bits/cycle.

For efficient data access, we propose a data configuration method with the following principles. First, the three views' image and disparity frames are separately stored into different banks, while their no-motion count and motion flag are placed in the same bank. Second, multiple image (disparity) segment rows are configured into the same memory row to avoid frequent row miss because image pixels (disparities) are accessed by raster-scan order and cross-multiple rows. Third, the image pixel and disparity pixel in low resolution are packed together for efficient access, as their access positions in image coordinates are identical.

With the above data configuration in external memory, we propose an access schedule as shown in Figure 24.8. This schedule is hierarchically defined from one frame to one tile access while considering the latencies of different row misses in DRAM (Micron, n.d.). For a tile access, the memory access follows the defined processing order. Each access block only uses the first cycles to fetch data and preserves other cycles to handle row misses. The data access in one tile is executed repeatedly for a whole frame. Note that the bank 7 is not scheduled here because of its infrequent access. In addition, the available block for occlusion is also not scheduled here owing to its different and longer access period.

To sum up, with the above data configuration and memory access schedule, our disparity estimation engine can support all the access requests from the main core and attain the target real-time throughput.

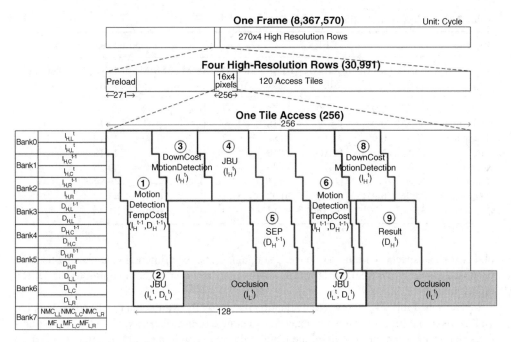

Figure 24.8 Proposed external memory access schedule where the number means the processing order

24.5 Experimental Results

24.5.1 Comparison of Disparity Quality

The simulation is set as follows. We use the three-view configuration for a nine-view display (MPEG, 2010) to generate three-view disparity maps by the proposed algorithm and the DERS algorithm, and synthesize virtual view videos. The most left and right synthesized videos in nine views are compared with the real captured videos by the peak signal-to-noise ratio (PSNR), the structural similarity (SSIM) (Wang *et al.*, 2004), and the temporal part of the peak signal-to-perceptible-noise ratio (T_PSPNR) (Zhao and Yu, 2009) in the objective evaluation. The former two methods evaluate the spatial distortion, and the latter method evaluates the temporal distortion.

Table 24.1 lists the evaluation results. Note that the DERS algorithm requires five view videos as input owing to its algorithm flow, and our algorithm only uses three view videos. Because of insufficient input views for some sequences, the results are not available. For the objective evaluation results, the proposed algorithm is better than the DERS algorithm in most of the sequences. In addition, Table 24.2 presents the effect of each proposed method. The downsampling and upsampling method drops quality in BookArrival and Newspaper due to detail loss in low resolution. For the temporal consistency enhancement methods, the SEP method improves the temporal quality but degrades the spatial quality. Finally, the quality of all the videos is improved by the occlusion handling because background regions are recovered. Figure 24.9 shows the disparity maps and synthesized frames. There is some pepper noise around the object boundary in the proposed algorithm due to its processing with small window size. On the other hand, the DERS algorithm suffers from distortion of the

Table 24.1 Objective evaluation for synthesis frames

Sequence	DERS			Proposed algorithm					
	PSNR (dB)	SSIM	T_PSPNR (dB)	PSNR (dB)	ΔPSNR (dB)	SSIM	ΔSSIM	T_PSPNR (dB)	ΔT_PSPNR (dB)
BookArrival	35.07	0.97	52.39	35.54	0.47	0.95	−0.02	53.26	0.87
LoveBird1	30.88	0.94	44.31	29.38	−1.50	0.93	−0.01	44.10	−0.22
Newspaper	30.70	0.99	45.65	30.81	0.12	0.99	0.00	45.19	−0.46
Café	N.A.	N.A.	N.A.	32.50	—	0.99	—	46.18	—
Kendo	N.A.	N.A.	N.A.	35.43	—	0.98	—	49.26	—
Balloons	N.A.	N.A.	N.A.	35.11	—	0.98	—	49.87	—
Champagne	24.76	0.97	34.39	30.07	5.31	0.97	0.00	43.79	9.40
Pantomime	35.56	0.97	50.15	36.15	0.59	0.97	0.00	51.36	1.21

object shape, especially in the sequence Champagne. The experimental results show that our proposed disparity estimation algorithm attains disparity quality comparable to the DERS algorithm.

In addition, Table 24.3 shows the evaluation results by the Middlebury test bed (Scharstien and Szeliski, n.d.) for all pixels with an error tolerance of one disparity. Note that the parameters in our algorithm are not fine tuned, and only two view images are inputted. Compared with the baseline algorithms (Scharstien and Szeliski, 2002), our algorithm can perform better than the 1D optimization algorithms.

In summary, the proposed algorithm performs better than the advanced DERS algorithm in the evaluation for the targeted 3DTV applications, but cannot deliver very accurate disparity maps in the disparity error evaluation. This implies the performance of targeted 3DTV applications depends mainly on the temporal consistency of disparity maps, instead of the error rate of disparity with ground truth.

24.5.2 Analysis of Sampling Factor

The selection of sampling factors highly affects the hardware cost and the disparity quality. Table 24.4 shows the evaluation results with different sampling factors. The final selected

Table 24.2 Objective evaluation for each proposed method

Method	BookArrival (100f)			Newspaper (300f)			LoveBird1 (300f)		
	PSNR (dB)	SSIM	T_PSPNR (dB)	PSNR (dB)	SSIM	T_PSPNR (dB)	PSNR (dB)	SSIM	T_PSPNR (dB)
Cost diffusion	35.44	0.95	48.88	29.91	0.98	39.57	28.90	0.93	39.42
+Downsample, JBU	35.25	0.95	41.59	29.95	0.93	39.65	29.16	0.98	40.84
+Temporal cost	35.51	0.95	49.99	30.15	0.98	41.24	29.31	0.93	41.51
+NMR	35.51	0.95	49.99	30.15	0.98	41.24	29.31	0.93	41.99
+SEP	34.80	0.95	50.65	29.15	0.98	41.78	28.96	0.93	43.55
+Occlusion handle	35.13	0.95	52.32	29.95	0.99	44.55	29.40	0.95	44.31

| DERS disparity maps | Proposed disparity maps | DERS synthesis frames | Proposed synthesis frames |

BookArrival

LoveBird1

Newspaper

Champagne Tower

Pantomime

Figure 24.9 Disparity maps and synthesized images of the DERS algorithm and the proposed algorithm

sampling factors in our algorithm keep the view synthesis quality for all resolutions, especially for the smaller size of 1024×768.

24.5.3 Implementation Result

The proposed design is implemented by Verilog and synthesized using the 90 nm CMOS technology process. Table 24.5 lists the implementation result. The proposed disparity estimation engine achieves a throughput of 95 fps for three-view HD1080p disparity maps with the logic cost of 1645K gate counts and an internal memory of 59.4 Kbyte.

Table 24.3 Disparity error rate by Middlebury test bed

	Tsukuba (%)	Venus (%)	Teddy (%)	Cones (%)
Proposed algorithm (1D opt.)	6.45	7.10	20.6	17.0
GC (2D opt.)	4.12	3.44	25.0	18.2
DP (1D opt.)	5.04	11.10	21.6	19.1
SO (1D opt.)	7.22	10.9	28.2	22.8

Table 24.4 Evaluation with different sampling factors in Y-PSNR (dB)

Horiz. factor	Vert. factor	BookArrival	LoveBird1	Newspaper	Café	Kendo	Balloons	Champagne	Pantomime	Avg.
1/2	1/2	36.40	30.72	30.77	N.A.	36.17	34.10	30.73	38.47	33.91
1/2	1/4	36.34	30.89	30.79	33.96	36.00	33.97	30.46	38.38	33.85
1/2	1/8	36.07	30.85	30.62	33.78	35.89	33.67	29.38	37.56	33.48
1/2	1/16	35.57	30.85	30.29	33.05	35.24	32.77	28.84	37.54	33.02
1/4	1/2	36.00	30.77	30.73	33.82	35.75	33.81	29.83	38.48	33.65
1/4	1/4	35.90	30.92	30.64	33.96	36.04	33.64	29.79	38.46	33.67
1/4	1/8	35.68	30.87	30.57	33.71	35.73	33.18	29.93	38.46	33.52
1/4	1/16	35.27	30.90	30.12	32.45	35.10	32.66	29.12	38.40	33.00
1/8	1/2	35.66	30.77	30.24	33.21	35.50	33.00	29.04	38.48	33.24
1/8	1/4	35.49	30.77	30.15	33.16	35.37	32.89	29.40	38.49	33.21
1/8	1/8	35.23	30.73	30.18	32.16	35.08	32.43	29.01	38.47	32.91
1/8	1/16	34.66	30.68	29.76	30.84	34.53	32.32	28.82	38.44	32.50
1/16	1/2	34.56	30.38	29.43	32.04	34.42	31.92	29.30	38.42	32.56
1/16	1/4	34.62	30.51	28.95	31.66	34.55	32.15	29.07	38.45	32.50
1/16	1/8	34.46	30.46	29.11	31.11	34.29	31.89	34.13	38.43	32.99
1/16	1/16	34.17	30.67	28.08	30.60	33.88	31.58	27.85	38.49	31.92

Table 24.5 Performance of the proposed disparity estimation engine

I/O function	
Input/output data	3-view HD1080p videos/disparity maps
Disparity range (pixel)	128
Frame rate (fps)	95
System	
External memory	8 DDR3 SDRAMs (800 MHz)
Bus width (bit)	128 (800 MHz)
Core	
Technology process	UMC 90 nm
Clock frequency	200 MHz
Power consumption (mW)	361.5
Gate-count (including memory)	2020K
Gate-count (excluding memory)	1645K
Internal memory (byte)	59.4K
Throughput (pixel-disparity/s)	75.64G

Table 24.6 Comparison of real-time disparity estimation implementation

	Yang et al. (2006)	Liang et al. (2011)	Diaz et al. (2007)	Chang et al. (2010)	Liang et al. (2011)	Our design
No. input/output view	2/1	2/1	2/1	2/1	2/1	3/3
Algorithm	Hierachical BP	Tile BP	Phase matching	MC ADSW	Tile BP	Cost diffusion
Frame size	800 × 600	450 × 375	1280 × 960	352 × 288	640 × 480	1920 × 1080
Frame rate (fps)	0.67	1.68	52	42	58	95
Disparity range (pixels)	300	60	29	64	64	128
Implementation method	GPU	GPU	FPGA	ASIC	ASIC	ASIC
	Nvidia 8800GTX	Nvidia 8800GTS	Xilinx Vertex-II	UMC 90 nm	UMC 90 nm	UMC 90 nm
Frequency(MHz)	—	—	65	95	185	200
Logic area (gate-count)	—	—	—	562 K	633K	1645K
Memory usage (gate-count)	—	—	—	—	1871K	375K
	(9 Mbtye)			(21.3 Kbyte)		(59.4 Kbyte)
Total area	—	—	—	—	2505K	2020K
Throughput (pixel-disparity/s)	96M	17M	1885M	272M	1146M	75 644M

Table 24.6 compares our design with previous implementations. Compared with the previous studies, our disparity estimation engine reaches the highest throughput and satisfies the requirement of HD 3DTV applications.

24.6   Conclusion

We have proposed a new hardware-efficient high-quality disparity estimation that adopts the downsampled matching cost to reduce the computational cost significantly and the cost diffusion algorithm to reduce the memory cost in disparity optimization. To further improve disparity quality, we proposed the NMR and SEP methods to deal with the temporal consistency problems and inter/intra-view-based occlusion handling for the occlusion problem. Moreover, a three-stage pipelining architecture is proposed for the disparity estimation algorithm. The final proposed disparity estimation engine achieves real-time speed for three-view HD1080p disparity maps, and its disparity quality is comparable to the DERS algorithm for view synthesis in 3DTV applications.

References

Bang, G., Lee, J., Hur, N., and Kim, J. (2009) ISO/IEC JTC1/SC29/WG11/M16070: *The Consideration of the Improved Depth Estimation Algorithm: The Depth Estimation Algorithm for Temporal Consistency Enhancement in Non-moving Background*, MPEG, Lausanne, Switzerland.

Boykov, Y. and Kolmogorov, V. (2004) An experimental comparison of min-cut/max-flow algorithms for energy minimization in vision. *IEEE Trans. Pattern Anal.*, **26**(9), 1124–1137.

Boykov, Y., Veksler, O., and Zabih, R. (2001) Fast approximate energy minimization via graph cuts. *IEEE Trans. Pattern Anal.*, **23**(11), 1222–1239.

Chang, N. Y.-C. and Chang, T.-S. (2007) A scalable graph-cut engine architecture for real-time vision. Proceedings of 2007 VLSI Design/CAD Symposium, Hualien, Taiwan.

Chang, N. Y.-C., Tsai, T.-H., Hsu, B.-H. *et al.* (2010) Algorithm and architecture of disparity estimation with mini-census adaptive support weight. *IEEE Trans. Circ. Syst. Vid.*, **20**(6), 792–805.

Cherkassky, B. V. and Goldberg, A. V. (1995) On implementing the push-relabel method for the maximum flow problem, in *Integer Programming and Combinatorial Optimization* (eds E. Balas and J. Clausen), Springer, Berlin, pp. 157–171.

Chou, C.-W., Tsai, J.-J., Hang, H.-M., and Lin, H.-C. (2010) A fast graph cut algorithm for disparity estimation, in *Picture Coding Symposium, 2008*, IEEE, pp. 326–329.

Delong, A. and Boykov, Y. (2008) A scalable graph-cut algorithm for N-D grids, in *IEEE Conference on Computer Vision and Pattern Recognition, 2008. CVPR 2008*, IEEE, pp. 1–8.

Diaz, J., Ros, E., Carrillo, R., and Prieto, A. (2007) Real-time system for high-image resolution disparity estimation. *IEEE Trans. Image Process.*, **16**(1), 280–285.

Egnal, G. and Wildes, R. P. (2002) Detecting binocular half-occlusions: empirical comparisons of five approaches. *IEEE Trans. Pattern. Anal.*, **24**(8), 1127–1133.

Felzenswalb, P. F. and Huttenlocher, D. P. (2006) Efficient belief propagation for early vision. *Int. J. Comput. Vision*, **70**(1), 41–54.

Ford, L. and Fulkerson, D. (1962) *Flows in Networks*, Princeton University Press.

Gong, M. (2006) Enforcing temporal consistency in real-time stereo estimation, in *ECCV'06 Proceedings of the 9th European Conference on Computer Vision – Volume Part III*, Springer-Verlag, Berlin, pp. 564–577.

Gong, M. and Yang, Y.-H. (2007) Real-time stereo matching using orthogonal reliability-based dynamic programming. *IEEE Trans. Image Process.*, **16**(3), 879–884.

Greisen, P., Heinzle, S., Gross, M., and Burg, A.P. (2011) An FPGA-based processing pipeline for high-definition stereo video. *EURASIP J. Image Vid. Process.*, **2011**, 18.

Hirschmüller, H. (2005) Accurate and efficient stereo processing by semi-global matching and mutual information, in *IEEE Computer Society Conference on Computer Vision and Pattern Recognition, 2005. CVPR 2005*, Vol. 2, IEEE Computer Society Press, Washington, DC, pp. 807–814.

Khoshabeh, R., Chan, S., and Nyuyen, T. Q. (2011) Spatio-temporal consistency in video disparity estimation, in *2011 IEEE International Conference on Acoustics, Speech, and Signal Processing (ICASSP)*, IEEE, pp. 885–888.

Kim, J. C., Lee, K. M., Choi, B. T., and Lee, S. U. (2005) A dense stereo matching using two-pass dynamic programming with generalized ground control points, in *CVPR '05 Proceedings of the 2005 IEEE Computer Society Conference on Computer Vision and Pattern Recognition*, Vol. 2, IEEE Computer Society Press, Washington, DC, pp. 1075–1082.

Klaus, A., Sormann, M., and Karner, K. (2006) Segment-based stereo matching using belief propagation and self-adapting dissimilarity measure, in *18th International Conference on Pattern Recognition, 2006. ICPR 2006*, Vol. **3**, IEEE Computer Society Press, Washington, DC, pp. 15–18.

Kolomogorov, V. and Zabih, R. (2001) Computing visual correspondence with occlusions using graph cuts, in *Eighth IEEE International Conference on Computer Vision, 2001. ICCV 2001*, Vol. **2**, IEEE, pp. 508–515.

Kopf, J., Cohen, M.F., Lischinski, D., and Uttyendaele, M. (2007) Joint bilateral upsampling. *ACM Trans. Graph.*, **26**(3), article no. 96.

Larsen, E. S., Mordohai, P., Pollefeys, M., and Fuchs, H. (2007) Temporally consistent reconstruction from multiple video streams using enhanced belief propagation, in *IEEE 11th International Conference on Computer Vision, 2007. ICCV 2007*, IEEE, pp. 1–8.

Liang, C.-K., Cheng, C.-C., Lai, Y.-C. *et al.* (2011) Hardware-efficient belief propagation. *IEEE Trans. Circ. Syst. Vid.*, **21**(5), 525–537.

Micron Inc. (n.d.) 1Gb DDR3 SDRAM: MT41J128M8JP-125 [online], http://www.micron.com/parts/dram/ddr3-sdram/mt41j128m8jp-125 [Accessed 2 January 2013].

Min, D., Yea, S., Arican, Z. and Vetro, A. (2010a) Disparity search range estimation: enforcing temporal consistency, in *2010 IEEE International Conference on Acoustic, Speech, and Signal Processing (ICASSP)*, IEEE, pp. 2366–2369.

Min, D., Yea, S. and Vetro, A., (2010b) Temporally consistent stereo matching using coherence function, in *3DTV-Conference: The True Vision – Capture, Transmission and Display of 3D Video (3DTV-CON), 2010*, IEEE, pp. 1–4.

MPEG (2010) ISO/IEC JTC1/SC29/WG11 W10925: Description of exploration experiments in 3D video coding, MPEG, Xian, China.

Ohta, Y. and Kanade, T. (1985) Stereo by intra- and inter- scanline search using dynamic programming. *IEEE Trans. Pattern Anal.*, **7**(2), 139–154.

Park, S. and Jeong, H. (2007) Real-time stereo vision FPGA chip with low error rate, in *International Conference on Multimedia and Ubiquitous Engineering, 2007. MUE '07*, IEEE, pp. 751–756.

Park, S., Chen, C. and Jeong, H. (2007) VLSI architecture for MRF based stereo matching, in *Embedded Computer Systems: Architectures, Modeling, and Simulation*, Lecture Notes in Computer Science, Vol. 4599, Springer, Berlin, pp. 55–64.

Riemens, A. K., Gangwal, O. P., Barenbrug, B., and Berretty, R.-P. M. (2009) Multi-step joint bilateral depth upsampling. *Visual Communications and Image Processing 2009* (eds R. L. Stevenson and M. Rabbani), Proceedings of the SPIE, Vol. **7257**, SPIE, Bellingham, WA, p. 72570M.

Scharstien, D. and Szeliski, R. (2002) A taxonomy and evaluation of dense two-frame stereo correspondence algorithm. *Int. J. Comput. Vision*, **47**(1)–(3), 7–42.

Scharstien, D. and Szeliski, R. (n.d.) Middlebury Stereo Evaluation – Version 2 [online], http://vision.middlebury.edu/stereo/eval/ [Accessed 15 May 2011].

Sun, J., Zhang, N.-N. and Shum, H.-Y. (2003) Stereo matching using belief propagation. *IEEE Trans. Pattern Anal.*, **25**(7), 787–800.

Szeliski, R., Zabih, R., Scharstein, D. *et al.* (2008) A comparative study of energy minimization methods for Markov random fields with smoothness-based priors. *IEEE Trans. Pattern. Anal.*, **30**(6), 1060–1080.

Tanimoto, M., Fujii, T., Tehrani, M. P., and Wildeboer, M. (2009) ISO/IEC JTC1/SC29/WG11 M16605: Depth estimation reference software (DESR) 4.0, MPEG, London.

Veksler, O. (2005) Stereo correspondence by dynamic programming on a tree, in *IEEE Computer Society Conference on Computer Vision and Pattern Recognition, 2005. CVPR 2005*, Vol. **2**, IEEE Computer Society Press, Washington, DC, pp. 384–390.

Van Meerbergen, G., Vergauwen, V., Pollefeys, M., and Van Gool, L. (2002) A hierarchical symmetric stereo algorithm using dynamic programming. *Int. J. Comput. Vision*, **47**(1–3), 275–285.

Wang, L., Liao, M., Gong, M. *et al.* (2006) High quality real-time stereo using adaptive cost aggregation and dynamic programming, in *Third International Symposium on 3D Data Processing, Visualization, and Transmission*, IEEE Computer Society Press, Washington, DC, pp. 798–805.

Wang, Z., Bovik, A. C., Sheikh, H. R., and Simoncelli, E. P. (2004) The SSIM index for image quality assessment [online], http://www.cns.nyu.edu/~lcv/ssim/ [Accessed 4 April 2011].

Yang, Q., Wang, L., Yang, R. *et al.* (2006) Real-time global stereo matching using hierarchical belief propagation. Proceedings of the 17th British Machine Vision Conference, Edinburgh.

Yang, Q., Wang, L., Yang, R. *et al.* (2009) Stereo matching with color-weighted correlation, hierarchical belief propagation and occlusion handling. *IEEE Trans. Pattern. Anal.*, **31**(3), 492–504.

Yang, Q., Wang, L., and Ahuja, N. (2010) A constant-space belief propagation algorithm for stereo matching, in *2010 IEEE Conference on Computer Vision and Pattern Recognition (CVPR)*, IEEE, pp. 1458–1465.

Zhang, K., Lu, J., and Lafruit, G. (2009) Cross-based local stereo matching using orthogonal integral images. *IEEE Trans. Circ. Syst. Vid.*, **19**(7), 1073–1079.

Zhao, Y. and Yu, L. (2009) ISO/IEC JTC1/SC29/WG11, M16890: PSPNR Tool 2.0, MPEG, Xian, China.

Index

Emerging Technologies for 3D Video: Creation, Coding, Transmission and Rendering, First Edition.
Frédéric Dufaux, Béatrice Pesquet-Popescu, and Marco Cagnazzo.
© 2013 John Wiley & Sons, Ltd. Published 2013 by John Wiley & Sons, Ltd.